D1669023

Greenhouse Gas Emissions from International Aviation

Greenhouse Gas Emissions from International Aviation

Legal and Policy Challenges

Alejandro José Piera

Published, sold and distributed by Eleven International Publishing
P.O. Box 85576
2508 CG The Hague
The Netherlands
Tel.: +31 70 33 070 33
Fax: +31 70 33 070 30
e-mail: sales@budh.nl
www.elevenpub.com

Sold and distributed in USA and Canada
International Specialized Book Services
920 NE 58th Avenue, Suite 300
Portland, OR 97213-3786, USA
Tel.: 1-800-944-6190 (toll-free)
Fax: +1 503 280-8832
orders@isbs.com
www.isbs.com

Eleven International Publishing is an imprint of Boom uitgevers Den Haag.

ISBN 978-94-6236-467-7
ISBN 978-94-6274-143-0 (E-book)

Printed in The Netherlands

Table of Contents

Preface		**xiii**
Acknowledgments		**xvii**
Abbreviations		**xix**
Introduction		**1**
1	**Setting the Scene**	**11**
1.1	Understanding the Problem	12
1.1.1	Aviation Market Outlook and Economic Contribution	12
1.1.2	Aviation under Siege	15
1.1.3	Climate Change	16
1.1.4	Aviation and Climate Change	19
1.1.5	The "Only 2 Percent" Exculpation Argument	23
1.1.6	The "Communication Problem" Defense	25
1.1.7	The All-Mighty Technological Efficiencies	25
1.1.8	Alternative Fuels	28
1.1.9	Aviation's Fuel Factor: Is Regulation Needed After All?	30
1.2	International Legal Regime	32
1.2.1	UNFCCC/Kyoto Protocol	32
1.2.2	The EU ETS	34
1.3	The Political Dimension: A Small Piece within a Bigger Puzzle	35
1.3.1	The Problems with Lack of Progress	36
1.3.2	Exogenous Threats	38
1.4	Conclusion	39
2	**Aviation and Climate Change: A Case of Fragmentation of International Law**	**41**
2.1	The Interaction between International Aviation and the Climate Change Regime	42
2.1.1	The Kyoto Protocol	42
2.1.2	ICAO and Climate Change: Some Unanswered Questions	43
2.1.3	The CBDR/Non-Discrimination Saga	47
2.2	From Theory to What Happens in Practice	50

2.2.1 Understanding Non-Discrimination 51
2.2.2 Understanding CBDR 53
2.3 A Look into Fragmentation of International Law 55
2.3.1 The ILC Report on Fragmentation of International Law 58
2.4 Applying *VCLT* Rules to the CBDR/Non-Discrimination Saga 62
2.4.1 Systemic Integration and ICAO 65
2.5 Attempts to Accommodate the Special Needs of Developing Countries 66
2.5.1 The De Minimis Principle 67
2.5.1.1 The 2010 De Minimis Proposal 67
2.5.1.2 The 38th Assembly: A Similar De Minimis *Proposal* 69
2.5.2 Reinventing CBDR: ICAO's SCRC 71
2.5.2.1 What Is the Difference? 72
2.5.2.2 Why Has SCRC Not Worked? 74
2.5.3 CBDR Finally Arrives at ICAO through the Back Door 75
2.6 Toward a New Approach: Reconciling Principles 76
2.6.1 CBDR Is Not Static 76
2.6.2 Avoiding Isolation 78
2.6.3 Reconciling CBDR with Non-Discrimination 78
2.6.4 From Theory to Practice: Some Design Elements to Consider 80
2.7 Conclusion 82

3 The International Civil Aviation Organization 85
3.1 ICAO's Institutional Setting 86
3.1.1 Objectives 87
3.1.2 Governing Structure 91
3.1.3 Constituency 94
3.1.4 ICAO's Committee on Aviation Environmental Protection (CAEP) 94
3.1.5 Industry Participation 98
3.1.6 NGO Participation 99
3.2 ICAO and Climate Change 100
3.2.1 Historical Background 100
3.2.2 The Long Road to the CO2 Standard 103
3.2.3 State Action Plans 105
3.2.4 Aspirational Goals 106
3.2.5 A Framework for MBMs 108
3.2.6 A Global Scheme for MBMs 113
3.3 Conclusion 116

4	**The Inclusion of International Aviation in the EU ETS**	**119**
4.1	The EU ETS	120
4.1.1	Background	120
4.1.2	EU ETS and Aviation	124
4.1.3	Calculation of Emissions	125
4.1.4	Distribution of Allowances: Surrendering of Allowances	126
4.1.5	Emission Credit Units (ECUs)	128
4.1.6	Generation and Use of Revenues	130
4.1.7	Cost or Windfall Gains?	133
4.2	International Opposition	135
4.2.1	Judicial Attacks: A4A's Legal Challenge	136
4.2.2	Political Statements: From New Delhi to Moscow	138
4.2.2.1	The New Delhi Declaration	138
4.2.2.2	ICAO Council Declaration of 2 November 2011	139
4.2.2.3	Moscow Declaration	140
4.2.3	The Nail in the Coffin: The Anti-EU ETS US Bill	141
4.3	Why Did Europe Decide to Fly Solo?	145
4.3.1	Rationale for Including Non-EU Aircraft Operators	148
4.3.2	The "Temporary" Suspension of the EU ETS	149
4.4	Conclusion	152
5	**Legal Challenges against the EU ETS: Extraterritoriality**	**155**
5.1	Principles of International Law on Jurisdiction	157
5.1.1	Territorial Principle	157
5.1.2	Nationality Principle	163
5.1.3	Passive Personality Principle	166
5.1.4	Protective Principle	167
5.1.5	Universal Jurisdiction	168
5.2	State Responsibility and Obligations *Erga Omnes*	172
5.3	The Effects Doctrine	177
5.4	The EU ETS and Extraterritoriality	181
5.4.1	The Commission's Approach	181
5.4.2	The *ATA Decision* and Extraterritoriality	182
5.4.2.1	Drawing Symbolism	182
5.4.2.2	Physical Presence Leads to Unlimited Jurisdiction	183
5.4.2.3	Events Partly Outside	186
5.4.2.4	A Condition to Access the Market	188
5.4.2.5	A Vehicle to Advance EU's Climate Change Policy: Where Extraterritorial Jurisdiction Meets Rational Choice	191

5.4.2.6 The Surreptitious Effects Doctrine 194
5.5 The EU ETS, Extraterritoriality, and Principles of International Law 195
5.5.1 The Nationality Principle 195
5.5.2 Territorial Principle: The National Airspace Approach 195
5.5.2.1 Flight Information Regions 197
5.5.2.2 The 38th Assembly and the Non-Recognition of the Airspace Approach 198
5.5.2.3 Universal Jurisdiction 202
5.6 Is the Extraterritoriality of the EU ETS Unprecedented? Other Examples of Extraterritorial Jurisdiction in International Civil Aviation 202
5.7 Can the Doctrine of State Responsibility Exonerate the EU ETS? 206
5.8 Conclusion 208

6 Additional Legal Issues Involving the EU ETS 211
6.1 Unilateral Action 212
6.1.1 The *Kyoto Protocol*, ICAO, and Unilateral State Action 213
6.1.2 Types of Unilateral State Actions 215
6.1.3 Where Does the EU ETS Stand? 219
6.1.4 The EU ETS under the Lenses of Bodansky's Balanced, Liberal Approach 221
6.2 The Tax Controversy 224
6.2.1 Understanding the Genesis of ETS: An Alternative to Taxes 225
6.2.2 Does Article 15 of the *Chicago Convention* Prohibit the EU ETS? 229
6.2.3 Jennison's Functional Equivalency Theory 230
6.2.4 Why the EU ETS Is Not a Tax 232
6.2.5 Does the EU ETS Contravene Article 24 of the *Chicago Convention*? 236
6.2.6 Is *Braathens* Relevant? 238
6.3 A Scheme Violating WTO Rules? 239
6.4 An Attack upon CBDR? 241
6.5 Does the *Chicago Convention* Matter? 244
6.5.1 Background 244
6.5.2 The Theory of Functional Succession 245
6.5.3 When *Strasbourg* Negated *Chicago* 247
6.5.4 Repercussions 249
6.6 Conclusion 251

7	**We Are All Ahead of the Curve**	**253**
7.1	The Theory of Norms and Norm Entrepreneurship	255
7.1.1	What Are Norms?	255
7.1.2	The Emergence of Non-State Actors	257
7.1.3	Who Are Norm Entrepreneurs?	259
7.1.4	Transnational Networks	262
7.1.5	What Drives Norm Entrepreneurs?	263
7.1.6	The Norm's Life Cycle	264
7.1.7	Conditions Inducing Norm Entrepreneurship	267
7.1.8	Influences Hampering Norm Entrepreneurship	268
7.1.9	Norm Internalization	270
7.1.10	The Relevance of Regime Architectural Design in the Formation of International Norms	273
7.1.10.1	Bottom-Up Norm Construction	273
7.1.10.2	Top-Down Norm Construction	275
7.2	Who Is the Real Norm Entrepreneur?	276
7.2.1	Europe	276
7.2.1.1	An Obstacle or a Building Block for an International Agreement?	276
7.2.1.2	Why Has the EU ETS Not Led to Norm Cascading?	282
7.2.1.3	EU ETS and Norm Internalization	285
7.2.1.4	EU ETS: Significant Contribution or Missed Opportunity?	286
7.2.2	IATA	286
7.2.2.1	A Reactive Response: The IATA Four-Pillar Strategy	286
7.2.2.2	The Fear of Patchwork Regulations	288
7.2.2.3	The Long Road to the IATA Industry Targets	288
7.2.2.4	A Global, Sectoral Approach	291
7.2.2.5	Why Not Binding Industry Self-Regulation?	293
7.2.2.6	The "Historic" Resolution: Shaping the Future?	295
7.2.2.7	Exogenous Influences: Some Problems for Sectoral Norm Entrepreneurs	298
7.2.2.8	An Assessment of IATA's Norm Entrepreneurship Role	300
7.2.3	ICAO	301
7.2.3.1	ICAO: A Norm Entrepreneur or an Institutional Platform?	301
7.2.3.2	Barriers to Norm Entrepreneurship	302
7.2.3.3	Assessing ICAO's Role	310
7.2.4	The United States	310
7.2.4.1	Not Just Another Player	310
7.2.4.2	The United States and Climate Change	311
7.2.4.3	The United States: Aviation and Climate Change	312

7.2.4.4 Changing the Approach: From Resistance to Endorsement 313
7.2.4.5 Assessing the United States' Role 313
7.3 Conclusion 315

8 The Way Ahead: Key Considerations in Addressing GHG Emissions from International Aviation in the Future 319
8.1 Addressing Some of the Required Design Elements of the Global MBM Scheme 320
8.1.1 Reconciling CBDR and Non-Discrimination 320
8.1.1.1 Route-Based/Phase-In Approach 320
8.1.1.2 The Potential Value of Voluntary Commitments 332
8.1.1.3 Potential Criticism 334
8.1.2 Redirecting Financial Flows 335
8.1.3 Addressing Fast Growers and Early Movers 337
8.2 The Challenge of Finding a Legal Vehicle to Enforce Compliance with the Global MBM Scheme 337
8.2.1 ICAO Standards 338
8.2.2 ICAO Assembly Resolution 340
8.2.3 International Convention 341
8.2.4 Enforcement 343
8.2.4.1 The Value of Transparency 343
8.2.4.2 External Enforcers 345
8.2.4.3 Reporting Non-Compliance to the Assembly: Article 54 (j) 346
8.3 Additional Actions 348
8.3.1 ICAO 348
8.3.2 IATA 350
8.3.3 Europe 351
8.4 Conclusion 351

9 Concluding Remarks 353
9.1 Setting the Aviation and Climate Change Discourse and the Research Problem 353
9.2 The Interaction between International Aviation and the Climate Change Regime 354
9.3 ICAO and Climate Change 355
9.4 The EU ETS 357
9.5 The Global MBM Scheme 359
9.6 The Role Played by Some of the Main Actors 361
9.7 The Road Ahead 361

Bibliography **363**

Preface

Imprimisque hominis est propria veri inquisitio
("The first duty of man is seeking after and investigating the truth")
Marcus Tullius Cicero, *De Officiis* I., 4. 18

Every author of a book would approve Cicero's adage that a primary human instinct is to seek truth. However, more often than not, there is no general agreement on what the "truth" is. Pontius Pilate, and surely many thinkers well before him, is recorded as having asked "*quid est veritas?*" (what is "truth"?). It takes much effort to seek the truth and great courage to defend it. This book is a result of much labor and great courage, and I was privileged to witness some of its early roots.

In the last year of the twentieth century, I was screening the list of applicants for admission to the graduate program at the Institute of Air and Space Law of McGill University. Among the 27 selected applicants my highest priority was Alejandro Piera, a young lawyer from Asuncion, Paraguay, with high academic standing, interesting airline and corporate experience, and clearly defined ambitions. He was the first citizen of Paraguay to come to the Institute since its foundation in 1951, and he soon proved his abilities, qualifying for very competitive scholarships. In the classroom, he was highly visible by his focused attention, copious note taking, and initiation of lively discussions. He was endlessly curious with a probing mind that identified problems for which the teacher often did not have a ready answer. Alejandro's graduate studies at McGill University in Montréal may have motivated him for legal research and possibly helped to further sharpen his skills for a critical analysis. For me – now an ageing former teacher – he was definitely a stellar student to be remembered as one of the most promising academic prospects of his generation. It has been a joy to follow his progress.

His professional progress after graduation was quite impressive. He spent some years as a partner in a law office in Asuncion, Paraguay, practicing in aviation law, as well as advising the Government on a proposed Aeronautical Code and in air transport negotiations with the United States, Panama, and Chile. Soon thereafter, he appeared on the international scene as Senior Legal Counsel of the International Air Transport Association (IATA) that brought him into working contacts with the ICAO and its Legal Committee, with regional aviation institutions, and he was involved in the studies and advisory functions on issues of aviation security, air transport policy, and safety oversight, among others. Later he was retained by the Government of the United Arab Emirates as a Permanent Advisor on legal, policy, and environmental issues. In that position, he prepared documentation and attended many ICAO meetings, including Assemblies, Diplomatic Conferences, and Sessions of the

Legal Committee and its Sub-Committees, and became a frequent speaker at air transport conferences and seminars. A notable acknowledgment of his professionalism was his appointment as a Rapporteur of the ICAO Legal Committee on the modernization of the Tokyo Convention of 1963. As a testimony of his significant contribution to ICAO's work in the legal field, Alejandro was also elected Vice-Chairman of the ICAO Legal Committee, a pinnacle of international recognition in the field of air law.

Now Alejandro presents a book on a difficult and controversial issue. Climate change and protection of the environment have been at the forefront of international attention for a long time, but the international community is far from any agreement on how to limit the carbon footprint generated by different sectors of the economy. The fate of the 1997 *Kyoto Protocol* shows a continuing deep cleavage of attitudes among States and – regardless of the differences between developed and developing countries – no State seems to be ready for any substantial sacrifices in the economy or level of employment to protect the environment. The issue is, first and foremost, a matter of policy of individual States. The influence of the powerful lobbies of multinational corporations cannot be disregarded.

Aviation does contribute to the general atmospheric pollution through its substantial emissions of carbon dioxide, but its share is only some 2 percent, far below other sectors, such as road transport or coal-generated power plants. Yet, there is pressure on one side that aviation has to reduce its possibly growing share in the overall carbon footprint while on the other side there is resistance to any such step. Who is right? Are the airlines at fault and should they be subject to tax/fee/charge/levy for any quantity of greenhouse gas emissions above a stated basis and should any credits/debits be traded within a market-based system? The airlines do not have a choice. They have to use the fuel currently available and operate with the best available engines the manufacturers can supply and that are the real sources of pollution. Should this not be the responsibility of fuel and aircraft suppliers to motivate them to market a better product?

At the last two to three Assemblies of ICAO, the issue of climate change was very high on the order of priorities and generated unprecedented confrontational discussions; for the first time in ICAO's history, many States attached reservations to adopted resolutions. The European Union with its controversial Emissions Trading Scheme (ETS) – fortunately shelved by 2012 pending progress in ICAO – perhaps gave impetus to the 38th Session of ICAO Assembly in 2013 to seek by 2016 a model of market-based system. Alejandro is courageous in addressing the conflicting policy issues and identifying a possible legal framework for an agreed solution. However, he does not overlook the absence of any specific mandate for ICAO in the *Chicago Convention* to guide the 191 contracting States on this matter. He leans towards the introduction of a market-based system but cannot decide on the legal form the ICAO rules should take:

- A possible Annex to the *Chicago Convention* (with all the weaknesses, giving States the right to file differences)?
- A resolution of the Assembly (that would have no legally binding force)?
- A new *Convention* (how long would it take to bring it into force and how many of the 191 ICAO Member States would ratify it)?

Alejandro made an audacious and innovative step in analyzing the problem and seeking the "truth" in particular solutions. Many will still keep asking "what is truth"? There is probably a long way to go to reach a practical solution. Confucius is recorded as having said that "a journey of 1000 miles begins with one step…" Alejandro Piera has made several thousands of steps on this road.

Professor Dr. Michael Milde
Emeritus Director, Institute of Air and Space Law, McGill University

Acknowledgments

I am indebted to Professor Jaye Ellis, for her invaluable support, constant encouragement, and guidance through the very lonely undertaking of pursuing this research. Professor Ellis saw in me the ability to conclude this book long before I could recognize it myself. I am privileged to have had the support and encouragement of Professors Fabien Gélinas, Ram Jakhu, and Paul Stephen Dempsey (as well as Professor Ellis). These Professors reviewed countless drafts, provided the most constructive criticism, and motivated me. I am most grateful for all their support.

Dr. Yaw Nyampong also deserves special recognition not only for having carefully edited various chapters of this manuscript but also for having challenged a number of my assumptions. This exercise, which at times I found extremely tedious, proved to be extremely helpful. By rebutting most of my main arguments, Dr. Nyampong's criticism significantly enhanced the analysis of my topic and refined the ideas advanced. Similarly, Mr. Norberto Luongo patiently went through the entire draft. His comments were most constructive to frame the real purpose of this book. In addition to his constructive criticism on the whole draft, my long-time friend and mentor, Professor Julian Hermida, provided invaluable assistance with the introduction and the conclusion. A number of friends also took the time to provide the most helpful comments to various drafts. These include Michael Gill, Giselle A. Deiró, Mark Glynn, Kuan Wei (David) Chen, Jae Woon (June) Lee, Enrique Boone Barrera, Melissa Martins Casagrande, Sandra Brand, Maria Esmeralda Moreno, and Andrea Marcora.

I was fortunate to discuss the topic of this book with a number of renowned experts, academics, negotiators, and policy makers around the world. These exchanges were instrumental in concluding this work. These include Professor Michael Milde, Professor Pablo Mendes de Leon, Dr. Ludwig Weber, Sandra Brand, Tetsuya Tanaka, Jane Hupe, Paul Steele, Rasa Sceponaviciute, Michael Gill, Andreas Hardeman, Majid Al Suwaidi, Mariam Al Balooshi, Kevin Welsh, Dr. Lourdes Maurice, Alfredo Iglesias, Dr. Arturo Benito, Scott Stone, Brian Pearce, Tim Johnson, Philip Good, David Batchelor, Kim Carnahan, Wan Ren, Timothy Fenoulhet, Paul Lamy, Carlos Vallarino, David Blanco, Raúl Romero, Jitendra Thaker, Jim Marriott, Anda Djojonegoro, John Augustin, Benoit Verhaegen, Armando Quiroz, Christopher Dalton, Lynette Lim, Paul FitzGerald, Henry Gourdji, Mohamed Elamiri, Vince Galotti, Carlos Tornero, Aurora Torres, Víctor Aguado, Theodore Thrasher, Blandine Ferrier, Iulia Bogatu, Stephan Sicars, Bernd Hackmann, Rob Bradley, Carlos Grau Tanner, Nancy Young, Lynette Lim, Gordana Milinic, Brian Day, Narjess Abdennebi, Juan Carlos Salazar, Jerome Simon, Annie Petsonk, Jenny Cooper, Arie Jakob, John Ilson, Jimena Blumenkron, Maria Liz Viveros, Constance O'Keefe, Bill Hemmings,

Sue-Ann Rapattoni, Auguste Hocking, Jongin Bae, Sara Harrison, Miguel Marin, Nathalie Herbelles, Andrew Parker, Paul Zissermann, Linden Coppell, Mark Antaki, Roderick MacDonald, and Niel Dickson.

My colleagues of the UAE Delegation to ICAO have been extremely supportive. I would like to recognize the support, understanding, and encouragement I have received from Capt. Aysha Al Hamili, Mr. Rashed Ali Al Kaabi, Ms. Sanaa Kiadi, and Ms. Lara Zidine. I also acknowledge the constant encouragement I have received throughout these years from my friends and colleagues Alejandro Guanes Mersan, Fernando Heisecke, and Jose Ignacio Santiviago. My long time friends Miguel Arnando Canale, Perla Alderete, Cynthia Fatecha, Patricia Vitale, Judith Gauto, Ariel Ojeda, Mirta Dos Santos have also been instrumental in my legal education.

Numerous people provided invaluable childcare assistance for my kids. This allowed the needed time for academic seclusion. Amongst others, these include Romina Romero, Diego Rodriguez, Carmen Casanova, Teresa Marmolejo, Gaston Iglesias, Paola Valdredi, and Raul Acosta. My parents-in-law (Hugo and Maria Ofelia) and my parents (Jesus and Maria Nydia) also came to Canada several times during the course of this academic endeavor. I cannot thank them enough. Their help was tremendous. I remain most obliged for the praying support of various members of my family, including Mark and Carol Loomis, Diosma Frescura, Chochita de Perán, and my beloved parents. My parents, along with my late grandfather, Joaquin Rogelio, have been extremely supportive.

Last but certainly not least, I will always be indebted for all the love, understanding, cheering, renunciations, and support I have been blessed with from my wife, Alicia, and my two wonderful kids, Octavio Augusto and Fabrizio José. They are the joy of my life and the source of all my inspiration. They have endured so much through this academic endeavor. I could have not done it without you!

Thank you all! The shortcomings remain all mine.

Abbreviations

AATS	Annex on Air Transport Services
ACI	Airport Council International
AEA	Association of European Airlines
AHWG	Ad Hoc Working Group on market-based measures
AOC	Aircraft Operator Certificate
API	Advance Passenger Information
APU	Auxiliary Power Unit
ASA	Air Services Agreements
AWG	Aviation Working Group
BCA	Border Carbon Adjustment
BSA	Burden Sharing Agreement
BTA	Border Tax Adjustment
CAB	Civil Aeronautics Board
CAEE	Committee on Aircraft Engine Emissions
CAEP	Committee on Aviation Environmental Protection
CAN	Committee on Aircraft Noise
CANSO	Civil Air Navigation Services Organization
CBDR	Common But Differentiated Responsibilities and Respective Capabilities
CDA	Continuous Descent Approach
CDM	Clean Development Mechanism
CERCLA	Comprehensive Environmental Responsive Compensation and Liability Act
CGPM	Conférence générale des poids et mesures/General Conference of Weights and Measures
CJEU	Court of Justice of the European Union
CMR	Convention on the Contract for the International Carriage of Goods by Road
CNG	Carbon Neutral Growth
EAG	Environment Advisory Group
EC SAFA	European Safety Assessment of Foreign Aircraft
ECAC	European Civil Aviation Conference
ECU	Engine Control Unit
EDF	Environmental Defense Fund
EDTO	Extended Division Time Operations

EEC	European Economic Community
EEZ	Exclusive Economic Zone
EPS	Emissions Performance Standards
ERU	Emission Reduction Unit
ETOPS	Extended Range Twin-Engine Operations
EU ETS	European Union Emission Trading Scheme
EUAAs	European Union Aviation Allowances
EUAs	European Union Allowances
FAA	Federal Aviation Administration
FIR	Flight Information Region
FTC	Federal Trade Commission
GATS	Agreement on Trade in Services
GATT	General Agreement on Tariffs and Trade
GDP	Gross Domestic Product
GHG	Greenhouse Gas/Greenhouse Gases
GIACC	Group on International Aviation and Climate Change
GLADs	Global Aviation Dialogues
HGCC	High Level Group on International Aviation and Climate Change
IASA	International Aviation Safety Assessment
IATA	International Air Transport Association
ICAO	International Civil Aviation Organization
ICJ	International Court of Justice
ICTY	International Criminal Tribunal for Yugoslavia
IEnvA	IATA's Environmental Assessment Program
IFSO	In-Flight Security Officers
IHR	International Health Regulations
ILC	International Law Commission
IMO	International Maritime Organization
ISO	International Organization for Standardization
IUU	Illegal, Unreported and Unregulated Stocks
JI	Joint Implementation
LLDCS	Least Developed, Land Locked States
MBM	Market-Based Measure
MEA	Multilateral Environmental Agreements
MIT	Massachusetts Institute of Technology
MRV	Monitoring, Reporting, and Verification
NAMAs	Nationally Appropriate Mitigation Actions

NGO	Non-Governmental Organization
PCIJ	Permanent Court of International Justice
PPM	Process Production Method
QELRCs	Quantified Emissions Limitation/Reduction Commitments
REDD-plus	Reducing Emissions from Deforestation, Degradation and Forest Enhancement
RTK	Revenue Tonne Kilometer
SAFA	Safety Assessment of Foreign Aircraft
SARPs	Standards and Recommended Practices
SCRC	Special Circumstances and Respective Capabilities
SIDS	Small Island Developing States
SPS	Sanitary and Phytosanitary Measures
TBT	Technical Barriers to Trade
TED	Turtle Excluder Device
TFEU	Treaty on the Functioning of the European Union
TSA	Transportation Security Administration
UNCLOS	UN Convention on the Law of the Sea
UNFCCC	United Nations Framework Convention on Climate Change
USAID	United States Agency for International Development
USOAP	ICAO's Universal Safety Oversight Audit Program
VCLT	Vienna Convention on the Law of Treaties
WGOG	Working Group on Governance
WTO	World Trade Organization

Introduction

Background

Despite the cyclical crises in international aviation, the sector is expected to grow annually by approximately 5 percent,[1] making it the fastest growing mode of transportation.[2] The sector's enormous contribution to the world's economy and society at large is undeniable;[3] however, international aviation's projected growth poses significant challenges to the environment and in particular to concerns over climate change.[4] At present, international aviation contributes approximately 2 percent of total global greenhouse gas (GHG) emissions.[5] If the international aviation industry were a State, it would rank seventh in the world in terms of global contribution of GHG emissions.[6] Aviation's carbon footprint is expected to increase 3 to 4 percent annually, leading the sector to introduce a number of technological and operational measures to reduce its negative environmental impact.[7] However, these initiatives will not offset the emissions expected to be generated as a result of its projected growth. At this pace and in the absence of regulation or massive deployment of alternative fuels, reports estimate that aviation's CO_2 emissions will rise fourfold by 2050.[8]

---

1 See Airbus, "Global Market Forecast: Future Journeys 2013-2032" *Airbus* online: <www.airbus.com/company/market/forecast/?eID=dam_frontend_push&docID=33752> at 13 [Airbus, Global Market Forecast 2013-2032].

2 See Bert Metz et al., *Climate Change 2007: Mitigation. Contribution of Working Group III to the Fourth Assessment Report* (Cambridge: Cambridge University Press, 2007) at 330 [IPCC/4].

3 See "The Economic and Social Benefits of Air Transport 2008", *ATAG* online: <www.iata.org/pressroom/facts_figures/fact_sheets/Pages/fuel.aspx?NRMODE=Unpublished> [ATAG, Economic and Social Benefits of Air Transport 2008].

4 See Brian Graham & Jon Shaw, "Low-Cost Airlines in Europe: Reconciling Liberalization and Sustainability" (2008) 39 Geoforum 1439.

5 See Jane Hupe, "Towards Environmental Sustainability" in ICAO, *2013 Environmental Report: Destination Green* (Montréal: ICAO, 2013) at 11.

6 See ICAO, HGCC/3-AIP/6 at 1 (Presentation of the International Coalition for Sustainable Aviation at the third meeting of the High Level Group on International Aviation and Climate Change, 25-27 March 2013) [HGCC/3-AIP/6].

7 See ICAO CAEP/8-WP/80, *Committee on Aviation Environmental Protection (CAEP) Eighth Meeting, Montreal, 1 to 12 February 2010 Agenda Item 1* at 1-32.

8 See "Hedegaard Sets out Conditions on ICAO Agreement as EU Legislators Approve EU ETS 'Stop the Clock' Measure", *Green Air Online* (17 April 2013) online: <www.greenaironline.com/news.php?viewStory=1681> ["Hedegaard Sets out Conditions on ICAO Agreement"]. ICAO's estimates are a bit more conservative. In 2013, ICAO forecasted that by 2050 GHG emissions from international aviation will increase between 2.8 and 3.9 times compared with 2010 levels. See Gregg Fleming and Urs Ziegler, "Environmental Trends in Aviation to 2050" in ICAO, *supra* note 5 at 22.

Those outside aviation circles have been unimpressed by the sector's achievements in confronting climate change. Critics argue that international aviation should be responsible for the environmental externality it generates.[9] In addition, a number of international institutions have singled out international aviation as a potential source of funds for mitigation and adaptation activities in the wider climate change context. Likewise, various States have levied international departure taxes (e.g. embarkation taxes) on the aviation sector for climate change purposes although it is not entirely clear whether these funds are allocated to climate change-related measures.[10] Increasingly, international aviation is facing mounting pressure to address its climate change impact.

In an effort to establish an overall legal framework to regulate anthropogenic GHG emissions, the international community adopted the *United Nations Framework Convention on Climate Change* (*UNFCCC*) in 1992.[11] In order to operationalize this framework, States adopted the *Kyoto Protocol* during the third Conference of Parties (COP/3) of the *UNFCCC* in 1997.[12] This instrument tasked developed States to limit or reduce GHG emissions from aviation by working through the International Civil Aviation Organization (ICAO).[13] In practice, this has been interpreted as an implicit mandate bestowed upon ICAO by the *UNFCCC* regime to handle the aviation sector's GHG emissions. However, the mandate is extremely vague and imprecise, allowing States to interpret it in furtherance of parochial national interests. This mandate has also made it extremely difficult to reconcile the *UNFCCC*'s cardinal principle of common but differentiated responsibilities and respective capabilities (CBDR) with the *Convention on International Civil Aviation's* non-discrimination principle.[14] In the *UNFCCC* context, on the basis of CBDR, only developed States bear quantified emissions limitation/reduction commitments (QELRCs). Developing countries ought to report emissions, maintain national inventories, and cooperate to achieve the regime's objectives. In international civil aviation, by invoking CBDR, developing countries argue that their aircraft operators should not be subject to emission reduction obligations. Developed States contend that CBDR cannot be applied in civil aviation, because it contravenes ICAO's non-discrimination principle and leads to market distortions.

---

9 See Joyce E Penner et al., eds, *Aviation and the Global Atmosphere*, Published for the Intergovernmental Panel on Climate Change (Cambridge: Cambridge University Press, 1999) at 341 [IPCC Aviation Report].

10 See ICAO, HGCC/3-IP/5, *Addressing Carbon Emissions From Aviation: Industry Views* (Presented by the Air Transport Action Group (ATAG) on behalf Airports Council International (ACI), Civil Air Navigation Services Organisation (CANSO), International Air Transport Association (IATA), International Business Aviation Council (IBAC), International Coordinating Council for Aerospace Industries Associations (ICCAIA)).

11 See *United Nations Framework Convention on Climate Change*, 9 May 1992, 1771 UNTS 107 [*UNFCCC*].

12 See *Kyoto Protocol to the United Nations Framework Convention on Climate Change*, 11 December 1997, UN Doc FCCC/CP/1997/7/Add.1, 37 ILM 22 (entered into force 16 February 2005) [*Kyoto Protocol*].

13 *Ibid* at art 2.2.

14 Also known as the *Chicago Convention*; see Chapter 2.

A long time has passed since the *Kyoto Protocol* was adopted, yet GHG emissions from international aviation continue to be essentially unregulated. ICAO has undertaken an impressive amount of technical work. It has produced myriad guidance material and recommendations on the subject for States.[15] Currently, however, there is no system in place to reduce or limit emissions. These continue to grow. Frustrated by the lack of progress at ICAO and the unwillingness of the international community, the European Commission unveiled a comprehensive plan to include foreign aircraft operators into its emissions trading scheme (EU ETS) in 2005.[16] To this end, the European Union adopted Directive 2008/101/EC in December 2008,[17] which indicated its intention to subject all European and foreign aircraft operators flying to and from airports situated in the EU to the EU ETS with effect from 1 January 2012. Under the Directive, emissions were to be computed for the whole duration of the flight even when portions thereof were produced in airspace that is outside European jurisdiction (i.e. over the high seas or over the airspace of non-EU States). Strong international opposition was assembled to challenge the scheme. As a result, in November 2012, Europe was forced to suspend the application of the scheme to foreign aircraft operators for a period of one year.[18] In addition, in April 2014, the European Parliament announced that the scheme will be put on hold until 2017.[19]

Ever since Europe announced its intention to incorporate aviation into its ETS, the airline industry, led by the International Air Transport Association (IATA), has been extremely active in addressing climate change issues. IATA has adopted non-binding, aspirational industry targets that have proved to be very influential in ICAO's work.[20]

In the meantime, at the 38th session of the ICAO Assembly held in October 2013, ICAO's Member States finally agreed to develop a global market-based measure (MBM) to regulate GHG emissions from international aviation.[21] Specifically, the Assembly tasked the Council to work on the design elements of a concrete MBM proposal to be presented for consideration at the 39th Assembly in September 2016.[22] At this stage, it is expected

15 See Chapter 3.

16 See EC, *Communication from the Commission to the Council, the European Parliament, the European Economic and Social Committee and the Committee of the Regions Reducing the Climate Change Impact of Aviation*, COM (2005) 459 (Brussels, 27 September 2005).

17 EC, *Parliament and Council Directive 2008/101 of 19 November 2008 Amending Directive 2003/87/EC so as to Include Aviation Activities in the Scheme for Greenhouse Gas Emission Allowance Trading Within the Community*, [2008] OJ, L 8/3 [Aviation Directive].

18 See EC, "Stopping the Clock of ETS and Aviation Emissions Following Last Week's International Civil Aviation Organization (ICAO) Council", MEMO/12/854 (Brussels, 12 November 2012).

19 See Chapter 4.

20 See Giovanni Bisignani, *Words of Change* (Geneva: IATA Corporate Communications, 2011) at 96-98 [Bisignani, *Words of Change*].

21 ICAO, A38-18, *Consolidated Statement of Continuing ICAO Policies and Practices Related to Environmental Protection: Climate Change* at preamble, para 18 [ICAO, A38-18].

22 *Ibid.* at para 19 (d).

that the Assembly will adopt such a scheme to be implemented as of 1 January 2020. At present, ICAO is working toward developing such a scheme.

The Research Problem

GHG emissions from international aviation continue to grow steadily. At their current rate of growth and projected future growth, technological and operational efficiencies cannot be relied on to offset emissions. Without regulation, the international aviation sector will not be in a position to reduce its emissions. It is certainly in the best interest of the sector to do so now since failure to tackle the issue will only result in the imposition of an enormous burden upon the sector by external forces. For instance, in an attempt to fund 100 billion US dollars annually until 2020 for climate change mitigation in developing countries, the UN Secretary-General's High Level Advisory Group on Climate Change Financing has suggested that international air transport could contribute between 1 and 6 billion US dollars annually through the establishment of levies.[23] Likewise, a World Bank report prepared for the G20 recommended levying 25 US dollars per tonne of CO_2.[24] This proposal could levy a fee of up to 12 billion US dollars per year for climate financing.[25] In this context, the sector can no longer postpone an appropriate regulatory mechanism to address GHG emissions from international aviation.

Research Objective

This book examines aspects of the legal framework underlying the aviation and climate change discourse with a view to providing some recommendations that may facilitate the adoption, implementation, and, ultimately, compliance with ICAO's global MBM to limit GHG emissions from international aviation. In other words, it seeks to analyze certain issues that have directly or indirectly influenced discussions on aviation and climate change and that have played or will play a significant role in ICAO's global MBM scheme. To this end, the book deals with five broad areas. First, it examines ICAO's relationship with the climate change regime, the implicit mandate provided by the *Kyoto Protocol*, and the interplay between the core principles of the *UNFCCC* regime and those of the international

23 See *Report of the Secretary-General's High-Level Advisory Group on Climate Change Financing* (5 November 2010) online: UN <www.un.org/wcm/webdav/site/climatechange/shared/Documents/AGF_reports/AGF%20 Report.pdf> at 28.

24 See International Monetary Fund and the World Bank, *Market-Based Instruments for International Aviation and Shipping as a Source of Climate Finance, Background Paper to the G20 on the Mobilizing Sources of Climate Finance*, (November 2011) online: IMF <www.imf.org/external/np/g20/pdf/110411a.pdf> at 5 [International Monetary Fund and the World Bank Report].

25 *Ibid.*

aviation regime. In this connection, the theory of fragmentation of international law is used as an analytical framework for the resolution of normative conflict. Second, the book explores ICAO's institutional setting and its suitability for handling climate change. Third, it comprehensively analyzes the EU ETS, its influence on the climate change discourse, as well as the legal issues associated with it. Fourth, by resorting to the theory of norm entrepreneurship, the book assesses the role of the major players involved in the climate change discourse. Fifth, in its closing chapters, the book highlights issues that should be carefully considered during the design of ICAO's global MBM. These considerations should not be construed as an attempt to develop a comprehensive global MBM scheme for international aviation. Such an endeavor requires substantial qualitative and quantitative research that is outside the scope of this project.

Relevance of the Subject Matter of the Research

Climate change exerts enormous pressure on States, industry stakeholders, non-governmental organizations (NGOs), and ICAO. Climate change has almost hijacked the last four sessions of the ICAO Assembly to the point where discussions on other aviation-related issues were put on hold until an agreement could be reached. This is a testimony to the subject's predominance.[26] In recent years, ICAO's budget for environmental protection has increased, whereas funds for activities of other more traditional sectors of the industry are diminishing.[27] There is no indication that the significance and importance accorded to the aviation and climate change issue is likely to dwindle in the near future, and it is up to the international aviation community to stand ready for the challenge.

Original Contribution to Legal Research and Knowledge

This book contributes to legal research and knowledge in at least five ways. First, commentators have been quick to point out ICAO's lack of achievement and slow progress on climate change issues. Yet, to date, there has not been a careful examination of ICAO's institutional setting and an assessment of whether the organization is the appropriate forum to address climate change issues. Also missing from the literature is an assessment of whether ICAO's institutional setting inhibits or promotes participation by Member States. This book tackles precisely this issue[28] by examining ICAO's constitutional framework and its recently adopted strategic objectives and proposing realistic corrective

26 See Michel Wachenheim, "Interview with Michel Wachenheim: President of the 38th Session of the ICAO Assembly" in *The European Civil Aviation Conference Magazine, ECAC News* 50 (Winter 2013) 10 at 11.

27 See ICAO Doc 10030, *Budget for the Organization 2014-2015-2016*, (2013) [ICAO, Budget 2014-2016].

28 See Chapter 1.

measures that may ultimately bear influence in the design and implementation of ICAO's global MBM.

Second, most of the legal scholarship on aviation and climate change primarily centers on the legality or illegality (depending on the reader's point of view) of the EU ETS. Perhaps this is understandable if one considers that, since it was first announced in 2005, this regional initiative has monopolized discussions on aviation and climate change. While this book explores some of the legal issues surrounding the EU ETS (such as its unquestionable extraterritoriality) in great detail, it goes further to analyze issues such as the contribution of the EU ETS to the ICAO process of regulating GHG emissions from international aviation and the negative impacts as well as the missed opportunities of the EU ETS. This book is innovative in the sense that it provides a critical assessment of the EU ETS rather than simply defending or attacking the scheme. It also draws lessons from the EU ETS that may enhance the design and implementation of ICAO's global MBM scheme and enable it to avoid certain pitfalls.[29]

Third, most of the political discourse and industry allegations on aviation climate change issues have been dominated by self-interested statements. Quite often, these have been taken for granted. By challenging these assumptions, this book not only identifies the merits and shortcomings of the activities carried out by the main actors involved but also identifies potential corrective measures that may contribute to the success of ICAO's global MBM scheme.[30]

Fourth, State representatives and scholars have argued that either CBDR or the non-discrimination principle should govern ICAO's work on climate change. This approach, which selects one principle over the other depending on the political perspective held by the proponent(s), has not been successful in achieving the intended results. I attempt to reconcile these seemingly divergent principles in a manner that is compatible with the aviation industry. In fact, it advances some concrete proposals as to how this reconciliation may be operationalized in the design of the ICAO's global MBM.[31] This may well be a fundamental issue for the success of the scheme, for it may facilitate broader acceptance.[32]

Fifth, although States, industry stakeholders, ICAO, and many authors have proposed ideas for the global MBM scheme, there has been no consideration of the legal vehicle to bring about the optimal adoption of such a scheme. If unaddressed, this may well prove to be the stumbling block. This book comprehensively analyzes the advantages and disad-

---

29 See Chapters 4-6.

30 See Chapter 7.

31 See Chris Lyle's perspective on Green Air Online, "Mitigating International Air Transport Emissions through a Global Measure: Time for Some Lateral Thinking" (2014) (noting that "ICAO equal application principle and the *UNFCCC*'s CBDR can be reconciled through transitional implementation of MBMs on a route or route group basis") online: Green Air Online <www.greenaironline.com/news.php?viewStory=1820> [Lyle, "Mitigating International Air Transport"].

32 See Chapter 8.

vantages of a number of legal vehicles that could potentially be used to adopt the scheme. Mindful of the significant pressure on ICAO to adopt the scheme by 2016 and the time constraints involved, I also explore some realistic and practical enforcement mechanisms.

METHODOLOGY

The book examines primary sources such as jurisprudence, legislation, regulations, and proposed amendments to existing legislation. It reviews secondary legal materials such as books, journal articles, and theses. Given their strong influence in the aviation and climate change discourse, the book also analyzes industry reports, studies, and publications in this field. To avoid a natural bias, such data is often contrasted with that provided by other organizations such as ICAO and NGOs.

In light of the fact that ICAO serves as the institutional platform where, for the most part, aviation and climate change discussions take place, the book devotes significant consideration to the organization's assembly resolutions, guidance materials, standards and recommended practices, manuals, and the minutes and the decisions of the Council. More specifically, the book tracks all Council discussions and Assembly resolutions on the issue of climate change since 1998, the year following the adoption of the *Kyoto Protocol*. This is intended to facilitate understanding of the different positions of States and also to document the progressive evolution of this subject matter at ICAO.

Although there are some scattered references to *UNFCCC* documents, for the most part, the book is not so much concerned with these sources. Admittedly, the *UNFCCC* process influences discussions at ICAO. However, much if not all of the aviation and climate change discussions have occurred at ICAO. It is also evident that Sates and industry stakeholders prefer to hold such deliberations under the auspices of ICAO. Similarly, the *Kyoto Protocol* expressly tasked ICAO to address GHG emissions from international aviation. At this stage, it is highly unlikely that the *UNFCCC* would be able and willing to address aviation and climate change issues.

Although this book primarily studies legal materials, it also relies on some political science and international relations literature. This is particularly the case in Chapter 7 where the issues of norm entrepreneurship and norms in general are considered.

ORGANIZATION OF CHAPTERS

The book comprises nine chapters. Chapter 1 describes the setting in which the aviation and climate change discourse evolves. In particular, it discusses the growth trends in the aviation sector, its expected contribution to global GHG emissions, as well as the technological and operational efficiencies introduced by the industry. In essence, this chapter makes the point that, given its projected growth trends in the next 20-30 years, technological and operational efficiencies will not be sufficient to limit international aviation's GHG emissions. The chapter also suggests that, at this stage, there is no indication that alternative fuels will be available in sufficient quantities and at reasonable prices to play a major role in reducing the sector's emissions – at least in the foreseeable future. In addition to briefly introducing the international legal regime, the chapter explains the risks that the sector is likely to face if it decides not to tackle its climate change impact.

Through the lens of the theory of fragmentation of international law, Chapter 2 looks into the interplay between the climate change regime and ICAO. More specifically, the chapter thoroughly investigates the implicit mandate that the *Kyoto Protocol* granted to ICAO to limit or reduce GHG emissions from international aviation. It examines the practical and operational implications of this mandate, studying the apparent conflict between the *UNFCCC*'s CBDR principle and the *Chicago Convention's* non-discrimination principle. Seeking to reconcile these two principles, the chapter explores the seminal work of the International Law Commission (ILC) on fragmentation and, more specifically, whether the rules contained in the *Vienna Convention on the Law of Treaties* (*VCLT*) governing conflicts between treaties and other legal obligations may provide a viable solution. The chapter advances an innovative approach to reconciling these principles; a central concern of this book is reconciling these two principles in the design of ICAO's global MBM. This should facilitate adherence to the scheme by a broader audience. In fact, it is submitted that one of the reasons why States have yet to implement a measure to limit or reduce GHG emissions from international aviation has been their inability to agree on how the relationship between these two principles should be handled.

Chapter 3 addresses ICAO's involvement in climate change. This assessment is necessary in order to better understand what the organization may achieve and what corrective measures may be implemented to contribute toward the success of the global MBM scheme. Before discussing the organization's specific achievements in this field, the chapter carefully examines ICAO's institutional setting and its suitability to handle climate change issues. Particular attention is paid to its governing structure and its institutional objectives, and whether these inhibit or foster participation by its Member States in the climate change discourse. Similarly, the chapter assesses the role of industry stakeholders and NGOs and the influence (or lack thereof) they exert in the formation of climate change policy at ICAO. It also examines the work of the Committee on Aviation Environmental Protection (CAEP) – ICAO's technical body in charge of environmental issues.

After providing a brief historical background, the chapter then examines ICAO's specific work on climate change with particular emphasis on the CO_2 standard, State actions plans, the framework for MBMs, and the agreement to develop a global MBM scheme.

Chapters 4 to 6 deal with the EU ETS. Considerable attention is devoted to the scheme's architectural design as well as the policy and legal issues involved. Although a number of legal arguments have been advanced against the EU ETS, the most serious legal allegation has focused on the issue of extraterritoriality. Because the original geographical scope of the scheme envisioned that fuel consumption would be computed for the whole duration of the flight, non-EU States did contend that it was tantamount to an unlawful exercise of extraterritorial jurisdiction. To examine these allegations, the book digs into the principles of international law on jurisdiction. It also explores whether the principles of State responsibility may exonerate the EU ETS.

A thorough analysis of the European scheme, how it came into existence, its struggles, and missed opportunities, is necessary not only to provide a clear understanding of how ICAO, its Member States, and the airline industry reached an agreement to develop a global MBM scheme but also to take advantage of the true potential of the scheme. In fact, the EU ETS is central to the purpose of this book for a number of reasons. First, it stands as a policy response to the unwillingness of the international aviation community to regulate GHG emissions from international aviation. Europe decided to include foreign aircraft operators into its ETS only after it became abundantly clear that States gathered at ICAO were unwilling to do so.

Second, the EU ETS prompted the formation of an unprecedented international coalition to challenge it. As described in Chapter 4, instead of working toward establishing a mechanism to regulate GHG emissions from international aviation at ICAO, States and the airline industry devoted an enormous amount of time, effort, and resources in a concerted attempt to derail the EU ETS.

Third, in spite of the dire opposition it ignited, the EU ETS has exerted considerable pressure on ICAO, its Member States, and industry stakeholders to take action. Had it not been for the EU ETS, ICAO would not have agreed to develop a global scheme for international aviation, and it is also doubtful that the airline industry would have ever developed a dedicated strategy to combat climate change. As explored in Chapter 7, the industry's engagement in climate change came about as a reactive response to the EU ETS. In addition, the fact that climate change now ranks very high on ICAO's policy agenda may, to a large extent, be attributed to the European normative push.

Fourth, the EU ETS represents the first robust normative approach to regulate GHG emissions from international aviation. Moreover, having been implemented by States representing almost 23 percent of the ICAO membership, the EU ETS serves as a fantastic learning experience for ICAO's global MBM scheme. For instance, it is highly likely that in developing standards for GHG emissions monitoring, reporting, and verification (MRVs), ICAO will resort to the same methodologies as those applied in the EU ETS. Further, given the animosity that the EU ETS has prompted, it is quite likely that the international community will adopt any other regulatory option apart from emissions trading when designing ICAO's global MBM scheme. As a policy instrument, emissions trading is viewed as a highly toxic regulatory option because it is so closely linked to Europe.

The climate change discourse is full of dogmatic statements. Most often, these are taken at face value without much corroboration. The industry contends that it has a comprehensive strategy to address climate change issues; it has set ambitious targets and is on course to achieve them. ICAO says that it is the first UN-specialized agency with a concrete global plan to address GHG emissions from the sector it regulates. Other States also claim that emissions generated by their respective aviation sectors have been dramatically reduced. If this is the case, then climate change should not pose a major challenge to international aviation. To assess the various actors' achievements, shortcomings, and missed opportunities, Chapter 7 relies heavily upon the theory of norm entrepreneurship. In so doing, the chapter identifies those actors who by challenging the status quo have pushed for normative change. The application of the norm entrepreneurship theory also enables us to identify the conditions under which these changes are likely to occur and the requirements for norm internalization (e.g. acceptance). In addition, this assessment facilitates the identification of corrective measures that may have to be taken into account for a successful global MBM scheme. It also provides an appropriate platform to assess the level of acceptance that the proposed normative change (e.g. proposal to adopt global MBM scheme) has received within the membership of ICAO and, should it be necessary, the required measures to introduce corrective action.

In Chapter 8 the book presents some issues that must be considered in the design of the global MBM scheme. Particular attention is paid to a practical way of integrating the principles of CBDR and non-discrimination. In this regard, Chapter 8 proposes the adoption of a route-based approach where different routes will be phased in accordance to a set of criteria. It also examines in detail the potential legal vehicles to be used in adopting such a global scheme, their respective benefits and drawbacks, as well as possible enforcement mechanisms. It highlights some additional roles that key actors should play that may contribute to the success of the global MBM scheme. Finally, Chapter 9 provides some concluding remarks. As this project deals with evolving issues, the reader should be aware that the facts and data presented herein are up to date as of November 2014.

1 Setting the Scene

This chapter provides context for international aviation and climate change discourse. It makes clear that as international aviation is expected to experience growth, GHG emissions will continue to increase. Despite the sector's introduction of technological and operational efficiencies, these measures will not be able to offset the growth in GHG emissions. Given that alternative fuels will not be available in sufficient and affordable quantities in the short to medium term, regulation will be required to limit or reduce international aviation's carbon footprint.

By examining various market outlooks, Section 1.1 describes international aviation's growth trends. Although traffic is expected to increase in all regions, this growth will predominantly take place in developing countries, underscoring the importance of developing countries' participation in addressing GHG emissions from international aviation. Section 1.1 also examines aviation's contribution to the world's economy and argues that its GHG emissions are a key consideration in policy measures that address climate change. As GHG emissions grow exponentially, they will become important to international efforts to reduce climate change. Section 1.1 also examines the defenses that the sector has developed in response to the criticism that it has failed to act on climate change. Both industry and ICAO have placed strong emphasis on aviation's small contribution to global GHG emissions (e.g. only 2 percent). They have also suggested that it is not that the sector has failed to achieve results in addressing climate change issues but that its efforts and achievements have not been properly communicated. In addition, they have overemphasized the role of technological improvements in reducing GHG emissions. In addressing these defenses, I dismiss the argument that, in light of ever-increasing fuel prices, regulation is not needed to limit GHG emissions in international aviation and provide concrete examples.

Section 1.2 briefly introduces the international legal regime applicable to aviation and climate change. The section highlights some of the main features of the *UNFCCC* regime and the *Kyoto Protocol*. It also discusses the interaction between the climate change regime and ICAO. As a major force in addressing aviation and climate change issues, the section also presents the EU ETS and the inclusion of international aviation.

Section 1.3 explains the political issues associated with aviation and climate change and the direct effect that discussions at ICAO may have in the *UNFCCC* context and vice versa. It also addresses the problems associated with the perception that ICAO has not been able to deliver substantial results.

Section 1.4 underlines the exogenous threats affecting international aviation in its quest to address climate change. The section analyzes the risks associated with the failure on the part of the sector to take action. It suggests that taking proactive action is in the sector's best interests, given that other proposals advanced outside aviation circles will be more onerous.

1.1 UNDERSTANDING THE PROBLEM

1.1.1 Aviation Market Outlook and Economic Contribution

Aviation is one of the most sophisticated sectors of the global economy. It is the fastest-growing mode of transportation.[1] As a result of economic growth, globalization, and progressive liberalization, which have gradually eradicated barriers to market access and lowered airfares, air transport has rapidly become more accessible, which has increased its demand.[2] Aviation promotes social inclusion by connecting people like no other industry in the world, and each day, more and more people travel by air.[3] While the aggregate economic activity of the world – measured by Gross Domestic Product (GDP) – has escalated at an average annual rate of 3.9 percent in real terms since 1960, the air transport industry has grown at a considerably greater pace.[4] Air traffic has increased

---

1 See IPCC Aviation Report, *supra* note 9 (Introduction) at 330. Furthermore, aviation has surpassed road transportation as the fastest growing mode of transportation. From 2003 to 2004 aviation grew at a rate double to that of road transport. See IEA, *World Energy Outlook 2006* (Paris: OECD/IEA, 2006) at 231 [IEA]. Please note that footnotes will begin at 1 for each chapter. Sources cited in previous chapters are referenced in subsequent citations (e.g. *supra* note 2 (Introduction) for a source cited in a subsequent chapter).

2 See Airbus, "Global Market Forecast 2013-2032," *supra* note 1 (Introduction); Annela Anger, "Including Aviation in the European Emissions Trading Scheme: Impacts on the Industry, CO_2 Emissions and Macroeconomic Activity in the EU" (2010) 16 J Air Transp Manag 100 [Anger, "Including Aviation in the EU ETS"] (pointing out that the sector's "rapid expansion and estimated future growth [is] caused by globalization, increasing GDP, liberalization of the air transport market combined with current matured technology and the appearance of the low fares business models" at 104); Jane Barton, "Including Aviation in the EU Emissions Trading Scheme: Prepare for Take-Off" (2008) 5:2 JEEPL 183 [Barton, "Including Aviation in the EU ETS"] (identifying as some of the causes of the sector's growth the "liberalisation of the air transport market, new leisure patterns, higher incomes and the increasing demand for and value of goods transported" at 183). According to statistics of the International Civil Aviation Organization (ICAO), in 2012, the world's airline traffic grew approximately 4.9 percent compared with 2011. See ICAO, Doc 10001, *Annual Report of the Council 2012*, at 1 [ICAO Council Annual Report, 2012].

3 See Bisignani, *Words of Change*, *supra* note 20 (Introduction) (noting that "[a]ir transport has never been more accessible" at 53).

4 See ICAO, Cir 313, *Air Transport Outlook to the Year 2025* (2007) at 1 [ICAO, *Air Transport Outlook*]; ICAO, Doc 10012, *Committee on Aviation Environmental Protection, Report of the Ninth Meeting, Montréal 4-15 February 2013* (2013) at 1-6 [ICAO, CAEP/9].

tenfold in the last 30 years.[5] Passenger traffic, expressed in revenue passenger kilometers,[6] has grown at a yearly rate of 9 percent since 1960, and only since 1997 has it slowed down to roughly 5 percent.[7] This is 2.4 times greater than the world's average GDP growth rate.[8]

Air travel statistics are projected to more than double over the next 15 years.[9] ICAO estimates that between 2005 and 2025, the world's GDP is expected to increase at an average annual rate of 3.5 percent. ICAO also projects that global air transport will grow at a rate of 4.9 percent annually.[10] International traffic is expected to grow faster at 5.1 percent, whereas the more mature domestic traffic will experience a 4.4 percent annual increase.[11] Aircraft manufacturers' forecasts are even more optimistic.[12] Reflecting a shift in the dynamics of the sector, most of this growth is taking place in developing countries.[13]

5 See Airbus, "Global Market Forecast 2009-2028," online: Airbus <www.airbus.com/en/corporate/gmf2009/> [Airbus, "Global Market Forecast 2009-2028"]. Boeing estimates that, from 2012 to 2032, while the world's GDP will grow 3.2 percent, airline's traffic will increase by approximately 5 percent per year. See Boeing, "Current Market Outlook 2013-2032," *Boeing* online: <www.boeing.com/boeing/commercial/cmo/> [Boeing, "Market Outlook 2013-2032"].

6 RPK or revenue passenger kilometer, which is one of the traffic measures used in international civil aviation, refers to one revenue paying passenger carried one kilometer.

7 See Airbus, "Global Market Forecast 2013-2032," *supra* note 1 (Introduction) at 13.

8 See IPCC Aviation Report, *supra* note 9 (Introduction).

9 See Airbus, "Global Market Forecast 2013-2032," *supra* note 1 (Introduction) at 3.

10 See ICAO, CAEP/9, *supra* note 4 at 1-6.

11 *Ibid.*

12 Airbus forecasts that "traffic growth is expected to average 4.7 [percent] per year." Airbus, "Global Market Forecast 2013-2032," *supra* note 1 (Introduction) at 13. In addition, the tourism industry – highly dependent on air transport – is rapidly developing into one of the largest service industries in most regions. See Ludwig Krämer, *EC Environmental Law* (London: Sweet & Maxwell, 2007) at 414.

13 Various market outlooks suggest that most of the sector's growth will take place in developing countries. For instance, Boeing points out that while, in 1992, the traffic between North America and Europe accounted for almost 70 percent of international traffic, by 2032 it is expected that it will drop to 39 percent. Aviation's growth will essentially take place in Asia Pacific, Latin America, and the Middle East regions. See Boeing, "Current Market Outlook 2013-2032," *supra* note 5. ICAO studies indicate that, while air traffic in Europe and North America increased at a rate of 3.9 and 1.3 percent as compared to 2011, respectively, Asia Pacific grew at 6.4 percent, Latin America and the Caribbean grew at 8.6 percent, Africa grew at 4.2 percent, and the Middle East posted a stunning 13.7 percent. All of these regions present higher growth rates than those of Europe and North America. Moreover, for the period 2013-2032, the European aircraft manufacturer forecasts that aircraft operators located in Asia Pacific, Latin America, and the Middle East will enjoy a growth rate of 7.1 percent, 5.5 percent, and 6 percent, respectively. Airbus, "Global Market Forecast 2013-2032," *supra* note 1 (Introduction) at 46. In addition, Airbus points out that of all the 29,226 aircraft to be delivered during that period, the Asia Pacific region will retain 36 percent of those aircraft, whereas Europe and North America's market share will only be 20 and 19 percent, respectively. Boeing's numbers are also very positive and suggest that approximately 35,380 new aircraft will be delivered during such period; Boeing predicts that almost 12,820 of those aircraft will be allocated within the Asia Pacific region. See Boeing, "Current Market Outlook 2013-2032," *supra* note 5.

Despite severe economic downturns, ICAO reports that, in 2013, aviation carried 3.1 billion passengers and fifty-one million tons of cargo.[14] The aircraft manufacturer Boeing predicts that by 2050 the air transport market will surpass the staggering milestone of 16 billion airline passengers,[15] more than ten times the number of passengers carried by IATA member airlines in 1999.[16] These are astonishing figures for an industry that as recently as 1945 – one year after ICAO was established – transported only 9 million passengers.[17] Few industries have experienced such growth.[18]

Industry estimates show that aviation's economic impact is roughly 7.5 percent of the world's GDP (3,560 billion US dollars). This includes air transport's direct, indirect, induced, and catalytic effects on the global economy.[19] Aviation generates 5.5 million jobs contributing approximately 408 billion US dollars to global GDP, and the aviation supply chain is responsible for an additional 6.3 million jobs.[20] The significance of its contribution to the global economy can hardly be questioned.

As a symbol of globalization and industrialization, aviation is a major driver of tourism and world trade. For a large number of countries, aviation serves as their main gateway to the world by supplying connectivity, bringing people together, and facilitating international commerce. It acts as a catalyst that fosters economic and social interaction and development.[21] Aviation's multiplier effect is unquestionable.[22] It is hard to imagine a properly functioning world without aviation. In the effort to address aviation's impact on climate change, any attempt to curb the growth of the industry must be carefully examined in connection with the unintended consequences that such measure may have, particularly

14 See "2013 ICAO Air Transport Results Confirm Robust Passenger Demand, Sluggish Cargo Market" online: ICAO <www.icao.int/Newsroom/Pages/2013-ICAO-AIR-TRANSPORT-RESULTS-CONFIRM-ROBUST-PASSENGER-DEMAND,-SLUGGISH-CARGO-MARKET.aspx>.

15 See Boeing, "Current Market Outlook 2009-2028," *Boeing* online: <www.boeing.com/commercial/cmo/pdf/Boeing_Current_Market_Outlook_2009_to_2028.pdf at 8>.

16 See IATA, *World Air Transport Statistics* (Montréal: IATA, 2000). See also IATA, News Release, PS/12/00, "2000 W.A.T.S. – More Passengers, Less Profits" (19 June 2000) online: IATA <www.asiatraveltips.com/travelnews2000/20June2000IATA.htm>.

17 See ICAO Circular 292-AT/124, *Economic Contribution of Civil Aviation* (2006) at I-1-3.

18 Earlier reports indicate that, in 1962, IATA member airlines had only 3,161 aircraft. That year, those airlines transported 136 million passengers. See Sir William Holdred, IATA Director General, "Extracts from the 1963 Annual Report" (1963) 30 J Air L Com 72 at 74.

19 See ATAG, Economic and Social Benefits of Air Transport 2008, *supra* note 3 (Introduction).

20 *Ibid.*

21 The United Nations World Tourism Organization (UNWTO) notes that "[a]ir passenger transport is closely associated with tourism, which generates a higher contribution of Gross Domestic Product, jobs and investment than most other economic activities. This is particularly the case in developing countries, where tourism is the principal service sector activity. At the same time, air passenger transport is the dominant, and a growing, contributor to global Greenhouse Gases (GHGs) generated by visitors." ICAO, A37-WP/174 EX/31, *Statement Regarding Mitigation of Greenhouse Gas Emissions from Air Passenger Transport* (presented by the World Tourism Organization, UNWTO).

22 See ICAO, ATConf/6-WP/33, *Views on Advancing ICAO's Work on Air Transport Liberalization* (presented by the United Arab Emirates) at 1-3.

on those countries whose development is highly dependent on aviation and tourism.[23] The ICAO Assembly has recognized "the vital role which international aviation plays in global economic and social development and the need to ensure that international aviation continues to develop in a sustainable manner."[24] This is a major concern of the industry in addressing aviation's high contribution to climate change.[25]

1.1.2 *Aviation under Siege*

The unconstrained growth of the air transport sector continues to generate increasing concern for the international community. In light of environmental externalities the industry casts upon society, commentators contend that such hasty expansion is incompatible with the medium- and long-term sustainability of the sector.[26] These externalities, which are costs borne by third parties as a result of the environmental damage caused by aviation activity, may include aircraft noise,[27] local air quality degradation produced in the vicinity of airports, and aircraft emissions.[28]

Although the airline industry argues that the sector is heavily taxed and it is the only industry that fully pays for the use of its own infrastructure,[29] there is a strong perception that the industry is heavily subsidized. The Intergovernmental Panel on Climate Change (IPCC) has noted that aviation does not bear its environmental cost to society, which constitutes an implicit governmental subsidy offered to the industry and its consumers.[30] Even ICAO has acknowledged that the aviation sector bears only direct costs of air travel but externalities "are not fully reflected in the economic pressures under which the industry operates."[31] For instance, scholars have contended that aircraft landing fees applicable for the use of international airports are generally structured on the basis of weight-based charges that do not take into account the environmental costs that society

23 See Ben Daley & Holly Preston, "Aviation and Climate Change: Assessment of Policy Options" in Stefan Gossling & Paul Upham, eds, *Climate Change and Aviation: Issues, Challenges and Solutions* (London: Earthscan, 2009) 347 at 347.

24 ICAO, A38-18, *supra* note 21 (Introduction).

25 See Chapter 3.

26 See Graham & Shaw, *supra* note 4 (Introduction).

27 See Peter Davies & Jeffrey Goh, "Air Transport and the Environment" (1993) 18 Air & Space L 123 (suggesting that, in the 1960s, noise was aviation's major environmental concern, at 125).

28 Generally speaking, while identifying environmental externalities caused by air transport, scholars point out to noise, air pollution, climate change, change of nature and landscape, and detrimental urban effects. See Werner Rothengatter, "Climate Change and the Contribution of Transport: Basic Facts and the Role of Aviation" (2010) 15 Transportation Research Part d 5 at 7. See also Gerd Winter, "On Integration of Environmental Protection into Air Transport Law: A German and EC Perspective" (1996) 21:3 Air & Space L 132.

29 See ATAG, "Economic and Social Benefits of Air Transport 2008," *supra* note 3 (Introduction).

30 See IPCC Aviation Report, *supra* note 9 (Introduction) at 341.

31 CAEP/3-WP/102, *Committee on Aviation Environmental Protection (CAEP) Third Meeting* at 2-B-2.

must bear as a result of the activities of the aviation industry, thereby leading to artificially low prices that induce economically unjustified growth in demand.[32] Also, it has been argued that despite "being the most environmentally damaging form of transport per passenger-kilometer," aviation does not properly account for the externalities it generates, creating economic inefficiency costs not borne by the sector that nevertheless benefit it directly.[33]

Considering aviation's vigorous growth, the failure to internalize its environmental cost can only be interpreted as explicitly placing the industry in a privileged position – an entitlement that most other sectors of the economy do not enjoy.[34] Arguably, this contradicts Principle 16 of the *Rio Declaration on Environment and Development*, which embraces the notion that the polluter must pay for the environmental damage caused to the natural environment as a result of his or her activities.[35] Public pressure is mounting, and the aviation industry should not remain immune to these concerns as it is undisputed that "there is a compelling and urgent need to address the environmental effects of air transportation."[36] The real policy challenge will be to balance internalizing these costs while facilitating access to air transport. Although the rise of air traffic primarily depends on the world's economy and trade growth, to a lesser but much more foreboding extent, the future growth of the air transport sector will be influenced by one key factor: aviation's response to environmental challenges.[37]

1.1.3 Climate Change

Climate change refers to a change in the statistical properties of the climate during an extended period of time usually decades or even longer.[38] Climate change is characterized

32 See Roberto Rendeiro Martin-Cejas, "Ramsey Pricing Including CO_2 Emission Cost: An Application to Spanish Airports" (2010) 16 J Air Transp Manag 45.

33 Graham & Shaw, *supra* note 4 (Introduction) at 1449.

34 Addressing whether States internalize environmental externalities, Eichenberg claims that "most economies do not account for climate-change-associated costs incurred by GHG emissions." Benjamin Eichenberg, "Greenhouse Gas Regulation and Border Tax Adjustments: The Carrot and Stick" (2010) 3 Golden Gate U Envtl LJ 283 at 287.

35 Principle 16 reads:
"Internalization of Environmental Costs
National authorities should endeavor to promote the internalization of environmental costs and the use of economic instruments, taking into account the approach that the polluter should, in principle, bear the cost of pollution, with due regard to the public interest and without distorting international trade and investment."
Rio Declaration on Environment and Development, 3-14 June 1992, UN Doc A/CONF.151/26 vol 1.

36 Ian Waitz et al, *Aviation & Environment: A National Vision Statement, Framework for Goals and Recommended Actions* (Cambridge: Massachusetts Institute of Technology, 2004) [MIT Report].

37 See ICAO, *Air Transport Outlook*, *supra* note 4 at 1.

38 See IPCC Aviation Report, *supra* note 9 (Introduction) at 30.

by rising sea levels, storm surges, melting glaciers, and changes in precipitation patterns,[39] making ecosystems, water resources, food security, human health, and settlements and society vulnerable.[40] Where the climate variation results from human activity, the phenomenon is referred to as anthropogenic climate change.[41] In fact, the *United Nations Framework Convention on Climate Change* (*UNFCCC*) seeks to address only those changes that may be directly or indirectly attributed to human activities that alter the composition of the global atmosphere over extended periods of time.[42]

Scientific evidence demonstrates that human activities have caused a substantial increase in the atmospheric concentrations of the so-called greenhouse gases (GHGs), which in turn produce drastic changes in climate patterns.[43] GHGs are "those gaseous constituents of the atmosphere, both natural and anthropogenic, that absorb and remit infrared radiation."[44] An increase in the concentration of GHGs in the atmosphere enhances the natural greenhouse effect, resulting in additional warming of the Earth's surface. Today, it is not contested that global warming is man-made. Other theories such as the Earth's long-term temperature oscillations have proven not to be persuasive[45] although skeptics still remain.[46] Similarly, the IPCC report underscores the fact that the Earth's warming is "unequivocal." Robust evidence confirms average increases in global temperatures,[47] causing a significant impact on natural ecosystems.[48] Sectors such as energy supply, transport, and industry are driving the global anthropogenic GHG footprint increase of 70 percent from 1970 to 2004.[49]

39 See Cleo Paskal, *Global Warring: How Environmental Economic and Political Crises Will Redraw the World Map* (Toronto: Key Porter Books Limited, 2010) at 26.

40 See IPCC, *Climate Change 2007: Synthesis Report, IPCC* (2007) online: IPCC <www.ipcc.ch/pdf/assessment-report/ar4/syr/ar4_syr.pdf> [IPCC4 "Synthesis Report"].

41 *Ibid.* See also National Snow and Ice Data Center, "Climate Change" *NSIDC* online: <http://nsidc.org/arcticmet/glossary/climate_change.html>.

42 See *UNFCCC, supra* note 11 (Introduction) at Art 3. At present, 194 states are parties to the *UNFCCC*. See *UNFCCC*, "Status of Ratifications of the Convention" *UNFCCC* online: <http://unfccc.int/essential_background/convention/status_of_ratification/items/2631.php>.

43 See generally Stuart Bell & Donald McGillivray, *Environmental Law,* 6th ed (London: Oxford, 2006); Richard Macrory, Ian Havercroft & Ray Purdy, eds, *Principles of European Environmental Law: Proceedings of the Avosetta Group of European Environmental Lawyers* (Groningen: Europa Law Publishing, 2004).

44 *UNFCCC, supra* note 11 (Introduction) at Art 2.

45 See Rothengatter, *supra* note 28 at 1.

46 Margaret Wente, "Whatever Happened to Global Warming," *The Globe and Mail* (6 January 2014) online: <www.theglobeandmail.com/commentary/whatever-happened-to-global-warming/article7725145/> (noting that "uncertainty on the future [of climate change] is rather great").

47 See IPCC4 "Synthesis Report," *supra* note 40 at 30.

48 *Ibid.* at 31. The IPCC notes that "[t]here is very high confidence, based on more evidence from a wider range of species, that recent warming is strongly affecting terrestrial biological systems, including such changes as earlier timing of spring events, such as leaf-unfolding, bird migration and egg-laying; and poleward and upward shifts in rages in plant and animal species." *Ibid.* at 33.

49 *Ibid.* at 36.

During this period, these sectors also prompted an 80 percent growth of CO_2 emissions, one of the most important GHGs derived directly from fossil fuel use.[50]

Climate change is arguably one of the most challenging and intricate policy issues of our times.[51] Part of the problem with climate change is that it is seen as an "amorphous" dilemma.[52] While it is intangible, people know it is out there.[53] Its effects, however, are perceived as a very distant phenomenon.[54] There is a tendency to acknowledge its existence, but since scientists are not yet able to link specific extreme climate events (e.g. Hurricane Katrina, tsunami, deadly Pakistani floods) to the accumulation of GHGs in the atmosphere, concrete actions have been rare.[55] Presently, there is public confusion about climate change[56] as society and governments struggle to implement policies that correct effects on the environment that may be felt at a much later time.[57] The IPCC stresses that with current mitigation policies, global GHG[58] emissions will continue to grow steadily.[59] Yet, since climate change evolves in an "open commons" setting beyond the jurisdiction of a given nation,[60] where strong incentives to free ride exist,[61] no single country or regional policy

---

50 *Ibid.*

51 Climate change has been described as a typical "tragedy of commons" of present times. See Barton H Thompson, Jr. "Tragically Difficult: The Obstacles to Governing The Commons" (2000) 30 Enviro L 241 at 253 [Thompson, Jr].

52 Brian Evans, "Principles of Kyoto and Emissions Trading Systems: A Primer for Energy Lawyers (2004) 42 Alberta L Rev 167 at 168.

53 *Ibid.*

54 As Suzi Kerr suggests, "[t]he people who bear the costs of abatement may not be those who would have borne the damages from climate change. This introduces not only discounting issues but also intertemporal equity issues." Suzi Kerr, ed, *Global Emissions Trading: Key Issues for Industrialized Countries* (Northampton: Edward Elgar 2000) 1 at 7. Posner and Weisback also note that the "benefits of emissions reductions will be enjoyed in the future rather than the present." Eric A Posner & David Weisbach, *Climate Change Justice* (Princeton: Princeton University Press, 2010) at 144 [Posner & Weisbach]. Similarly, Thompson, Jr. describes the challenges of climate change as a tragedy of commons involving intertemporal trade-offs. See Thompson Jr, *supra* note 51 at 262.

55 See Gregory C Unruh, "Escaping Carbon" (2002) Ener Pol'y 30 317 at 323. Sachs, on the other hand, claims that "climate change has real victims who can be located thousands of miles away from the major emissions sources." Noah Sachs, "Beyond the Liability Wall: Strengthening Tort Remedies in International Environmental Law" (2008) 55 UCLA L Rev 837 at 901.

56 See Evans, *supra* note 52 at 167.

57 See generally Scott Barrett, *Why Cooperate? The Incentive to Supply Global Public Goods* (Oxford: Oxford University Press, 2007) (stressing that "different countries will be affected in different ways" at 5).

58 Economist Nicolas Stern has described GHGs as a type of externality forming the "biggest market failure the world has ever seen." Nicholas Stern, "The Economics of Climate Change" (2008) 98 American Economic Review: Papers & Proceedings 2 at 1. In a similar vein, Ghosh & Gray point out that this negative externality requires a huge "global cooperative response." Koushik Ghosh & Peter Gray, "Rushing to Copenhagen? Is Cap-and-Trade the Answer?" (2010) Challenge (January/February) 5 at 6.

59 See IPCC4 "Synthesis Report," *supra* note 40 at 44.

60 See Diana S Solomon & Kenneth FD Hughey, "A Proposed Multi Criteria Analysis Decision Support Tool for International Environmental Policy Issues: A Pilot Application to Emissions Control in the International Aviation Sector" (2007) Environmental Science and Pol'y 645 at 653.

61 Michael P Vandenbergh "Order without Social Norms: How Personal Norm Activation Can Protect the Environment" (2005) 99 Nw UL Rev 1101 at 1128.

tool could achieve meaningful and lasting results.[62] The protection of the commons requires collective policy action. The question then is who should do what and when.[63]

1.1.4 *Aviation and Climate Change*

In 1999, at ICAO's request, the IPCC carried out the first sector-specific assessment of aviation and its impact on climate change.[64] To ICAO's credit, aviation is the only sector that has been thoroughly studied.[65] Today, no one disputes the fact that aviation contributes at least marginally, if not significantly, to the degradation of the atmosphere.[66] The IPCC report highlighted the fact that during the course of their operations, aircraft discharging of gases into the upper troposphere and lower stratosphere at very high altitudes (8 to 13 kilometers) affects the atmospheric composition and concentration of an array of GHGs.[67] This impact is estimated to represent approximately 3.5 percent of the total radiative forcing[68] from all human activities.[69] Similarly, the Massachusetts Institute of Technology's (MIT) report on aviation and the environment underlines that "scientific assessments also suggest that the resulting chemical and physical effects due to aviation are such that aviation may have a disproportionate effect on climate per unit of fuel burned when compared to terrestrial sources."[70] The MIT report also points out that "[t]hese estimates reflect a finding that per unit of fuel burned, radiative forcing from aircraft is expected to be approximately double that of land-based use of hydrocarbon fuels."[71] Therefore, when considering aviation's contribution to climate change, one must not only factor in CO_2

62 See Daniel Bodansky, *The Art and Craft of International Environmental Law* (Cambridge: Harvard University Press, 2010) at 11 [Bodansky, *The Art and Craft of International Environmental Law*].

63 As Joyeeta Gupta points out, part of the problem with addressing climate change is the "lack of statesmanship." No country wants to take measures, unless others are also engaged. For instance, developed countries such as the United States are unwilling to make substantial commitment, unless major developing economies are also on board. See Joyeeta Gupta, "Climate Change and Shifting Paradigms" in Duncan French, ed, *Global Justice and Sustainable Development* (Leiden: Martinus Nijhoff Publishers, 2010) 167 at 174. Jeffrey L Dunoff writes that "the tragedy of commons illustrates how self-interested actors left to their own devices will frequently generate individually and collectively sub-optical outcomes." Jeffrey L Dunoff, "Levels of Environmental Governance" in Daniel Bodansky, Jutta Brunnée & Ellen Hey, eds, *The Oxford Handbook of International Environmental Law* (Oxford: Oxford University Press, 2006) 85 at 93.

64 See IPCC Aviation Report, *supra* note 9 (Introduction).

65 See IPCC/4, *supra* note 2 (Introduction) at 331.

66 See generally Michael Faure, John Vervaele & Albert Weale, eds, *Environmental Standards in the European Union in an Interdisciplinary Framework* (Antwerpen: Maklu, 1994). See also DS Lee et al, "Transport Impacts on Atmosphere and Climate: Aviation" (2010) 44 Atmospheric Environment 4678; *Beattie v. United States*, 756 F 2d 91 (DC Cir 1984) (describing Antarctica as a "global common" at 99).

67 See IPCC Aviation Report, *supra* note 9 (Introduction) at 3.

68 Radiative forcing is a common metric to describe "warming influence." See IPCC/4, *supra* note 2 (Introduction); IPCC "Synthesis Report," *supra* note 40 at 36.

69 *Ibid.*

70 MIT Report, *supra* note 36.

71 *Ibid.*

generation but also the fact that aviation has a larger impact on radiative forcing.[72] Aviation-released GHGs, which change the composition of the atmosphere and perturb the Earth's balance of radiation, include carbon dioxide (CO_2), ozone (O_3), methane (CH_4), water vapor (H_2O), nitrogen oxides (NO_x), sulfur oxides (SO_x), and soot particles.[73] Despite their negative effects on global climate change,[74] GHGs stimulate the formation of condensation trails and boost cirrus clouds;[75] however, there is still little understanding of their impact on radiative forcing.[76] CO_2 is the GHG of greatest concern for aviation because it persists for a long time in the atmosphere and is released in large quantities.[77]

According to the IPCC, the highly petroleum-dependent transport sector, whose growth rate is the highest among the end-user sectors, is responsible for 23 percent of the global anthropogenic GHG CO_2 footprint, generating 6.3 Gt. of emissions annually.[78] By 2030, these emissions are expected to boost an 80 percent rise in GHGs as compared to 2002 levels.[79] While road transport accounts for 77 percent of CO_2 emissions from the transport sector, aviation is responsible for 13 percent.[80] As such, aviation is the second largest transport sector contributor to climate change.[81] Under a business-as-usual scenario, CO_2 emissions from air transport will grow to 23 percent of the transportation sector's footprint by 2050.[82] Although the IPCC estimates that aviation emissions currently account for (only) 2 percent of total global CO_2 emissions,[83] the latter will experience average annual

72 NGOs point out that international aviation is responsible for "5 [percent] of cumulative global warning." See "Global Deal or no Deal: Your Free Guide to ICAO's 38th Triennial Assembly?" *Transport & Environment* online: <www.transportenvironment.org/sites/te/files/publications/2013%2009%20Your%20Guide%20to%20ICAO_final.pdf> [Transport & Environment, "Global Deal or No Deal"].

73 See MIT Report, *supra* note 36.

74 See David Woolley et al, eds, *Environmental Law* (New York, Oxford: Oxford University Press, 2000).

75 See IPCC Aviation Report, *supra* note 9 (Introduction) at 3.

76 See Paul Peeters and Victoria Williams, "Calculating Emissions and Radiative Forcing" in Stefan Gossling & Paul Upham, eds, *Climate Change and Aviation: Issues, Challenges and Solutions* (London: Earthscan, 2009) 69 at 84. These authors argue that if aviation CO_2 emissions increase by more than 2 percent per year, the total radiate forcing will increase even faster due to an increase in the radiative forcing index. *Ibid.*

77 Although aviation's impact on climate change involves a number of GHGs, this book places emphasis on CO_2 emissions.

78 See IPCC/4, *supra* note 2 (Introduction). See also Robert L Hirsch, Roger Bezdek, & Robert Wendling, "Peaking of World Oil Production and Its Mitigation" in Daniel Sperling & James S Cannon, eds, *Driving Climate Change: Cutting Carbon from Transportation* (Amsterdam: Elsevier, 2007) at 9 (claiming that "world oil demand will continue to grow, increasing approximately 2 percent per year, driven primarily by the transportation sector" at 9).

79 *Ibid.* at 357.

80 See ICAO CAEP/8-WP/80, *Committee on Aviation Environmental Protection (CAEP) Eighth Meeting, Montréal, 1 to 12 February 2010,* at 1-38. See also IEA, *supra* note 1 at 233.

81 See Daniel Rutherford, "The Role of Aviation Alternative Fuels in Climate Change Mitigation" (paper delivered at the ICAO Alternative Fuels Workshop, Montréal, Quebec, 10-12 February 2009) [unpublished].

82 See OECD, *Transport Outlook 2008* (Paris: OECD, 2008).

83 By way of comparison, in 2007, shipping represented approximately 2.7 percent of total CO_2 emissions. See Haifeng Wang, "Economic Cost of CO_2 Emissions Reduction for Non-Annex I Countries in International Shipping" (2010) 14 Energy for Sustainable Development 280 [Wang, "Economic Cost"].

growth rates of 3.5 percent to 2036.[84] It is striking to learn that the world's energy usage for transportation purposes will climb yearly at 2 percent,[85] a rate which is lower than that projected for the growth of aviation's CO_2 footprint.[86]

Technology improvements in fuel productivity explain the lower CO_2 emission growth rates projected for the transportation sector as a whole vis-à-vis those projected for the air transport sector.[87] Between 1991 and 2003, aviation's CO_2 footprint went up 87 percent.[88] Bearing in mind aviation's projected annual growth rate of nearly 5 percent, this CO_2 emission growth rate is expected to continue for at least the next twenty years.[89] While emissions from most other sectors remain stable, the aviation sector's CO_2 emissions will rise fourfold by 2050[90] with its global footprint share escalating from 2 to 3 percent[91] if no remedial efforts are implemented. To put things into perspective, global energy-related CO_2 emissions will have increased by 55 percent between 2004 and 2030 (equivalent to 1.7 percent per year),[92] and aviation's footprint is a significant contributor to this projected increase. In Europe, for instance, international aviation emissions account for approximately 3 percent of global CO_2 inventories.[93] These emissions will rise between 6 and 7 percent per year.[94]

---

84 See ICAO CAEP/8-WP/80, *Committee on Aviation Environmental Protection (CAEP) Eighth Meeting, Montréal, 1 to 12 February 2010 Agenda Item 1* at 1-32 [ICAO CAEP/8-WP/80].

85 See IPCC/4 *supra* note 2 (Introduction) at 325.

86 European States estimate that by 2020 aviation emissions will grow 70 percent compared to 2005 levels. See *Document Presented by Belgium, Denmark France, Germany, Italy, Slovenia, Spain, and the United Kingdom* for consideration of the Council when discussing Item No 25 of the Work Programme of the Council for the 194th Session ("European Emissions Trading Scheme (ETS)") [unpublished, archived on file with the author]. See ICAO, C-MIN 194/2, *Council, 194th Session, Minutes of the Second Meeting.* See also "UK climate advisers recommend international aviation emissions be included in national carbon budgets," *Green Air Online* (5 April 2012) online: <www.greenaironline.com/news.php?viewStory=1447> (discussing that UK emissions from international aviation doubled from 16 Mt. CO_2 in 1990 to 32 Mt. CO_2 in 2010, compared to a sharp fall in other sectors).

87 See ICAO, *Air Transport Outlook, supra* note 4 at 17.

88 See Rothengatter, *supra* note 28 at 11.

89 See IPCC/4, *supra* note 2 (Introduction) at 326; ICAO, *Air Transport Outlook, supra* note 4.

90 See ICAO, DGCIG/1-IP8. See also "Hedegaard Sets Out Conditions on ICAO Agreement," *supra* note 8 (Introduction).

91 See IPCC Aviation Report, *supra* note 9 (Introduction) at 6. See also WWF, "Aviation Report: Market Based Mechanisms to Curb Greenhouse Gas Emissions from International Aviation" *WWF* online: <http://awsassets.panda.org/downloads/aviation_main_report_web_simple.pdf> at 6 [WWF, Aviation Report] (claiming that international aviation emissions are "predicted to increase above 660 [$MtCO_2$] per annum, reaching 800 [$MtCO_2$] by 2025, 980 [$MtCO_2$] by 2030 and more than 2150 [$MtCO_2$] by 2050").

92 See IEA, *supra* note 1 at 41.

93 See Anger, Including Aviation in the EU ETS, *supra* note 2 at 100.

94 See Alice Bows, Kevin Anderson, & Anthony Footitt, "Aviation in a Low-Carbon EU" in Stefan Gossling & Paul Upham, eds, *Climate Change and Aviation: Issues, Challenges and Solutions* (London: Earthscan, 2009) 89 at 106. Conversely, the road transport sector is accountable for a third of the total of Europe's CO_2 emissions. See Charlotte Streck & David Freestone, "The EU and Climate Change" in Richard Macrory, ed, *Reflections on 30 Years of EU Environmental Law: A High Level of Protection* (Groningen: Europa Law Publishing, 2006) 87 at 102.

As Gossling and Upham indicate, "there is no other human activity pushing individual emission levels as fast and as high as air travel."[95] To put these figures into context, by 2020 international aviation's CO_2 footprint will be larger than the overall emissions of Germany.[96] Without strong policy intervention, aviation CO_2 emissions "will cancel out more than a quarter of the reductions" that European countries have undertaken under international agreements.[97]

International aviation produces a great deal of its CO_2 emissions over the high seas beyond the territorial jurisdiction of any state. Monitoring emissions from stationary sources is a rather easy task since there are usually no doubts about where such emissions are produced.[98] However, it is much more difficult to trace and monitor emissions from mobile sources.[99] Aviation defies standard CO_2 emissions accounting rules that attribute emissions to states. According to these rules, a country is only responsible for emissions that "result directly from activities that occur within its territory."[100] The case of oil-producing countries is highly illustrative. These countries are not responsible for the fossil fuels they export. These emissions will only be computed when the fuel is burned and attributed to the country where such action takes place.[101]

Aviation's transboundary nature makes its GHG emissions international in nature.[102] ICAO estimates that international traffic is responsible for 62 percent of aviation's global CO_2 footprint, while domestic traffic only accounts for 38 percent.[103] Environmental problems that transcend traditional national boundaries require a much higher degree of

---

95 Stefan Gossling and Paul Upham, "Introduction: Aviation and Climate Change in Context" in Gossling and Upham, eds, *Climate Change and Aviation: Issues, Challenges and Solutions* (London: Earthscan, 2009) 1 at 5.

96 See IPCC/4, *supra* note 2 (Introduction).

97 EC, *Communication from the Commission to the Council, the European Parliament, the European Economic and Social Committee and the Committee of the Regions Reducing the Climate Change Impact of Aviation*, COM (2005) 459 (Brussels, 27 September 2005) at 2. Southgate explains that international aviation carbon footprint grew from 250 million tonnes of CO_2 in 1990 to 455 million in 2008. See David Southgate, "Aviation Carbon Footprint: Global Scheduled International Passenger Flights 2012" online: <www.scribd.com/doc/137044034/Aviation-Carbon-Footprint-Global-Scheduled-International-Passenger-Flights-2012> at 6. See also Christian Brand & Brenda Boardman, "Taming of the Few: The Unequal Distribution of Greenhouse Gas Emissions from Personal Travel in the UK" (2008) 36 Energy Pol'y 224.

98 See Andreas Hardeman & Kalle Keldusild, "Overview of ICAO Guidance on Emissions Trading" *in ICAO Environmental Report 2007* (Montréal: ICAO, 2007) at 149.

99 See ICAO, "Shared Vision on International Aviation and Climate Change Poznan, Poland, 1-10 December 2008" online: ICAO <www.icao.int/environmental-protection/Documents/STATEMENTS/Awglca4_2008_Submission_SharedVision.pdf> at 3 [ICAO SBSTA Submission 2008].

100 Andrew Macintosh, "Overcoming the Barriers to International Aviation Greenhouse Gases Emissions Abatement" (2008) 33:6 Air & Space L at 405.

101 *Ibid.*

102 See Solomon & Hughey, *supra* note 60 at 653.

103 See ICAO CAEP/8-WP/80, at 1-36.

"international cooperation at a scale [at] which we [have] not [yet been] accustomed."[104] Thus, measures to address international aviation must also factor in a number of legal and policy issues that are not present in the context of domestic aviation.[105] A global approach may be desirable, for unilateral measures increase the risk for normative arbitrage that allows regulated actors to move to less environmentally stringent landscapes.[106]

In analyzing the broader climate change problem, Bodansky has suggested that "just twelve states are responsible for more than 80 percent of global greenhouse gas emissions."[107] To this end, he argues that "broad participation is not essential to solve the climate change problem. A smaller regime, with relatively few members, still has the potential to address the problem effectively."[108] This proposition suggests that forming consensus within a "club" or a group of major emitting States would be far more effective than trying to obtain broader participation. Although Bodansky's views are valid for addressing climate change in the broader context, they are not necessarily applicable to international civil aviation. Although the top 21 States are responsible for roughly 80 percent of international aviation activity[109] and one can deduce that this is directly linked to GHG emissions, aircraft operators from these States compete head-to-head with others who fall below such threshold. Excluding those States and their aircraft operators may lead to significant market distortions. Certainly, this increases the complexity of the problem.[110] As addressed in subsequent chapters, addressing GHG emissions from international aviation requires collective action, albeit in a gradual and incremental manner.

1.1.5 The "Only 2 Percent" Exculpation Argument

The disproportionate visibility of the air transport industry makes it an easy target for government regulation and prompts attacks from environmental advocates and politicians.[111] The industry has responded to these attacks by asserting that aviation's climate

104 Lawrence Summers, "Foreword" in Joseph E Aldy & Robert N Stavins, eds, *Architectures for Agreement: Addressing Global Climate Change in the Post-Kyoto World* (Cambridge: Cambridge University Press, 2007) xviii at xxi.

105 See Andrew Macintosh & Lailey Wallace "International Aviation Emissions to 2025: Can Emissions be Stabilized Without Restricting Demand?" (2009) 37 Energy Pol'y 264 at 264 [Macintosh & Wallace].

106 Stepano Clò, "Grandfathering, Auctioning and Carbon Leakage: Assessing the Inconsistencies of the New ETS Directive" (2010) 38 Energy Pol'y 2420 at 2430.

107 Bodansky, *The Art and Craft of International Environmental Law*, *supra* note 62 at 175.

108 *Ibid.*

109 See "Civil Aviation: 2012 International RTK by State of Air Operator Certificate (AOC)," online: ICAO <www.icao.int/Meetings/a38/Documents/International%20Scheduled%20RTK%20(Annual%20Report).PDF>.

110 Bodansky discusses that a higher number of participants in a scheme to address climate change issues raise transactional costs. As such, the scheme is more likely to experience free riding issues. See Bodansky, *The Art and Craft of International Environmental Law*, *supra* note 62 at 50.

111 See Bisignani, *Words of Change*, *supra* note 20 (Introduction) at 97.

change contribution must be balanced against the role the sector plays in society's socioeconomic development.[112] The industry repeatedly describes its 2 percent share of the global CO_2 footprint as being relatively small.[113] The industry is also quick to point out that it contributes 8 percent of the world's GDP.[114] Similarly, ICAO has exhibited a tendency to always remind its Member States that aviation's share of GHG emissions is only 2 percent.[115] In doing this, ICAO has also been keen to compare aviation's 13 percent climate change contribution within the transport sector to that of road transportation, which is 74 percent.[116] Intentionally or unintentionally, this action has not only downplayed aviation's true contribution to climate change but has also served to shift blame and to provide the perfect recipe for collective inaction. Following this line of reasoning, the benefits of aviation far exceed its impacts on the climate. If there is something that the great majority of ICAO's Member States remember about aviation's climate change impact, it is the fact that its global footprint is only 2 percent. Arguably, in communicating this message to its membership, ICAO has placed very little emphasis on the unimpeded projections of aviation's CO_2 emissions and the consequences of not taking responsible action.[117]

Commentators rightly characterize the industry's disaggregating of its participation in climate change to the least minimum contribution as "dangerously misleading."[118] Beijing, New York, New Delhi, and Paris are also responsible for "only" less than 2 percent of the global CO_2 footprint.[119] So is the case of Canada, the United Kingdom, and Australia.[120] As one NGO has noted, "[i]f global aviation was a country, its emissions would be ranked seventh between Germany and South Korea on CO_2 alone."[121] If major metropolitan centers and countries also avail themselves of the "only 2 percent" dogmatic defense, that serves to maintain the status quo, since each contribution would be so minimal that it would not

---

112 See Andy Kershaw & Mark Watson, "Global Sectoral Approach for International Aviation Emissions" (presentation delivered at the ICAO Colloquium on Aviation and Climate Change, 13 May 2010) online: ICAO <www.icao.int/CLQ10/Docs/5_Kershaw-Watson.pdf>. See also ICAO, A37-WP/217 EX/39, *Development of a Global Framework for Addressing Civil Aviation CO_2 Emissions* (presented by the International Air Transport Association (IATA), on behalf of ACI, CANSO, IATA, IBAC, and ICCAIA, referred to hereafter as the "aviation industry") [ICAO, A37-WP/217 EX/39].

113 See Andreas Hardeman, "A Common Approach to Aviation Emissions Trading" (2007) 32:1 Air & Space L 1 at 3 [Hardeman, "Common Approach"].

114 See Bisignani, *Words of Change*, *supra* note 20 (Introduction) at 71.

115 ICAO, A38-18, *supra* note 21 (Introduction); ICAO, A37-WP/402 P/66, *Report of the Executive Committee on Agenda Item 17 (Section on Climate Change)* (presented by the Chairman of the Executive Committee). See also ICAO SBSTA Submission 2008, *supra* note 99 at 2.

116 See Jane Hupe, "Report Overview" in ICAO, *Environmental Report 2007* (Montréal: ICAO, 2007) at 2.

117 See generally ICAO SBSTA Submission 2008, *supra* note 99 at 2.

118 Bows et al, *supra* note 94 at 105.

119 *Ibid.*

120 *Ibid.*

121 Transport & Environment, Global Deal or No Deal, *supra* note 72. See also ICAO, HGCC/3-AIP/6, *supra* note 6 (Introduction) at 1.

be worth taking any action.[122] This situation compares closely to a prisoner's dilemma where actors misleadingly justify their respective inactions with their net gains. ICAO's determination to downplay aviation's contribution to global GHG emissions has also made it much more difficult for its Member States to understand the urgency to act.[123]

1.1.6 The "Communication Problem" Defense

ICAO, some of its Member States, and industry stakeholders have said that the sector has made substantial progress in addressing GHG emissions from international aviation. The problem has, however, been that both ICAO and the industry have not adequately "communicated" their achievements in this field. There has been a communication glitch. A poor and weak outreach campaign has misled those outside aviation circles and created a false impression that the sector is not doing enough to address the problem.[124] This has resulted in exogenous attacks. In this regard, some years ago, Giovanni Bisignani, former Director General of IATA, suggested that the "[aviation industry has] been silent in [its] success and now [it has] a reputation crisis. That makes [the industry] an easy target for politicians who think green and see cash."[125] This perception will have profound implications for the behavior of actors who show a remarkable tendency to over-emphasize environmentally related achievements. Arguably, the emphasis on the lack of effective communication of the sector's environmental achievements diverts attention from the core issues and removes the urgency to act.

1.1.7 The All-Mighty Technological Efficiencies

As the aviation industry has downplayed its contribution to climate change, it has shown a remarkable ability to exaggerate the significance of technological efficiencies introduced over the years. It is undisputed that very few industries can match aviation's track record

122 Chris Lyle, "Mitigating International Air Transport Emissions through a Global Measure: Time for Some Lateral Thinking," *Green Air Online* (6 February 2014) online: <www.greenaironline.com/news.php?viewStory=1820> (discussing that when addressing climate change, ICAO has carried out a "defensive approach, effectively protecting the aviation industry") [Lyle, "Mitigating International Air Transport Emissions through a Global Measure"].

123 See Chapter 7.

124 See ICAO, C-MIN 196/7, *Council, 196th Session, Decision of the Seventh Meeting* at 19 (Statement of the Russian Federation) [ICAO, C-MIN 196/7]; ICAO, C-MIN 196/8 (Statement of Paraguay) (noting that "if it was repeatedly stated that ICAO was not making any efforts to reduce CO_2 emissions from civil aviation, then the public would start to believe it"); ICAO, C-MIN 196/8 at 12 (Statement of Singapore) (underscoring that "ICAO [has] indeed done much work in the field of environmental protection. [The organization was just not] very good at publicizing it, not just to the public but even within Council").

125 Bisignani, *Words of Change*, *supra* note 20 (Introduction) at 97.

of efficiency improvements. Noise from aircraft has been reduced by 75 percent,[126] and fuel efficiency and environmental performance of modern generation aircraft have improved by a margin of about 70 percent over the last four decades.[127] From 1970 to 1990, fuel efficiency improved at a yearly rate of between 3 and 4 percent.[128] Aircraft delivered in 2050 will be 40 to 50 percent more fuel efficient than those manufactured today.[129]

IATA claims that their 1.5 percent annual fuel efficiency goal translates into a CO_2 emission reduction of about 2.2 billion tonnes.[130] Similarly, the trade association of US airlines, Airlines for America (A4A), has noted that US airlines have "improved fuel efficiency 120 percent since 1978, reducing carbon emissions by 3.3 billion metric tons, savings equivalent to taking 22 million cars off the road every year."[131] Some States have also noted that an "overall increase in civil aircraft fuel efficiency of approximately 70 percent over the last 40 years has significantly reduced aviation greenhouse gas emissions."[132]

These milestone achievements have been well below traffic growth rates.[133] As one commentator notes, "history has shown that demand growth generally outstrips emission intensity gains."[134] The growth in demand for air transport simply outpaces efficiencies gained by the deployment of technological and operational advancements.[135] In addition, it is noteworthy that in the future, annual improvements in aircraft fuel efficiency will only be in the order of 1 to 2 percent per annum.[136] Therefore, unless there is policy intervention, these energy intensity gains will be outperformed by unswerving yearly air traffic growth of around 5 percent. This will lead to an annual increase of aviation's CO_2 footprint of 3 to 4 percent per year.[137]

---

126 See Hupe, *supra* note 116 at 2.

127 See Airbus, Global Market Forecast 2013-2032, *supra* note 1 (Introduction) at 8.

128 See ICAO, CAEP/2-WP/73, *Committee on Aviation Environmental Protection (CAEP) Second Meeting* [ICAO, CAEP/2-WP/73].

129 See IPCC/4, *supra* note 2 (Introduction) at 353.

130 See ICAO, HGCC/3-IP/5 at 1.

131 Nancy Young, "Letter: Airlines Getting Greener" *The New York Times* (3 February 2013) online: <www.nytimes.com/2013/02/04/opinion/airlines-getting-greener.html?_r=1&>.

132 New Delhi Joint Declaration of 30 September 2011 [New Delhi Declaration]; Joint Declaration of the Moscow Meeting on the Inclusion of International Civil Aviation in the EU ETS of 20-21 February 2012 [Moscow Declaration]. See also ICAO, Declaration Adopted by the Council of the International Civil Aviation Organization (ICAO) at the Second Meeting of the 194th Session on 2 November 2011[ICAO Council Declaration of 2 November 2011].

133 See Lyle, "Mitigating International Air Transport Emissions through a Global Measure," *supra* note 122; Benoît Mayer, "Case -366/10, Air Transport Association of America and Others *v.* Secretary of State for Energy and Climate Change" (2012) 49:3 Common Market Law Review 1113 at 1118.

134 Macintosh & Wallace, *supra* note 105 at 271.

135 See Simon Eggleston, "IPCC Guidelines for Estimating National Greenhouse Gas Inventories" in *ICAO Environmental Report 2007* (Montréal: ICAO, 2007) at116.

136 IPCC/4, *supra* note 2 (Introduction) at 326.

137 *Ibid.*

As some scholars have said, if air transport is to be developed in a sustainable manner, "there is a need to reduce CO_2 emissions at a rate equal [to] or greater than the rate of increase of traffic."[138] At this stage, there is no evidence to suggest that technological efficiencies will offset growth.[139] As Giovanni Bisignani put it, "the [airline] industry needs a wakeup call. [Its] carbon footprint is growing and that is not politically acceptable for any industry."[140] Yet the industry is not alone in overemphasizing the role of technological efficiencies.[141] Critics have said that ICAO's initiatives aimed at reducing GHG emissions are mainly geared toward technical and operational measures.[142] The ICAO Assembly has also repeatedly drawn attention to these technological efficiencies in its environmental resolutions, but it has mysteriously failed to make any reference to CO_2 emission trends and the fact that the growth of the latter far outstrips that of the former.[143] Unfortunately, this oversight sends a misleading message to ICAO's constituency, erroneously prompting them to downplay aviation's contribution to climate change.

---

138 Sgouris Sgouridis, Philippe A Bonnefoy, & R John Hansman, "Air Transportation in a Carbon Constrained World: Long-Term Dynamics of Policies and Strategies for Mitigating the Carbon Footprint of Commercial Aviation" (2011) 1 Transportation Research Part 1077 at 1078.

139 See Kathryn Kisska-Schulze & Gregory P Tapis, "Projections for Reducing Aircraft Emissions" (2012) 77 J Air L & Com 701 at 745.

140 Bisignani, *Words of Change, supra* note 20 (Introduction) at 97. See also ICAO, A38-WP/176 EX/67, *Expectations and Desirable Objectives of the 38th Session of the Assembly Relating to International Aviation and Climate Change: Perspective of the Kingdom of Saudi Arabia* (presented by the Kingdom of Saudi Arabia) (suggesting that climate change is "one of the most important issues of discussions in ICAO") [A38-WP/176 EX/67].

141 The industry's high expectations on technology are best illustrated in the words of Mr. Tony Tyler, IATA's Director General:
"We fully expect that technology, operations and infrastructure measures alone will provide the long-term solution for aviation's sustainable growth. As such MBMs will be a temporary gap-filling measure until the full impact of new technologies and sustainable biofuels can be realized."
IATA, Press Release, "Remarks of Tony Tyler at the Greener Skies Conference in Hong Kong," *IATA* (26 February 2013) online: IATA <www.iata.org/pressroom/speeches/Pages/2013-02-26-01.aspx> [IATA, "Remarks of Tony Tyler"].

142 See Anming Zhang, Sveinn Vidar Gudmundsson, & Tae H Oum, "Air Transport, Global Warming and the Environment" (2009) Transport Research Part D.

143 See ICAO, A38-18, *supra* note 21 (Introduction) (stressing that "the significant technological progress made in the aviation sector, with aircraft produced today being about 80 percent more fuel efficient per passenger kilometer than in the 1960's" at preambular clauses). While elsewhere the Assembly Resolution timidly acknowledges that "CO_2 emissions are projected to grow as a result of the continued development of the sector," such reference is not linked to the lower rate of gains provided by technological improvements. *Ibid.* This makes it much more difficult for the intended audience to understand that, despite the significant technological efficiencies achieved by the sector, given its current growth trends, GHG emissions will not be reduced unless regulation is put in place or massive quantities of alternative fuels are deployed.

1.1.8 Alternative Fuels

The aviation community has great expectations for alternative fuels as potential vehicles for substantially reducing the prevailing mitigation gap between the growth in the sector's emissions and the efficiency improvements achieved. Even ICAO acknowledges the need "to promote improved understanding of the potential use, and the related emissions impacts, of alternative aviation fuels."[144] A number of airlines have already launched test flights using some form of alternative fuels. The three most promising types of alternative fuels are jatropha, camelina, and algae,[145] none of which require structural changes to aircraft or the fuel delivery chain.[146] However, at present these fuels are not produced in sufficient quantities to cover global demand for aviation fuel.[147] High prices are also a concern.[148]

This scenario is very unlikely to change in the next decade. Lack of standardization, cost, and sufficient availability of alternative fuels for aviation serve as barriers to a scale-up of their production. Another key challenge for alternative fuels is the lack of a definition of their sustainability criteria to establish what feedstocks should be used.[149] Testing and production infrastructure will demand steep investments.[150] Currently, it is not clear how and who is going to bear the exorbitant financial cost involved in the construction of refineries and the production of feedstock necessary to meet demand.[151]

---

144 ICAO, A36-22, *Consolidated Statement of Continuing ICAO Policies and Practices related to Environmental Protection* at Appendix I [ICAO, A36-22]. See also ICAO, A38-18, *supra* note 21 (Introduction) (suggesting that "the use of sustainable alternative fuels for aviation, particularly the use of drop-in fuels in the short to midterm [may be] an important means of reducing aviation emissions" at Preamble).

145 See Charles E Schlumberger, "Are Alternative Fuels an Alternative? A Review of the Opportunities and Challenges of Alternative Fuels for Aviation" (2010) XXXV Ann Air & Sp L 119 at 151.

146 See ICAO, A37-WP/23 EX/6, *Aviation and Alternative Fuels* (presented by the Council of ICAO) [A37-WP/23].

147 *Ibid.*

148 See Tommi Ekholm et al, "Effort Sharing in Ambitious, Global Climate Change Mitigation Scenarios" (2010) 38 Ener Pol'y 1797 at 1802; Peter Forsyth, "Environmental and Financial Sustainability of Air Transport: Are They Incompatible?" (2011) 17 J Air Transp Manag 27 (suggesting that it will take years to deploy alternative fuels) at 29.

149 See ICAO, A37-WP/23, *supra* note 146. The National Resource Defense Council has stated the following: "[T]he aviation industry has a responsibility to use biofuels that are certified as sustainable because the sector's massive purchasing power has the potential to reshape the supply chain. Poorly sourced biofuels - produced using the wrong feed stocks or employing damaging management practices - could drive deforestation, food insecurity, and carbon pollution, and thereby damage brand value. Sustainable aviation biofuels, on the other hand, could create wealth and jobs, help provide energy security, and reduce carbon pollution." NRDC, "Aviation Biofuel Sustainability Survey" *NRDC* online: <www.nrdc.org/energy/aviation-biofuel-sustainability-survey>. Even the United States, a strong proponent of alternative fuels, recognizes that there still remains some uncertainty on GHG life cycles. See ICAO, A37-WP/185 EX/34, *Sustainable Alternative Aviation Fuels* (presented by the United States of America).

150 See ICAO, A37-WP/23, *supra* note 146.

151 See Curt Epstein, "Aviation Still Faces Hurdles in Environment," *Aviation International News* (27 August 2010) online: <www.ainonline.com/aviation-news/aviation-international-news/2010-08-27/alternative-fuels-still-face-hurdles>.

Even if existing barriers are removed, alternative fuels are unlikely to become available in sufficient quantities for the needs of all modes of transportation.[152] Various studies suggest that, if available, alternative fuels will primarily be used to meet the global needs of road transportation.[153] One could expect fierce competition among different sectors to lay hands on the Holy Grail. In addition, since aviation is a relatively minor consumer of the world's fuel supply, it would not be a surprise if alternative fuel producers would initially target much larger markets, such as road transportation.[154] ICAO has therefore called for regulatory and financial frameworks to ensure that alternative fuels become available in sufficient quantities for aviation.[155]

The airline industry estimates that alternative fuels could decrease aviation's carbon footprint by 80 percent.[156] But even if alternative fuel production were to scale up, they would only replace approximately 30 percent of all commercial jet fuel by 2030.[157] Most of the fuel consumed will still be fossil fuel. Thus, there is little indication that alternative fuels will reduce aviation's CO_2 emissions in the short to medium term.[158] Although it is worth conducting further studies on the feasibility of such fuels to reduce aviation's contribution to global climate change and to assign financial resources for research and development,[159] it is simply unrealistic to expect any substantial gain from alternative fuels in the near future.[160] Furthermore, these technologies will require significant investments from the public sector.[161] Alternative fuels are thus not the cure of all of aviation's environmental evils.

152 Some authors also suggest that the introduction of alternative fuels may also require significant changes in the supporting infrastructure of the aviation supply chain, such as airports and suppliers. See Robert Kivits, Michael B Charles, & Neal Ryan, "A Post-Carbon Aviation Future: Airports and the Transition to a Cleaner Aviation Sector" (2010) 42 Futures 199 at 209.

153 See IEA, *supra* note 1 at 44.

154 See ICAO, A37-WP/23, *supra* note 146.

155 *Ibid.* See also Aysha Al Hamili, "Can ICAO Act as a Catalyst in the Development of Alternative Fuels Among Member States?" (2012) 1 ICAO J (stressing that need to adopt "research and development legal frameworks" to ensure that alternative fuels are allocated in sufficient quantities to aviation at 30-31).

156 See Giovanni Bisignani, "State of the Air Transport Industry" (presentation delivered at the 66th IATA Annual General Meeting, World Air Transport Summit, Berlin 6-8 June 2010) online: IATA <www.iata.org/pressroom/speeches/pages/2010-06-07-01.aspx>, citing Michigan Technological University, "Life Cycle Assessment of Green Jet from Oils and Tallow: Comparison to Petroleum Jet Fuel." February 2009 [Bisignani, "State of the Air Transport Industry"].

157 See Airbus, "Global Market Forecast 2009-2028," *supra* note 5.

158 See Rutherford, *supra* note 81.

159 See generally Zack Colman, "White House Extends Aviation Biofuel Program," *The Hill* (15 April 2013) online: <http://thehill.com/blogs/e2-wire/e2-wire/293951-white-house-extends-aviation-biofuel-program #ixzz2SQL8wP3e>.

160 See Emma Nygyren, Kjell Aleklett, & Mikael Hook, "Aviation Fuel and Future Oil Production Scenarios" (2009) 37 Ener Pol'y 4003 at 4010.

161 See Boeing, "Current Market Outlook 2013-2032," *supra* note 5.

1.1.9 *Aviation's Fuel Factor: Is Regulation Needed After All?*

The International Energy Agency (IEA) estimates that aviation uses 6.3 percent of the world's total refinery production, which is equivalent to 243 million tonnes.[162] Each tonne of fuel that aviation burns releases a fixed amount of CO_2. Since aviation's CO_2 footprint is proportional to the amount of fuel it burns,[163] fuel consumption serves as the proper indicator to determine the sector's CO_2 output.[164]

During the summer of 2008, the price of aviation fuel skyrocketed to nearly 150 US dollars per barrel,[165] reaching an all-time record of 40 percent of most airlines' operating costs.[166] According to IATA estimates, the average fuel price for 2013 was 126.40 US dollars per barrel.[167] High fuel prices serve as a very strong incentive for airlines to reduce excess capacity, and it is not surprising that fuel prices have also become a major consideration in aircraft fuel efficiency.[168] It has been argued that market signals are an effective way to reduce aviation's carbon footprint since airlines have strong financial incentives to reduce fuel consumption.[169] This laissez-faire view has been particularly well received in the United States where aviation's CO_2 emissions are mostly unregulated despite the fact that the United States is responsible for over one-third of aviation's global footprint.[170] On the strength of the argument that the invisible hand of the market will correct any deficiencies, regulatory intervention has received very little support in the United States thus far.

---

162 See IEA, *supra* note 1. Yet if aviation continues growing at a yearly rate of nearly 5 percent, its share of the world refinery production will reach 13.7 percent by 2030.

163 1 tonne of fuel is equivalent to 3.16 tonnes of CO_2.

164 See ICAO, CAEP/2-WP/73, *supra* note 128. ICAO forecasts that aviation's fuel consumption in 2006 was approximately 187 metric tons (Mt) of fuel. This means a CO_2 footprint of roughly 591 Mt CO_2. With aircraft fuel consumption expected to grow at between 3.0 and 3.5 percent per year, aviation's aggregate CO_2 output is estimated at between 1,450 and 1,710 Mt CO2 by 2036.

165 The 2008 fuel spike was due not only to market speculations but primarily to low inventories, supply disruptions, low refining capacity, and extremely high demands from China and India. See generally Charles E Schlumberger, "The Oil Price Spike of 2008" (2009) XXXIV Ann Air & Sp L 114.

166 See IATA, "Striking Oil – Understanding Oil Price Volatility" *IATA* (October 2009) online: <www.iata.org/publications/airlines-international/october-2009/Pages/2009-10-07.aspx>. At present, fuel accounts for roughly 33 percent of an airline's cost. IATA, "Remarks of Tony Tyler," *supra* note 141; Boeing, "Current Market Outlook 2013-2032," *supra* note 5 (stressing that "[f]uel is now the largest component of an airline's cost structure"); See IEA, *supra* note 1 at 232.

167 IATA, Price Fuel Analysis, *IATA* online <www.iata.org/publications/economics/fuel-monitor/Pages/price-analysis.aspx>.

168 See IEA, *supra* note 1 at 232.

169 See Carl Burleson, "International Aviation and Emissions Trading: A US Perspective" (presentation delivered at the ICAO Colloquium on Aviation and Climate Change, 13 May 2010) online: ICAO <www.icao.int/Meetings/EnvironmentalColloquium/Documents/2010-Colloquium/5_Burleson_Faa.pdf>; Carl Burleson, "The EU Emissions Trading System Proposal" (2007) 21:3 Air & Space L; Ray Lahood, "Statement of the Honorable Ray Lahood, Secretary of Transportation before the Committee on Commerce, Science and Transportation of the United States Senate" (6 June 2012) (discussing that US aviation emissions have declined 12 percent since 2000 through 2010).

170 See IEA, *supra* note 1 at 234.

Some commentators consider any attempt to regulate aviation's carbon footprint as an "impediment" for the sector.[171] In this vein, US airlines have long argued that environmental regulatory measures will only add a financial burden that will in turn make it very difficult to purchase new (aircraft) technology to improve fuel efficiency and reduce their carbon footprint.[172] An argument could be made that regulation is not warranted since airlines already have all incentives in the world to reduce fuel consumption and become more efficient. This is driven not only by the environmental pressure but also and most importantly by high oil prices.[173] In other words, reducing GHG emissions, it is not a matter of corporate social responsibility or good stewardship, but, rather, it is a matter of corporate profitability.

Fuel prices have recently dropped, the result of a number of factors, including the US credit crunch following the housing market failure. In 2010, aviation fuel averaged 88 US dollars per barrel.[174] Air traffic returned to pre-recession levels.[175] Aviation is once again expanding. Therefore, despite the volatility of the price of fuel, the sector shows a historical ability to rebound and recover from the most challenging crises.[176] Relying solely on market signals may not be the best policy alternative to addressing aviation's ever-growing CO_2 footprint.

The argument that regulation is not needed also ignores the fact that technological and operational efficiencies are not, by themselves, sufficient to offset emissions produced as the result of the sector's remarkable growth. For example, Emirates, one of the largest and most sophisticated airlines in the world, has a fleet of more than 200 aircraft.[177] The average age of its fleet is roughly 6.5 years,[178] one of the youngest within the airline industry.

---

171 Tate L Hemingson, "Why Airlines Should be Afraid: The Potential Impact of Cap and Trade and Other Carbon Emissions Reduction Proposals on the Airline Industry" (2010) 75 J Air L & Com 741 at 743.

172 Nicholas E Calio and Lee Moak, "Why U.S. Airlines Oppose EU's Emissions Tax Scheme," *USA Today* (27 March 2012) online: <www.usatoday.com/news/opinion/forum/story/2012-03-27/airlines-eu-planes-emissions-tax/53805612/1>; "US legislation to prohibit airlines from joining EU ETS moves a step closer as Senate bill receives bipartisan support," *Green Air Online* (8 March 2012) online: <www.greenairon-line.com/news.php?viewStory=1436>. See also See Forsyth, *supra* note 148 at 27.

173 See Peter Morrel, "The Potential for European Aviation CO_2 Emissions Reduction through the Use of Larger Jet Aircraft" (2009) 15 J Air Transp Manag 151.

174 See IATA, "Fuel Monitor" *IATA* online: <www.iata.org/whatwedo/economics/fuel_monitor/Pages/index.aspx>.

175 See Bisignani, "State of the Air Transport Industry 2010," *supra* note 156.

176 In addition to the 2008 oil spike, these crises include the 2001 9/11 terrorist attacks in the United States, the 2003 severe acute respiratory syndrome (SARS), and the 2010 Icelandic volcanic ash eruption. As Boeing has noted, "[c]ommercial aviation has weathered many downturns in the past. Yet recovery has followed quickly as the industry reliably returned to its long-term growth rate of approximately 5 percent per year." Boeing, "Current Market Outlook 2013-2032," *supra* note 5. See also Airbus, "Global Market Forecast 2013-2032," *supra* note 1 (Introduction) at 8.

177 In 2013 alone, Emirates received 34 brand new wide-body aircraft. See Emirates Group, *Annual Report 2012-2013, Emirates Group* online: <http://content.emirates.com/downloads/ek/pdfs/report/annual_.report_2013.pdf>. [Emirates, "Annual Report 2012-2013"].

178 See Emirates Group, "Environment Report 2012-2013" online: <http://content.emirates.com/downloads/ek/

According to the carrier, its "flights are 16.6 [percent] more efficient in terms of emissions per passenger-kilometre than the industry average."[179] The carrier is well known across the industry for having implemented numerous operational initiatives to reduce fuel consumption.[180] Over the past decade, its growth rate has remained in double figures above the industry average.[181] With the largest fleet of A-380s and B-777s, the carrier is flying the most fuel-efficient aircraft known to date.[182] Despite these efficiencies implemented by Emirates, its GHG emissions' growth rate far exceeds the offsets made by fuel efficiencies. Emirates reports that its 2012-2013 emissions increased by 15.9 percent as compared to 2011-2012 levels.[183] The airline recognizes that the increase in CO_2 emissions has been due to its remarkable growth.[184] Given current growth trends in the international air transport sector, the case of Emirates speaks for the industry at large: technological and operational efficiencies will not be sufficient. Without new regulations, there is simply no way to reduce or limit the sector's carbon footprint.

1.2 International Legal Regime

Having addressed air transport's growth trends, the economic and environmental impact of aviation, and its contribution to climate change, I now provide a brief overview of the international legal regime.

1.2.1 UNFCCC/Kyoto Protocol

In 1992, States adopted the *United Nations Framework Convention on Climate Change* (*UNFCCC*).[185] This instrument seeks to establish an overall legal framework to stabilize anthropogenic GHG "concentrations in the atmosphere at a level that would prevent dangerous anthropogenic interference with the climate system."[186] As its name indicates, the *UNFCCC* is a framework agreement. It neither sets emission reduction targets for individual countries nor contains an enforcement mechanism.[187] It serves, however, as the

pdfs/environment/environment_report_2012_13_locked.pdf> [Emirates, "Environmental Report 2012-2013"] at 6.

179 *Ibid.*

180 *Ibid.*

181 See Emirates, "Annual Report 2012-2013," *supra* note 177 at 6.

182 *Ibid.*

183 Emirates reports that in 2012-2013, it emitted approximately 22.4 million tonnes of CO_2, as opposed to only 19.3 million in 2011-2013. *Ibid.* Fuel efficiency, however, has improved 0.79 percent. *Ibid.*

184 *Ibid.* at 20.

185 See *UNFCCC*, *supra* note 11 (Introduction).

186 *Ibid.* at 2.

187 *Ibid.*

legal skeleton for States to exchange information on GHGs, establish national or regional policies for mitigation purposes, develop national inventories, and cooperate in preparing for adaptation to the impacts of climate change.[188]

In order to make the *UNFCCC*'s objectives operational, States negotiated and adopted the *Kyoto Protocol* in 1997.[189] This instrument, which entered into force on 16 February 2005, imposes specific emission targets with respect to four GHGs for industrialized countries over a five-year commitment period. By assigning specific quantified emission limitation/reduction commitments (QELRCs), the *Kyoto Protocol* required these countries – Annex 1 Parties – to reduce their overall emissions of GHGs to at least 5.2 percent below 1990 levels in the period 2008-2012.[190] On the other hand, non-Annex 1 Parties (developing States) did not have to pursue quantified emission reductions but were called upon to undertake mitigation and adaptation measures in accordance with their respective capabilities.[191] In essence, developing countries were only called upon to undertake voluntary mitigation measures. To achieve its objectives, the *Kyoto Protocol* contemplated three options: Emissions Trading Systems (ETS), Joint Implementation (JI), and the Clean Development Mechanism (CDM).[192] Under the *Kyoto Protocol*, developed States undertook to pursue limitations or reductions of GHGs from aviation and marine bunker fuels by working through the International Civil Aviation Organization (ICAO) – the United Nations specialized agency in charge of civil aviation – and the International Maritime Organization (IMO), respectively.[193]

188 *Ibid.*, Arts 3, 4, and 5.

189 See *Kyoto Protocol, supra* note 12 (Introduction).

190 See David Freestone, "The UN Framework Convention on Climate Change, the Kyoto Protocol, and the Kyoto Mechanism" in David Freestone & Charlotte Streck, eds, *Legal Aspects of Implementing the Kyoto Protocol Mechanisms: Making Kyoto Work* (Oxford: Oxford University Press, 2005) 3 at 9.

191 See *UNFCCC, supra* note 11 (Introduction) at Art 4.

192 See Bodansky, *The Art and Craft of International Environmental Law, supra* note 62 at 34. In very simple terms, under the CDM mechanism, a firm from a developed country develops an emission reduction project in a developing country. Such firm bears the cost associated with such initiatives but generates certified emission reduction units (CERs) that may be used to meet their reduction targets. These units are also tradable. The developing country benefits from the transfer of technology and know-how that stems from carrying out such project. The project also achieves emission reductions in the developing country. It is a win-win situation for both developed and developing countries. See Graciela Chichilnisky & Kristen A Sheeran, *Saving Kyoto: An Insider's Guide to the Kyoto Protocol, How it Works, Why it Matters and What it Means for the Future* (London: New Holland, 2009) at 121; Jon Birger Skjærseth & Jørgen Wettestad, "Implementing EU Emissions Trading: Success or Failure?" (2008) 8 Int Enviro Agreements 275 at 285; Friedrich Soltau, *Fairness in International Climate Change Law and Policy* (Cambridge: Cambridge University Press, 2009) at 83; Axel Michaelowa et al, "The Market Potential of Large-Scale Non-CO_2 CDM Projects" in W Th Douma, L Massai, & M Montini, eds, *The Kyoto Protocol and Beyond: Legal and Policy Challenges of Climate Change* (The Hague: TMC Asser Press, 2007) at 59.

193 See *Kyoto Protocol, supra* note 12 (Introduction) at Art 2.2.

1.2.2 The EU ETS

By joining the *Kyoto Protocol*, Member States of the European Union undertook to reduce their GHGs by 8 percent at 1990 levels by 2012.[194] EU Member States agreed to share the allocation of that overall target through the so-called Burden-Sharing Agreement (BSA).[195] On 13 October 2003, the European Parliament and Council passed Directive 2003/87 establishing a community trading system to reduce CO_2 emissions.[196] Known as the EU ETS, this is a market-based policy instrument seeking to achieve emission reductions in a cost-effective and economically efficient manner.[197] To date, and in spite of its unquestionable difficulties, the EU ETS is the world's largest initiative to limit emissions and combat climate change.[198]

Ever since the 1960s when ICAO started addressing aviation's environmental impact, European States have been the most fervent advocates.[199] These States have pushed the environmental agenda at ICAO's Committee on Aviation Environmental Protection (CAEP) on myriad issues such as aircraft noise and other environment-related charges like no other group of States.[200] Europe's interest in climate change issues at ICAO has not been any different. It should not come as a surprise then that when the EU launched its

---

194 See John Vogler & Charlotte Bretherton, "The European Union as a Protagonist to the United States on Climate Change" (2006) 7 Int'l Stud Perspectives 1 at 3. Sebastian Oberthür and Claire Dupont point out that the 8 percent reduction target is the highest among industrialized nations. See Sebastian Oberthür & Claire Dupont, "The Council, the European Council and International Climate Policy: From Symbolic Leadership to Leadership by Example" in Rüdiger KW Wurzel & James Connelly, eds, *The European Union as a Leader in International Climate Change Politics* (London: Routledge, 2011) at 77. Although the *Kyoto Protocol* is perceived as being essentially flawed for not including the major emitters, namely, the United States and China, part of its true legacy lies in the fact that it has not only embraced economic instruments to reduce emissions, but it has also paved the way for measures such as the EU ETS. See Dieter Helm, "Climate-Change Policy: Why Has to Little Been Achieved?" (2008) 24:2 Oxford Rev Econ Pol'y 211 at 212.

195 See Leonardo Massai, "Legal Challenges in European Climate Policy" in W Th Douma, L Massai, & M Montini, eds, *The Kyoto Protocol and Beyond: Legal and Policy Challenges of Climate Change* (The Hague: TMC Asser Press, 2007) at 13. For instance, Germany and Denmark agreed to reduce GHG emissions by 21 percent below 1990 levels.

196 EC, *Parliament and Council Directive 2003/87 of 13 October 2003 on Establishing a Scheme for Greenhouse Gas Emission Allowance Trading Within the Community and Amending Council Directive 96/61/EC*, [2003] OJ, L 275/32 [General EU ETS Directive]. See also Anthony Hobley, "The UK Emissions Trading System: Some Legal Issues Explored" in Julian Boswall & Robert Lee, eds, *Economics, Ethics and the Environment* (London, Sydney: Cavendish Publishing Limited, 2002) at 61.

197 See General EU ETS Directive, *supra* note 196 at Art 1. See also Jon Birger Skjærseth & Jørgen Wettestad, "Fixing the EU Emissions Trading System? Understanding the Post-2012 Changes" (2010) 10 Global Environmental Politics 4 at 10; Robert W Hahn, "Greenhouse Gas Auctions and Taxes: Some Political Economy Considerations" (2009) 3:2 Rev Enviro Econ & Pol'y 167 at 179 [Hahn, "Greenhouse Gas"].

198 The purpose of the ETS is to stabilize GHGs in order for the rate of increase of overall annual temperatures not to exceed 2° C. See EC, *Parliament and Council Directive 2009/29 of 23 April 2009 Amending Directive 2003/87/EC so as to Improve and Extend the Greenhouse Gas Emission Allowance Trading Scheme of the Community* [2009] OJ, L 140/63 [Revised Directive].

199 See Davies & Goh, *supra* note 27 at 125.

200 See Chapter 3.

sixth community environmental action program in 2002, the set target was clearly to "reduce [GHG] emissions from aviation if no such action [was] agreed within [ICAO]."[201] Frustrated by the lack of concrete progress and the almost nonexistent commitment on the part of other ICAO Member States, the European Commission decided in 2005 to take the "bull by the horns" and announced its intention to incorporate international aviation into the general ETS regime.[202]

On 19 November 2008, after a consultation process that lasted more than 3 years, the European Parliament and the Council of the European Union passed Directive 2008/101/EC, which included international aviation activities into the community's emissions trading regime.[203] Although the Commission expected to cut CO_2 emissions by 70 million tonnes per year by 2020[204] and in spite of the fact that analysts predict that aviation emissions will roughly double from 2005 levels, the inclusion of international aviation into the European scheme has created one of the most highly fought controversies in the annals of international aviation.[205]

1.3 THE POLITICAL DIMENSION: A SMALL PIECE WITHIN A BIGGER PUZZLE

Although the analysis of legal issues surrounding the aviation and climate change discourse is of utmost importance, it is also necessary to examine the political, economic, as well as other sectorial interests. These exert an enormous influence in addressing GHG emissions from international aviation. In this respect, it is noteworthy that some States tend to approach aviation and climate change issues from a sectorial perspective. As the aviation industry would advocate, this requires "an aviation solution to a climate problem [and] not a climate solution to an aviation problem."[206] Other States, on the other hand, consider the broader policy implications of the problem. For these States, the concern is not neces-

201 EC, *Decision 1600/2002/EC of the European Parliament and of the Council of 22 July 2002 laying down the Sixth Community Environment Action Programme* [2002] OJ, L 242/1 at 7. See also M Vittoria Giugi Carminati, "Clean Air & Stormy Skies: The EU ETS Imposing Carbon Credit Purchases on United States Airlines" (2010) 37 Syracuse J Int'l L & Com 127 at 137.

202 See Richard Smithies, "Regulatory Convergence – Extending the Reach of EU Aviation Law" (2007) 72 J Air L & Com 3 at 18.

203 Aviation Directive, *supra* note 17 (Introduction). The EU ETS requires that EU Member States transpose the directive into national legislation within a year. See also Ulrich Steppler & Angela Klingmüller, "EU Emissions Trading Scheme and Aviation: Quo Vadis?" (2009) 34:4/5 Air & Space L at 253.

204 See Artur Runge-Metzger, "Aviation and Emissions Trading" ICAO Council Briefing (29 September 2011) [unpublished, archived on file with the author] at 33.

205 See Annela Anger & Jonathan Köhler, "Including Aviation Emissions in the EU ETS: Much ado About Nothing?" (2010) 17 Transp Pol'y 38 at 42. As a central part of this book, policy and legal implications of the EU ETS are thoroughly addressed in Chapters 4, 5, and 6.

206 Paul Steele, "IATA AGM Resolution Climate Change," Informal Briefing Delivered to the Council of ICAO (11 June 2013) [unpublished, archived on file with the author].

sarily what is best for their aviation sectors, but rather how addressing the problem may spill over and affect other aspects of their national interests (e.g. economic development).

Addressing the climate change concerns associated with a particular sector is in and of itself a very challenging undertaking. Yet, the endeavor becomes even more complex when actions taken to tackle the sector's carbon footprint in one forum may be perceived as potentially having knock-on effects in another regime. In other words, decisions in one forum may either prejudge discussions or set a precedent in the other. Some States would like to completely disassociate ICAO from the *UNFCCC* process. Others strongly advocate that ICAO's action in handling this issue should be strictly in line with the *UNFCCC* regime and the principles thereof. In spite of the fact that ICAO Assembly resolutions addressing climate change issues have repeatedly stated that discussions at ICAO do not prejudge the outcome of the *UNFCCC* regime,[207] the fact of the matter is that a large number of States are concerned with potential spillovers.

One of the main concerns of a group of developing countries in accepting non-binding, non-attributable aspirational goals for international civil aviation has been that such goals may set a precedent for binding goals in the future not only within the context of aviation discussions but also with regard to the *UNFCCC*.[208] Likewise, these States have objected to the EU ETS, given that such a regime does not take into account the *UNFCCC*'s CBDR principle. In other words, it does not establish a differentiated scheme between aircraft operators from developed and developing countries. Some developed States have also strongly opposed any attempt to incorporate such principle into ICAO's discussions. This serves to illustrate that international aviation is but one small piece of a bigger puzzle that comprises climate change as a whole.

1.3.1 The Problems with Lack of Progress

GHG emissions from international aviation have not been reduced or limited, as the *Kyoto Protocol* had instructed.[209] Critics have said that ICAO's work in addressing the aviation

207 See ICAO, A38-18, *supra* note 21 (Introduction) (noting that the ICAO Resolution on climate change "does not [in theory] set a precedent for or prejudge the outcome of negotiations under the UNFCCC and its Kyoto Protocol nor represents the position of the Parties to the *UNFCCC* and its Kyoto Protocol" at preambular clauses). See also ICAO, A37-19, *Consolidated Statement of Continuing ICAO Policies and Practices related to Environmental Protection: Climate Change* at preambular clauses [ICAO, A37-19].

208 In 2006, while at the 12th meeting of the *UNFCCC*'s Conference of the Parties (COP/12), a group of developing countries rejected a proposal that developing countries may undertake voluntary commitments to reduce GHG emissions. For these countries, as Friedrich Soltau notes, voluntary commitments may represent a "slippery slope to binding commitments, as well as underpinning [*UNFCCC*'s] principle of common but differentiated responsibility." Soltau, *supra* note 192 at 94.

209 See *Kyoto Protocol*, *supra* note 12 (Introduction) at 2.2.

sector's carbon footprint has not gone beyond mere "symbolic action."[210] The lack of a system in place that actually reduces or limits the sector's GHG emissions has produced two different results.

First, some question ICAO's ability to handle the issue.[211] Reducing the sector's carbon footprint does not necessarily form part of its core business.[212] Under this line of reasoning, ICAO has only engaged in aviation and climate change discussions as a result of two factors: the possibility that the *UNFCCC* may take over the regulation of GHG emissions from international aviation[213] and the threat posed by unilateral EU action.[214] For instance, just months after the adoption of the *Kyoto Protocol*, the 32nd Assembly acknowledged that "other international organizations [were] becoming involved in activities relating to environmental policies affecting air transport."[215] The Assembly also instructed the Council to "not leave [developing policy guidance initiatives] to other organizations [such as the *UNFCCC*]."[216]

Second, a lack of progress at ICAO has served as an opportune justification for the EU to unilaterally include foreign aircraft operators into its ETS. To ICAO's defense, the fact that at present there is no system in place to reduce emissions is the result of the lack of political will of its Member States and not fault of the organization. Other international organizations assigned similar tasks under the *Kyoto Protocol* such as the International Maritime Organization (IMO) have also not been successful.

---

210 Sebastian Oberthür, "The Climate Change Regime: Interactions with ICAO, IMO and the EU Burden-Sharing Agreement" in Sebastian Oberthür & Thomas Gehring, eds, *Institutional Interaction in Global Environmental Governance: Synergy and Conflict among International and EU Policies* (Cambridge: Massachusetts Institute of Technology, 2006) 53 at 66 [Oberthür, "Climate Change Regime"].

211 *Ibid.*

212 *Ibid.* at 132.

213 Allen Pei-Jan Tsai & Annie Petsonk claim that the *UNFCCC* "retains legal competence to review the work that ICAO and IMO undertake pursuant to Article 2.2. of the Kyoto Protocol." Allen Pei-Jan Tsai & Annie Petsonk, "Tracking the Skies: An Airline-Based System for Limiting Greenhouse Gas Emissions from International Civil Aviation" (1999) 6 Envtl Law 763 at 782.

214 See Oberthür, "Climate Change Regime," *supra* note 210 at 60; Gabriel S Sanchez, "In Defense of Incrementalism for International Aviation Emissions Regulation" (2012) 53:1 Va J Int'l L 1 (commenting that ICAO's sudden push to address international aviation's carbon footprint through a multilateral global scheme is primarily due to the "fear of having its limited custody over aviation-emissions issues displaced by the EU's unilateralism" at 8); Kati Kulovesi, "Addressing Sectoral Emissions Outside the United Nations Framework Convention on Climate Change: What Roles for Multilateralism, Minilateralism and Unilateralism" (2012) 21 RECIEL 193 at 197 [Kulovesi, "Addressing Sectoral Emissions"]. Sebastian Oberthür is of the view that, although weak, the *UNFCCC* has served as a regulatory competitor inducing the work of ICAO. *Ibid.* at 66. Havel and Sanchez go even further by saying that "ICAO's direct authority to regulate aviation emissions can be characterized as weak or, more realistically, nonexistent." Brian F Havel & Gabriel S Sanchez, "Toward a Global Aviation Emissions Agreement" (2012) 36 Harv Envtl L R 352 at 358 [Havel & Sanchez, "Toward a Global Aviation Emissions Agreement"].

215 ICAO, A32-8, *Consolidated Statement of Continuing ICAO Policies and Practices related to Environmental Protection*, Appendix A, General Policy at preambular Clauses [ICAO, A32-8].

216 *Ibid.* at para 1.

1.3.2 Exogenous Threats

Outside observers and critics remain unimpressed with ICAO and industry achievements toward reducing GHG emissions. It is clear that combating climate change will require substantial financial investment, particularly for developing countries.

In this context, international civil aviation has been identified as a feasible financial source to bridge the gap between developed and developing countries in addressing climate change issues. For instance, in an attempt to find 100 billion US dollars per year until 2020 for climate change mitigation in developing countries, the UN Secretary-General's High-Level Advisory Group Climate Change Financing has suggested that international air transport could contribute between 1 and 6 billion US dollars per year through the establishment of levies.[217] Likewise, a World Bank report prepared for the G20 also recommended levying a fee of 25 US dollars per tonne of CO_2.[218] This proposal could generate up to 12 billion US dollars per year for climate financing.[219] Environmental NGOs have highlighted the fact that, given that international aviation is under-taxed, a "global levy on fuel" may provide up to 5 billion US dollars for climate change adaptation and mitigation measures primarily destined for developing countries.[220] Similarly, the European Commission has noted that "[a]t the global level, international maritime and aviation transport could be promising new sources for raising climate finance."[221] Although supportive of the overall goal to reduce GHG emissions, the ICAO Assembly has expressed serious concerns regarding "the use of international aviation as a potential source for the mobilization of revenue for climate finance to the other sectors."[222] In addition, the airline industry frequently complains that numerous countries have adopted international departure taxes for environmental purposes, yet it is not clear that funds so collected from such levies have been used for their intended purposes.[223] The ICAO Secretariat, Member States, and the airline industry do not want taxes imposed on the sector.[224]

---

217 See "Report of the Secretary-General's High-Level Advisory Group on Climate Change Financing," *supra* note 23 (Introduction) at 28.

218 See International Monetary Fund and the World Bank Report, *supra* note 24 (Introduction) at 5.

219 *Ibid.*

220 International Maritime Emissions Reduction Scheme, "Rebate Mechanism for Fair and Global Carbon Pricing of International Transport" *IMERS* online: <http://imers.org/docs/RM_Aviation_Fact_Sheet.pdf>.

221 EC, *Commission Staff Working Document, Scaling Up International Climate Finance After 2012*, SEC (2011) 487 final (Brussels, 8 April 2011).

222 ICAO, A38-18, *supra* note 21 (Introduction) at para 30. Moreover, the ICAO Assembly urged its Member States to raise such concern within the UNFCCC process.

223 Industry has identified the following taxes: (i) UK Air Passenger Duty, (ii) Germany Air Passenger Tax, (iii) Ireland Air Travel Tax, (iv) Australia Carbon Tax, (v) Norway Carbon Tax, (vi) Austria Carbon Tax, (vii) South Africa Carbon Tax, and (vii) Brazil Environmental Tax. See ICAO, HGCC/3-WP 5 at Appendix A.

224 In 2005, France proposed "taxing airline tickets to finance development" in accordance with the UN Millennium Development Goals. Although a number of States recognized the value in providing financial contributions for development purposes, the vast majority of States did not find appropriate targeting international

Targeting international aviation as a vehicle to provide additional funds for climate change is a threat to the sector. These proposals are thought to be far more financially burdensome for the sector than an industry-led proposal to be implemented under the auspices of ICAO. Failing to address the issue will only encourage other international institutions to seek additional funds from international aviation for climate change purposes. It may also provide the perfect justification to impose international departure taxes for environmental purposes. Establishing a system to reduce or limit the sector's GHG emissions is clearly in the best interest of the airline industry and ICAO. Failure to do so will be more costly for the sector and will significantly affect ICAO's credibility.

1.4 Conclusion

Aviation unquestionably represents a catalytic force that fosters economic development and facilitates global connectivity. The modern world could simply not function without aviation. Yet its consistent growth poses significant challenges for the environment.[225] Concerns over aviation's impact on the climate continue to escalate. Although the sector has introduced a number of technological and operational efficiencies, such measures by themselves will not offset the CO_2 emissions expected to be generated as a result of its projected growth. The sector represents a "growing source of CO_2 emissions."[226] Likewise, there is little indication that alternative fuels will reduce aviation's carbon footprint in the short to medium term.

Those defending the aviation sector argue that the real problem has been the lack of an adequate communications campaign to explain the sector's actual achievements. They are quick to point out that aviation's carbon contribution is only 2 percent of global GHG emissions. This approach has also been extremely detrimental in triggering concrete remedial actions, for it has given the false impression that the problem is either insignificant or that it will not make any real contribution to the larger climate change issue.

---

aviation as a means to fund such initiatives. While noting the proposal, the ICAO Council did not endorse it. See ICAO, C-WP/12498, *Taxes on Airline Tickets to Finance Development* (presented by France). See ICAO, C-MIN 176/13, *Council, 176th Session, Minutes of the Thirteenth Meeting*; ICAO, C-MIN 176/14, *Council, 176th Session, Minutes of the Fourteenth Meeting*; ICAO, C-MIN 175/15, *Council, 175th Session, Minutes of the Fifteenth Meeting*; ICAO, C-DEC 176/14, *Council, 176th Session, Minutes of the Fourteenth Meeting.*

225 See John Broderick, "Voluntary Carbon Offsetting for Air Travel" in Stefan Gossling & Upham, eds, *Climate Change and Aviation: Issues, Challenges and Solutions* (London: Earthscan, 2009) 329 at 341.

226 ICAO, A37-WP/108 EX/26, *Addressing Aviation's Environmental Impacts Through: A Comprehensive Approach* (presented by Belgium on behalf of the European Union and its Member States and by the other States Members of the European Civil Aviation Conference and by Eurocontrol). Environmental non-governmental organizations (NGOs) estimate that by 2050 the sector's fuel consumption will increase approximately 250 percent compared to 2006 levels. See WWF, Aviation Report, *supra* note 91 at 3. This NGO also points that "there will be a 274 [percent] increase in fuel used by airlines in the next 38 years." *Ibid.*

Although some commentators and sectors within the industry contend that regulation is not warranted, the fact of the matter is that without it aviation will not be in a position to either reduce or limit its carbon footprint.

Hence, at the international level, States adopted the *UNFCCC* as an over-arching regime to address anthropogenic climate change. A few years later, the *Kyoto Protocol* put some of the precepts of the *UNFCCC* into operation. The *Kyoto Protocol* mandated that developed States take steps to limit or reduce GHG emissions from aviation working through ICAO. Unimpressed with progress made at ICAO thus far, European States resorted to unilateral action at the regional level to address the problem by including foreign aircraft operators into its emissions trading scheme. This has created an epic international dispute.

As pressure to address climate change mounts, it is in the aviation sector's best interest to tackle its carbon footprint as soon as possible. Failure to do so may be extremely financially burdensome in the short and long run. Various international organizations have already singled out international aviation as a potential source to finance climate change adaptation and mitigation projects. The following chapters discuss policy and legal issues that may facilitate addressing GHG emissions from international aviation.

2 Aviation and Climate Change: A Case of Fragmentation of International Law

The *Kyoto Protocol* tasks developed countries to pursue the limitation or reduction of GHG emissions from aviation by working through ICAO.[1] Despite its prima facie rationale as an international mechanism to reduce GHG emissions, *Kyoto* has resulted in one of the most frequently disputed normative conflicts. There appears to be an incompatibility between the *UNFCCC* regime and its "common but differentiated responsibilities and respective capabilities" (CBDR) principle on the one hand and the *Chicago Convention* and its non-discrimination principle on the other hand. Through the lens of fragmentation of international law, this chapter analyzes the relationship between the two regimes and the scope of the conflict, offering proposals aimed at reconciling the principles.

I first examine the international aviation sector's interaction with the global climate change regime. In particular, I explore the provisions of the *Kyoto Protocol* that involve ICAO in the climate change regime. It has been suggested that these provisions give an implicit mandate to ICAO to address GHG emissions from international aviation; however, these provisions are extremely ambiguous, which helps explain the fraught relationship between the *Chicago Convention* and the *UNFCCC* regime. As I show, one of the salient characteristics of fragmentation is the emergence of multiple regimes whose principles may be in conflict or in tension.

I then explore fragmentation as a phenomenon of international law. In addition to pointing out the risks and advantages of fragmentation, the section reviews the seminal International Law Commission (ILC) report on fragmentation. In evaluating the merits and shortcomings of the ILC report, significant attention is devoted to the ILC's consideration of the rules on successive treaty interpretation prescribed by the Vienna Convention on the Law of Treaties (*VCLT*), as well as the examination of the principles of systemic integration of international law.

I move on to apply the *VCLT* rules on successive treaty interpretation to the CBDR/non-discrimination saga. The purpose of this exercise is to explore whether the ILC report is sufficient to address this case of normative conflict or whether other alternatives must be sought.

My next section highlights various attempts to accommodate the special needs of developing countries when addressing aviation and climate change. Through the adoption of exclusion (e.g. de minimis) provisions, the section explains ICAO's unsuccessful attempts

1 See *Kyoto Protocol*, *supra* note 12 (Introduction) at Art 2.2.

to engage broader participation from developing countries. It also points out the reasons for their failure and explores the unsuccessful attempt to replace CBDR in the aviation context with the notion of special circumstances and respective capabilities (SCRC). Considerable attention is paid to CBDR and its implications for the future design of a global market-based measure (MBM) for international civil aviation.

Finally, this chapter proposes a new approach that reconciles the principles of CBDR and non-discrimination. I argue that the principles should not be interpreted as static and immutable, but evolving and adaptable. I draw support from my argument from comparative debates in international law that assert the ways international organizations act within the context of interactions with other institutions and can find common ground where principles do not overtly conflict. The reconciliation of the CBDR and non-discrimination principles is central to the success of the global MBM scheme.

2.1 THE INTERACTION BETWEEN INTERNATIONAL AVIATION AND THE CLIMATE CHANGE REGIME

2.1.1 The Kyoto Protocol

The *Kyoto Protocol* explicitly provides that developed countries must pursue the limitation or reduction of GHG emissions from aviation by "working through [ICAO]."[2] In essence, the third Conference of Parties (COP/3) of the *UNFCCC* decided that GHG emissions from the aviation sector would be handled at another forum, namely, ICAO.[3] As Koremenos et al. point out, States may opt to move one issue to another forum that offers "better institutional design."[4]

Under significant pressure, COP/3 adopted the *Kyoto Protocol* at the eleventh hour. Although the *UNFCCC* did not track the minutes of COP/3, the literature points out that this was the result of the difficulties encountered in allocating GHG emissions from international aviation to national inventories.[5] As seen in Chapter 1, a great deal of GHG

2 *Ibid.*

3 It is not unusual that States gathered at one forum decide that one issue will be handled by another international organization. For instance, the Annex on Telecommunications of the General Agreement on Trade and Services, which forms part of the Uruguay Round Agreement, provides that States will address "standards for global compatibility and inter-operability of telecommunication networks and services" through "the International Telecommunication Union and the International Organization for Standardization." See *General Agreement on Trade in Services*, 15 April 1994, 1869 UNTS 183 at Annex on Telecommunications, Art 7 (a) (entered into force 1 January 1995) [*GATS*].

4 Barbara Koremenos, Charles Lipson, & Duncan Snidal, "The Rational Design of International Institutions" (2001) 55:4 International Organization, 761 at 767 [Koremenos et al].

5 See Paul Stephen Dempsey, *Public International Air Law* (Montréal: Institute and Center for Research in Air & Space Law, 2008) (discussing that international aviation emissions are not part of the *Kyoto Protocol*

emissions from international aviation occurs over the high seas beyond the territorial jurisdiction of any State or over the territories of several States in succession. Unlike emissions from traditional stationary sources, GHG emissions from international aviation are not confined to the geographical boundaries of the States over which they initially occur.[6]

Given ICAO's unquestionable technical credentials, the decision to transfer the handling of aviation's GHG emissions to ICAO appeared reasonable at the time.[7] In addition, it appears that those present at COP/3 did not critically consider the legal, policy, political, and operational repercussions of the decision.[8] ICAO's subsequent interaction with the climate change regime raises a number of questions.[9] These are examined below.

2.1.2 *ICAO and Climate Change: Some Unanswered Questions*

It is clear that the *Kyoto Protocol* triggers ICAO's interaction with the climate change regime.[10] However, its language lacks precision, which has caused a number of problems. First, as Dempsey has noted, Article 2.2 of the *Kyoto Protocol* speaks about "emissions from aviation rather than emissions from international aviation."[11] A literal interpretation

"due to the difficulty of attributing these emissions to a single State or dividing the emissions between States" at 450).

6 See Chris Lyle, "Rio, Kyoto, Brussels and Chicago: Reconciling Principles Related to International Air Transport Emissions," *Green Air Online* (27 July 2012) online: <www.greenaironline.com/photos/Rio_Kyoto_Brussels_and_Chicago_Chris_Lyle_July_2012.pdf>. (suggesting that the mandate entrusted to ICAO has permitted to provide more technical expertise and the peculiarities of the aviation industry, but it has also resulted in a form of protection) [Lyle, "Rio, Kyoto, Brussels and Chicago]; Fariborz Zelli, "The Fragmentation of the Global Climate Governance Architecture" (2011) 2 Wires Clim Change 255 at 262.

7 See Kathryn Harrison, "The United States as Outlier: Economic and Institutional Challenges to US Climate Policy" in Kathryn Harrison & Lisa McIntosh Sundstrom, eds, *Global Commons, Domestic Decisions: The Comparative Politics of Climate Change* (Cambridge: MIT Press, 2010) 67 at 80; Chichilnisky & Sheeran, *supra* note 192 (Chapter 1) at 72. See also Jin Liu, "The Role of ICAO in Regulating the Greenhouse Gas Emissions of Aircraft" (2011) 4 Carbon & Climate L Rev 417 (arguing that, at the time, in light of its technical expertise, it was thought that ICAO was the appropriate international organization to handle this matter) at 418.

8 Jenks brilliantly writes that "[o]ne of the most serious sources of conflict between law-making treaties is the imperfect development of the law governing the revision of multipartite instruments and defining the legal effect of such revision." C Wilfred Jenks, "The Conflict of Law-Making Treaties" (1953) 30 Brit YB Int'l L 401 at 403.

9 The literature points out that the interaction of the climate change regime with institutions dealing with international transport, such as ICAO and IMO, is not at all unique. In fact, the *Kyoto Protocol*, which is one of the main instruments forming the climate change regime, "has an enormous scope, overlapping and interacting with a multitude of other issue areas and institutions in various ways." Sebastian Oberthür & Thomas Gehring, "Institutional Interaction: Ten Years of Scholarly Development" in Sebastian Oberthür & Olav Schram Stokke, eds, *Managing Institutional Complexity: Regime Interplay and Global Environmental Change* (Cambridge: Massachusetts Institute of Technology, 2011) 25 at 28.

10 See Oberthür, "Climate Change Regime," *supra* note 210 (Chapter 1) at 61.

11 Dempsey, *supra* note 5 at 450.

of this provision would mean that States working with ICAO must tackle GHG emissions from both domestic and international aviation. In theory, this could present a problem, given that ICAO mainly deals with international not domestic civil aviation.[12] This interpretation seems to disregard the fact that, under *UNFCCC* rules, emissions from domestic aviation form part of States' national inventories.[13] Also, as explained above, the rationale behind tasking this issue to ICAO was precisely the problem encountered in allocating GHG emissions from *international* aviation. Therefore, one may reasonably conclude that, in spite of the drafting oversight, the drafters' intention was geared toward emissions from international aviation. Coincidentally, this is how ICAO has interpreted Article 2.2 of the *Kyoto Protocol*.[14]

Second, a literal reading of the *Kyoto Protocol* suggests that only some States are obliged to address GHG emissions from international aviation. In this regard, Denys Wibaux, former Director of ICAO's Legal Bureau, recognized that this cannot be construed as an obligation bestowed upon ICAO.[15] Under this line of thinking, ICAO is simply a forum. Failure to reduce GHG emissions from international aviation cannot be attributed to ICAO. It is up to Member States to develop the required policies.[16] In the absence of political will, ICAO can just sit and wait. This position is a formalistic interpretation of the *Kyoto Protocol*.

Although it is undeniable that ICAO is not a party to the *Kyoto Protocol* and thus unbound by its provisions, the interpretation is not illustrative of ICAO's own subsequent practice in response to the *Kyoto Protocol*, a practice to which its membership has acquiesced.[17] As described throughout this book, States and industry stakeholders regard ICAO

---

12 *Ibid.*

13 See ICAO, A33-7, *Consolidated Statement of Continuing ICAO Policies and Practices related to Environmental Protection* [ICAO, A33-7] (noting that "domestic aviation emissions are included in national targets" at Appendix H, preambular Clauses).

14 *Ibid.*

15 See ICAO- C-MIN 181/21, *Council, 181st Session, Minutes of the Twenty-First Meeting* (the Statement of Germany recognized that the *Kyoto Protocol* "imposes obligation to States (some) not ICAO. ICAO is not obliged to develop policy nor is it prevented from doing so. In this respect, Kyoto is not an obstacle") at 256.

16 *Ibid.* (noting that "[t]here was not an obligation put on ICAO, and it was therefore up to its Member States. ICAO did not have an obligation as much as an opportunity") at 257.

17 ICAO does not expressly regard Art 2.2 of the *Kyoto Protocol* as granting a "mandate" to act. However, an examination of all the ICAO Assembly resolutions dealing with environmental protection from 1998 onward illustrates that the *Kyoto Protocol* has been the starting point that triggered action on aviation and climate change issues. For instance, less than 10 months after the *Kyoto Protocol's* adoption, the 32nd Assembly instructed the Council to "pursue all aviation matters related to the environment and also maintain the initiative in developing policy guidance on these matters, and not leave such initiatives to other organizations." The Assembly made explicit reference to the *Kyoto Protocol*. ICAO, A32-8, *supra* note 215 (Chapter 1) at Appendix A, para 2. In 2001, the Assembly tasked the Council "to study policy options to limit or reduce the environmental impact of aircraft engine emissions and to develop concrete proposals." ICAO, A33-7, *supra* note 13 at Appendix H, para 3 (b). In 2004, the 35th Assembly agreed that the organization "will strive to... limit or reduce the impact of aviation greenhouse gas emissions on the global climate." Notice the similar language of the *Kyoto Protocol*. See ICAO, A35-5, *Consolidated Statement of Continuing ICAO Policies*

as the appropriate forum to address GHG emissions from international aviation. Part of the controversy with regard to the EU ETS stems from the argument that only ICAO has jurisdiction to address GHG emissions from international aviation. Therefore, it seems quite appropriate to conceptualize the *Kyoto Protocol* as granting an implicit mandate to ICAO to handle GHG emissions from international aviation.[18] Without expressly acknowledging it, ICAO's own subsequent practice also suggests that the organization has interpreted this as an implicit mandate.[19]

Third, the *Kyoto Protocol* does not provide much guidance on how its implicit mandate should be implemented. It does not shed light on what the limitations and exceptions are, nor does it set a time frame to reduce or limit GHG emissions from international aviation.[20] For instance, what happens if States are not able to agree on measures at ICAO? In the absence of an ICAO solution, can States adopt unilateral measures or should the implicit mandate be construed as being exclusive to ICAO?[21] If the mandate is exclusive to ICAO and no action takes place at ICAO, would States' inability to act not contradict the objectives of the *UNFCCC*? If, on the other hand, the implicit mandate is to be regarded as non-exclusive and a State adopts a measure, should such State only target its own aircraft operators, or should it also target foreign aircraft operators?[22] Should the measures only

---

and Practices related to Environmental Protection at Appendix A, para 1 [ICAO, A35-5]. See also *Kyoto Protocol, supra* note 12 (Introduction) at Art 2.2. The Assembly "[emphasized] the importance of ICAO taking a leadership role on all civil aviation matters related to the environment." *Ibid.* at Appendix A, para 2. The Assembly also "[endorsed] the further development of an open emissions trading system for international aviation." *Ibid.*, Appendix I, para 2 (c) (1). In 2007, the 36th Assembly ordered the Council to "ensure that [the organization undertakes] continuous leadership on environmental issues in relation to international civil aviation, including GHG emissions." ICAO, A36-22, *supra* note 144 (Chapter 1) at Appendix J, para 1 (a). The Assembly also "[requested] the Council to examine the potential for carbon offset mechanism." *Ibid.* at Appendix L, para 1 (c) (1). Thereafter, in 2010, in addition to agreeing to aspirational goals, the 37th Assembly decided to "develop a framework for market-based measures (MBM), and explore the feasibility of a global MBM scheme." See ICAO, A37-19, *supra* note 207 (Chapter 1) at paras 6, 13, and 18. Finally, in 2013, the 38th Assembly agreed to develop a global MBM scheme for international aviation. See ICAO, A38-18 *supra* note 21 (Introduction) at para 18.

18 See generally ICAO, C-MIN 181/21, *supra* note 15 at 253 (the Statement of India argued that the *Kyoto Protocol* grants a mandate to ICAO).

19 See Lyle, "Mitigating International Air Transport Emissions through a Global Measure," *supra* note 122 (Chapter 1) (noting that "ICAO received a mandate for the reduction of GHG emissions from international aviation through the UNFCCC's Kyoto Protocol of 1997").

20 See *Kyoto Protocol, supra* note 12 (Introduction) at Art 2, para 2.

21 *The Vienna Convention for the Protection of the Ozone Layer*, for instance, clearly states that parties retain the right to adopt domestic measures, in addition to those of the treaty, provided that such measures are (i) in accordance with international law and (ii) are not incompatible with their obligations under such instrument. See *Vienna Convention for the Protection of the Ozone Layer*, 22 March 1985, 1513 UNTS 323, 26 ILM 1529 (1987) at Art 2, para 3 [*Vienna Convention*].

22 To better understand this problem, let us take the example of Germany, a well-established developed country. Under the *Kyoto Protocol*, Germany should pursue measures to reduce GHG emissions from international aviation by working through ICAO. Having said this, the *Kyoto Protocol* does not specify whether Germany is supposed to only target its own air carriers or whether it could also cover those foreign aircraft operators flying to and from its territory. If the *Kyoto Protocol* is to be construed as giving Germany only the authority

cover emissions generated over implementing States' own airspace, or could they also include emissions occurring over the high seas or over third States? These questions are at the heart of the legal challenges advanced against the EU Emissions Trading Scheme (EU ETS).[23]

Fourth, in what is arguably the most serious challenge for international aviation, on the basis of CBDR, the *Kyoto Protocol*'s implicit mandate strictly follows the differentiated, bipolar system that animates the *UNFCCC* regime in which only developed countries bear quantified emissions reduction obligations. Applied in the context of international aviation, this would mean that, for instance, Germany would have to take steps to reduce or limit its GHG emissions from aviation, but this would not be the case for the United Arab Emirates – one of the fastest-growing aviation markets and currently home to some of the world's most successful and fastest-growing air carriers. This may lead to serious discriminatory treatment among States and aircraft operators. In fact, those who drafted the mandate not only failed to take into account the specific characteristics of international aviation, but they also did not envision the tension created in the CBDR and non-discrimination relationship.[24]

Fifth, just as is the case with many other international instruments, the *Kyoto Protocol* does not contain explicit provisions to address conflicting principles.[25] Moreover, the *Kyoto Protocol* does not address whether States attempting to address GHG emissions from international emissions at ICAO should be bound by the principles[26] of the *UNFCCC*, whether they may simply opt to disregard them or whether they should attempt to reconcile them with those of the *Chicago Convention*. The mandate is completely silent on this issue,

---

to impose mitigation obligations on its own air carriers, then in practice it also renders it inapplicable. German air carriers compete head-to-head on numerous routes with air carriers from developing countries. Given that these carriers would not bear any mitigation obligations, they would stand to benefit from a significant competitive advantage. This would in turn create considerable market distortions. If, on the other hand, the *Kyoto Protocol* is to be interpreted as conferring Germany the right to include foreign air carriers, this is also problematic because most of their emissions are produced outside German jurisdiction, thereby raising issues of extraterritorial application of law. As will be discussed in Chapters 4, 5, and 6, this has been the EU ETS's Achilles heel.

23 These issues are discussed in detail in Chapters 4, 5, and 6.

24 See Md Saiful Karim & Shawkat Alam, "Climate Change and Reduction of Emissions of Greenhouse Gases from Ships: An Appraisal" (2011) 1:1 Asian J Int'l L 131 at 134.

25 Rosendal discusses that many times international instruments are negotiated without explicit provisions to address conflicting norms. See G Kristin Rosendal, "Impacts of Overlapping International Regimes: The Case of Biodiversity" (2001) 7 Global Governance 95.

26 Dworkin conceives a principle as a "standard that is to be observed, not because it will advance or secure an economic, political, or social situation deemed desirable, but because it is a requirement of justice or fairness or some other dimension of morality." Ronald M Dworkin, "The Model of Rules" (1967) 35 U Chi L Rev 14 at 23. For Dworkin, rules are also standards that should be observed, but they "are applicable in an all-or-nothing fashion." *Ibid.* at 25. In other words, if a rule is valid, it should be applied. Principles on the other hand have a different dimension not present in rules. This is what Dworkin calls the "dimension of weight and importance." *Ibid.*

and States have interpreted this lacuna to their advantage. As will be explained below, developed States contend that CBDR has no place at ICAO's discussions on climate change issues, for it contradicts the non-discrimination principle. On the other hand, developing States argue that any scheme addressing GHG emissions from international aviation must take into account CBDR. In other words, aircraft operators from developing countries should not be subject to mitigation obligations.

2.1.3 *The CBDR/Non-Discrimination Saga*

Scholars point out that a conflict occurs when a State party to two regimes cannot simultaneously comply with the obligations it has undertaken under such treaties.[27] Conflict also exists where one treaty "may frustrate the goals of another treaty without there being any strict incompatibility between their provisions."[28] In embracing a broad characterization of the phenomenon, the ILC has stated that a conflict arises "where two rules or principles suggest different ways of dealing with a problem."[29] Conflicts appear to be "a natural consequence of how international law is developed" within specialized or self-contained regimes.[30] In this regard, a conflict between principles of different international legal regimes is not unique to the relationship between international civil aviation and climate

27 See Harro van Asselt et al., "Global Climate Change and the Fragmentation of International Law" (2008) 20:4 L & Pol'y 429 [van Asselt et al, "Climate Change"]. Jenks notes that "[a] conflict in the strict sense of direct incompatibility arises only where a party to the two treaties cannot simultaneously comply with its obligations under both treaties." Jenks, *supra* note 8 at 426. See also "Draft Report of the Study Group of the International Law Commission, Fragmentation of International Law: Difficulties Arising from the Diversification and Expansion of International Law, Fifty-eighth Session, 1 May-9 June and 3 July-11 August 2006," UN Doc A/CN.4/L.682 (13 Apr 2006, as corrected UN Doc A/CN.4/L.682/Corr.1 (11 Aug. 2006) (finalized by Martti Koskenniemi) at 19 [ILC, Fragmentation Report]; Anja Lindroos & Michael Mehling, "Dispelling the Chimera of Self-Contained Regime' International Law and the WTO" (2005) 16 EJIL 857 (holding that a measure adopted by one party in the performance of an international obligation under one regime may constitute a breach of another obligation under another international legal regime) at 861. Rosendal suggests that "[c]onflict is more likely to result when the overall policy objectives as well as the obligations emanating from overlapping international agreements fail to complement and enhance each other – or worse, when they are mutually exclusive." Rosendal, *supra* note 25 at 97.

28 See ILC, Fragmentation Report, *supra* note 27 at 19. Jenks reports that early international air law instruments, such as the *Convention for the Regulation of Aerial Navigation* of 1919 [*Paris Convention*] and the *Havana Convention on Commercial Aviation* of 1928, were classical examples of conflictive international rules. See Jenks, *supra* note 8 at 410.

29 ILC, Fragmentation Report, *supra* note 27 at 19. Pauwelyn distinguishes between vertical and horizontal conflicts. Vertical conflicts involve conflicts between international instruments and domestic law, whereas horizontal conflicts deal with norms of international law. See Joost Pauwelyn, *Conflict of Norms in Public International Law: How WTO Law Relates to Other Rules of International Law* (Cambridge: Cambridge University Press, 2003) at 11.

30 See Yuka Fukunaga, "Civil Society and the Legitimacy of the WTO Dispute Settlement System" (2008) 34 Brook J Int'l L 85 at 109. See also Samantha A Miko, "Norm Conflict, Fragmentation and the European Court of Human Rights" (2013) 36 BC Int'l & Comp L Rev 1351 at 1363.

change.[31] There is also a conflict between the international trade regime and a number of multilateral environmental agreements (MEAs).[32] In fact, scholars claim that tension exists between the WTO regime and the environment.[33] Also, according to the WTO, there are at least 20 MEAs containing trade clauses.[34]

There are at least three ways to categorize the interplay between the *Chicago Convention* and the *Kyoto Protocol* and the relationship between the principles of non-discrimination and CBDR. First, this relationship can be conceived as a normative conflict.[35] A literal interpretation of the relevant provisions of both the *Kyoto Protocol* and the *Chicago Convention* leads to some State parties to both instruments being unable to comply with their obligations. Although both of these principles will be thoroughly examined below, it suffices to say here that, in the climate change context, CBDR differentiates developed from

31 See Kristen E Boon, "Regime Conflicts and the U.N. Security Council: Applying the Law of Responsibility" (2010) 42 Geo Wash Int'l L Rev 787 (arguing that "regime conflicts are no longer a rarity.") at 832. The Convention on the Biological Diversity (CBD) and the Agreement on Trade-Related Aspects of Intellectual Property (TRIPS) may also have conflicting norms and objectives. See *Convention on Biological Diversity*, 5 June 1992, 1760 UNTS 79, 31 ILM 818 (1992) [*CBD*]; *Agreement on Trade-Related Aspects of Intellectual Property Rights*, 14 April 1994, 1869 UNTS 299, 33 ILM 1197 [TRIPS]. Rosendal writes that the CBD is "concerned with conservation of biological diversity and equitable sharing of benefits derived from the world's genetic resources." Rosendal, *supra* note 25 at 95. On the other hand, TRIPS seeks to create a minimum global standard for the protection of intellectual property rights around the world. The conflict arises because plant varieties may fall under TRIPS' patentable subject matter. In other words, TRIPS requires that State parties adopt patent legislation or through other "effective sui generis system or by any combination thereof" to provide for intellectual property protection of such plant varieties. *Ibid.* at Art 27, para 3 (b). See also G Kristin Rosendal, "The Convention on Biological Diversity: Tensions with the WTO TRIPS Agreement over Access to Genetic Resources and the Sharing of Benefits" in Sebastian Oberthür & Thomas Gehring, eds, *Institutional Interaction in Global Environmental Governance: Synergy and Conflict among International and EU Policies* (Cambridge: MIT Press, 2006) 79 at 88.

32 See W Bradnee Chambers, "International Trade Law and the Kyoto Protocol: Potential Incompatibilities" in W Bradnee Chambers, ed, *Inter-Linkages: The Kyoto Protocol & the International Trade & Investment Regimes* (New York: United Nations University Press, 2001) 88 at 88.

33 See John S Dryzek, "Paradigms and Discourses" in Daniel Bodansky, Jutta Brunnée, & Ellen Hey, eds, *The Oxford Handbook of International Environmental Law* (Oxford: Oxford University Press, 2006) 44 at 48. See also John Vogler, "The European Contribution to Global Environmental Governance" (2005) 81:4 Int'l Affairs 835 (suggesting that the EU was the first global actor to recommend that WTO discussions take into account environmental issues) at 845.

34 See Gabrielle Marceau, "Conflicts of Norms and Conflicts of Jurisdiction: The Relationship between the WTO Agreement and MEAs and other Treaties" (2001) 35:6 J World Trade 6 1081 at 1095 [Marceau, "Conflicts"].

35 See Coraline Goron, "The EU Aviation ETS Caught between Kyoto and Chicago: Unilateral Legal Entrepreneurship in the Multilateral Governance System" *GREEN-GEM Doctoral Working Papers Series* (2012) online: University of Warwick <www2.warwick.ac.uk/fac/soc/csgr/green/papers/workingpapers/gem/no._2_c._goron.pdf> at 2; Md Saiful Karim, "IMO Mandatory Energy Efficiency Measures for International Shipping: The First Mandatory Global Greenhouse Gas Reduction Instrument for an International Industry" (2011) 7 Macquarie J Int'l & Comp Envtl L 111 (discussing that "given the unique characteristic of [international maritime], strict application of CBDR may not be viable") at 112; Doaa Abdel Motaal, "Curbing CO_2 Emissions from Aviation: Is the Airline Industry Headed for Defeat?" (2012) 3 Climate Change 1 (noting that the "CBDR principle clashes head on with a number of provisions of the Chicago Convention") at 6.

developing countries where only the former bear emission reduction/limitation obligations.[36] On the other hand, one of the cardinal principles of international civil aviation is to avoid discrimination. A scheme to regulate GHG emissions from international aviation implemented by one State and a group of States or, at a global level, a scheme that only applies to aircraft operators from developed States contradicts this principle and may create significant market distortions. Although the *Kyoto Protocol* requires developed countries to address the issue at ICAO, it does not prevent developing countries from doing so if they so wish. The *Kyoto Protocol* does not establish quantified mitigation obligations for developing countries, but this should not be interpreted as justifying a permanent do-nothing attitude of these States. The obvious criticism to this approach is that it is difficult to conceptualize as a case of normative conflict two principles in abstract.

In light of *Kyoto's* imprecision, a second approach would be to regard the relationship with the *Chicago Convention* as encompassing completely independent principles. The ICAO Secretariat has taken this view. Its Legal Bureau has said that CBDR "[is] not binding for the organization."[37] ICAO has the option of either embracing or completely discarding CBDR. This principle could be a policy option, but it is certainly not a legal obligation.[38] In practice, however, ICAO has opted to discard CBDR in its work on climate change issues. In fact, as will be seen below, ICAO has, with limited success, developed the notion of special circumstances and respective capabilities (SCRC) as an alternative principle. Although one could defend ICAO's formalistic stance, disassociating from CBDR has proven unfruitful. In fact, this stance has been detrimental to the regulation of GHG emissions from international aviation.

A third approach would be to conceptualize the relationship between the two regimes as one that leads to the significant tension of two different principles. Here, it is immaterial whether a conflict exists in strict legal terms or whether ICAO is bound by the *Kyoto Protocol* principles. The practical effect is that there is a tremendous political disagreement over this issue. This approach would seem to be more in line with what has occurred in the aftermath of *Kyoto*. I now explain this approach in more detail.

---

36 See *Convention on International Civil Aviation*, 7 December 1944, 15 UNTS 295, ICAO Doc 7300/06 [*Chicago Convention*] at Arts 7, 9, 11, 15, and 44 (g) [*Chicago Convention*].

37 ICAO-C-MIN 181/22, *Council, 181st Session, Minutes of the Twenty-Second Meeting*. ICAO's Legal Bureau's reasoning resembles the stance taken by the Arbitral Body (AB) in the Beef Hormone trade dispute. Here the AB ruled that the precautionary principle was not binding on the WTO regime. Although the AB justified its position in the fact that such principle has yet to achieve the status of customary international law, in essence the AB was reluctant to incorporate foreign, non-trade principles. See WT/DS26/AB/R (13 February 1998) (confirming the panel's view that "the precautionary principle does not override the provisions of Articles 5.1 and 5.2. of the SPS Agreement" at para 125). See also Martti Koskenniemi & Päivi Leino, "Fragmentation of International Law? Postmodern Anxieties" (2002) 15 Leiden J Int'l L 553 at 572. Similarly, in the Tuna II case, the panel did not resort to other environmental treaties to interpret GATT's Art XX. See United States – Restrictions on Imports of Tuna, 10 June 1994, not adopted, DS 29/R at para 5.19.

38 See ICAO-C-MIN 181/22, *supra* note 37 at 264.

2.2 From Theory to What Happens in Practice

On numerous occasions, developed countries have argued that CBDR cannot be applicable in the context of international aviation.[39] The United States has said that CBDR's differential treatment is "inconsistent with ICAO's non-discrimination principle."[40] According to the United States, CBDR may provide an "unequal treatment of different States" with potential "spill over into other areas such as safety and security."[41] Given that in the broader climate change context CBDR is considered a threat to the competitiveness of US industries,[42] the United States has been firm in maintaining the position that the *Chicago Convention* not only governs climate change discussions at ICAO but also trumps *UNFCCC* rules. Under this view, international civil aviation would be a "self-contained regime" completely isolated from other international regimes and principles of international law.[43]

Hardeman writes that "an orthodox application of the [CBDR] concept spilling over into the aviation policy arena, as advocated by a number of developing country governments in the climate negotiations, would bring about an unjustified differentiation between economic actors in global aviation markets."[44] In fact, some commentators have described CBDR as having a "paralyzing effect" in advancing ICAO's work on aviation and climate change.[45] In the words of Tony Tyler, Director General of the International Air Transport Association (IATA), CBDR "has made finding a common way forward a difficult challenge."[46] Concerned about the potential market distortions, IATA has opted not to address CBDR and has left it for States to decide.[47]

39 *Ibid.*; Letter from Joseph L Novak to Raymond Benjamin, "Statement of Reservation of the United States regarding Resolution 17/2 of the 38th ICAO Assembly Resolution: Consolidated Statement of Continuing ICAO Policies and Practices related to Environmental Protection – Climate Change" online: ICAO <www.icao.int/Meetings/a38/Documents/Resolutions/United_States_en.pdf>; ICAO, C-MIN 194/2, *supra* note 86 (Chapter 1) at para 101 (Statement of the United States).

40 ICAO-C-MIN 181/22, *supra* note 37 at 263 (Statement of the United States).

41 *Ibid.* The United States has also resisted the incorporation of the CBDR principle in other *fora*. For instance, in the context of the RIO + 20 framework, the United States has pushed to delete any reference to CBDR. See Third World Network, "Common but Differentiated Responsibilities under Threat," *TWN* (13 June 2012) online: <www.twnside.org.sg/title2/rio+20/news_updates/TWN_update2.pdf>.

42 See Bernd G Janzen, "International Trade Law and the "Carbon Leakage" Problem: Are Unilateral U.S. Import Restrictions the Solution?" (2008) 8 Sustainable Dev L & Pol'y 22.

43 While describing self-contained regimes, the ILC has said that these regimes "[come] with [their] own principles, its own form of expertise and its own ethos" not necessarily identical to the "ethos of neighboring specializations." See ILC, Fragmentation Report, *supra* note 27 at 14.

44 Andreas Hardeman, "Reframing Aviation Climate Politics and Policies" (2011) XXXVI Ann Air & Sp L 1 at 12 [Hardeman, "Reframing Aviation Climate Politics and Policies"].

45 See Motaal, *supra* note 35 at 7; Stephanie Koh, "The Case against Extending the EU Emissions Trading Scheme to International Aviation" (2012) 30 Sing L Rev 125 [Stephanie Koh] (characterizing the CBDR principle as an obstacle that needs to "be overcome") at 144.

46 IATA Press Release, "Remarks of Tony Tyler," *supra* note 141 (Chapter 1).

47 See IATA Resolution on the Implementation of the Aviation CNG 2020 Strategy, Appendix 1 [IATA, CNG Resolution].

A contrario, for a long time, developing countries such as Argentina, Brazil, China, India, and Saudi Arabia have strongly argued that CBDR should guide ICAO's work on climate change issues.[48] India has underscored that, given that the *Kyoto Protocol* entrusted a mandate to ICAO to handle GHG emissions from international aviation, CBDR must be observed.[49] To this end, these developing countries have argued that their airlines should not be subject to any emission reduction obligations.[50] Such obligations should only target aircraft operators from developed States. Yet, a cardinal principle of the climate change regime such as CBDR cannot be integrated into another system to defeat the purpose of the latter.

I contend that the *Kyoto Protocol*'s interplay with the *Chicago Convention* regime involves principles that are not irreconcilable. What is necessary is to integrate them in a manner compatible with their respective environments.

2.2.1 Understanding Non-Discrimination

Although it is not defined in the *Chicago Convention*, non-discrimination is a cornerstone principle of international civil aviation. In fact, one of ICAO's objectives is precisely to "avoid discrimination between [States parties]."[51] Several provisions of the *Chicago Convention* illustrate this principle. For instance, in exercising jurisdiction over the arrival and departure of aircraft from its territory, a State may not discriminate between aircraft on the basis of nationality.[52] With respect to access to public airports, provision of air navigation facilities, radio and meteorological services, and the levying of aeronautical and air navigation charges, States cannot provide more favorable treatment to national aircraft operators.[53] If military or safety reasons so require, a State may establish prohibited areas within its

48 See ICAO, A36-WP/88 EX/36, *Viewpoint of the Arab Republic of Egypt as a Developing Country on Emissions Trading for Civil Aviation* (presented by Egypt) at 4; ICAO, HLM-ENV/09-WP/28, *African Position on the GIACC Programme of Action* (presented by Nigeria on behalf of African States) (discussing that ICAO should "adhere to the UNFCCC fundamental principle of [CBDR]") at 3; ICAO, A38-WP/272 EX/92, *Position of African States on Climate Change* (presented by 54 African States) (noting that "in developing a scheme for MBMs, ICAO should take into consideration the principle of [CBDR]") at 1; ICAO, C-MIN 194/2, *supra* note 86 (Chapter 1) (Statement of Saudi Arabia discussing that the "principle of CBDR should be emphasized") at para 68; ICAO, C-MIN 194/2, *supra* note 86 (Chapter 1) (Statement of Cuba discussing that measures to combat international aviation's impact on climate change "should be based on the *UNFCCC* principles of equity and common but differentiated responsibilities (CBDR) and should promote an open international economic system that would lead to the development of all parties, particularly developing countries") at para 58.

49 See ICAO- C-MIN 181/21, *supra* note 15 at 253 (Statement of India).

50 See WWF, Aviation Report, *supra* note 91 (Chapter 1) at 4.

51 *Chicago Convention*, *supra* note 36 at Art 44 (g).

52 *Ibid.*, Art 11.

53 *Ibid.*, Art 15. See also ICAO Doc 9626, *Manual on the Regulation of Air Transport*, 2nd edn (2004) at 4.10.

airspace.[54] However, if this is the case, the prohibitions and/or restrictions must apply equally to foreign and local aircraft; neither foreign nor local aircraft operators may fly over such areas.[55] States are also forbidden from granting exclusive cabotage rights.[56] Mendes de Leon argues that if a State nonetheless grants such rights and if another State so requests them, the granting State cannot afford a discriminatory treatment with regard to the requesting State which came in second.[57]

Many States have brought claims alleging discriminatory treatment against their airlines before ICAO's dispute settlement mechanism in light of the provisions of the *Chicago Convention* and the Air Transit Agreement.[58] In 1952, India challenged the establishment of a prohibited air zone by Pakistan alleging that this amounted to discriminatory treatment especially since aircraft operators from Iran were allowed to operate over such area.[59] In 1996, Cuba filed a complaint against the United States alleging that the prohibition of Cuban-registered aircraft from overflying the US airspace was discriminatory.[60] In 2000, in the so-called Hushkit dispute, the United States challenged a European noise regulation which was perceived as discriminatory against US interests.[61] Because all of these disputes were settled through diplomatic negotiations, the ICAO Council never had a chance to rule on the merits of the discriminatory allegations. In any case, the allegations were founded upon the non-discrimination principle. [62]

In the context of international civil aviation, the non-discrimination principle manifests itself in the form of a prohibition that enjoins States to refrain from arbitrarily and capri-

54 See *Chicago Convention*, *supra* note 36 at Art 9.

55 *Ibid.*

56 Cabotage refers to the right to transport passenger, cargo, and mail between two points within a State. Most states restrict these rights to aircraft operators registered in such States. See *Chicago Convention*, *supra* note 36 at Art 7.

57 See Pablo Mendes de Leon, *Cabotage in Air Transport Regulation* (Leiden: M Nijhoff, 1992).

58 See *International Air Services Transit Agreement*, 7 December 1944, 84 UNTS 389 (entered into force 30 January 1945).

59 See Michael Milde, *International Air Law and ICAO*, 2nd edn (The Hague: Eleven International Publishing, 2012) at 200; Ludwig Weber, *International Civil Aviation Organization: An Introduction* (The Netherlands: Kluwer Law International, 2007) at 42. See also ICAO, C-WP/1169, *Request of the Government of India to the Council of ICAO*.

60 See Milde, *supra* note 59 at 203; Weber, *supra* note 59 at 42.

61 See Memorial of the United States of America, *Disagreement Arising under the Convention on International Civil Aviation done at Chicago on December 7, 1944* (14 March 2000).
See also EC, *Council Regulation (EC) No 925/1999 of 29 April 1999 on the Registration and Operation within the Community of Certain Types of Civil Subsonic Jet Aeroplanes which Have been Modified and Recertificated as Meeting the Standards of Volume I, Part II, Chapter 3 of Annex 16 to the Convention on International Civil Aviation, Third Edition* (July 1993) [1999] OJ, L 115/1; EC, *Directive 2002/30/EC of the European Parliament and of the Council of 26 March 2002 on the Establishment of Rules and Procedures with Regard to the Introduction of Noise-related Operating Restrictions at Community Airports* [2002] OJ, L 85/40.

62 See Milde, *supra* note 59 at 200.

ciously applying differential treatment to aircraft operators on the basis of nationality.[63] Such treatment is not justifiable on any objective basis. More specifically, in the context of designing a scheme to address GHG emission from international aviation, the non-discrimination principle means that aircraft operators flying on the same route should be subject to the same rules.

2.2.2 Understanding CBDR

The climate change regime embraces CBDR.[64] Although States disagree on its definition[65] and scope and whether, in its current form, it should remain part of the climate change regime,[66] the principle bears the following salient characteristics. First, all States bear a common, inter-generational obligation to protect the climate system as a whole.[67] In this respect, Stavins claims that "all nations should engage in the solution (because of the global-commons nature of the problem)."[68]

Second, while recognizing that they are responsible for the largest share of historical emissions, developed countries should take the lead in mitigation efforts.[69] In order to take into account the special circumstances of developing countries, on the basis of equity[70]

63 Partsch discusses that "the rule of non-discrimination is nothing more than the negative form of the rule of equality connected with a specification of certain illegal criteria." Karl Josef Partsch, "Discrimination Against Individuals and Groups" in R Bernhardt & Max Planck Institute for Comparative Public Law and International Law eds, *Encyclopedia of Public International Law*, vol 1 (Amsterdam: North Holland, 1992) 1079 at 1079.

64 See *UNFCCC, supra* note 11 (Introduction) at Art 3, para 1.

65 See Douglas Bushey & Sikima Jinnah, "Evolving Responsibility? The Principle of Common but Differentiated Responsibility in the UNFCCC" (2010) 6 Berkeley J Int'l L Publicists 1 (noting that CBDR "has never been susceptible to precise definition" at 1).

66 See Gupta, *supra* note 63 (Chapter 1) (noting that although it is undisputed that CBDR has been embraced in different international legal regimes, the principle is extremely controversial,) at 170. See also Daniel Bodansky, Jutta Brunnée, & Ellen Hey, "International Environmental Law: Mapping the Field" in Daniel Bodansky, Jutta Brunnée, & Ellen Hey, eds, *The Oxford Handbook of International Environmental Law* (Oxford: Oxford University Press, 2006) 1 at 13 [Bodansky et al, *Oxford Handbook*].

67 See Rüdiger Wolfrum, "International Environmental Law: Purpose, Principles and Means of Ensuring Compliance" in Fred L Morrison & Rüdiger Wolfrum, eds, *International, Regional and National Environmental Law* (The Hague: Kluwer Law International, 2000) 3 at 26 [Wolfrum, "International Environmental Law"].

68 See Robert N Stavins, "An International Policy Architecture for the Post-Kyoto Era" in Ernesto Zedillo, ed, *Global Warming: Looking Beyond Kyoto* (Washington: Brookings Institution Press, 2008) at 146 [Stavins, "Post-Kyoto Era"].

69 See Wolfrum, *supra* note 67 at 27.

70 Eichenberg describes CBDR as "an equity principle to balance the burdens of environmental protection." Benjamin Eichenberg, "Greenhouse Gas Regulation and Border Tax Adjustments: The Carrot and Stick" (2010) 3 Golden Gate U Envtl LJ 283 at 291.

and fairness,[71] CBDR differentiates responsibilities.[72] In practical terms, the *Kyoto Protocol* imposes specific emission targets for four GHGs upon developed countries over a five-year commitment period. By assigning specific quantified emission limitation/reduction commitments (QELRCs) for the first commitment period, the *Kyoto Protocol* requires that developed countries reduce their overall GHG emissions to at least 5.2 percent below 1990 levels.[73] On the other hand, developing countries do not have to pursue quantified emission reductions, but they are required to maintain national inventories and cooperate with the international climate change regime.[74] In essence, the international climate change regime has been divided in two: those with reduction obligations and those without such obligations.

Sterns point out that it is unfair to demand that developing countries lower the rate of their economic growth since they are not primarily responsible for the climate change dilemma that persists today.[75] Shielding themselves behind CBDR, several major developing countries have made financial assistance and technology transfer from developed States a precondition to assuming commitments to participate in any emission reduction scheme.[76] In the context of international aviation, for instance, China has clearly stated that "developed countries should take the lead in taking reduction measures in order to offset the growth of emissions from international aviation of developing countries."[77] Developing countries would argue that "developed States [must] take more ambitious [mitigation] actions [to]

71 The definition of fairness may also be problematic. In the climate change context, some states argue that all major emitters should be subject to mitigation obligations. Others, on the other hand, contend that any mitigation commitment should be in strict adherence to the State's historical emissions. See Friedrich Soltau, *Fairness in International Climate Change Law and Policy* (Cambridge: Cambridge University Press, 2009) at 4.

72 See Philippe Sands, "The Greening of International Law: Emerging Principles and Rules" (1994) 1:2 Ind J Global Legal Stud 293 at 307.

73 See Jason N Glennon, "Directive 2008/101 and Air Transport: A Regulatory Scheme Beyond the Limits of the Effects Doctrine" (2013) 78:3 J Air L & Com 479 at 481.

74 See *UNFCCC*, *supra* note 11 (Introduction) at art 4. See also Hardeman, "Reframing Aviation Climate Politics and Policies," *supra* note 44 at 12.

75 See Cameron Hepburn & Nicolas Stern, "A New Global Deal on Climate Change" (2008) 34:2 Oxford Rev Econ Pol'y 259 at 270. See also Robert W Hahn, "Climate Policy: Separating Fact from Fantasy" (2009) 33 Harv Envtl L Rev 557 at 571 (arguing that major emitters such as China and India reject the idea of binding emission reductions because of the limitations for the growth of their economies). See also Robert Mendelsohn, "The Policy Implications of Climate Change Impacts" in Ernesto Zedillo, ed, *Global Warming: Looking Beyond Kyoto* (Washington: Brookings Institution Press, 2008) 82 at 87.

76 See Jiahuan Pan, "Common but Differentiated Commitments: A Practical Approach to Engaging Large Developing Emitters under L20" (presentation delivered at the Commissioned Briefing Notes for the CIGI/CFGS L20 Project, 20-21 September 2004) online: <http://goo.gl/H8TTFz>.

77 Reservation of China to ICAO A38-18, *Consolidated Statement of Continuing ICAO Policies and Practices related to Environmental Protection: Climate Change*, online: ICAO <www.icao.int/Meetings/a38/Documents/Resolutions/China_en.pdf>.

offset an increase in emissions from the growth of air transport in developing States."[78] Developed countries would, on the other hand, respond that emissions from developing countries grow at a very high rate.[79] In fact, Stavins warns that "developing countries are likely to account for more than half of global emissions by the year 2020, if not before."[80] Various models confirm the view that developing countries will soon rank among the largest emitters.[81] Korea, for instance, is the world's fourteenth largest economy and the tenth largest GHG emitter. Yet, it is not subject to any emission reduction target(s) because of its developing country status.[82]

Finally, CBDR also touches upon the often forgotten "respective capabilities of States" – an element that a number of developing countries consciously opt to disregard.[83] In theory, at least a literal interpretation of CBDR would suggest that the differentiation treatment should not be applicable for those States that are "capable" of contributing more to climate change, regardless of the fact that they may be labeled as developing countries.[84] For instance, with the highest per capita income, the State of Qatar is in a much better position to tackle climate change than the small island State of Seychelles.[85]

2.3 A Look into Fragmentation of International Law

Over the last few decades, in an effort to handle the surge of increasingly complex global issues, the international community has developed myriad different international legal

78 ICAO, A38-WP/427 EX/142, *Proposed Amendments for the Draft Consolidated Statement of Continuing ICAO Policies and Practices Related to Environmental Protection: Climate Change* (presented by Argentina, China, Cuba, India, Islamic Republic of Iran, Pakistan, Peru, the Russian Federation, Saudi Arabia and South Africa) at 2.

79 See Chichilnisky & Sheeran, *supra* note 7 at 121 (Chapter1).

80 Stavins "Post-Kyoto Era," *supra* note 68 at 147. As Helm explains, China on the other hand claims that per capita emissions are still very low and that most of its growth in GHG emissions is due to the increase of consumption from its exports that are destined to both Europe and the United States. Dieter Helm, "Climate Change Policy: Why Has so Little Been Achieved?" (2008) 24:2 Oxford Rev Econ Pol'y 211 at 234 [Helm, "Climate Change Policy"].

81 See Andrew E Dessler & Edward A Parson, *The Science and Politics of Global Climate Change: A Guide to the Debate*, 2nd edn (Cambridge: Cambridge University Press, 2009) at 122. See also Alex Bowen & James Rydge, "The Economics of Climate Change" in David Held, Angus Hervey, & Marika Theros, eds, *The Governance of Climate Change: Science, Economics, Politics & Ethics* (Cambridge: Polity Press, 2011) 68 at 82.

82 Jae-Seung Lee, "Coping with Climate Change: a Korean Perspective" in Antonio Marquina, ed, *Global Warming and Climate Change: Prospects and Policies in Asia and Europe* (London: Palgrave Macmillan, 2010) 357 at 357.

83 Wolfrum, *supra* note 67 at 26.

84 See Soltau, *supra* note 71 at 195.

85 See World Bank, "Income per Capita" *World Bank* online: <http://data.worldbank.org/indicator/NY.GDP.PCAP.CD>.

regimes,[86] and it has established numerous international institutions[87] as well as international courts.[88] The rise in the number of international instruments may lead to what Brown Weiss describes as "treaty congestion."[89] While describing new trends of international environmental law, Brown Weiss explains that since the 1970s a large number of international environmental agreements have been adopted.[90] These include bilateral, multilateral, binding, and non-binding agreements.[91] The high quantity of instruments places a significant burden on states, some of which may not be in a position to manage them all.[92] In Brown Weiss's own words, "[w]ith such a large number of international agreements, there is a great potential for the additional inefficiency of overlapping provisions in agreements, inconsistencies in obligations, significant gaps in coverage, and duplication of goals and responsibilities."[93] Similarly, Bodansky notes that treaty congestion may "[create] the potential for duplication of efforts, lack of coordination, and even conflict between different [legal] regimes."[94]

International agreements tend to evolve as self-contained systems isolated from international law.[95] Functional specialization leads to agreements that adopt their own principles and rules.[96] Lindroos points out that "[n]orms are created by the subjects of international law themselves in a variety of *fora*, many of which are disconnected and independent from each other, creating a system different from the more coherent domestic legal order."[97] Often, the multiplicity of international instruments leads to normative conflicts that, in

86 "Report of the Study Group of the International Law Commission, Fragmentation of International Law: Difficulties Arising from the Diversification and Expansion of International Law, Fifty-eighth Session, 1 May-9 June and 3 July-11 August 2006," UN Doc A/CN.4/L.702 (18 July 2006) [ILC, Fragmentation Report Summary] (noting that "the scope of international law has increased dramatically... it has expanded to deal with the most varied kinds of international activity, from trade to environmental protection, from human rights to scientific and technological cooperation") at 3.

87 *Ibid.*

88 See Francis N Botchway, "Threads in Fragments" (2012) Cardozo J Int'l & Comp L 639 (underscoring the "exponential growth in international law" and "international adjudicatory institutions") at 709.

89 Edith Brown Weiss, "International Environmental Law: Contemporary Issues and the Emergence of a New World Order" (1993) 81 Geo LJ 675 at 687.

90 *Ibid.* at 679.

91 *Ibid.*

92 *Ibid.* at 698. See also Djamchid Momtaz, "The United Nations and the Protection of the Environment: from Stockholm to Rio de Janeiro" (1996) 15:3/4 Political Geography 261 at 269.

93 Brown Weiss, *supra* note 89 at 699.

94 Bodansky, *The Art and Craft of International Environmental Law*, *supra* note 62 (Chapter 1) at 35.

95 See Karen N Scott, "International Environmental Governance: Managing Fragmentation through Institutional Connection" (2011) 12 Melbourne J Int'l Law 177 at 178. See also ILC, Fragmentation Report, *supra* note 27 (discussing that these regimes "have their basis in multilateral treaties and acts of international organizations [where] specialized treaties and customary patterns...are tailored to [their] needs and interest ... but rarely take account of the outside world") at 245.

96 See Rossana Deplano, "The Fragmentation and Constitutionalisation of International Law: A Theoretical Inquiry" (2013) 6:1 European J Leg Stud 67 at 68.

97 Anja Lindroos, "Addressing Norm Conflicts in a Fragmented Legal System: The Doctrine *of Lex Specialis*" (2005) 74 Nordic J Int'l L 27 at 28.

the absence of a central, hierarchical authority, are not easy to resolve.[98] This phenomenon amounts to fragmentation.

Matz-Lück writes that fragmentation includes both conflicts of norms between regimes but also those conflicts involving institutions.[99] Adopting a less fatalistic position, Simma conceives fragmentation as "nothing but the result of a transposition of functional differentiations of governance from the national to the international plane; which means that international law today increasingly reflects the differentiation of branches of the law which are familiar to us from the domestic sphere."[100] In this regard, fragmentation is not at all a new phenomenon, but rather a natural consequence of a globalized world where complex issues require specialized treatment.[101]

Hafner claims that fragmentation may lead to "frictions and contradictions between the various legal regulations and creates the risks that States even have to comply with mutually exclusive obligations."[102] Hafner further notes that "the international legal system cannot avoid normative conflicts and inhomogeneous application because it lacks clear legal guidance for the resolution of conflict and norms. This threatens the unity of the international legal system."[103] As the ILC has rightly pointed out, "fragmentation does create the danger of conflicting and incompatible rules, principles, rule-systems and institutional practices."[104] Fragmentation may lead to forum shopping,[105] "regime shifting,"[106] or an "a la carte approach to international law."[107] Other commentators also note that, while fragmentation may be detrimental to the least developed countries, it is most

98 See Carmen Pavel, "Normative Conflict in International Law" (2009) 46 San Diego L Rev 883; Frank Biermann et al., "The Architecture of Global Climate Governance: Setting the Stage" in Frank Biermann, Philipp Pattberg, & Fariborz Zelli, eds, *Global Climate Governance Beyond 2012: Architecture, Agency and Adaptation* (Cambridge: Cambridge University Press, 2010) 15 at 17 [Biermann et al., "Architecture of Global Climate"].

99 See Nele Matz-Lück, "Structural Questions of Fragmentation" (2011) 105 Am Soc'y Int'l L Proc 123 at 125.

100 Bruno Simma, "Universality of International Law from the Perspective of a Practitioner" (2009) 20:2 EJIL 265 at 270.

101 See Koskenniemi & Leino, *supra* note 37 at 556.

102 Gerhard Hafner, "Risks Ensuing from Fragmentation of International Law," Official Records of the General Assembly, Fifty-fifths Sessions, Supplement No 10 (A/55/10) at Annex at 144.

103 See Gerhard Hafner, "Pros and Cons Ensuing From Fragmentation of International Law" (2004) 25 Mich J Int'l L 849 at 854.

104 See ILC, Fragmentation Report, *supra* note 27 at 14. See also N Jansen Calamita, "Countermeasures and Jurisdiction: Between Effectiveness and Fragmentation" (2011) 42 Geo J Int'l L 233 at 237.

105 See Fariborz Zelli et al, "The Consequences of a Fragmented Climate Governance Architecture: A Policy Appraisal" in Frank Biermann, Philipp Pattberg, & Fariborz Zelli, eds, *Global Climate Governance Beyond 2012: Architecture, Agency and Adaptation* (Cambridge: Cambridge University Press, 2010) 25 at 31.

106 Tomer Broude, "Principles of Normative Integration and the Allocation of International Authority: The WTO, the Vienna Convention and the Law of Treaties, and the Rio Declaration" (2008) 6 Loy U Chi Int'l L Rev 173 (discussing that States favor one particular regime because they expect it be more advantageous to their interests,) at 184.

107 See Daniel Moeckli, "The Emergence of Terrorism as a Distinct Category of International Law" (2008) 44 Tex Int'l LJ 157 at 182. See also Christian Leathley, "An Institutional Hierarchy to Combat the Fragmentation of International Law: Has the ILC Missed an Opportunity?" (2007) 40 NYU J Intl'l L & Pol 259 at 271.

beneficial to the most powerful states.[108] Yet, some authors also point out some positive aspects of fragmentation, arguing generally that it brings "diversity" to international law.[109] For Coyle, specialization, which is one of the salient characteristics of fragmentation, offers greater clarity and efficiency.[110] Arguably, specialized regimes are better suited to provide adequate responses to issues within their domains.[111]

Fragmentation as a phenomenon of international law is characterized by a significant surge in the number of specialized international regimes that, in some circumstances, lead to normative conflict or significant tension. Although, in some cases, there is interaction among different international organizations in charge of these specialized regimes,[112] some of them tend to act as silos, in isolation from the rules of other organizations. This explains why normative conflicts are not unusual. Although these conflicts are a central component of fragmentation, it is useful to note that synergies between different regimes and norm reconciliation are also important aspects of fragmentation.[113]

2.3.1 *The ILC Report on Fragmentation of International Law*

Fragmentation has long been a matter of concern for many international law scholars and practitioners. In 2000, during a speech delivered to the UN General Assembly, Gilbert Guillaume, former President of the International Court of Justice (ICJ), warned about the "undesirable effects" and the "dangers of legal fragmentation." Shortly thereafter, the ILC included the issue of fragmentation in its work program.[114] Later, in 2002, a Study Group

---

108 See Eyal Benvenisti & George W Downs, "The Empire's New Clothes: Political Economy and the Fragmentation of International Law" (2007) 60 Stan L Rev 595 at 598. See also Andries Hof, Michel den Elzen, & Detlef van Vuuren, "Environmental Effectiveness and Economic Consequences of Fragmented versus Universal Regimes: What Can We Learn from Model Studies" in Frank Biermann, Philipp Pattberg, & Fariborz Zelli, eds, *Global Climate Governance Beyond 2012: Architecture, Agency and Adaptation* (Cambridge: Cambridge University Press, 2010) 35 at 55.

109 See Biermann et al., "Architecture of Global Climate," *supra* note 98 at 15.

110 See John F Coyle, "The Treaty of Friendship, Commerce and Navigation in the Modern Era" (2013) 51 Colum J Transnat'l L 302 at 347.

111 See ILC, Fragmentation Report Summary, *supra* note 86 at 5.

112 Palmer et al., for instance, note that the WTO "interacts with many international environmental regimes, which also regulate international trade. The WTO is often a source of the interaction, invoking reactions from international environmental regimes in the design and implementation of rules that are responsive to WTO prescriptions." Alice Palmer, Beatrice Chaytor, & Jacob Werksman, "Interactions between the World Trade Organization and International Environmental Regimes" in Sebastian Oberthür & Thomas Gehring, eds, *Institutional Interaction in Global Environmental Governance: Synergy and Conflict among International and EU Policies* (Cambridge: MIT Press, 2006) 181 at 181.

113 See Harro van Asselt, "Legal and Political Approaches in Interplay Management: Dealing with the Fragmentation of Global Climate Governance," in Sebastian Oberthür & Olav Schram Stokke, eds, *Managing Institutional Complexity: Regime Interplay and Global Environmental Change* (Cambridge: Massachusetts Institute of Technology, 2011) 59 at 60 [van Asselt, "Legal and Political Approaches"].

114 ILC, Fragmentation Report, *supra* note 27 at 8.

was established.[115] In 2006, under the chairmanship of renowned Finish scholar, Martti Koskenniemi, the ILC presented a seminal report on fragmentation to the UN General Assembly.[116]

Although fragmentation involves both "institutional and substantive [normative] problems,"[117] the ILC deliberately decided not to address institutional issues – that is to say the potential conflict between international institutions.[118] The ILC sought to find legal techniques "to establish meaningful relationships between [potentially conflicting] rules and principles so as to determine how they should be used."[119] The ILC reasoned that the application of international law triggers the interaction of two types of rules: on the one hand, those seeking to facilitate the interpretation of rules and, on the other hand, those guiding the resolution of conflicting norms.[120] The ILC identified four types of normative conflict. These include (i) relations between special and general law, (ii) relations between prior and subsequent law, (iii) relations between laws at different hierarchical levels, and (iv) relations of law to its "normative environment more generally."[121] Moreover, the ILC focused its attention on normative conflict rules such as *lex posterior* and *lex specialis.*[122] For the ILC, these are conflict techniques that "enable seeing a systemic relationship between two or more rules, and may thus justify a particular choice of the applicable standards and a particular conclusion."[123] Yet, none of these collision rules can be applied in abstracto.[124] They "should be decided contextually."[125]

The ILC suggested that the constitutional frameworks of specialized or "self-contained" regimes, such as those of ICAO or the *UNFCCC* may be regarded as *lex specialis.*[126] The latter may be used to "clarify or set aside general law,"[127] but general law principles may

115 *Ibid.* at 1. See also Michael J Matheson, "The Fifty-Sixth Session of the International Law Commission" (2005) 99 Am J Int'l L. 211 at 219.

116 See ILC, Fragmentation Report, *supra* note 27 at 1.

117 ILC, Fragmentation Report Summary, *supra* note 86 at 4.

118 *Ibid.* at 5; ILC, Fragmentation Report, *supra* note 27 at 247. See also Sean D Murphy, "Deconstructing Fragmentation: Koskenniemi's 2006 ILC Project" (2013) Temp Int'l & Comp LJ at 2; Chen Yifeng, "Structural Limitations and Possible Future Work of the International Law Commission" (2010) 9 Chinese J Int'l L 473 at 481.

119 ILC, Fragmentation Report Summary, *supra* note 86 at 6-7.

120 *Ibid.* at 7-8.

121 See ILC Fragmentation Report, *supra* note 27 at 16.

122 ILC suggests that, although *lex specialis* does not have any formal superiority over other sources of law, States tend to prefer it over general law. See ILC, Fragmentation Report Summary, *supra* note 86 (suggesting that "whenever two or more norms deal with the same subject matter, priority should be given to the norm that is more specific") at 8. See also Michael J Matheson, "The Fifty-Seventh Session of the International Law Commission" (2006) 100 Am J Int'l L 416 at 422.

123 ILC Fragmentation Report, *supra* note 27 at 25.

124 *Ibid.* at 24.

125 See ILC, Fragmentation Report Summary, *supra* note 86 at 9.

126 *Ibid.* at 11.

127 *Ibid.* at 9.

be used as a gap filler.[128] In addition, the ILC also looked into the "settlement of disputes within and across regimes" and the so-called inter se agreements or an instrument amending multilateral instruments.[129] It also examined the higher hierarchical order of *jus cogens* norms, obligations *erga omnes*, and obligations of States under the Charter of the United Nations.[130] Similarly, the ILC thoroughly examined the relevant conflict provisions of the *VCLT*.[131] According to the ILC, the "*VCLT* provides the normative basis – the 'tool-box' for dealing with fragmentation."[132] None of these types of "collision rules,"[133] however, may trump *jus cogens* norms.[134]

The ILC report on fragmentation points out that Article 30 of the *VCLT* establishes a system to resolve normative conflicts. However, as the ILC itself has recognized, this is subject to a restrictive condition.[135] The conflict must necessarily involve "successive treaties relating to the same subject-matter."[136] If this is the case, provisions of *lex posterior* prevail over those of prior treaties to the extent that the later treaty "is not to be considered as incompatible with the earlier treaty."[137] This embraces the principle that *lex posterior derogat legi priori.* If the same States are parties to both the earlier and the later treaties and unless the conclusion of the later treaty implies the termination of the earlier treaty, the *VCLT* provides that the provisions of the earlier treaty will apply only to the extent that they are compatible with those of the later treaty.[138]

Yet, perhaps one of the most interesting aspects of the ILC's work on fragmentation is its examination of the systemic integration of treaties. In this respect, the ILC paid close attention to Article 31(1)(c) of the *VCLT*. This provision states that, in interpreting treaties to resolve a given conflict, the interpreter must take into account "any relevant rules of international law applicable in the relations between the parties."[139] This may include customary international law,[140] principles of international law, and conventional interna-

128 *Ibid.* at 12.

129 *Ibid.* at 19.

130 *Ibid.* at 20-21.

131 See *Vienna Convention on the Law of Treaties*, 23 May 1969 UNTS 331, 8 ILM 679, 63 AJIL 875 (entered into force 27 January 1980) [*VCLT*].

132 See ILC, Fragmentation Report, *supra* note 27 at 250. See also ILC, Fragmentation Report Summary, *supra* note 86 (describing the *VCLT* as the "framework through which [fragmentation] may be assessed and managed in a legal-professional way") at 6.

133 See ILC, Fragmentation Report, *supra* note 27 at 250.

134 See ILC, Fragmentation Report Summary, *supra* note 86 at 10.

135 *Ibid.* at 17.

136 See *VCLT*, *supra* note 131 at Art 30.

137 *Ibid.* at Art 30, para 2. See also EW Vierdag, "The Time of the 'Conclusion' of a Multilateral Treaty: Article 30 of the Vienna Convention on the Law of Treaties and Related Provisions" (1988) 59 BYBIL at 99 (suggesting that *VCLT* "contains conditions that must be satisfied if it is to be operative") at Art 30.

138 See *VCLT*, *supra* note 131 at Art 30, para 3.

139 *Ibid.*, Art 31 (3) (c).

140 See ILC, Fragmentation Report, *supra* note 27 at 233.

tional law.[141] This is so because international treaties "receive their force and validity from general [international] law."[142] Along with customary international law, the rights and obligations stemming from these treaties are part of the system as a whole.[143] In essence, systemic integration refers to an interpretive process where norms are not considered in isolation, but rather "by reference to their normative environment."[144] As the ILC has brilliantly put it:

> [t]o hold those institutions as fully isolated from each other and as only paying attention to their own objectives and preferences is to think of law only as an instrument for attaining regime-objectives. But law is also about protecting rights and enforcing obligations, above all rights and obligations that have backing in something like a general, public interest. Without the principle of 'systemic integration' it would be impossible to give expression to and to keep alive, any senses of the common good of humankind, not reducible to the good of any particular institution or 'regime'.[145]

The main problem with the ILC's approach to fragmentation is that the collision rules force the interpreter to pick one norm over the other. Implicitly, it establishes a hierarchy between norms where one wins and another loses.[146] The ILC itself recognized that in some cases a "mutually supportive solution" may be more convenient.[147] In fact, the ILC acknowledged the limitations of the *VCLT* approach.[148] In the words of Jenks, the main problem with these collision rules is that they do not have "any absolute validity [n]or can [they] be applied automatically and mechanically to any particular class of case."[149] Murphy claims that the ILC's work on fragmentation falls short because it does not establish which rules resolve which conflicts.[150] In fact, by placing too much emphasis on a conflict resolution mechanism, this approach implicitly discards the "idea that norms may also reinforce each other."[151] There is no attempt to reconcile different norms. As Sydnes underscores,

141 *Ibid.* at 237; ILC, Fragmentation Report Summary, *supra* note 86 at 15.
142 *Ibid.*, Fragmentation Report, *supra* note 27 at 208.
143 *Ibid.*
144 *Ibid.* at 208 and 216.
145 *Ibid.* at 244.
146 According to Broude, the ICL's report focuses only on "principles of normative integration." Broude, *supra* note 106 at 174.
147 ILC, Fragmentation Report, *supra* note 27 at 131.
148 *Ibid.* (acknowledging that *VCLT*'s treatment "of bilateral and multilateral treaties through identical rules seems unsatisfactory" at 250).
149 See Jenks, *supra* note 8 at 453.
150 See Murphy, Deconstructing Fragmentation, *supra* note 118 at 3.
151 Harro van Asselt, "Managing the Fragmentation of International Environmental Law: Forests at the Intersection of the Climate and Biodiversity Regimes" (2012) 44 NYU J Int'l L & Pol 1205 at 1256.

the ILC report also ignores the political dimensions of the conflict, given that "in many cases the overlap between regimes is not resolved by legal rules or codification, but is subject to [heavily interested] political interpretations and [extensive] negotiations."[152] As Carlarne puts it, fragmentation should not be "separated from the social [and political] context which gives rise to it."[153] The relationship between CBDR and non-discrimination presents an enormous political element. Fry argues that the ILC "ignores the spontaneous reconciling, and even borrowing, of norms between specialized regimes, which the reference to autonomous self-contained regimes within the definition of fragmentation would seem to not allow."[154] As will be seen below, the ILC's approach to resolving normative conflicts and the *VCLT* collision rules do not provide an adequate response to resolve the CBDR/non-discrimination problem.

2.4 Applying *VCLT* Rules to the CBDR/Non-Discrimination Saga

I now attempt to apply the *VCLT* collision rules to the relationship between the *UNFCCC*'s CBDR and the *Chicago Convention*'s non-discrimination. Given that the ILC suggests that these rules should not be applied in abstracto, let us imagine this hypothetical scenario:

It is September 2016. The 39th Assembly has convened. Following the decision of the previous Assembly, the Council presents a proposal urging the Assembly to consider adopting an MBM for international aviation. The proposed measure is a global emissions trading scheme (Global ETS). As of 1 January 2020, all the airlines of the world will be subject to the scheme. Applying the *VCLT* rules, the Assembly must consider how to integrate *UNFCCC/Kyoto Protocol's* CBDR and the *Chicago Convention*'s non-discrimination principles. For the sake of simplification, I assume that all States are parties to the *Kyoto Protocol*, the *UNFCCC*, and the *Chicago Convention*.

The first thing the interpreter ought to establish is whether the successive treaties in question address the "same subject matter." In addition to establishing ICAO, the *Chicago Convention* sets the framework for the safe and orderly development of international civil aviation. The *UNFCCC* on the other hand seeks to "[stabilize GHG] concentrations in the atmosphere at a level that would prevent dangerous anthropogenic interference with the

152 See Are K Sydnes, "Overlapping Regimes: The SPS Agreement and the Cartagena Biosafety Protocol" in Oran R Young et al., *Institutional Interplay* (New York: United Nations University Press, 2008) 71 at 74. See also Vierdag, *supra* note 137 (stressing that *VCLT*'s rules will not resolve normative conflict between international instruments, but rather the matter will be settled through political channels, at 111).

153 See Cinnamon Piñon Carlarne, "Good Climate Governance: Only a Fragmented System of International Law Away?" (2008) 30 L & Pol'y 450 at 474-475.

154 See James D Fry, "International Human Rights Law in Investment Arbitration: Evidence of International Law's Unity" (2007) 18 Duke J Comp & Int'l L 77 at 135.

climate system."[155] The *Kyoto Protocol* operationalizes the objectives of the *UNFCCC*.[156] The ILC has indicated that "treaties that are institutionally linked or otherwise intended to advance similar objectives" would fall under the same subject matter.[157] Although, as explained above, there is a clear interplay between the *UNFCCC* regime and the *Chicago Convention*, given that they have different objectives and that they do not form part of the same regime, one would tend to conclude that, following the *VCLT*, these instruments do not deal with the same subject matter.

The ILC has nonetheless favored a broader interpretation. In the ILC's opinion, "[t]he criterion of 'same subject-matter' seems already fulfilled if two different rules or sets of rules are invoked in regard to the same matter, or if, in other words, as a result of interpretation, the relevant treaties seem to point to different directions in their application by a party."[158] Vierdag, whom the ILC cites, argues that "[i]f an attempted simultaneous application of two rules to one set of facts or actions leads to incompatible results it can safely be assumed that the test of sameness is satisfied."[159] According to the ILC, the characterization of a treaty into a predetermined field of expertise (e.g. trade, environment) follows a set of interests.[160] Therefore, this should not be relevant to establish whether conflicting rules are part of the same subject matter in order to apply the *VCLT* provisions. Otherwise, a different characterization would simply exclude the application of the *VCLT*.[161] In addition to leaving most issues outside of the scope of the *VCLT*,[162] this may lead to absurd results.[163]

In my view, the ILC's broad interpretation leads to the opposite result. Just about the provisions of all regimes may be considered as being part of the same subject matter. I, therefore, do not agree with ILC's approach. However, for the sake of discussion and to test the potential applicability of the *VCLT* rules to the *UNFCCC/Chicago Convention* relationship, I assume that these two different regimes address the same subject matter. This assumption requires a determination about which is the earlier instrument. Here, the answer is clear. Adopted in 1944, the *Chicago Convention* precedes both the *UNFCCC* and the *Kyoto Protocol*. The former is *lex prior* and the latter regime is *lex posterior*. It also must be established whether the *VCLT*'s Article 30(2) may govern the relationship between these instruments. The interpreter must consider whether any of the treaties expressly establish that it is subject to the other treaty or that is not to be deemed incompatible. As

155 *UNFCCC*, *supra* note 11 (Introduction) at Art 2.
156 See *Kyoto Protocol*, *supra* note 12 (Introduction).
157 ILC, Fragmentation Report Summary, *supra* note 86 at 18.
158 ILC, Fragmentation Report, *supra* note 27 at 18.
159 See Vierdag, *supra* note 137 at 100.
160 ILC, Fragmentation Report, *supra* note 27 at 17.
161 *Ibid.* (suggesting that "[t]he qualifications do not link to the nature of the instrument but the interest from the perspective for which the instrument is assessed by the observer") at 130.
162 *Ibid.* at 129.
163 *Ibid.* at 18.

previously mentioned, the *Kyoto Protocol* specifically refers to ICAO not to the *Chicago Convention*. Therefore, this rule is of no help.

I then proceed to examine the collision rules of Article 30(3). In this respect, the interpreter must establish whether the subject matter involves the same parties. With some notable exemptions, the *Chicago Convention*, the *UNFCCC*, and the *Kyoto Protocol* have almost identical membership.[164] For the sake of simplicity and for the purposes of this exercise, I assume that the same States are parties to all three instruments. The interpreter must also consider whether the operation of the *lex posterior* terminates or suspends *lex prior*.[165] It is clear that neither the *Kyoto Protocol* nor the *UNFCCC* in any way suspends or terminates the application of the *Chicago Convention*. The *VCLT* then says that the *lex prior* only applies to the extent that its provisions are compatible with those of the *lex posterior*.[166] In other words, *lex posterior* governs the relationship between the norms, unless the interpreter can demonstrate that *lex prior* is compatible with *lex posterior*. Assuming that there is an incompatibility between the two regimes, the *UNFCCC/Kyoto Protocol* would trump the *Chicago Convention*. That is to say that CBDR prevails over non-discrimination.

Going back to the Global ETS, this would mean that such a scheme could only be made applicable to developed States. In all likelihood, those designing the Global ETS would then have two options. The first option would be to include only aircraft operators from developed countries in the scheme. The second option would be to include only routes to and from developed States. Either option leads to enormous market distortions and provides an unfair advantage to aircraft operators from developing countries. By opting for one rule over the other, the *VCLT* does not adequately resolve, but it rather aggravates the relationship between the two principles. The *VCLT* collision rules should not be used to defy some of the cardinal objectives of one of the regimes that are subject to the dispute. The *VCLT* rules are of no use in resolving the CBDR non-discrimination saga.

---

164 At present, 191 States are parties to the *Chicago Convention*, 195 to the *UNFCCC*, and 192 to the *Kyoto Protocol*. The only *UNFCCC* parties that are not parties to the *Chicago Convention* are Dominica, the EU, Liechtenstein, Niue, and Tuvalu. The only parties to the *Chicago Convention* that are not parties to the *Kyoto Protocol* are Andorra, South Sudan, the United States, and Canada. The only parties to the *Kyoto Protocol* that are not parties to the *Chicago Convention* are Dominica, the EU, Liechtenstein, Niue, and Tuvalu. See *Chicago Convention*, *supra* note 36; *UNFCCC*, "Parties to the Convention and Observer States" *UNFCCC* online: <http://unfccc.int/parties_and_observers/parties/items/2352.php>.

165 See *VCLT*, *supra* note 131 at Art 30, para 3.

166 *Ibid.*

2.4.1 Systemic Integration and ICAO

The need to take into account rules from other systems is not unusual in normative conflict interpretation.[167] In the US-Shrimp trade dispute, the Arbitral Body relied on non-trade rules to explain the meaning of GATT's Article XX (General Exceptions).[168] Without necessarily being a case of conflict interpretation, an analysis of ICAO's long-standing practices reveals that, on numerous occasions, the organization has resorted to rules from other sectors. As such, the idea that ICAO, as a self-contained regime, should simply ignore rules of other organizations with which it interacts does not seem plausible. For instance, according to Annex 9 to the *Chicago Convention*, the disinfection of aircraft is subject to the rules prescribed by the World Health Organization (WHO).[169] The handling of consumable products and waste on board aircraft and at international airports is subject to WHO's requirements and those of the Food and Agriculture Organization.[170] ICAO has through the adoption of a standard in Annex 9 made it a mandatory requirement[171] for its Member States to comply with WHO's 2005 International Health Regulations (IHR).[172] With regard to units of measurement, international civil aviation relies on the system established by the General Conference of Weights and Measures (CGPM).[173] Through its World Radiocommunication Conferences, the International Telecommunication Union

167 See Palmer et al., *supra* note 112 (discussing that "[t]he WTO agreements also anticipate the need to take into account other existing international agreements, such as MEAs, and other relevant state practices. Both the [Agreement on Sanitary and Phytosanitary Measures (SPS Agreement)] and the [Agreement on Technical Barriers to Trade] (TBT Agreement) make reference to international standards developed by competent international organizations operating outside the WTO system" at 183).

168 See *United States – Import Prohibitions of Certain Shrimp and Shrimp Products, Recourse to Article 21.5 of the DSU by Malaysia* (22 October 2001), WT/DS58/AB/RW (Appellate Body Report) online: WTO <www.wto.org/english/tratop_e/dispu_e/58abrw_e.doc> at para 129. See also Yuka Fukunaga, "Civil Society and the Legitimacy of the WTO Dispute Settlement System" (2008) 34 Brooklyn J Int'l L 85 at 104; Palmer et al, *supra* note 112 (discussing that WTO's dispute settlement bodies have "taken into account existing international agreements and state practice when clarifying relevant provisions of the GATT" at 187). There have also been cases where WTO adjudicatory bodies have been reluctant to embrace foreign norms. For instance, in the 1998 Beef Hormone case, a WTO AB rejected the application of the precautionary principle to a trade-related dispute. See ILC, Fragmentation Report, *supra* note 27 at 34.

169 See Annex 9 to the *Chicago Convention* Facilitation, at standard § 2.25 [Annex 9]. See also Jenks, *supra* note 8 at 417.

170 Annex 9, *supra* note 169 at standard § 6.39 and 6.41.

171 Although ICAO's standard setting function will be explained in Chapter 3, it suffices to say here that, as part of its core functions, the organization adopts technical standards and recommended practices (SARPs). As far as the organization is concerned, compliance with the standards is mandatory to the extent that a State does not file a difference. That is to say a formal notification that such a State finds it impracticable to implement such standard. See *Chicago Convention*, *supra* note 36 at Arts 37 and 38. Whether these standards are in fact mandatory has been subject to interesting academic debates. For the purpose of this chapter, however, we will follow ICAO's interpretation that compliance with such standards is mandatory unless a State notifies a difference thereto.

172 See Annex 9, *supra* note 169 at standard § 2.5.

173 See Annex 5 to the *Chicago Convention*, Units of Measurement to be Used in Air and Ground Operations, standard § 3.1.1 [Annex 5].

allocates the radio-frequency spectra for international civil aviation for both air-ground communications and radionavigation.[174] Similarly, a number of standards establish mandatory observance of the technical specifications of the International Organization for Standardization (ISO).[175] Signage used at airports is subject to ICAO-IMO recommendations.[176] These examples reinforce the notion that ICAO – the UN specialized agency in charge of international civil aviation – does not remain isolated.

These examples arguably pose a substantial difference with regard to ICAO's rejection of CBDR, namely, they are not in direct conflict with the *Chicago Convention*. As such, there is no impediment to adopting them, given that these foreign norms contribute to achieving the *Chicago Convention*'s main objectives. CBDR, on the other hand, as applied in the *UNFCCC* context, threatens to destroy the basic premises upon which international civil aviation has been structured. This stance, however, presupposes that there is an inherent conflict between CBDR and the *Chicago Convention*. Assuming that both principles are mutually exclusive, the analysis does not attempt to integrate both principles in a reconciliatory manner. As I explain below, although the principles are clearly in tension, it is in fact feasible to reconcile them.

2.5 Attempts to Accommodate the Special Needs of Developing Countries

In this section, I discuss some attempts by ICAO to incorporate a special recognition for developing countries in the climate change discourse without creating the bipolar differentiation present in the CBDR principle. Although these initiatives have been unsuccessful, they are central to the argument of this book because they demonstrate that the reconciliation of CBDR and non-discrimination principles in a manner compatible with the aviation environment can no longer be postponed.

---

174 See ICAO Doc 9718, *Handbook on Radio Frequency Spectrum Requirements for Civil Aviation* at v. See also Annex 10 to the *Chicago Convention*, Aeronautical Telecommunications, vol III, standard § 4.3.1.1 [Annex 10] (establishing that "[w]hen providing AMS(R)S communications, an AMS(R)S system shall operate only in frequency bands which are appropriately allocated to AMS(R)S and protected by the ITU Radio Regulation"). Jenks also claims that "[a]eronautical telecommunications are governed partly by [ITU] and partly by [ICAO]." Jenks, *supra* note 8 at 16.

175 See Annex 14 to the *Chicago Convention* Aerodromes, vol 1, standard § 9.2.18 [*Annex 14*]; Annex 16, Aircraft Engine Emissions, Vol II, standard § 2.3 (smoke analysis system) [Annex 16].

176 See Annex 9, *supra* note 169 at recommended practice § 6.9.

2.5.1 The De Minimis Principle

2.5.1.1 The 2010 De Minimis Proposal

During the 37th Assembly held in 2010, Canada, Mexico, and the United States proposed an exclusion threshold to determine which States should submit action plans to address aviation's impact on climate change.[177] While seeking to capture those States to whom the vast majority of emissions from international aviation may be attributed, the proponents intimated that the suggested threshold should be based on international aviation activity measured in revenue tonne kilometers (RTKs).[178] More specifically, they proposed that those countries whose aviation activity accounted for less than 0.5 percent of total RTKs should be excluded.[179] This would have captured the top 34 countries and covered roughly 90 percent of emissions from international aviation.[180] Given that an RTK-based threshold takes into account aviation activity, the list would have included both developed and developing States, but it would still have excluded a large number of developing countries. Arguably, proponents of the proposal - which came to be known as the "de minimis" proposal - sought to obtain support from developing countries without creating the *UNFCCC/Kyoto Protocol* type of differentiation enshrined in CBDR. As originally proposed, it was only intended to exclude States from having to develop action plans. As such, its proponents did not envisage any potential risks for market distortions.[181]

African States then took up the essence of the idea embedded in the proposal but introduced a significant change. Under the African-led version of the de minimis proposal, the exclusion would not only be applicable to the submission of action plans, but it would also seek to ensure that aircraft operators from African States would be exempted from the "application of MBMs that are established on national, regional and global levels."[182] At the time, South Africa was the only African State with the highest level of international aviation activity, ranking above 0.5 percent but below 1 percent of international RTKs. Because of this, African States proposed an upward revision of the de minimis ceiling to

177 See ICAO, A37-WP/186 EX/35, *A More Ambitious, Collective Approach to International Aviation Greenhouse Gas Emissions* (presented by Canada, Mexico and the United States) at 3.

178 *Ibid.*

179 *Ibid.*

180 This would have captured the United States, China, Germany, the United Kingdom, UAE, France, Republic of Korea, Netherlands, Singapore, Japan, Ireland, Canada, Australia, Thailand, Spain, Qatar, Malaysia, Russian Federation, India, Turkey, Luxembourg, Switzerland, Italy, New Zealand, Saudi Arabia, Brazil, Israel, Mexico, South Africa, Chile, Portugal, Colombia, Finland, and Austria. See ICAO Doc 9952, *Annual Report of the Council: 2010*, Appendix at 8 [ICAO Doc 9952]. See also ICAO, States Ranking of International Aviation Traffic (RTK in 2009) [unpublished, archived on file with the author].

181 See Hardeman, "Reframing Aviation Climate Politics and Policies," *supra* note 44 (noting that only 22 ICAO Member States are above the de minimis threshold, at 9).

182 ICAO, A37-19, *supra* note 207 (Chapter 1) at para 15 (a).

1 percent.[183] With this proposal, African States also sought to send a strong political message that their aircraft operators should be exempted from the EU ETS. Given that Europe never amended its legislation, this de minimis proposal did not have much practical application with regard to the EU ETS. However, as will be seen below, it did set an ill-fated precedent.

At the 37th Assembly, Europe and the United States remained divided on a variety of issues. Likewise, major developing countries rejected a number of provisions, including those establishing non-binding, global, aspirational goals for international aviation. In addition, African States conditioned their support for the climate change resolution upon the inclusion of the de minimis proposal. Given that ICAO was under tremendous pressure to make progress and that not having a resolution could have had much worse consequences, the Assembly arrived at a political compromise in which it reluctantly accepted the African-led 1 percent de minimis threshold without any technical justification.[184]

The de minimis proposal was problematic for at least four reasons. First, it suggested that States implementing MBMs should exclude aircraft operators of those States that fall below the 1 percent RTK threshold. If implemented, this may lead to significant market distortions. It would basically mean that on a route between Santiago, Chile, and Miami, Florida, American Airlines (a US aircraft operator) would be subject to an MBM, whereas LAN (Chile's aircraft operator) would be exempted. This is simply a flagrant violation of ICAO's non-discrimination principle since aircraft operators serving the same route would be subject to different environmental requirements. It should, therefore, come as no surprise that 57 States filed reservations against the African-led de minimis proposal.[185]

Second, because the de minimis proposal was incorporated as part of a political decision that prompted discontent among numerous States, the Assembly had no other option but to task the Council to conduct an assessment to establish its potential "[impact] on the aviation industry and markets."[186] This was rather symbolic, given that those opposing the

183 This captured the United States, China, Germany, the United Kingdom, UAE, France, Republic of Korea, the Netherlands, Singapore, Japan, Ireland, Canada, Australia, Thailand, Spain, Qatar, Malaysia, Russian Federation, India, Turkey, Luxembourg, and Switzerland. See ICAO Doc 9952, *supra* note 180. See also ICAO, States Ranking of International Aviation Traffic (RTK in 2009) [unpublished, archived on file with the author].

184 See ICAO, A38-WP/258 EX/85, *UAE's Views on Aviation and Climate Change* (presented by the United Arab Emirates) [ICAO, A38-WP/258 EX/85]. See also ICAO, C-MIN 199/13, *Council, 199th Session, Decision of the Meeting* at 9 (Statement of India) [ICAO, C-MIN 199/13]. Likewise, the Republic of Korea correctly underscored that there was no "scientific or economic rationale to justify the de minimis threshold. *Ibid.* at 14.

185 See "Reservations to A37-19," online: ICAO <www.icao.int/Meetings/AMC/Assembly37/Documents/ReservationsResolutions/10_reservations_en.pdf>.

186 ICAO, A37-19, *supra* note 207 (Chapter 1) at para 16.

de minimis proposal knew that it would lead to substantial market distortions.[187] The Secretariat then hired a consulting firm to unequivocally reconfirm the suspicions.[188]

Third, an RTK-based exclusion threshold was not necessarily the best approach to measure the level of aviation activity of any given State. To calculate a State's share of international aviation activity, ICAO takes into account all traffic generated by aircraft operators to whom such State has issued an aircraft operator certificate (AOC). The statistics are distorted where foreign airlines dominate the traffic to and from that State as a result of the small market share of local airlines. This explains why, despite being major developing aviation markets, Argentina, Indonesia, and Nigeria feature at the bottom of ICAO's RTK rankings.[189]

Fourth, while the de minimis proposal ensured that African States would support ICAO's 2010 climate change resolution, it did not resolve the conflict between CBDR and non-discrimination.[190] Not only did it stand as another missed opportunity to address the conflict, but it made matters more complicated. In practice, it was almost impossible to implement the de minimis proposal without causing significant market distortions. However, it created among African States in particular a sense of entitlement or a perception that exempting African aircraft operators from MBM measures was in fact a vested right.[191] The fact that, in 2013, the 38th Assembly also resulted in an RTK-based de minimis resolution certainly reinforces this argument.[192]

2.5.1.2 The 38th Assembly: A Similar De Minimis *Proposal*

On 4 September 2013, at the last Council meeting held before the commencement of 38th Assembly, African States insisted that they "were prepared to live with other provisions [of the proposed draft Assembly resolution on climate change] as long as [the de minimis

187 See ICAO, A38-WP/258 EX 85, *supra* note 184 at 7.

188 See ICAO, WP/13798, *Study on Market-Based Measures (MBMs)* (presented by the Secretary General) at 3.2.4 and 3.2.5.

189 According to ICAO's 2012 international RTK list, Argentina and Nigeria rank fiftieth and eighty-third, respectively. See "Civil Aviation: 2012 International RTK by State of Air Operator Certificate (AOC)," online: ICAO <www.icao.int/Meetings/a38/Documents/International%20Scheduled%20RTK%20(Annual%20Report).PDF>.

190 See Hardeman, "Reframing Aviation Climate Politics and Policies," *supra* note 44 at 10.

191 The problem does not lie in providing some headroom for some States for a defined period of time. Difficulties arise when this is perceived as an automatic entitlement without technical justification. This makes the designing of any scheme to address GHG emissions much more challenging. For instance, in examining the broader climate change context, Heller questions whether the Russian Federation and Ukraine would agree to a second commitment period of the *Kyoto Protocol* unless significant headroom is provided to them. It is worth remembering that these countries already enjoyed considerable concessions in the first commitment period of the *Kyoto Protocol*. See Thomas Heller, "Climate Change: Designing an Effective Response" in Ernesto Zedillo, ed, *Global Warming: Looking Beyond Kyoto* (Washington: Brookings Institution Press, 2008) 115 at 122.

192 See ICAO, A38-18, *supra* note 21 (Introduction) at para 16 (b).

proposal] was adopted."[193] During the ensuing discussions at the Assembly, it was once again evident that a resolution on climate change would necessarily require the support of African States, for whom the de minimis proposal was a must. Mindful of the previous criticisms based on the likelihood of market distortions, the 2013 de minimis proposal excluded from new and existing MBMs "routes to and from developing States whose share of international civil aviation activities is below the threshold of 1 [percent] of total revenue ton[ne] kilometres of international civil aviation activities, until the global scheme is implemented."[194]

Although not expressly mentioned in the resolution, the de minimis provision would, in principle, exclude both aircraft operators from developed and developing countries serving routes to or from developing countries that fall under the specified threshold. As such, the issue is not whether the routes involve destinations with low levels of emissions, but rather whether they are situated in developing countries. Ethiopia serves as an apt example. It is a least developed African country that hosts one of Africa's most vibrant and sophisticated airlines: Ethiopian Airlines. According to ICAO's latest statistics, Ethiopia, whose share of international RTKs is roughly 0.60 percent, ranks 29th in the world.[195] As such, Ethiopia's level of international aviation activity places it below the de minimis threshold. In practice, this would mean that both Air France and Ethiopian Airlines should be excluded from an MBM that applies on routes to and from Ethiopia (e.g. Paris to Addis Ababa and Addis Ababa to Paris). It would also mean that a segment such as Nairobi to Hong Kong via Addis Ababa on Ethiopian Airlines would not be subject to an MBM scheme. Conversely, a similar segment Nairobi to Hong Kong via Doha with Qatar Airways would be subject to the MBM scheme for the sole reason that the State of Qatar ranks above the threshold. Although the de minimis provision applies equal treatment to aircraft operators flying on the same route, it ignores the competition between different hubs. Depending on the cost of carbon, this may expose carriers to market distortions. Again, not a single technical study was conducted prior to the adoption by the Assembly of this

193 ICAO, C-MIN 199/13, *supra* note 184 at 7 (Statement of Cameroon). See also ICAO, C-DEC 199/13. Similarly, Nigeria indicated that although African States had a number of reservations on the proposed Assembly resolution, they would be in a position to support it to the extent that the African-led de minimis would be retained. ICAO, C-MIN 199/13, *supra* note 184 at 9 (Statement of Nigeria). South Africa also warned that if the de minimis was not retained, the whole resolution would be "unacceptable" to African States. *Ibid.* at 15. Uganda said that "it could live with" the text, as long as the de minimis was part of the package. *Ibid.* at 17.

194 ICAO, A38-18, *supra* note 21 (Introduction) at para 16 (b).

195 See ICAO, Civil Aviation: 2012 International RTK by State of Air Operator Certificate (AOC) online: ICAO <www.icao.int/Meetings/a38/Documents/International%20Scheduled%20RTK%20(Annual%20Report).PDF>.

de minimis proposal to discard or confirm the possibility of market distortions.[196] Politics trumped reason.[197]

What is perhaps most dangerous about an RTK-based de minimis provision is the fact that it may prejudge future exclusion thresholds that may be needed for the purpose of designing the ICAO global MBM.[198] In other words, the de minimis provision will most likely serve as the starting point for negotiations. Arguably, African States will once again push for an absolute and permanent exemption for their aircraft operators. Nobody can challenge the fact that those aircraft operators with an insignificant level of emissions should be excluded from the application of an MBM. The administrative burden of running an MBM scheme certainly does not justify chasing insignificant emitters. The cost far outweighs the gains. However, it is quite different to suggest that a whole region should be permanently exempted from an MBM on the basis of a political decision with no technical justification other than the fact that such a compromise (de minimis) was required in order to save ICAO's climate change resolutions. As will be explained in Chapter 8, during the initial stages of ICAO's global MBM, the vast majority of African aircraft operators should be exempted. The level of their emissions still remains quite low. However, this does not mean that all African aircraft operators may be regarded as insignificant emitters. The level of development and sophistication and growth rates achieved by South African Airways and Ethiopian Airlines are not the same as those of State-owned Air Namibia operating only a handful of long-haul flights.

2.5.2 *Reinventing CBDR: ICAO's SCRC*

Since 2007, as a result of the pressure exerted by some developing countries, ICAO has included hortatory references to CBDR in its Assembly resolutions on climate change but only in preambular clauses.[199] However, to offset the potential market distortions that CBDR may create if applied in the aviation environmental context, reference to the principle has always been either preceded or followed by ICAO's non-discrimination principle. Until 2013, CBDR never featured in the operative clauses of ICAO's Assembly resolutions. The intention behind this was to acknowledge in a declaratory manner that although CBDR guided the *UNFCCC* process, it did not have any practical application within the international civil aviation regime. Again, a number of developed States and industry stakeholders

---

196 See also ICAO, C-MIN 199/13, *supra* note 184 at 15 (Statement of Spain).

197 See Lyle, "Mitigating International Air Transport Emissions through a Global Measure," *supra* note 122 (Chapter 1) (noting that "[t]he 1 [percent] concept is not only solely aviation determined, [but rather] also somewhat arbitrary") at 2.

198 See ICAO, A38-WP/258 EX/85, *supra* note 184 at 7.

199 See ICAO, A36-22, *supra* note 17 at Appendix L.

have strongly argued that because it conflicts with non-discrimination, CBDR has no place in international civil aviation context.

In 2010, while still replicating both the CBDR and the non-discrimination principles in the preambular clauses of its climate change resolution, the ICAO Assembly attempted to move away from the political divide and potential operational repercussions embedded in CBDR by developing the new concept of "special circumstances and respective capabilities" (SCRC).[200] As such, in working toward achieving ICAO's global aspirational goals, the 37th Assembly suggested that States and international organizations should take into account SCRC.[201] Originally, the scope of SCRC was exclusively restricted to developing countries. In 2013, however, the Assembly made several references to the principle without restricting it to developing countries.[202] As its name suggests, this principle seeks to accommodate those States whose specific circumstances may require special consideration. In other words, SCRC seeks to accord a degree of deference to some Sates in order for them to fully participate in ICAO's initiatives to tackle GHG emissions.

2.5.2.1 What Is the Difference?

On the basis of fairness and equity, both CBDR and SCRC seek to address the challenges that some (developing) countries face in their efforts to combat climate change. It is clear that not all countries are equipped with the same resources to perform this task. These principles however differ in at least four key respects.

First, there is a clear difference in their political perception. As embedded in the climate change regime, CBDR implies that only developed countries bear mitigation obligations. If one were to follow this interpretation, applying it in the ICAO context would pose operational challenges and increase the political risk of perpetuating a differentiated scheme in another forum beyond the context of the climate change regime. On the contrary, foregoing CBDR sets a precedent that developing countries may not be willing to accept: a climate change regime without differentiated obligations. CBDR is also perceived as the perfect excuse for inaction because (some) developing countries will continue to argue that they should not bear any obligation in spite of the fact that their airlines compete head-to-head with airlines from developed States and quite often have been much more successful. It is not unusual for airlines from developing States to enjoy a significantly larger share of the market on certain routes as compared to their peers from developed States. In addition, as mentioned in Chapter 1, the rate of growth of air transport in many

200 See ICAO, A37-19, *supra* note 207 (Chapter 1).

201 *Ibid.* at para 6 (a).

202 For instance, the 38th Assembly agreed to develop a global MBM for international aviation "taking into account the special circumstances and respective capabilities of States, in particular, developing States, while minimizing market distortion." ICAO, A38-18, *supra* note 21 (Introduction) at paras 19 (b) and 20. Similarly, ICAO's global aspirational goal must also take into account SCRC. *Ibid.* at para 7.

developing countries is now higher than it is in developed States. This explains why developed States and the aviation industry argue that CBDR should not exist at ICAO. On the other hand, SCRC is an ICAO creation, although its wording would seem to be a reformulation of existing *UNFCCC* language.[203]

Second, as mentioned above, CBDR refers to the "common but differentiated responsibilities and *respective capabilities*" of States.[204] However, when developing countries such as Brazil, China, and India refer to CBDR, they purposely ignore the last part. This is clearly self-serving. SCRC, on the other hand, emphasizes this forgotten part of the CBDR language. In theory, this would allow differentiating among developing countries, a concept that some States have strongly resisted within the *UNFCCC* context. For instance, for the purpose of *UNFCCC*'s classifications, Mauritania, Qatar, UAE, and Yemen are all developing countries. However, Qatar and UAE have achieved a significantly higher level of aviation development. In fact, the growth of aviation in these two countries poses a real and tangible threat to well-established aircraft operators in developed States. This is certainly not the case for civil aviation in Mauritania and Yemen. Given that their "respective capabilities" are substantially different, it does not seem fair to treat all of these States in the same manner.

Third, some States are of the view that when dealing with environmental issues, ICAO should assess the impact of proposed measures for both developed and developing countries.[205] In this context, SCRC in its latest form does not preclude the extension of special consideration to developed States whose capabilities may justify a request for assistance under a given set of circumstances. One could argue reasonably that, in spite of being a developed State and Member of the European Union, Greece may be going through some special circumstances. Some States have expressed the view that SCRC should only be applicable to developing countries.[206]

Fourth, as applied in context of the *UNFCCC*, CBDR very clearly takes into account the historical responsibilities of developed States. Because of their large share of historical GHG emissions, developed countries have taken the lead with regard to mitigation obliga-

203 It is worth noting that Art 3, para 2, of the *UNFCCC* speaks about "[t]he specific needs and special circumstances of developing country parties." *UNFCCC, supra* note 11 (Introduction) at Art 3, para 2.

204 *UNFCCC, supra* note 11 (Introduction) at Art 3, para 1.

205 See ICAO, C-MIN 181/22, *supra* note 37 at 262 (Statement of the United States).

206 See ICAO, C-MIN 199/13, *supra* note 184 at 6, 9, and 11 (Statements of Peru, India, and Cuba, respectively). India also noted that SCRC should not apply to "fast growing airlines and new entrants in the air transport industry." *Ibid.* See *contra Ibid.* at 11 (Statement of Malaysia raising no objection to SCRC being extended to cover developed countries); ICAO, A38-WP/424 EX/139, *Consolidated Statement of Continuing ICAO Policies and Practices Related to Environmental Protection: Climate Change* (presented by Argentina, Brazil, China, Cuba, Guatemala, India, Islamic Republic of Iran, Pakistan, Peru, Russian Federation, Saudi Arabia, South Africa) [ICAO, A38-WP/424 EX/139].

tions. At least as currently construed at ICAO, SCRC does not weigh in these two key elements.[207]

2.5.2.2 Why Has SCRC Not Worked?

SCRC represents a laudable attempt to incorporate a workable principle into the ICAO domain that would recognize the different circumstances and level of development of States while minimizing market distortions. It has certainly sought to bridge the political divide between die-hard CBDR adherents and non-discrimination fanatics. Having said this, judging from the unprecedented number of reservations filed against the last two ICAO Assembly resolutions on climate change, one could speculate that SCRC has not necessarily brought the divergent views of both developed and developing States any closer.[208] A number of reasons may explain why this has been the case. SCRC evolved from the premise that CBDR leads to market distortions. As such, CBDR conflicts with the *Chicago Convention*. This however holds if one follows CBDR's conceptualization as applied in the *UNFCCC* context. CBDR is not implementable in international aviation. However, as will be seen below, CBDR is not set in stone. Nothing prevents States from reformulating CBDR in a manner that allows them to take into account the particular characteristics of the aviation sector. However, developed States, industry, and the ICAO Secretariat did not attempt to adapt CBDR to the international aviation environment, but rather, under the pretext that CBDR does not exist at ICAO, they opted to replace it completely with SCRC. In the end, this not only alienated major developing countries but also contributed to creating a sense of distrust. It is true though that, to some extent, as Brazil has put it, while some States have "accepted the euphemism of SCRC in the name of consensus and to accommodate the concerns of certain States, it was not reasonable to distort the concept and essence of CBDR."[209]

The situation has been exacerbated by the fact that SCRC has never been made operational. It still remains an abstract concept.[210] In other words, almost 17 years after the *Kyoto Protocol* entrusted ICAO with the mandate to tackle GHG emissions from international aviation, it is not clear how the principle will be put in practice to accommodate developing countries or those states that may require special treatment. I argue that the approach should have been not to reject but reformulate CBDR in a manner compatible with ICAO's practices. It is evident that SCRC has not achieved the desired results. In the remaining

207 See ICAO, C-MIN 199/13, *supra* note 184 at 10 (Statement of Brazil).

208 At the 37th Assembly, 56 States filed reservations to ICAO's climate change resolutions. See "Reservations to Resolution A37-19," online: ICAO <www.icao.int/Meetings/AMC/Assembly37/Documents/ReservationsResolutions/10_reservations_en.pdf>. Similarly, at the 38th Assembly, 62 States filed reservations. See "Reservations to Resolution A38-18," online: ICAO <www.icao.int/Meetings/a38/Pages/resolutions.aspx>.

209 ICAO, C-MIN 199/13, *supra* note 184 at 10 (Statement of Brazil).

210 While stressing the importance of the SCRC principle, Colombia underscored that it "should be more specific about financial assistance, technology transfer and all other forms of assistance." *Ibid.* at 4.

portions of this chapter, I explain why CBDR requires a new conceptualization and advance ways to apply this new conceptualization to international civil aviation.

2.5.3 *CBDR Finally Arrives at ICAO through the Back Door*

One of the most interesting aspects of the outcome of the 38th Assembly on climate change is the fact that, for the first time ever, ICAO acknowledged CBDR within the operative clauses of its Assembly resolution.[211] This will frame future discussion on the subject.[212] As mentioned above, in the past, references to CBDR were only included in the preambular clauses. CBDR is no longer just a symbolic principle included for rhetorical purposes, but it is now a concept which, along with other principles, should be put into practice.

During ensuing discussions of the Assembly, a group of influential developing countries tabled the proposal to include CBDR.[213] Developed States suggested the inclusion of the principle of non-discrimination. Singapore proposed compromise language on SCRC, and China advanced the principle of fair and equal opportunity.[214] The result is a complete

211 See A38-18, *supra* note 21 (Introduction) at Annex (p).

212 See Lyle, "Mitigating International Air Transport Emissions through a Global Measure," *supra* note 122 (Chapter 1).

213 ICAO, A38-WP/425 EX/140 *Proposed Amendments for the Draft Consolidated Statement of Continuing ICAO Policies and Practices Related to Environmental Protection: Climate Change* (presented by Argentina, Brazil, China, Cuba, Guatemala, India, Islamic Republic of Iran, Pakistan, Peru, Russian Federation, Saudi Arabia, South Africa).

214 According to ICAO, "fair and equal opportunity" refers to a general principle that appears on bilateral air services agreements (ASAs). See ICAO Doc 9626, *supra* note 53 at 2.2. In principle, it seeks to "ensure against discrimination or unfair competitive practices affecting" aircraft operators designated under such bilateral arrangements. *Ibid.* In practice, however, States interpret it differently. For States desirous to gain market access and expand international routes for their aircraft operators, this principle seeks to ensure competition and prevent discriminatory treatment. On the other hand, for conservative States, the principle provides a justification to impose market restrictions in order to protect their designated aircraft operators. If, for instance, designated aircraft operators of State A can only fly twice a week to cities in State B, State B's designated aircraft operators will only be entitled to two frequencies per week, regardless of their capacity to fly daily. Under this conception, it is not fair to grant State B unrestricted market access. This will be limited and subject to the capabilities of those operators of State A. See also Dempsey, *supra* note 5 at 639-640. In the last three ICAO resolutions dealing with climate change issues, fair and equal opportunity has been linked to the principle of non-discrimination, as if they were one sole concept. See ICAO, A36-22, *supra* note 17 at preambular clauses ("acknowledging the principles of non-discrimination and equal and fair opportunities to develop international civil"); ICAO, A37-19, *supra* note 207 (Chapter 1) ("[a]lso acknowledging the principles of non-discrimination and equal and fair opportunities to develop international aviation set forth in the Chicago Convention" at preambular clauses); ICAO, A38-18, *supra* note 21 (Introduction) ("[a]lso acknowledging the principles of non-discrimination and equal and fair opportunities to develop international aviation set forth in the Chicago Convention" at preambular clauses). Although the preamble of the *Chicago Convention* recognizes that "air transport services may be established on the basis of equality of opportunity," it does not automatically follow that "equal and fair opportunities," as incorporated in the ICAO climate change resolutions, form an integral part of the non-discrimination principle. It is unquestionable however that both principles present some similarities. For instance, both of them are part of ICAO's objectives. See *Chicago Convention*, *supra* note 36 ("insure that... every [State] has a fair opportunity to

mélange: in developing its global MBM scheme for international aviation, ICAO must now take into account the CBDR, SCRC, non-discrimination, and equal and fair opportunity principles.[215]

The inclusion of CBDR into the Assembly resolution bears remarkable consequences for ICAO's future work in addressing climate change. First, although 50 States filed reservations against it,[216] the principle now forms an integral part of ICAO's guiding principles on MBMs. For States abhorring CBDR, it will be very difficult to preclude its consideration as one of the principles that must guide the design of a global MBM.

Second, it will also be an uphill battle to continue arguing that CBDR does not exist in the international civil aviation context. CBDR has been upgraded from the preambular to the substantive clauses. Arguably, for the vast majority of Member States, CBDR deserves not only to be recognized but also be made operational in any global MBM scheme. It is illusory to expect that, in 2020, once ICAO's global scheme kicks off, all States will be treated equally.

Third, the Assembly resolution lists three different principles. It is therefore now evident that CBDR and SCRC are different principles with different meanings. In the past, some States, the Secretariat, and the industry argued that SCRC was the manner in which international civil aviation applied CBDR at ICAO. The challenge of course will be how to reconcile these two principles. I discuss some proposals below.

2.6 TOWARD A NEW APPROACH: RECONCILING PRINCIPLES

2.6.1 CBDR Is Not Static

Principles of international law are not immutable or incapable of being adapted to changing circumstances. Many decades ago, in the Namibia case, the ICJ acknowledged

operate international airlines" at Art 44 (f)) ("avoid discrimination between [States]" at Art 44 (g)). In spite of this, the principles are essentially different. In the *Chicago Convention*, non-discrimination is constructed as a prohibitive rule against arbitrary treatment of both States and aircraft operators. The use of the word "shall" suggests a mandatory nature of the norm. See *Ibid.* at Arts 11 and 15. On the other hand, "equality of opportunity" denotes an ideal that should be met to operate air transport services. In other words, it seeks to ensure access to these services by all States. The reader will note the use of the word "may." See *Ibid.* at preamble. Although in some cases such access may be denied as a result of discriminatory treatment, the scope of the principle is much broader. As mentioned above, the main problem with the principle relates to its disparate interpretation. For the purpose of this chapter, we do not consider it to be part of the non-discrimination principle, but rather a serious drafting oversight.

215 See ICAO, A38-18, *supra* note 21 (Introduction) at Annex (o).

216 See "Reservations to Resolution A38-18" online: ICAO <www.icao.int/Meetings/a38/Pages/resolutions.aspx>.

that provisions of international treaties are not "static."[217] On the contrary, they are "by definition evolutionary."[218] The ICJ also said that "an international instrument has to be interpreted and applied within the framework of the entire legal system prevailing at the time of the interpretation."[219] In interpreting the term "natural resources" in the US-Shrimp trade dispute, the Appellate Body applied exactly the same rationale.[220] This presupposes that norms contained in international treaties may in fact evolve. It is beyond doubt that different regimes may have different objectives. In fact, some principles may conflict with those of other regimes. Having said this, quite often this is due to the manner in which these rules are interpreted.[221]

The conceptualization of CBDR within the wider climate change regime cannot constitute its only possible form of interpretation. Its recognition by numerous international instruments other than the *UNFCCC* and the *Kyoto Protocol* suggests that CBDR may evolve.[222]

It may certainly be re-interpreted in a manner that is consistent with international civil aviation. CBDR should not be applied in a manner that completely disregards the principles and objectives of the climate change regime; neither should it be interpreted to contravene the essential nature of international air transport.[223] CBDR and non-discrimination are not by definition mutually exclusive. As Lyle aptly notes, "[the international climate change regime] does not preclude alternative forms of differentiation in future agreements."[224]

---

217 *Legal Consequences for States of the Continued Presence of South Africa in Namibia (South West Africa) notwithstanding Security Council Resolution 276 (1970), Advisory Opinion,* [1971] *ICJ Rep 16.*

218 *Ibid.*

219 *Ibid.*

220 *United States – Import Prohibition of Certain Shrimp and Shrimp Products* (12 October 1998), WT/DS58/AB/R Appellate Body Report), online: WTO <https://docs.wto.org/dol2fe/Pages/FE_Search/FE_S_S006.aspx?Query=(@Symbol=%20wt/ds58/ab/r*%20not%20rw*)&Language=ENGLISH&Context=FomerScriptedSearch&languageUIChanged=true#> at para 130 [US-Shrimp trade dispute]. See also Marceau, "Conflicts," *supra* note 34 at 1089.

221 See Sydnes, *supra* note 152 (explaining that the tension between the Cartagena Protocol and the WTO regime is the result of "the scope of interpretation of both rule systems rather than from obvious incompatibilities of their rules") at 84.

222 The differentiation and special recognition of developing countries are not exclusive to the climate change regime. See Christopher D Stone, "Common but Differentiated Responsibilities in International Law" (2004) 98 Am J Int'l L 276.

223 In a similar vein, examining the interplay between the WTO and MEAs, Marceau claims that "further accommodation of environmental measures in the WTO is possible without undermining its open, non-discriminatory character." Gabrielle Marceau, "A Call for Coherence in International Law: Praises for the Prohibition against Clinical Isolation in WTO Dispute Settlement" (1999) 33:5 J World Trade 87 at 106 [Marceau, "Coherence"].

224 Lyle, "Rio, Kyoto, Brussels and Chicago," *supra* note 6. An analysis of the wider climate change negotiations reveals that different States have different views with regard to establishing differentiated schemes for regulating GHG emissions. Vogler and Bretherton provide an excellent explanation to the different approaches major players have taken with regard to CBDR and the possibility to develop a differentiated regime, albeit for a limited time, to address GHG emissions. According to these authors, the United States and Europe are divided on this issue. While Europe would be willing to explore a differentiated regime with concrete time

2.6.2 Avoiding Isolation

Nothing prevents ICAO or the *UNFCCC* from taking into account principles of international law or principles developed by the other regimes to the extent that they do not purport to undermine its objectives. In fact, the principle of systematic integration of the *VCLT* supports this view. As seen above, practice at ICAO reveals a long tradition of incorporating rules from other regimes. *UNFCCC* and ICAO norms should be integrated in a coherent manner.[225] As the ILC has put it, "legal rules rarely if ever appear alone, without relationship to other rules."[226] A regime's norms and principles should "not be interpreted in clinical isolation from public international law."[227]

2.6.3 Reconciling CBDR with Non-Discrimination

In the *Gabcíkovo-Nagymaros Project* case, Judge Weeramantry stated that the right of development "does not exist in the absolute sense, but is relative always to its tolerance by the environment."[228] In the words of Judge Weeramantry, "[e]ach principle cannot be given free rein, regardless of the other. The law necessarily contains within itself the principle of reconciliation."[229] Accordingly, the question is not whether one principle prevails over the other and whether one is applied and the other discarded. The central issue is that they do coexist and they do require a form of integration. This is achieved through reconciliation.

---

frames, the United States, on the other hand, supports full participation from major developing countries' emitters, such as Brazil, China, and India. Vogler & Bretherton, *supra* note 194 (Chapter 1) at 17.

225 See Marceau, "Coherence," *supra* note 223 (advocating the need for "coherence between the areas of trade, development and environment") at 87. Marceau also argues that non-WTO principles may bring coherence in addressing challenges posed on trade and environment. *Ibid.* at 89. See also Jenks, *supra* note 8 (claiming that "[n]o particular principle or rule can be regarded as of absolute validity. There are a number of principles and rules which must be weighted and reconciled in the light of the circumstances of the particular case") at 407; ILC, Fragmentation Report, *supra* note 27 (arguing that "only a coherent legal system treats legal subjects equally") at 248.

226 ILC, Fragmentation Report, *supra* note 27 at 19.

227 *United States – Standards for Reformulated and Conventional Gasoline* (20 May 1996), WTO Doc WT/DS2/9 (Appellate Body Report), online: WTO <www.wto.org/english/tratop_e/dispu_e/2-9.pdf> at para 17 [*US Gasoline*]; US-Shrimp trade dispute, *supra* note 220; *United States – Import Prohibition of Certain Shrimp and Ship Products* (12 October 1998), WT/DS58/AB/R Appellate Body Report), online: WTO <https://docs.wto.org/dol2fe/Pages/FE_Search/FE_S_S006.aspx?Query=(@Symbol=%20wt/ds58/ab/r*%20not%20rw*)&Language=ENGLISH&Context=FomerScriptedSearch&languageUIChanged=true#>.

228 *Case concerning the Gabčikovo-Nagymaros Project (Hungary v Slovakia)* [1997] ICJ Rep 7, Separate Opinion of Vice-President Weeramantry at 92 [*Gabcíkovo-Nagymaros Project*].

229 *Ibid.* at 90. For Judge Weeramantry, the principle of sustainable development serves to balance the right to development and the protection of the environment. *Ibid.* at 88. See also Rio Declaration, *supra* note 35 (Chapter 1) (stressing that "in order to achieve sustainable development, environmental protection shall constitute an integral part of the development process and cannot be considered in isolation from it") at Principle 4.

Drawing from Judge Weeramantry's views, I consider reconciliation as a complex exercise involving various elements. First, the emphasis is not on normative conflict, but rather how to create synergies.[230] The purpose should not be to establish a hierarchical order in a Kelsian sense.[231] As Carlarne suggests, climate change-related issues "require policymakers to develop linkages between normally compartmentalized systems of law."[232] Interaction between aviation and climate change may also lead to synergies.[233] Because principles will coexist, they should supplement and support each other.

Second, as Dworkin suggests, the principles involve a dimension of "weight and importance."[234] Dworkin writes that "[w]hen principles intersect, one who must resolve the conflict has to take into account the relative weight of each."[235] This involves balancing principles that appear to be in conflict or that are at least in tension. Principles are not absolute. I do not consider that CBDR is more important than non-discrimination or vice versa. Because both principles are equally relevant, they must be reconciled.

Third, this approach also requires a less rigid interpretation of each principle involved. This presupposes that their interpretation may evolve over time. A lack of flexibility may not only perpetuate the risk of conflicting norms, but it may also threaten the objectives of one of the regimes.

Fourth, although principles should be integrated in a manner consistent with the environment in which they will be made operational, they should also tolerate small variations in order to accommodate each other. In other words, while reconciling CBDR with non-discrimination, it is unrealistic to expect a situation where not even a single market distortion will arise.[236] In any event, market distortions do exist in international aviation even in the absence of CBDR. The purpose should then be to minimize such market distortions.

Fifth, because principles are not static, reconciliation also involves a continuous revision of such principles over time, in light of the objectives that the system purports to achieve.

---

230 See van Asselt, *supra* note 151 (suggesting that "different regimes and norms could work to support each other" with the goal of "[achieving] synergies" at 1208).

231 See, e.g., Lindroos, *supra* note 97 at 27.

232 Carlarne, *supra* note 153 at 472.

233 See Oberthür, "Climate Change Regime," *supra* note 210 (Chapter 1) at 68. See also Joyeeta Gupta, "Developing Countries and the Post-Kyoto Regime: Breaking the Tragic Lock-in of Waiting for Each Other's Strategy" in W Th Douma, L Massai, & M Montini, eds, *The Kyoto Protocol and Beyond: Legal and Policy Challenges of Climate Change* (The Hague: TMC Asser Press, 2007) 161 (commenting that treaties within the same field, such as climate change-related instruments, may "sometimes [have] synergetic [but also] conflicting goals" at 163).

234 Dworkin, *supra* note 26 at 27.

235 *Ibid.*

236 It is worth noting that the ICAO guiding principles on the implementation of MBMs suggest that these measures should "minimize" as opposed to avoid "market distortions." This would suggest that States were cognizant that eliminating market distortions completely is simply not possible. Market distortions will always exist to a certain degree. See ICAO, A38-18, *supra* note 21 (Introduction) at Annex (g).

In other words, if the objective is to implement a global MBM to regulate GHG emissions from international aviation and the system incorporates both CBDR and non-discrimination during implementation, such principles should be subject to a process of continuous examination to establish whether further changes are warranted.

Finally, the process of reconciliation must be cognizant of the political context in which discussions take place. Failure to do so may render the entire process futile.

Reconciling CBDR with non-discrimination is an issue central to tackling GHG emissions from international aviation. This should no longer be postponed under the pretext that CBDR is a foreign concept to the international civil aviation regime and that ICAO is not bound by the *Kyoto Protocol*. In addition to disregarding the principle of systemic integration, it fails to achieve the desired results. Arguably, the aviation-specific SCRC has not been much more successful. As illustrated in Chapter 3, one of aviation's biggest challenges in addressing climate change is how to engage a broader membership. In this regard, CBDR may certainly be an enabler (as opposed to a stumbling block) in attracting participation and engaging States that have been passive on aviation and climate change issues.

To reconcile these principles, one must first reformulate them in light of current circumstances and the objective sought to be achieved. The objective is very clear: to develop a scheme to limit or reduce GHG emissions from international aviation. This also sets out the environment in which these principles will coexist. None of the principles should be articulated in a manner that runs afoul of the other regime's rules. Next, the basic elements of both principles are to be identified: (i) CBDR (these are equity, historical responsibility, and the ability to respond to one's own capabilities)[237] and (ii) non-discrimination (the prohibition against treating States and aircraft operators arbitrarily). As a basic notion, aircraft operators flying on the same route should be subject to the same rules irrespective of their nationality.

2.6.4 From Theory to Practice: Some Design Elements to Consider

While designing a scheme to address GHG emissions from international aviation is beyond the scope of this book, I propose some key elements that can be taken into consideration. Although these issues are further explored in Chapter 8, it suffices to say here that the articulation of these principles may take many forms. A number of commentators have suggested that equity can be addressed in a number of ways.[238] For instance, if a system

237 In addressing GHG emissions from international aviation, any scheme may recognize equity and the issue of historical responsibility by providing some degree of deference to developing countries.

238 See Jennifer Morgan, "The Emerging Post-Cancun Climate Regime" in Jutta Brunnée, Meinhard Doelle, & Lavanya Rajamani, eds, *Promoting Compliance in an Evolving Climate Regime* (Cambridge: Cambridge

designed to address GHG emissions generates revenues, such revenues may be rechanneled into developing or least developed countries.[239] In addition, if the system requires the purchase of emission credit units (ECUs) through the implementation of an emissions trading or offsetting scheme, the design of such a scheme could mandate that a significant portion of these units be purchased from climate change-related projects carried out in developing countries, such as occurs under the Clean Development Mechanism (CDM) and Reducing Emissions from Deforestation, Degradation and Forest Enhancement (REDD-plus) schemes. Both of these mechanisms are addressed in more detail in the chapters that follow. The scheme may also include concrete provisions on financial assistance, technical cooperation, and technology transfer.[240]

Incorporating historical emissions and the greater contribution of developed countries requires considering the principle of "incrementalism" along the lines of the Montreal Protocol.[241] This would require the adoption of a route-based/phase-in approach where routes to and from different States will join at different times.[242] A key question is which States will join the scheme and when. Criteria could be developed to phase in States; for instance, flight routes from States could be phased in taking into account a State's level of development or economic growth.[243] These criteria may utilize Gross Domestic Product (GDP), level of aviation activity, or regional market share.[244] The underlying idea is that outgoing and incoming routes from some States may be given a temporary grace period

---

University Press, 2012) 17 at 35; Stavins, "Post-Kyoto Era," *supra* note 68 (discussing that "cost-effectiveness and distributional equity could both be addressed") at 148.

239 See Karim & Alam, *supra* note 24 at 135.

240 See Hardeman, "Reframing Aviation Climate Politics and Policies," *supra* note 44 at 27. See also Gilbert Bankobeza, "Compliance Regime of the Montréal Protocol" in Donald Kaniaru, ed, *The Montréal Protocol: Celebrating 20 Years of Environmental Progress* (London: Cameron May Ltd, 2007) at 75 (stressing the importance of financial assistance and compliance mechanism in the Montréal Protocol, at 83).

241 See Penelope Canan & Nancy Reichman, "Lessons Learned" in Donald Kaniaru, ed, *The Montréal Protocol: Celebrating 20 Years of Environmental Progress* (London: Cameron May Ltd, 2007) 107 at 107; Havel & Sanchez, *supra* note 214 (Chapter 1) at 372.

242 Andreas Tuerk et al, "Emerging Carbon Markets: Experiences, Trends, and Challenges" *Climate Strategies* (1 January 2013) online: Climate Strategies <www.climatestrategies.org/research/our-reports/category/63/370.html> at 1; Harald Winkler, "An Architecture for Long-term Climate Change: North-South Cooperation Based on Equity and Common but Differentiated Responsibilities" in Frank Biermann, Philipp Pattberg, & Fariborz Zelli, eds, *Global Climate Governance Beyond 2012: Architecture, Agency and Adaptation* (Cambridge: Cambridge University Press, 2010) (arguing that "multi-stage becomes a pathway towards a global regime in which developing countries participate in a commitments regime in several stages") at 106; Bodansky, *The Art and Craft of International Environmental Law*, *supra* note 62 (Chapter 1) (suggesting that a system that is developed "in an incremental fashion" may induce participation,) at 183; Havel & Sanchez, Toward a Global Aviation Emissions Agreement, *supra* note 214 (Chapter 1) at 372.

243 See Stavins, "Post-Kyoto Era," *supra* note 68 (analyzing, in the broader climate change regime, the possibility of "targets that become more stringent for individual developing countries as those countries become more wealthy") at 147.

244 See Lyle, "Mitigating International Air Transport Emissions through a Global Measure," *supra* note 122 (Chapter 1).

during which they will remain outside the scheme. Equity could also be achieved by incorporating exemption provisions that exclude those actors with insignificant levels of GHG emissions. These considerations are further elaborated in Chapter 8.

2.7 Conclusion

Fragmentation is a phenomenon of international law characterized by the emergence of multiple international institutions and the adoption of a multiplicity of various international treaties. It presents both advantages and risks. On the one hand, specialized regimes tend to respond to specific sectoral needs. On the other hand, in some cases, these regimes isolate themselves and disregard other norms of international law. This threatens the coherence of international law. Given the enormous amount of existing international instruments, normative conflicts are also likely to occur in the absence of centralized hierarchical rules. In fact, this is not unusual at all in international law. As such, aviation and climate change is but one example where rules from two different regimes are in tension.

The *Kyoto Protocol* sets out ICAO's interaction with the climate change regime. Unfortunately, aside from failing to shed light on the particularities of this relationship, the *Kyoto Protocol* also left a number of unanswered questions. States have exploited this ambiguity in support of their own parochial interests. Given that ICAO considers that the *Kyoto Protocol* does not bind it, it has opted to discard CBDR. Instead, it has developed an alternative principle: SCRC. Because this has not yet been made operational, the SCRC has not led to the desired results either. In order to gather support and to ensure that ICAO's work on climate change continues, the last two ICAO Assemblies have embraced a de minimis provision to exclude the vast majority of developing countries from the scope of application of potential MBMs. Arguably, the de minimis proposals are short-term fixes that do not address the cause of the problem: the integration of principles from different regimes.

In an attempt to solve the CBDR and non-discrimination dispute, I have resorted to the *VCLT* collision rules. Unfortunately, these rules aggravate rather than resolve the problem because they force the interpreter to pick one rule over the other. Even when this approach may be a viable solution to resolve normative conflicts in bilateral instruments, it is much less likely to work in a multilateral setting where the political component and the interests of States are much more complex. The rules on systemic integration of norms seem much more appropriate given that they reinforce the argument that regimes do not reside in isolation.

In this context, the articulation of CBDR and non-discrimination requires a different approach. Taking inspiration from Judge Weeramantry's enlightening conceptualization in *Gabcíkovo-Nagymaros*, it is suggested that the purpose should be to reconcile principles that are in tension instead of choosing between them. This could only be achieved by accepting the proposition that norms are dynamic and not static. One cannot simply adopt the interpretation of CBDR as originally conceived in 1992. The reconciliation of CBDR with non-discrimination is in fact feasible, but it requires innovative thinking. This chapter has identified the design elements that must be implemented in order to achieve this objective, a sine qua non requirement for the success of ICAO's global MBM scheme. The big challenge now is to put them in practice. In Chapter 8, I advance a concrete proposal to reconcile these principles.

3 THE INTERNATIONAL CIVIL AVIATION ORGANIZATION

The success of international aviation has not been without cost. As the sector continues to grow, concerns over its environmental impact have risen in tandem. Although ICAO has discussed environmental protection in connection with aviation since the 1970s, climate change issues are relatively new. Ever since the *Kyoto Protocol* entrusted ICAO to handle GHG emissions from international aviation, the organization has been at the center of the storm. This chapter seeks to explore ICAO's involvement in climate change issues and its strengths and shortcomings, as well as to identify better ways for the organization to handle GHG emissions from international aviation, in particular bearing in mind the recent agreement to develop a global MBM scheme. Understanding ICAO's constraints is central to determining its limitations and establishing realistic corrective actions to facilitate not only its adoption and implementation but, more importantly, participation in ICAO's global MBM.

To this end, the chapter discusses two main topics. Firstly, it analyzes the suitability of ICAO's institutional setting to handle climate change issues. It digs into the organization's aims and objectives as set out in the *Chicago Convention* and the recently adopted strategic objectives. It examines the lack of a specific reference to "environmental protection" in the organization's constitutional framework as well as ICAO's governing structure and considers whether it facilitates or inhibits participation by Member States. As mentioned in Chapter 1, broad participation and Member State engagement are desired to effectively address climate change. The nature of the organization's constituency, including the influence of industry stakeholders in the formulation of environmental policy and the absence of representatives of international civil society, is also an important consideration. I pay close attention to the work of ICAO's Committee on Aviation Environmental Protection (CAEP), the merits and shortcomings thereof, as well as perceived structural deficiencies. CAEP has been at the forefront in the development of environmental regulations applicable to international aviation and will continue to play a leading role in addressing GHG emissions given that it handles most of the technical work.

Secondly, through an examination of all relevant Assembly resolutions and discussions within the Council, this chapter studies ICAO's specific involvement in climate change issues. I focus on five key aspects: (i) the CO_2 standard, (ii) State action plans, (iii) the aspirational goals, (iv) the framework for MBMs, and (v) the global scheme. These topics adequately illustrate the challenges faced and the political implications involved. Similarly, they highlight unresolved issues and various opportunities for improvement that lie ahead.

These will be enormously valuable in designing and bringing into operation the global MBM scheme for international aviation that the 38th Assembly has decided to pursue.

3.1 ICAO's Institutional Setting

Adopted on 7 December 1944 and entered into force on 4 April 1947, the *Chicago Convention* serves as the constitutional framework for international air transport.[1] In addition to setting out basic rules to promote the safe and orderly development aviation,[2] the *Chicago Convention* established ICAO.[3] One hundred and ninety-one States are party to the *Chicago Convention* and are, therefore, members of ICAO.[4] Through the adoption of its standards and recommended practices (SARPs),[5] this specialized UN agency has been instrumental

1 See *Chicago Convention, supra* note 36 (Chapter 2).

2 These rules address issues such as the principle of exclusive sovereignty over a State's airspace, exclusion of military aircraft, prohibition on the use of weapons against civil aviation, right of non-scheduled flights, cabotage, requirement of prior approval of the State concerned for pilotless aircraft, provisions on prohibited areas, customs and immigration, measures to prevent spread of communicable diseases, airport and air navigation charges, nationality of the aircraft, search and rescue, fuel tax exemptions, and adoption of standards and recommended practices. See *Ibid.* at Arts 1-42.

3 *Ibid.* at Art 43.

4 *Ibid.*

5 The *Chicago Convention* entrusted to ICAO a quasi-legislative function to develop standards and recommended practices (SARPs) to regulate an array of aviation-related issues. These have ranged from air navigation procedures to aviation security and to environmental matters. See *Chicago Convention, supra* note 36 (Chapter 2) at Art 37, para 2. Standards are incorporated into Annexes to the *Chicago Convention*. Thus far, ICAO has developed 19 Annexes. Annex 16, for instance, addresses environmental protection, covering aircraft noise and aircraft engine emissions. See ICAO, Annex 16, *Environmental Protection, Volumes I & II - Aircraft Engine Emissions*, 6th edn (2011) [Annex 16]. Technical working groups or panels elaborate the draft provisions of the Annexes and their amendments. They are adopted with the two-third majority vote of the Council. An Annex or any amendment to an Annex becomes effective within three months after the date of its submission to ICAO Member States, unless a majority of ICAO Member States register their disapproval with the Council. To date, this disapproval notice or veto power to impede the enactment of an Annex or an amendment to an Annex has never been exercised, reflecting a "compromise-oriented consultative process which precedes [their] adoption." Thomas Buergenthal, *Law-Making in the International Civil Aviation Organization* (Syracuse: Syracuse University Press, 1969) at 68. The *Chicago Convention* sets out an opt-out provision whereby States can file a difference within sixty days of the standard's adoption. The notification obligation only applies to standards not recommended practices. See *Chicago Convention, supra* note 36 at Art 38. See also Weber, *supra* 59 (Chapter 2) at 35. As far as ICAO is concerned, if a State does not file a difference to a standard, implementation of such standard becomes an international obligation and its observance is therefore mandatory. There are no penalties for States failing to notify such differences. See Dempsey, *supra* note 5 (Chapter 2) at 79. ICAO's position is understandable. Broad implementation of such standards is desirable for the advancement of international civil aviation. In spite of this, a close reading of the Chicago Convention may suggest otherwise. In fact, according to the *Chicago Convention*, Member States have only undertaken to "collaborate in securing the highest practicable degree of uniformity in regulations, standards, and procedures." *Chicago Convention, supra* note 36 (Chapter 2) at Art 37. Such language does not necessarily resonate as being "mandatory." Dempsey, however, opines that standards de facto operate as "hard law." *Ibid.* at 79-80.

in laying down the foundational grounds that has allowed international civil aviation to evolve from perilous beginnings to the safest mode of transportation.

3.1.1 Objectives

The *Chicago Convention* sets out ICAO's main objectives.[6] These may be summarized as follows: (i) develop principles to advance international air transport, (ii) ensure safe and orderly growth, (iii) encourage infrastructure expansion, (iv) avoid discrimination, and (v) play a vigilant role to ensure that the rights of all Member States are respected.[7] These objectives were established in the early days of aviation when the organization's institutional framework served to encourage the growth of the sector and to ensure its stability in the post-World War II era. In anticipation of aviation's potential benefits of enhancing connectivity and encouraging economic prosperity,[8] the drafters of the *Chicago Convention* wanted to guarantee that international air transport would develop in a "safe, regular, efficient, and economical" manner to "meet the needs of the people of the world."[9] Never in their wildest dreams could they have envisioned that the tiny air transport sector would have reached its current dimensions. It is not surprising that the word "environment"[10] is not mentioned once in the *Chicago Convention*; the text pays much more attention to "growth" and "development," which were main concerns of the drafters. As China has noted, "the development of international air transport is [ICAO's] first priority."[11]

This historical context explains why ICAO's institutional framework heavily favors the expansion of air transport.[12] Some commentators therefore see the organization's objectives as "seemingly weighted against the needs of the climate system."[13] Similarly, others claim that ICAO's own legal framework and its institutional setting are not necessarily conducive to address environmental issues such as climate change.[14]

6 See *Chicago Convention*, *supra* note 36 (Chapter 2) at Art 44.

7 *Ibid.*

8 For instance, the *Chicago Convention's* preamble recognizes that air transport's development may "greatly help to create and preserve friendship and understanding among the nations and people of the world." *Ibid.* at Preamble.

9 *Ibid.* at Art 44 (d).

10 Nor does the word environment appear in GATT's original text adopted in the 1940s. See Bruce Neuling, "The Shrimp-Turtle Case: Implications for Article XX of GATT and the Trade and Environment Debate" (1999) 22 Loy LA Int'l & Comp L Rev 1 at 13.

11 ICAO, A37/WP-181 at 2.4. In spite of this emphasis on developing air transport, Heather Miller notes that ICAO has "a history" of tackling environmental issues. See Heather L Miller, "Civil Aircraft Emissions and International Treaty Law" (1997) 63 J Air L Com 697 at 712.

12 See Transport & Environment, "Global Deal or no Deal," *supra* note 72 (Chapter 1) (suggesting that aware of the "sector's environmental impacts" since the 1970s, ICAO has demonstrated "no willingness to constrain growth").

13 Macintosh, *supra* note 100 (Chapter 1) at 411.

14 See Kulovesi, "Addressing Sectoral Emissions," *supra* note 214 (Chapter 1) at 198.

ICAO's structural characteristics and its natural tendency to promote the growth and development of the air transport sector pose major challenges when the organization considers policies oriented at curbing aviation's environmental externalities.[15] Environmental policy at ICAO must be limited to measures that reduce aircraft engine emissions without constraining the growth of air transport in any way.[16] Above all, these are the organization's core aims and objectives.[17]

ICAO's constitutional framework places severe limitations on what the organization can actually accomplish in addressing climate change. There might be times when specific measures ought to be put in place in certain regions or locations seeking precisely to restrict growth in order for international civil aviation to develop in a sustainable manner. Yet these measures would directly conflict with the aims and objectives of ICAO as set out in the *Chicago Convention*. It is on the basis of this contradiction that ICAO's constituents will most likely argue against or simply reject growth-limiting proposals.[18] By way of illustration, some States have argued that emission-related charges are incompatible with the aims of the *Chicago Convention* since they place incremental burdens on the orderly development of air transport.[19]

15 See Graham & Shaw, *supra* note 4 (Introduction) at 1448.

16 The 38th Assembly has "[reiterated] that... emphasis should be on those policy options that will reduce aircraft engine emissions without negatively impacting the growth of air transport especially in developing economies." ICAO, A38-18, *supra* note 21 (Introduction) at para 3 (b).

17 This focus on the organization's aims, and inattention to advancing environmental values, is not unique to ICAO. Commentators have noted that WTO's appellate panels tend to favor free trade over environmental concerns. See Mark Edward Foster, "Making Room for Environmental Trade Measures within the GATT" (1998) 71 S Cal L Rev 393 at 395.

18 Climate change activists from outside aviation circles often suggest that what is really needed to address aviation's CO_2 footprint is to focus on curbing air traffic's demand. However, it is highly improbable that ICAO would undertake initiatives to seriously consider this alternative, at least as an option of first resort. ICAO's balanced approach to aircraft noise exemplifies a superb case study on how the organization proposes various other options before considering placing restrictions on aircraft operations when dealing with another major environmental issue of significant local/regional concern. The balanced approach to aircraft noise examines various measures available to managing noise on an airport-by-airport basis to achieve environmental benefits in the most cost-effective way, while preserving gains from aircraft operations. See ICAO, Doc 9829 AN/451, *Guidance on the Balance Approach to Aircraft Noise Management*, 2nd edn, 2008. These procedures may include, for instance, continuous descent approach (CDA) which decreases aircraft noise by reducing aircraft thrust during initial descent operations. *Ibid.* Only after other measures have been exhausted are operational restrictions, such as enforcing aircraft curfews, considered. See ICAO, "Aircraft Noise Overview" in ICAO *Environmental Report 2007* 20 at 21. Within ICAO, there is sufficient consensus that operational restrictions should not be applied as a measure of first resort to deal with aircraft noise issues. The foregoing case study is highly indicative of ICAO's indisposition to consider operational restrictions as a measure of first resort.

19 See ICAO, CSG-LAEC/1, *Council Special Group on Legal Aspects of Emissions Charges, Montréal, 6-9 September 2005* at 3-1.

Within ICAO, there is also a strong perception that the organization's core business is to deal with issues of air navigation, safety, and security.[20] For instance, while pointing out that ICAO has now devoted too much attention to climate change issues, one State "expressed doubt that [climate change] carried the same weight as safety and security and other tasks."[21]

In an effort to remedy the absence of a clear environmental objective in the *Chicago Convention*, ICAO has, through various proposals developed by the Secretary-General with input from the Council, adopted a number of different strategic objectives addressing environmental issues. For instance, for the first time ever, ICAO conducted a reassessment of its main objectives in 1997.[22] Following this exercise, the Council approved an action plan with strategic objectives addressing issues such as SARPs, strengthening the legal framework, air navigation, and technical cooperation.[23] Although the growing concern over aviation's environmental impact was noted in the action plan, ICAO did not establish a specific objective for environmental protection.[24] In addition to laying down the organization's vision and mission, ICAO developed a new set of strategic objectives in 2005.[25] At this time, "[m]inimizing the adverse effect of global civil aviation on the environment" became one of the aims of the organization.[26] For the 2010-2013 triennium, ICAO adopted three strategic objectives, one of which dealt with "environmental protection and sustainable development of air transport."[27] The objective was to "[f]oster [a] harmonized and economically viable development of international civil aviation, [while] not unduly [harming] the environment." Finally, in 2013, the 38th Assembly endorsed a proposal of the Council to adopt new strategic objectives for the 2013-2016 triennium. While addressing key areas of concern to ICAO such as safety, security, and air navigation, this latest iteration of the objective delinks air transport from environmental protection.[28] In other words, the objective seeks to "[m]inimize the adverse environmental effect of civil aviation activities" while "[fostering] ICAO's leadership in all aviation-related environmental activities."[29]

---

20 See Lionel Alain Dupuis, Discours d'adieu du Représentant permanent du Canada au Conseil de l'OACI, Doyen des Membres du Conseil (29 June 2011) [unpublished, archived on file with the author] (noting that safety and air navigation are ICAO's priorities).

21 ICAO, C-MIN 199/13, *supra* note 184 (Chapter 2) at 15.

22 "Guiding International Civil Aviation into the 21st Century," online: ICAO <www.icao.int/Documents/strategic-objectives/sap1997_en.pdf>.

23 *Ibid.*

24 *Ibid.*

25 "Strategic Objectives of ICAO for 2005-2010," online: ICAO <www.icao.int/Documents/strategic-objectives/strategic_objectives_2005_2010_en.pdf>.

26 *Ibid.*

27 "Strategic Objectives 2011-2012-2013," online: ICAO <www.icao.int/Documents/strategic-objectives/strategic_objectives_2011_2013_en.pdf>.

28 See ICAO Doc 10030, *Budget of the Organization 2014-2015-2016* at 24 [ICAO, *Budget 2014-2016*].

29 "Strategic Objectives 2013-2016," online: ICAO <www.icao.int/about-icao/Pages/Strategic-Objectives.aspx>.

The fact that environmental protection now forms part of ICAO's strategic objectives is very indicative of the relevance of this issue to the organization's activities. There is no doubt that ICAO is much more engaged with environmental issues than it previously was and it has allocated significantly more financial resources to the issue.[30] However, these strategic objectives do not amount to an amendment or a re-writing of the organization's aims as set out in the *Chicago Convention*. In practice, the fact that the *Chicago Convention* does not spell out "environment" as one of ICAO's main objectives does pose a problem. Arguably, in cases of conflict, the development of air transport would trump environmental protection. For instance, one State has said that "ICAO should put the development of international air transport as its first priority bearing in mind the mandate of Article 44 on the aims and objectives of ICAO in the *Chicago Convention*."[31] In spite of the foregoing, it seems very unlikely that, at least in the near future, the *Chicago Convention* will be amended to correct these deficiencies.[32]

---

30 While ICAO's budget for the triennium 2008-2010 only provided for 5.1 million dollars (Canadian) for the strategic objective of environmental protection, this jumped to 8.5 million dollars (Canadian) for the following triennium (2011-2013). See ICAO Doc 9895, *Budget for the Organization 2008-2009-2010* at 11; ICAO Doc 9955, *Budget for the Organization 2011-2012-2013* at 12. For the triennium 2014-2016, the 38th Assembly approved a budget of roughly 14 million dollars (Canadian) for environmental protection activities. See ICAO, *Budget 2014-2016*, *supra* note 28 at 10. While environmental protection still represents less than 2 percent of ICAO's overall budget, safety accounts for roughly 10 percent and air navigation for 9.5 percent. *Ibid.* What is interesting to note, however, is that of all the strategic objectives established for the 2014-2016 triennium, environmental protection showed the largest increase compared to the previous triennium. *Ibid.*

31 See ICAO, A37-WP/181 EX/32, *Addressing Global Climate Change within the Framework of Sustainable Development of International Aviation* (presented by the People's Republic of China) [ICAO, A37-WP/181 EX/32].

32 In 2007, at the 36th Assembly, a group of States led by Canada, India, and the United Kingdom proposed to review the adequacy of the *Chicago Convention*. See ICAO, A36-WP/284 EX/91 Rev 1, *Proposal for a Study of Policy and Programme with Respect to Examining the International Governance of Civil Aviation* (presented by Antigua and Barbuda, the Bahamas, Barbados, Belize, Canada, Dominica, Grenada, Guyana, Haiti, Hungary, India, Jamaica, Pakistan, the Republic of Korea, Saint Kitts and Nevis, Saint Lucia, Saint Vincent and the Grenadines, South Africa, Suriname, Trinidad and Tobago, the United Arab Emirates, and the United Kingdom) [ICAO, A36-WP/284 EX/91 Rev 1]. These States stressed that not all of the strategic objectives that ICAO had agreed on for the period 2005-2010 can "relate back" to the *Chicago Convention*. *Ibid.* at 2. They pointed out that the words "environment" and "security" are nowhere mentioned in the instrument. *Ibid.* While not making a decision on the proposal, the Assembly tasked the Council to review it. In 2009, Council's Working Group on Governance (WGOG) undertook such delicate task. See ICAO, C-WP/13416, *Review of International Governance* (*Chicago Convention*) (presented by the Chair of the Working Group on Governance (Policy) (WGOG)). Unable to achieve sufficient consensus, WGOG simply reported back to the Council the divergence of views of its members. In particular, WGOG noted that, in light of the flexibility provided by the *Chicago Convention*, some States thought that changes were not warranted. *Ibid.* at 5. Others, on the other hand, opined that the instrument required a "progressive adaptation" and that "[s]ecurity and environment [should] also be added to ICAO's competencies." *Ibid.* On 30 October 2009, the Council considered WGOG's report. See ICAO, C-MIN 188/6, *Council, 188th Session, Minutes of the Sixth Meeting* [ICAO, C-MIN 188/6]. Mexico, for instance, underscored that "the [Chicago] Convention had enough flexibility to be able to govern all situations inherent to international civil aviation." *Ibid.* at 70. Mexico stressed that "Article 44 (Objectives)[of the *Chicago Convention*] was sufficiently broad in scope to allow the Council to deal with any new or emerging threats to international civil aviation and any new

3.1.2 Governing Structure

ICAO consists primarily of an Assembly, a Council, a Secretariat, and other technical bodies that report to the Council such as the Air Navigation Commission (ANC).[33] This project focuses on the Assembly and the Council.

As the supreme body of the organization, the Assembly is responsible for (i) electing States to the Council, (ii) approving the budget, (iii) considering proposals to amend the *Chicago Convention*, (iv) delegating prerogatives to the Council, and (v) considering reports from the Council.[34] Initially, the *Chicago Convention* stipulated annual Assembly meetings.[35] However, on 14 June 1956, the Assembly adopted a protocol to amend the *Chicago Convention* and changed the annual frequency of its meetings to no less than once every three years (in practice this has meant a meeting every three years).[36]

---

challenges." *Ibid.* Similarly, Nigeria suggested that the "*Chicago Convention* had been sufficiently flexible to deal with developments in the aviation sector... that a new Annex 16 had been adopted to cover environmental issues and that various international air law instruments had been adopted to address aviation security issues." *Ibid.* at 71. On the other hand, the United Kingdom, supporting the need for change, pointed that "the only reason that the *Chicago Convention* worked was because half of it was ignored and the other half was dodged." *Ibid.* at 70. Given that the vast majority of States were skeptical about the need to amend the *Chicago Convention*, the Council opted for not pursuing the issue any further. See ICAO, C-DEC 188/6, *Council, 188th Session, Decision of the Sixth Meeting* at 3; ICAO, C-MIN 188/6, *Council, 188th Session, Minutes of the Sixth Meeting*. Similarly, renowned scholars contend that "ICAO has an usual capacity for adapting its constitutive instrument (the *Chicago Convention*) to current demands" making unnecessary any amendment to its constitutional framework. Buergenthal, *supra* note 5 at 228. Discussions on whether the *Chicago Convention* should be amended are extremely interesting. Two aspects are germane for the purpose of this book. The first of these aspects deals with the belief supported by those against change that an amendment is not necessary because the Convention itself provides sufficient flexibility. For instance, since the 1970s, ICAO has been dealing with aviation security and environmental issues. Yet these are not part of the *Chicago Convention*. The argument is true to some extent. However, it chooses to ignore a major problem. Although the Chicago Convention has been extremely flexible, areas not spelt out in Art 44 as part of the organization's objectives do not enjoy the same hierarchical preeminence. Given the post 9/11 environment, this does not so much affect aviation security. Endogenous and exogenous pressures make aviation security an overarching concern for the organization. This however is not the case for environmental issues. Those not so keen in embracing environmental initiatives often point out that ICAO's aims should be those of the Chicago Convention, nothing more. The environment is notoriously absent. The second relevant aspect for this book is who were in favor and who were against the proposal to amend the Chicago Convention. Those in favor included mostly European States and Canada. On the other hand, Latin American, African, Asia, and Middle Eastern States, as well as the United States, were against it. It is no surprise that, for decades, European States have been strong advocates of environmental issues at ICAO.

33 See *Chicago Convention*, *supra* note 36 (Chapter 2) at Arts 48, 50, 56, 58, and 59.

34 *Ibid.* at Art 49.

35 *Ibid.* at Art 48 (original).

36 See *Protocol Relating to the Amendments of Articles 48(a), 49(e), and 61 of the Convention on International Civil Aviation, 12 December 1956*, ICAO DOC 7300. The Protocol entered into force on 12 December 1956 and thus far has attained 143 ratifications. This means that, at least in theory, for the forty-eight Member States that have yet to accede to this instrument, the Assembly should still hold annual meetings. See ICAO, *Protocol Relating to Certain Amendments to the Convention on International Civil Aviation – Articles 48(a), 49(e), and 61, 14 June 1954.*

This amendment marks one of the most unnoticed but profound structural changes that ICAO has ever undertaken. Less frequent sessions have diminished the Assembly's role[37] while "strengthening the position and influence of [the Council]."[38] Milde claims that, with this change, "the powers of the Assembly have eroded."[39] The change has progressively dissuaded active participation of Member States other than those fortunate enough to have Council representation. Although Member States do attend ICAO's triennial Assembly sessions, they do so in a rather passive manner. The tremendous backlog of massive working papers makes it impossible for delegates to really grasp what the Assembly is supposed to consider in a short two-week session. In addition, the more infrequent these sessions have become, the farther Member States have gotten from the day-to-day activities of the organization. Commentators also point out that this structural change in ICAO has led to the "majority of States simply [not contributing] to the advancement of the international aviation."[40] In 2013, the 38th Assembly considered a proposal to hold its meetings once every two years. The Council reported that such a proposal would represent an additional cost of US$ 2.4 million over a period of six years. To no one's surprise, the Assembly quickly discarded the proposal.[41]

Initially, the Council, the organization's permanent political body, consisted of 21 States.[42] At present, however, the Council has 36 States.[43] As one environmental non-governmental organization puts it, the Council "has been carefully nurtured over the years to call all the shots."[44]

The Council's functions may roughly be summarized as follows: (i) executing the mandates given by the Assembly,[45] (ii) managing the organization's finances, (iii) appointing the Secretary-General, (iv) adopting standards and recommended practices (SARPs), and (v) reporting infractions to the convention.[46] The fact that extremely important functions such as the adoption of SARPs (ICAO's uncontested core activity)

37 See Michael Milde, "*Chicago Convention* - 50 Years Later: Are Major Amendments Necessary or 'Desirable'?" (1994) XIX: 1 Ann Air & Space L 401 at 429.

38 See Peter Ateh-Afac Fossungu, A Critique of the Powers and Duties of the Assembly of the International Civil Aviation Organization (LL M Thesis, McGill University Institute of Air & Space Law, 1996) [unpublished] at 21.

39 Milde, *International Law and ICAO*, *supra* note 59 (Chapter 2) at 130.

40 Peter Ateh-Afac Fossungu, "999 University, Please Help the Third World (Africa) Help Itself: A Critique of Council Elections" (1999) 64:2 J Air L & Com 339.

41 See ICAO, A38-WP/18 EX 1, *Declaration on Aviation Security and the ICAO Comprehensive Aviation Security Strategy (ICASS)* (presented by the Council of ICAO).

42 Membership to the Council increased from 21 States to 27, 30, 33, and finally to 36 States.

43 See *Chicago Convention*, *supra* note 36 (Chapter 2) at Art 50. See Milde, *International Law and ICAO*, *supra* note 59 (Chapter 2) at 315.

44 See Transport & Environment, "Global Deal or no Deal," *supra* note 72 (Chapter 1).

45 What is perhaps most interesting about the tasks that the Assembly entrusts the Council to carry out is that most of them stem from proposals that the Council itself tables to the Assembly.

46 See *Chicago Convention*, *supra* note 36 (Chapter 2) at Art 54.

and the appointment of the Secretary-General are vested in the Council is very indicative of its influential role. As Milde points out, it is the Council which is without a doubt "the real focus of the ICAO decision-making" process.[47] Some have rightfully referred to the Council as a genuine "executive committee."[48]

The foregoing may explain why States continue to suggest that Council membership should be increased. In 2007, at the 36th Assembly, 15 Arab States presented a proposal to increase Council membership from 36 to 39 States.[49] These States argued that the "increasing importance of civil aviation in the Arab region, particularly in the Gulf area, [justified] a higher weight of participation in supervisory bodies in the field of international civil aviation."[50] They also contended that "many regions of the world are over-represented compared to the 22 Arab States, such as the Northern European States as well as Central and Latin America."[51] Three years later, in 2010, at the 37th Assembly, Saudi Arabia tabled a similar proposal.[52] At the time, the Assembly referred the matter to the Council for consideration. Although noting "the need and growing desire by many States to be represented on the Council," the Council observed that "increasing membership may not be the most appropriate means to address this matter."[53] The Council therefore did not endorse an increase of its membership to 39.[54] In 2013, the 38th Assembly followed the Council's recommendation, and the proposal was abandoned.[55]

Although both of these proposals ultimately failed, they echo a growing feeling within Member States that in order to participate in the organization's activities, one must be a member of the Council. If not, their contributions to the organization become much less significant. Some States with significant aviation activity are currently not members to the Council. These include, among others, Indonesia, Thailand, Qatar, Turkey, Ireland, Lux-

---

47 See Milde, *International Law and ICAO*, *supra* note 59 (Chapter 2) at 130.

48 IH Ph Diederiks-Verschoor, *An Introduction to Ari Law*, 8th edn (The Netherlands: Kluwer Law International, 2006) at 45.

49 See ICAO, A36-WP/258 EX 86, *Increasing ICAO Council Membership to a Minimum of 39 Seats* (presented by the Arab Civil Aviation Commission (ACAC)).

50 *Ibid.* at 3.

51 *Ibid.*

52 See "Assembly 37th Session – Working Papers by numbers," online: ICAO <www.icao.int/cgi/a37.pl?wp;LE>.

53 ICAO, A38-WP/17 EX 12, *Proposal to Amend Article 50 a) of the Convention on International Civil Aviation so as to Increase the Membership of the Council to 39* (presented by the Council of ICAO).

54 *Ibid.*

55 In 2013, the State of Qatar shocked the world of civil aviation when it formally launched a bid to host ICAO's headquarters in an attempt to gain greater participation. See ICAO, PRES RK/2166. In other words, Qatar sought to move the ICAO Secretariat from Montréal to Doha. *Ibid.* The proposal itself involved a generous offer to provide state-of-the art headquarter facilities in Doha for the organization, as well as privileges and immunities for the Secretariat and foreign delegations substantially better than those presently provided by Canada. *Ibid.* What the proposal did not mention is that Qatar, among other things, also sought to obtain a permanent seat in the Council, given that, by conventional practice, the host State has always maintained de facto membership in the Council.

embourg, Ethiopia, Taiwan, Finland, the Philippines, Israel, and Colombia.[56] The existing structure poses significant barriers to participation and engagement from Member States, a sine qua non for addressing aviation and climate change issues. The more disengaged and the less informed the majority of Member States are on this issue, the stronger their inclination to reject collaborative action and block global proposals. Participation is a key element in the development of a global MBM scheme to address GHG emissions from international aviation.[57]

3.1.3 *Constituency*

Most attendees to ICAO events are mostly delegates from civil aviation authorities and ministries of transport of Member States. In fact, it is rare to find someone with a background other than aviation at an Assembly. Council composition is not much different. Permanent representatives to the Council are either civil servants from civil aviation authorities or career diplomats attached to their ministries of foreign affairs. Likewise, with very few exceptions, most staff members of the ICAO Secretariat possess a technical aviation background. It is unlikely that a predominantly aviation-oriented constituency will weigh in heavily on environmental considerations. In this context, it should not be a surprise that the ICAO constituency has the tendency to examine climate change issues from an aviation perspective and not vice versa.[58]

3.1.4 *ICAO's Committee on Aviation Environmental Protection (CAEP)*

On 5 December 1983, the Council merged the Committee on Aircraft Noise (CAN) and the Committee on Aircraft Engine Emissions (CAEE) and thereby established the Committee on International Aviation Environmental Protection (CAEP), a body of interdisci-

56 According to ICAO statistics, these countries rank in the top 36 in terms of international aviation activity, albeit not Council members. See "Civil Aviation: 2012 International RTK by State of Air Operator Certificate (AOC)," online: ICAO <www.icao.int/Meetings/a38/Documents/International%20Scheduled%20RTK%20(Annual%20Report).PDF>.

57 Milde argues that if ICAO is really serious about reaching out to a broader audience of Member States, the right step is not to increase the number of Council seats, but rather to enhance the role of the Assembly, first by re-establishing its annual meetings and second by making the Council a non-permanent body of the organization. Milde also notes that, at present, the Council holds three sessions (winter, spring, and fall) for a total of no more than 12 weeks during the whole course of the year. For Milde, the Council imposes an unnecessarily high administrative load on the Secretariat, which could be substantially decreased, should Council only meet for one or two weeks every year. See Milde, *International Air Law and ICAO*, *supra* note 59 (Chapter 2) at 207. This however is very unlikely to take place in the near future.

58 As one environmental NGO says, "[t]ransport ministries predominate at ICAO. They are not specialists on climate change questions. Rather, they see their role most often as the guardians of aviation's future." Transport & Environment, "Global Deal or no Deal," *supra* note 72 (Chapter 1).

plinary experts to formulate recommendations on issues involving technical, economic, social, and policy aspects of aviation and the environment.[59] CAEP is first and foremost a committee of the Council. The Council approves its meetings, terms of reference, agenda, and work program. The CAEP work program is carried out over a cycle of three years during which various steering and working groups meet regularly. At the end of the cycle, a two-week meeting is convened in Montréal to prepare a final report with proposed recommendations for consideration and approval by the Council.[60]

CAEP has been instrumental in ICAO's work on environmental issues. Its initial focus was geared toward standard setting on aircraft engine emissions, air quality, technical aspects, cost-effectiveness, and vapor displacement from fuel tanks.[61] Later, just before its second meeting, the Council refined CAEP's terms of reference to specifically tackle issues relating to the control of aircraft noise and gaseous emissions from aircraft engines.[62] It was only after the conclusion of the 1992 *United Nations Conference on Environment and Development* that the 31st Assembly decided to entrust CAEP with a very broad mandate to expand its work plan to include climate change issues associated with aviation and to work closely with other organizations such as the *UNFCCC* and the IPCC.[63]

CAEP recommendations are adopted following a fourfold test that includes technical feasibility, environmental effectiveness, economic reasonableness, and the interdependencies of measures.[64] Under technical feasibility, CAEP seeks to ensure that a recommended environmental regulation is viable from a technical perspective in order not to jeopardize aviation safety. This part of the test has been criticized as favoring technical conservatism where market forces – often summarized as the views of the industry – and not ICAO regulations determine what is technically feasible.[65] Under environmental effectiveness, CAEP seeks to ensure that the proposal will in fact produce concrete environmental benefits. In considering the economic reasonableness of recommendations, CAEP looks into

59 See ICAO, C-WP/13520, *Membership in the Committee on Aviation Environmental Protection (CAEP)* (presented by the Secretary-General) at 1.1 and 1.2 [ICAO, C-WP/13520].

60 Acting as a genuine consultative expert group subordinate to the Council, CAEP does not have any authority to introduce changes to existing environmental standards or recommended practices, for it is only allowed to make technical recommendations for the Council's consideration.

61 See ICAO, CAEP/1-WP/97, *Committee on Aviation Environmental Protection (CAEP) First Meeting* at 4-1 [ICAO, CAEP/1-WP/97].

62 See ICAO, CAEP/2-WP/1, *Committee on Aviation Environmental Protection (CAEP) Second Meeting* [ICAO, CAEP/2-WP/1].

63 At present, experts from the following states are members to CAEP: Argentina, Australia, Brazil, Canada, China, Egypt, France, Germany, India, Italy, Japan, the Netherlands, Poland, Russian Federation, Singapore, Spain, Sweden, Switzerland, South Africa, Tunisia, Ukraine, the United Kingdom, and the United States. The following States serve as observers: Greece, Indonesia, New Zealand, Norway, Turkey, and United Arab Emirates. See "CAEP – Members and Observers," online: ICAO <www.icao.int/ENVIRONMENTAL-PROTECTION/Pages/CAEP.aspx>.

64 CAEP Terms of Reference," online: ICAO <www.icao.int/environmental-protection/Documents/CAEP/Images/CAEPToR.jpg>.

65 See Elizabeth Duthie, "ICAO Regulation: Meeting Environmental Need?" (2001) 3:3/4 Air & Sp Europe 27.

the most cost-effective ways to carry them out. There have been concerns that economic reasonableness considerations include complex cost-benefit analyses that do not account for the cost of internalizing externalities that aviation creates for the environment. In other words, the test does not consider the application of the polluter pays principle through the imposition of an environmental externality on airlines and consumers to be economically reasonable.[66] In connection with interdependencies of measures, CAEP assesses the link between the different components of environmental protection to avoid situations in which, for instance, a CO_2 reduction measure may generate an increase in NO_X emissions.[67] Yet perhaps the most serious criticism that CAEP's four-pronged test has received relates to the fact that it fails to recognize the cost to society of aviation's development.[68]

Another criticism that CAEP has attracted relates to its imbalanced membership. At CAEP/1, the original roster included experts from thirteen ICAO Member States, of which seven were European States.[69] Later, at CAEP/3 membership was increased to 15 States with two new European additions.[70] At CAEP/5 held in 2001, nearly eighteen years after its formation, membership was expanded to eighteen ICAO Member States in order to include experts from Singapore, Egypt, and South Africa.[71] For the first time, regional diversity became a relevant factor in determining CAEP's membership. An expert from China, the fastest-growing domestic aviation market in the world, was only admitted in 2007 during the CAEP/7 cycle. Later, experts from Nigeria and Ukraine became members in 2009.[72] At present, CAEP is composed of experts from twenty-three ICAO Member States with an additional 6 experts having observer status.[73] European countries account for roughly 41 percent of CAEP's membership including observer States, whereas developing

66 *Ibid.* at 28.

67 For instance, Lee & Sausen note that any initiative to reduce aviation's CO_2 emissions must also factor the effect that such measure may have on radiative forcing, for it may produce an increase of the latter. See DS Lee & R Sausen, "New Directions: Assessing the Real Impact of CO_2 Emissions Trading by the Aviation Industry" (2000) 34 Atmospheric Environment 5337 at 5338.

68 *Ibid.*

69 CAEP's original members were Australia, Brazil, Canada, Denmark, France, Germany, Italy, Japan, the Netherlands, Sweden, USSR, the United Kingdom, and the United States. See ICAO CAEP/1-WP/97, *supra* note 61.

70 Although Denmark dropped out from CAEP at its third meeting, Poland, Spain, and Switzerland joined. The Council also decided to add Norway as an observer State. This produced a net get gain for Europe of two additional Member States and one observer State. At the time, no other country from any other of ICAO's geographical regions became a CAEP member. See ICAO, CAEP/3, *Committee on Aviation Environmental Protection (CAEP) Third Meeting.*

71 See ICAO, CAEP/5-WP/86, *Committee on Aviation Environmental Protection (CAEP) Fifth Meeting* [ICAO, CAEP/5-WP/86].

72 See ICAO, CAEP/8-WP/80, *Committee on Aviation Environmental Protection (CAEP) Eighth Meeting, Montréal, 1 to 12 February 2010 Agenda Item 1* at 1 [ICAO, CAEP/8-WP/80].

73 See "CAEP Members and Observers," online: ICAO <www.icao.int/ENVIRONMENTAL-PROTECTION/Pages/CAEP.aspx>.

countries only account for 34.4 percent.[74] While six industry trade associations provide experts to CAEP as observers, only one environmental NGO is allowed to participate.[75]

CAEP's notorious imbalance in favor of a heavily represented European constituency may be attributed to the fact that at the time of its origin the predominant issue in its work program referred to aircraft noise, a subject about which the European countries were profoundly concerned. Progressive European liberalization policies prompted rapid expansion of air transport, causing severe issues of airport congestion. Countless local areas neighboring airport premises were affected, provoking frequent public outcry. At the time, aircraft noise became the major aviation environmental issue in Europe. It was therefore the European push which forced the Council to include noise-related topics in CAEP's agenda. However, in regions where aircraft noise was not the subject of much attention, the environmental issues surrounding aviation seemed mostly remote. Countries from these regions did not see an acute need to participate in CAEP; neither did ICAO encourage them to join. In ICAO's defense, it is also true that these countries did almost nothing to engage in CAEP's discussions. It is not surprising that, during the early days of CAEP, Brazil, an aircraft manufacturing country, was the only developing country to take part in its proceedings. Japan was the only representative from Asia, and there were no members from Africa. This historical imbalance also helps to explain why aviation's environmental problems, including climate change, are to some extent perceived as a European phenomenon, from which developing States distance themselves to the greatest extent possible.

CAEP will continue to play a leading role in the formulation of ICAO standards and guidance material, as well as in providing policy advice on delicate issues. CAEP's work will be instrumental in the design of the global MBM scheme for international civil aviation. In fact, in November 2013, the Council tasked CAEP to develop sustainability criteria for offsets of carbon credits eligible for any such scheme.[76] As mentioned in Chapter 2, establishing the sustainability criteria for these credits may be one of the ways to reconcile ICAO's non-discrimination principle with the *UNFCCC*'s CBDR principle. The Council also entrusted CAEP with the mandate to work on monitoring, reporting, and verification (MRV) standards.[77] Because of the significant role CAEP is likely to have in the development of these standards, as well as other tasks that the Council may later assign, the imbalance in CAEP's membership and the participation of its experts should be revisited.

74 *Ibid.* On the other hand, experts from developed States account for 65.5 percent of CAEP's membership.

75 *Ibid.*

76 See ICAO, C-DEC 200/4, *Council, 200th Session, Decision of the Fourth Meeting* [ICAO, C-DEC 200/4]; ICAO, MIN-200/4, *Council, 200th Session, Minutes of the Fourth Meeting*; ICAO, *Committee on Aviation Environmental Protection – CAEP – Informal Briefing to the Council* (31 January 2014) [unpublished, archived on file with the author].

77 *Ibid.*

3.1.5 *Industry Participation*

Nobody can question the value of industry involvement in an international organization's rule-making process.[78] At ICAO, industry stakeholders play a significant role. Yet, outsiders have already pointed out that the organization echoes the interests of industry stakeholders such as air carriers far too strongly.[79] Some years ago, this was subject to noteworthy scrutiny particularly in the case of ICAO's Legal Committee. In June 2009, the Working Group on Governance and Policy (WGOG) recommended an amendment to the committee's rules of procedure with regard to the participation of (industry) observers.[80] The WGOG report argued that, in pursuit of consistency with the procedures of other United Nations specialized agencies, a clear differentiation was needed between Member States and observers, since "[observers] have neither the same responsibilities nor the same duties [as Member States]."[81] Although the Legal Committee did not agree with WGOG's recommendation to change its rules of procedure to restrict participation of industry stakeholders, the proposal serves to illustrate the discontent of some States.[82]

Industry observers bring invaluable technical expertise and advice to ICAO. Their participation helps to ensure that rules are drafted in a manner that recognizes the practicalities of the market and the realities of a sector that is technical by definition. International rule making should not adopt the form of a unilateral set of commands which fail to take into account the views of those who are the subject of intended regulations. As Fuller notes, this should be an interactive process where all parties involved actively participate as opposed to "only a unidirectional exercise of authority."[83] It is precisely this interaction between lawgivers (ICAO as the regulating agency) and law subjects (industry stakeholders as their ultimate recipients) that "gives meaning to this law[-making]" process.[84]

Having said this, it is also uncontested that this interaction cannot be to the detriment of the participation of Member States, ICAO being first and foremost an organization of sovereign States. Industry's overwhelming involvement in CAEP, for instance, exemplifies

78 See Canan & Reichman, *supra* note 241 (Chapter 2) (highlighting that strong industry involvement was important for the successful adoption of the Montréal Protocol on Ozone Depleting Substance,) at 114; Paul S Horwitz, "Harnessing the Power of the Montréal Protocol to Deliver Even More Climate Change Benefits" in Donald Kaniaru, ed, *The Montréal Protocol: Celebrating 20 Years of Environmental Progress* (London: Cameron May Ltd, 2007) at 187 (suggesting that "the climate process may benefit from an explicit understanding of the importance of engaging industry in the process" at 198).

79 See Richard Janda, "Passing the Torch: Why ICAO Should Leave Economic Regulation of International Air Transport to the WTO" 21:1 Ann Air & Sp (1995) 409 at 413.

80 See ICAO, C-WP/13399, *Legal Committee: Participation of Observers and Election of Officers* (presented by the Chair of the Working Group on Governance (Policy) (WGOG)).

81 *Ibid.*

82 See ICAO, Doc 9926-LC/194, *Legal Committee – Report of the 34th Session, Montréal 9-17 September 2009* (2009) at 3.1.

83 Lon L Fuller, *The Morality of Law* (New Haven: Yale University Press, 1964) at 223.

84 *Ibid.* at 195.

what ought not to be the standard.[85] Should this trend continue, it will only perpetuate a practice that thus far has only disengaged State participation.

Complex issues such as aviation and climate change may only accentuate this imbalance where the asymmetry of information gives a significant advantage to industry. Perhaps there is the need to mark the boundaries within which industry stakeholders may participate at ICAO, but restricting industry participation will not resolve the dilemma. Without industry input, ICAO will quickly lose touch with the practical needs and realities of the market on a number of regulatory issues including climate change.

3.1.6 NGO Participation

It is interesting to compare the role of industry stakeholders with that of organizations representing civil society at large. Arguably, industry enjoys greater access to the ICAO process than environmental NGOs. It is not unusual that industry is invited to participate in ICAO's discussion, whereas NGOs are not. For instance, when the 38th Assembly agreed to develop a global MBM for international aviation, it tasked the Council to "finalize work on the technical aspects…of the possible options for a global MBM scheme…taking into account…the proposal of the aviation industry."[86] There is no reference to proposals from those sectors representing civil society.

NGOs bring a completely different perspective on climate change issues which may be worth considering. Perhaps this surreptitious bias has been due to the fact that, historically, the ties between ICAO and some industry trade organizations have been very strong. For instance, one of IATA's own objectives is precisely to "cooperate with [ICAO]."[87] The fact that IATA is headquartered in Montréal is because of ICAO. The Airport Council International (ACI), the international trade association of the world's airports, has also relocated its headquarters to Montréal. The industry-ICAO symbiosis may also be explained from the perspective that as a specialized institution, ICAO deals with technical issues on which input from industry stakeholder is usually extremely important.[88]

---

85 Since its early beginnings, CAEP has been characterized by an overwhelming industry participation, which ranged from 33 percent of the total number of delegates at CAEP/4 to 65 and 63 percent during CAEP/7 and CAEP/8, respectively. See ICAO Doc 9886, *Committee on Aviation Environmental Protection, Report of the Seventh Meeting, Montréal 5-16 February 2007* (2007); ICAO Doc 9938, *Committee on Aviation Environmental Protection, Report of the Eighth Meeting, Montréal 1-12 February 2010* (2010).

86 ICAO, A38-18, *supra* note 21 (Introduction) at para 19 (a).

87 *Articles of Association*, online: IATA <www.iata.org/about/Documents/articles-of-association.pdf> Art 4 (3).

88 Lyle, "Mitigating International Air Transport Emissions through a Global Measure," *supra* note 122 (Chapter 1) (stating that "ICAO has also not done itself or the process any favors in the last few years by a lack of transparency, mostly working behind closed doors and excluding substantive dialogue with its main actuators – tourism, trade and finance – or with increasingly concerned and active NGOs").

Yet, this should not be translated into a perpetual advantage. Climate change is a complex issue characterized by different interests. It should be in the best interest of the whole process that all parties, including those representing civil society at large, are given a fair opportunity to participate. Restricting access to these organizations or not providing them with a level playing field may reflect an unintentional bias which may be detrimental to the rule-making process.

3.2 ICAO and Climate Change

ICAO has an impressive track record when it comes to the production of guidance materials directly or indirectly relating to aviation and climate change issues.[89] Yet, the main criticism against the work of the organization is that GHG emissions from international aviation continue to grow in spite all its efforts. To date, there is no system in place to reduce or limit these emissions. This section describes the evolving challenges that ICAO has faced and focuses on key elements that are expected to play a significant role in the organization's future work on GHG emissions.

3.2.1 Historical Background

In 1998, less than a year after the adoption of the *Kyoto Protocol*, the 32nd Assembly tasked the Council to explore "policy options to limit or reduce greenhouse gas emissions from civil aviation, taking into account the IPCC special report and the requirements of the *Kyoto Protocol*."[90] CAEP then established a working group to look into policy options such as charges, fuel taxes, offsetting schemes, and emissions trading.[91] Three years later, the

89 See ICAO Doc 9949, *Scoping Study of Issues Related to 'Linking' Open Emissions Trading Systems Involving International Aviation* 1st edn (2011); ICAO Doc 9931 AN/476, *Continuous Descent Operations (CDO) Manual* 1st edn (2010); ICAO Doc 9574 AN/934, *Manual on a 300 m (1000 ft) Vertical Separation Minimum between FL 290 and FL 410 Inclusive* 3rd edn (2012); ICAO Doc 9885, *Guidance on the Use of Emissions Trading for Aviation* 1st edn (2008); ICAO Doc. 9977 AN/489, *Manual on Civil Aviation Jet Fuel Supply* 1st edn (2012); ICAO Doc 9988, *Guidance on the Development of States' Action Plans on CO_2 Emissions Reductions Activities* 1st edn (2013); ICAO Doc. 9501 AN/929, *Environmental Technical Manual* 1st edn (2010); ICAO Doc 10018, *Report of the Assessment of Market-based Measures* 1st edn (2013) [ICAO, Global MBM Assessment]; ICAO Doc 9997 AN/498, *Performance-based Navigation (PBN) Operational Approval Manual* 1st edn (2013); ICAO Doc 9993 AN/495, *Continuous Climb Operations (CCO) Manual* 1st edn (2013); ICAO Doc 9992 AN/494, *Manual on the Use of Performance-based Navigation (PBN) in Airspace Design* 1st edn (2013); ICAO Circular 134-AN/94, *Control of Aircraft Engine Emission* 1st edn (1977); ICAO Circular 334-AN/184, *Guidelines for the Implementation of Lateral Separation Minima*; ICAO Circular 303 AN/176, *Operational Opportunities to Minimize Fuel Use and Reduce Emissions* (2003).

90 ICAO, A32-8, *supra* note 215 (Chapter 1) at Appendix F, para 4.

91 See ICAO, MBM Activities Chronology [unpublished, archived on file with the author] [ICAO, MBM Chronology].

33rd Assembly noted CAEP's recommendation that an "open emissions-trading system"[92] was a "cost effective measure to limit or reduce" GHG emissions from international civil aviation.[93] Due to the lack of political will, the Assembly did not adopt such a system. Instead, the Assembly encouraged States to address the aviation sector's carbon footprint through the adoption of voluntary measures.[94] It also tasked the Council to "develop guidance for States on the application of [MBMs]."[95] For the next 3 years, CAEP conducted a number of studies on several subjects, including the impact of environmental levies on developing countries, emissions trading, and the possibility of implementing voluntary agreements.[96] In 2004, the 35th Assembly, "[endorsed] the further development of an open emissions trading system."[97] It also tasked the Council to continue carrying out feasibility studies.[98] At that time, the fact that the Assembly was unable to implement concrete measures to address the sector's GHG emissions but rather continue in a "studying mode" sine die was perceived as a de facto rejection of emissions trading.[99] CAEP nevertheless developed guidance on emissions trading.[100] ICAO also formed a group of legal experts to examine the legality of emission-related levies and GHG charges.[101]

The next session of the Assembly took place in September 2007,[102] more than two years after of the European Union announced its intention to rope foreign aircraft operators into its emissions trading scheme (EU ETS). Although this issue is examined in detail in Chapters 4 through 6, it suffices to mention here that this announcement ignited fierce international opposition never seen in the context of international civil aviation. To no one's surprise, the deliberations at the 36th Assembly were clearly aimed at sending a strong message to Europe that the international community will not tolerate the inclusion of foreign aircraft operators in such a unilateral regime. The Assembly stated categorically that a State should not include foreign aircraft operators into its emissions trading scheme unless it has previously sought and obtained consent from home State(s) of the foreign

92 ICAO defines an open emissions trading as the "system where allowances can be traded in and outside the given scheme or sector." ICAO Doc 9885, *Guidance on the Use of Emissions Trading for Aviation* at xiv [ICAO, Doc 9885]. In other words, the aviation sector would be permitted to purchase allowances from other industry sectors.

93 ICAO, A33-7, *supra* note 13 (Chapter 2) at Appendix I, Preamble.

94 *Ibid.* at Appendix I, para 2 (1).

95 *Ibid.* at Appendix I, para 1.

96 See ICAO, MBM Chronology, *supra* note 91.

97 ICAO, A35-5, *supra* note 17 (Chapter 2) at Appendix I, para 4 (c) (1).

98 *Ibid.* at Appendix I, para 4 (c) (2).

99 See Transport & Environment, "Global Deal or no Deal," *supra* note 72 (Chapter 1) (suggesting that the 35th Assembly turned down emissions trading as a policy option).

100 See ICAO Doc 9885, *supra* note 92; ICAO, *Council Special Group on Legal Aspects of Emissions Charges*, Montréal, 6-9 September 2005 [ICAO, CSG-LAEC/1].

101 See ICAO, MBM Chronology, *supra* note 91.

102 See *Assembly 36th Session*, online: ICAO <www.icao.int/Meetings/AMC/36th/Pages/default.aspx>.

aircraft operators.[103] It is an understatement to say that the issue of mutual consent dominated the Assembly's discussions on climate change. Moreover, in light of the significant disagreements and the failure to adopt a system to regulate GHG emissions, the Assembly tasked the Council to establish a high-level Group on International Aviation and Climate Change (GIACC) to provide policy recommendations on what the Assembly described as "an aggressive Program of Action."[104]

After numerous meetings, GIACC provided three key recommendations, namely, the need to develop (i) a CO_2 aircraft emission standard, (ii) a framework for MBMs, and (iii) State action plans.[105] In October 2009, ICAO convened a high-level meeting on International Aviation and Climate Change (HLM-ENV/09).[106] Building upon GIACC's work, HLM-ENV/09 adopted a non-binding declaration suggesting that as part of ICAO's action plan on climate change, States should embrace an aspirational, non-attributable goal "to achieve a global annual average fuel efficiency improvement of 2 percent over the medium term until 2020."[107] HLM-ENV/09 also recommended that a framework for MBMs should be adopted and that States should develop action plans.[108] The following year, in 2010, signaling progress in three specific areas, the 37th Assembly instructed the Council to (i) develop a CO_2 standard for aircraft emissions,[109] (ii) develop a framework for MBMs,[110] and (iii)

103 See ICAO, A36-22, *supra* note 17 (Chapter 2), Appendix L, para 1 (b) (1).

104 *Ibid.* at Appendix K, para 2 (a).

105 See ICAO, *Group on International Aviation and Climate Change (GIACC) Report* (1 June 2009) at 5-6 [GIACC Report].

106 See ICAO Doc 9929, *Report of the High-Level Meeting on International Aviation and Climate Change, Montréal 7-9 October 2009* [HLM-ENV/09 Report]. Seventy-three States and sixteen international organizations attended HLM-ENV/09 representing roughly 38 percent of the ICAO membership, which at the time was formed by 190 States. It is interesting to compare how HLM-ENV/09 ranked with other major meetings hosted by ICAO in recent times. Since 2009, ICAO has held two high-level meetings and two major aviation-related conferences. The High-Level Safety Conference took place in Montréal, from 29 March through 1 April 2010. This event was attended by delegates of 117 States and observes from 32 international organizations. See ICAO Doc 9935, *High-Level Safety Conference 2010* at ii-1 [HLSC 2010]. Likewise, from 12-14 September 2012, ICAO convened a High-Level Conference on Aviation Security. The event was attended by delegates from 132 States and observers from 23 international organizations. See ICAO Doc 9990, *High-Level Conference on Aviation Security* at ii-1 [*HLCAS 2012*]. Later, the Twelfth Air Navigation Conference – ICAO's major event in this field in the last ten years – took place in Montréal, from 19 to 30 November 2012. Delegates from 120 States and 29 international organizations attended the conference. See ICAO Doc 10007, *Twelfth Air Navigation Conference* [AN-Conf/12] at iv-1. Finally, the Sixth Worldwide Air Transport Conference (18-22 March 2013) was attended by 131 States and 39 international organizations. See ICAO Doc 10009, *Sixth Worldwide Air Transport Conference: Sustainability of Air Transport*, [AtConf/6] at 1. Attendance in all these events was significantly higher than that of the HLM-ENV/09. If attendance reflects the level of interest among States, then it is clear that for ICAO Member States environmental protection, and in particular climate change, is not perceived as important as safety, aviation security, air navigation, and air transport issues.

107 See HLM-ENV/09 Report, *supra* note 106 at 20, para 2.

108 *Ibid.* at 20, para 5, 7.

109 See ICAO, A37-19, *supra* note 207 (Chapter 1) at para 24 (e).

110 *Ibid.* at para 13. The Assembly also incorporated numerous overarching principles for the implementation of MBMs. Among others, these principles seek to ensure that MBMs are implemented in a cost-effective

explore the feasibility of a global scheme for MBMs.[111] At the time, the possibility of developing a global scheme seemed extremely remote. Three years later, however, the 38th Assembly finally agreed to develop a global MBM scheme for international aviation.[112]

3.2.2 *The Long Road to the CO_2 Standard*

The process leading to the development of the CO_2 standard for aviation emissions has involved some highly controversial environmental policy debates at ICAO. The case is illustrative of the challenges that an international organization such as ICAO faces when industry is not necessarily on board or convinced that regulation is warranted. As will be seen below, it is also symptomatic of the limitations embedded in the regulation.

In simple terms, a CO_2 standard refers to the minimum certification requirements applicable to aircraft engine emissions. Based on type certification, these standards apply solely to aircraft engines.[113] To date, some countries have established minimum fuel consumption standards for road vehicles. However, there are no fuel efficiency standards in existence in any country for purposes of aircraft certification. As a result of successful industry lobbying efforts, market forces helped determine fuel efficiency and CO_2 emissions.[114] For a long time, aviation industry advocates, with consistent support from free-market champions, have strongly opposed emission standards on the grounds that market forces have already ensured aircraft fuel efficiency. Aircraft and engine manufacturers have sufficient incentives to produce the most fuel-efficient aircraft. This is precisely what their customers (mostly commercial airlines) demand. With ever-increasing fuel prices and with fuel being the major element in the cost structure of most airlines, reducing fuel consumption is not just an environmental crusade for airlines but also a significant part of their everyday survival mode. Should inefficiencies arise, free-market advocates argue that the market's invisible hand will correct them. They point to the US capacity cutbacks in 2008 that were the result of the fuel spike, which translated into lower fuel consumption levels than those of the year 2000.[115] Free-market advocates claim that since CO_2 emissions are directly proportional to the amount of fuel consumed, there is no need to develop such a standard.[116] Adhering strictly to this rationale, CAEP/5 completely ruled out the possibility

manner, avoid duplicative measures, provide guidance on the use of revenues, suggest exclusion thresholds, minimize market distortions, recognize early movers, and promote transparency. *Ibid.* at Annex.

111 *Ibid.* at para 18.

112 *Ibid.* at para 18.

113 See Duthie, *supra* note 65 at 27.

114 See IPCC/4, *supra* note 2 (Introduction) at 353.

115 See Maryalice Locke, "Aviation & the Environment: Issues for Considering Global Aviation Emissions" (presentation delivered at the ICAO Workshop on Aviation Carbon Markets, 19 June 2008) online: ICAO <www.icao.int/Meetings/EnvironmentalWorkshops/Documents/WACM-2008/5_Locke.pdf>.

116 See ICAO, CAEP/5-WP/86, *supra* note 71 at 1-2.

of developing a CO_2 standard in 2001.[117] While this line of reasoning may be true, it also fails to recognize that even if market forces may drive fuel efficiency standards to some extent, this only applies to the acquisition of new equipment.[118]

ICAO's long-standing reluctance to adopt a CO_2 standard gradually came to an end when GIACC recommended the development of the standard in its final report. However, the recommendation was limited in scope in that it was applicable exclusively to new aircraft types.[119] On 30 June 2009, the Council accepted the GIACC's recommendation and directed the Secretariat to work through CAEP on establishing such a standard.[120] Later, in October 2009, ICAO's HLM-ENV recommended that the organization should "seek to develop a global CO_2 standard for new aircraft types"[121] and implicitly reaffirmed the Council's earlier decision. At this stage, pressure was mounting. Aircraft manufacturers could no longer resist, and it was evident that the standard will have to be developed. A year later, as mentioned above, the 37th Assembly formally instructed the Council to develop the CO_2 standard.[122]

According to CAEP's reports, ICAO is pursuing three major objectives in establishing a CO_2 standard. Firstly, the standard seeks to provide additional incentives to improve aircraft fuel efficiency performance.[123] Secondly, the standard will measure such performance across different aircraft types.[124] And finally, ICAO intends to minimize counter-productive incentives and adverse interdependency effects.[125] In early 2013, CAEP finally agreed on a common metric for the standard.[126] It is expected that Council will be in a position to adopt the standard in or about 2016.

Critics would highlight a number of shortcomings in the CO_2 standard. In all likelihood, the standard will only be applicable to new aircraft types, that is, to the new Airbus XXX[127] and not for a new unit of its A-380 model. Likewise, it will have almost no effect on aircraft in service or those that are already on the production line. In practical terms, from an environmental perspective, the effect of the CO_2 standard will be limited. In fact, the CO_2 standard is nothing more than what industry is already doing or willing to concede. In

117 *Ibid.*

118 See Duthie, *supra* note 65 at 27.

119 See ICAO, C-WP/13385 (Progress Report on the Group on International Aviation and Climate Change – GIACC).

120 See ICAO, C-DEC 187/14, *Council, 187th Session, Decision of the Fourteenth Meeting.*

121 HLM-ENV/09 Report, *supra* note 106 at 21.

122 See ICAO, A37-19, *supra* note 207 (Chapter 1) at para 24 (e).

123 See ICAO, CAEP/8, *supra* note 85 at 2-18.

124 *Ibid.*

125 *Ibid.*

126 See ICAO, COM 4/13, "ICAO Environmental Protection Committee Delivers Progress on Aircraft CO_2 and Noise Standards" (14 February 2013) online: ICAO <www.icao.int/Newsroom/Pages/ICAO-environmental-protection-committee-delivers-progress-on-new-aircraft-CO2-and-noise-standards.aspx>.

127 The letters "XXX" refers to a new aircraft type which has yet to be launched.

this respect and as will be discussed below, the case of the CO_2 standard is very similar to that of the global MBM scheme.

It is important to recognize the enormous opposition that this issue has attracted. For almost 10 years, industry and some States were reluctant to even entertain the idea of a CO_2 standard, let alone its metric or scope of applicability. The fact that ICAO is now on track to adopt such a standard should be seen as a major accomplishment. If there is the necessary political will, ICAO and its Member States could expand the scope of the CO_2 standard to aircraft in production and, perhaps one day, to aircraft in service.

3.2.3 *State Action Plans*

In 2009, GIACC suggested that ICAO "should encourage States to develop action plans."[128] In an apparent follow-up effort on this suggestion in 2010, the 37th Assembly "[encouraged] States to submit their action plans [by June 2012] outlining their respective policies and actions, and annual reporting on international aviation CO_2 emissions to ICAO."[129] As a tool for the achievement of ICAO's aspirational goals, action plans essentially seek to induce States to monitor their aviation emissions and, as well, to identify measures to address their carbon footprint.[130] To this end, the Assembly suggested that States should establish a baseline or counterfactual basis to assess their progress.[131] In 2013, at the 38th Assembly, ICAO reported that 61 States had submitted action plans.[132] To date, 24 States have made their action plans public.[133] The 38th Assembly continued to encourage States to submit action plans but this time once every three years.[134]

Although it is not yet possible to determine actual GHG emission reductions as a result of the action plans, this has arguably proved to be one of ICAO's major achievements in addressing climate change issues. It is true that the development of these action plans is still undergoing a trail-and-error phase, but it has been a gigantic step toward instilling among Member States the culture of addressing GHG emissions from international aviation. Even if one remains skeptical about the content of these action actions (e.g. accuracy of data, overemphasis on technological and operational measures, or, in some cases, lack of evidence that concrete measures have been put in place to limit or reduce GHG emissions),

128 See GIACC Report, *supra* note 105.
129 ICAO, A37-19, *supra* note 207 (Chapter 1) at para 9.
130 "Climate Change: Action Plans," online: ICAO <www.icao.int/environmental-protection/Pages/action-plan.aspx>.
131 See "Guidance Material for the Development of States' Action Plans," online: ICAO <www.icao.int/environmental-protection/Documents/GuidanceMaterial_DevelopmentActionPlans.pdf>.
132 See ICAO, A38-WP/30 EX/25, *States' Action Plans for CO_2 Emissions Reduction Activities* (presented by the Council of ICAO).
133 "Action Plans," online: ICAO <www.icao.int/environmental-protection/Pages/action-plan.aspx>.
134 See ICAO, A38-18, *supra* note 21 (Introduction) at paras 11 and 12.

the sole fact that States discuss measures and options to address climate change issues at large is a major accomplishment. Here, one must bear in mind that the predominant transport-oriented culture within the vast majority of ICAO Member States is not necessarily conducive to tackling climate change issues.

State action plans have been an invaluable tool to engage a broader audience. The fact that the ICAO Secretariat has attempted to link technical assistance for some States to the preparation of these action plans is a step in the right direction. Obviously, here the availability of additional funds and resources will be of paramount importance. For the future, ICAO should consider whether the submission of these action plans should be made mandatory, possibly through the adoption of a special ICAO standard. The action plans could also form part of the ICAO audit process. At present, ICAO only audits States on the basis of their level of compliance with aviation safety and security standards.[135]

3.2.4 *Aspirational Goals*

In 2010, moving beyond HL-ENV/09, the 37th Assembly adopted three aspirational goals. First, the sector should strive to achieve "a global annual average fuel efficiency improvement of 2 percent."[136] Second, this goal should also be maintained from 2020 to 2050.[137] And finally, from 2020 onward, the Assembly also decided to cap the growth of aviation's emissions at 2020 levels.[138] This is what industry refers as carbon neutral growth (CNG).[139] In working toward CNG, the Assembly called upon States to take into account a number of factors, such as "the special circumstances and respective capabilities [SCRC] of developing countries, the maturity of aviation markets, [and the fact] that some States may take more ambitious actions prior to 2020, which may offset an increase in emissions from the growth of air transport in developing states."[140]

Although no State objected to the fuel efficiency goals, 51 States filed reservations against the CNG goal. For 44 European States and the United States, such a goal was not stringent enough.[141] Instead of CNG, European States had proposed a 10 percent reduction

135 At present, the ICAO safety audits only include airworthiness protocol-related noise certification standards of Annex 16 (Environmental Protection). See generally, ICAO Doc 9735 AN/960, *Universal Safety Oversight Audit Program Continuous Monitoring Manual* 3rd edn (2011); ICAO Doc 9859 AN/474, *Safety Management Manual (SMM)* 3rd edn (2013); ICAO, *Universal Security Audit Program* 2nd edn (2010).

136 ICAO, A37-19, *supra* note 207 (Chapter 1) at para 4.

137 *Ibid.*

138 *Ibid.* at para 6.

139 See ICAO, A37-WP/217 EX/39, *Development of a Global Framework for Addressing Civil Aviation CO_2 Emissions* (presented by the International Air Transport Association (IATA), on behalf of ACI, CANSO, IATA, IBAC and ICCAIA, referred to hereafter as the "aviation industry") [ICAO, A37-WP/217 EX/39].

140 ICAO, A37-19, *supra* note 207 (Chapter 1) at para 6(a), (c), and (d).

141 See "Reservations to A37-19," online: ICAO <www.icao.int/Meetings/AMC/Assembly37/Documents/ReservationsResolutions/10_reservations_en.pdf> [ICAO, Reservations to A37-19].

in emissions by 2020 but, with a 2005 baseline, a significantly more aggressive goal.[142] The United States, along with Canada and Mexico, in turn, had proposed CNG from 2020 onward with a 2005 baseline.[143] On the other hand, six States claimed that it was unfair for developing countries to commit to such goals.[144] According to them, only developed States should take the lead and implement CNG.[145] Given the divergence in views and how far apart the positions were, it is in fact remarkable that the Assembly was able to adopt such goals – albeit subject to numerous reservations. To a large extent, these goals are strongly industry oriented.[146]

These goals are non-binding and are not assigned or "attributed" to States or aircraft operators. All States are encouraged to collectively strive to achieve them. This was done to delink ICAO's goals from the binding emission reduction commitments under the *UNFCCC/Kyoto* regime. For some developing States, the fact that these goals were "global" and did not set different obligations for developed and developing countries made them unacceptable. This explains why a number of developing countries filed reservations. For other developed states, the suggestion by the Assembly that some States may take action to offset the growth of emissions in developing countries was perceived as a camouflaged incarnation of the CBDR principle at ICAO. The United States, Canada, and Australia expressed strong reservations on this issue.[147] It would seem that, for both developing and developed States, these discussions at ICAO bear implications for the broader climate change negotiations at the *UNFCCC*.[148]

---

142 See ICAO, A37-WP/108 EX/26, *Addressing Aviation's Environmental Impacts Through: A Comprehensive Approach* (presented by Belgium on behalf of the European Union and its Member States, by the other States Members of the European Civil Aviation Conference, and by Eurocontrol).

143 See ICAO, A37-WP/186 EX/35, *A More Ambitious, Collective Approach to International Aviation Greenhouse Gas Emissions* (presented by Canada, Mexico and the United States).

144 See ICAO, Reservations to A37-19, *supra* note 141.

145 See ICAO, A37-WP/181 EX/32, *supra* note 31. Developing countries argue that it is unfair to constrain the growth of their aviation sector that still needs significant room to develop. Aviation has already reached a point of maturity in developed countries. This is not the case in developing countries. Climate change is the result of the accumulation of GHGs in the atmosphere over a very long period of time. This has been caused by the process of developed States' industrialization, not developing countries. During this period, no limitation on the sector's growth of developed countries was ever put in place. Therefore, addressing GHG emissions from international aviation is first and foremost a responsibility of developed States. From a fairness perspective, although one cannot deny the legitimacy of this argument, as seen in Chapter 1, it chooses to ignore that all air transport outlooks unequivocally point that the sector's growth is and will primarily take place in developing countries. Any system that permanently leaves out developing countries will be meaningless.

146 See ICAO, A37-WP/217 EX/39, *supra* note 139. With regard to the aspirational goals, the 37th Assembly adopted all but one of the aspirational goals proposed by the industry. More specifically, the Assembly did not endorse the industry proposal to half emissions by 2050. *Ibid.*

147 See ICAO, Reservations to A37-19, *supra* note 141.

148 See ICAO C-MIN 197/6, *Council, 197th Session, Minutes of the Sixth Meeting* at 8 (Statement of India) (suggesting that climate change issues involving international aviation cannot be disassociated from *UNFCCC* developments) [ICAO, C-MIN 197/6].

ICAO's goals have been the subject of severe criticism.[149] Some commentators doubt that they will in fact lead to meaningful emission reductions.[150] ICAO itself admits that fuel efficiency improvement by itself will be insufficient.[151] In spite of this, one must also recognize that the goals have been instrumental in setting a reference point, which also serves as a basis for ICAO's work on climate change. This is important in determining whether the implementation of technical and operational measures will be sufficient or whether MBMs ought to be developed. Even if these goals prove later to be insufficient, it would be almost impossible to make any progress without them. One could argue that, in light of the current growth trends of the air transport sector, ICAO should consider more aggressive goals. To date, however, there is no political will to do so and is likely that these goals will continue to be non-binding.[152]

3.2.5 A Framework for MBMs

Ever since Europe announced its intention to include foreign aircraft operators into its ETS, one of the industry's greatest concerns has been the potential emergence of a patchwork of multiple regimes attempting to address the sector's GHG emissions.[153] In the words of Giovanni Bisignani, IATA's former Director General, "a global industry [requires] a global solution."[154] The industry's preference for a coordinated approach was self-evident. At this juncture, however, the possibility of adopting a single global scheme seemed a far-fetched proposition.[155] Cognizant of the industry's concerns, both GIACC and HLM-

149 See Lyle, "Mitigating International Air Transport Emissions through a Global Measure," *supra* note 122 (Chapter 1) (describing ICAO's global aspirational goals as "tenuous").

150 See "Climate Researchers Find even Carbon-Neutral Growth from 2020 will not be enough to Stave off Climate Impacts," *Green Air Online* (3 October 2013) online: <www.greenaironline:.com/news.php?viewStory=1761>.

151 See ICAO, A37-19, *supra* note 207 (Chapter 1) at Preamble; ICAO, A38-18, *supra* note 21 (Introduction) at Preamble.

152 Although there appears to be an intuitive preference for binding norms when addressing climate change issues, such as a treaty-based regime, commentators also point out the benefits of systems constructed on the basis of "soft law" or "non-mandatory accords." Heller notes that "[v]oluntary agreements that induce members to shift behavior toward collective goals are most often carefully structured around reporting, aspirational but realistic targets, the diffusion of best practices through benchmarking, and positive incentives and aid for compliance action." Heller, *supra* note 191 (Chapter 2) at 131. Goldsmith and Posner suggest that soft law or "non-legal agreements" also "play an important role in international politics." Jack L Goldsmith & Eric Posner, *Limits of International Law* (Oxford: Oxford University Press, 2005) at 82. Similarly, Keohane points out that international codification "is not necessarily an effective way to obtain [for instance] real change in State behavior." Robert O Keohane, "When Does International Law Come Home" (1998) 35 Hous L Rev 699 at 701. Hudson suggests that voluntary mechanisms may facilitate participation of some States in climate change issues. See Blake Hudson, "Climate Change, Forests, and Federalism: Seeing the Treaty for the Trees" (2011) 82 U Colo L Rev 363.

153 See Bisignani, *Words of Change*, *supra* note 20 (Introduction) at 52.

154 *Ibid.*

155 In fact, "GIACC acknowledged that the implementation of a unique global sectoral system would face major challenges, particularly in the short and medium-term." GIACC Report, *supra* note 105 at 16.

ENV/09 recommended that a framework for MBMs should be pursued.[156] The framework was perceived as a tool to induce coordination among different regimes, avoid market distortions, and shield the industry from additional financial burdens and redundant compliance.[157] At this stage, there was not much information as to what the framework should be; neither was there any indication as to whether it should be adopted through a binding or non-binding instrument.

In 2010, the 37th Assembly tasked the Council "to develop a framework for [MBMs]."[158] Unlike many other provisions of the 2010 ICAO climate change resolution, this task did not receive a single reservation.[159] Arguably, States saw in the framework a tool to avoid uncoordinated approaches and, for some, an instrument to send a strong political message to Europe in connection with its decision to include foreign aircraft operators in its ETS.[160]

After the Assembly session, much of the work on the framework remained dormant for more than a year following the Council's declaration of a one-year cooling-off period.[161] ICAO resumed discussions on the framework in 2012. At this time, all non-EU States joined in the effort to stop the application of the EU ETS. This issue will be discussed at length in Chapter 4. In this context, it seemed reasonable to push for the development of the framework. At the invitation of the United States, a group of States opposing the EU ETS met in Washington, DC, on 31 July and 1 August 2012 to discuss international aviation emissions.[162] These States concluded that ICAO's work on the framework should be prioritized.[163] Other political factors also explain the preference for a framework as opposed to a global scheme. For instance, the US presidential elections were scheduled for November 2012. Seeking re-election and given tremendous difficulties encountered with the US Congress on other domestic issues, the Obama administration was very concerned about any potential spillover effects that could be triggered if it expressed support for the concept of a global scheme.

The framework must be clearly distinguished from a global MBM scheme.[164] In essence, it is only a set of common, non-binding principles[165] aimed at guiding those States or group

156 See *Ibid.* at 5-6; HLM-ENV/09 Report, *supra* note 106 at 20, para 5.

157 *Ibid. at* GIACC Report, *supra* note 105 at 16.

158 ICAO, A37-19, *supra* note 207 (Chapter 1) at para 13.

159 See ICAO, Reservations to A37-19, *supra* note 141.

160 See ICAO, HGCC/2-WP/9, *Suggested Elements of the Framework for MBMs* (presented by the United States).

161 See ICAO, C-DEC 192/6, *Council, 192nd Session, Decision of the Sixth Meeting* at 2.

162 See Meeting on International Aviation Emissions, Chair's Summary, 31 July-1 August 2012, Washington, DC [unpublished, archived on file with the author [Washington Meeting]].

163 *Ibid.* at 2.

164 See ICAO, C-WP/13861, *Market-Based Measures (MBMs)* (presented by the Secretary-General) at 3-4 [ICAO, C-WP/13861].

165 Although States did discuss different instruments (e.g. convention, Assembly resolutions, standard), they have always assumed that the framework will be adopted through an Assembly resolution. See ICAO, HGCC/2-WP/6.

of States that voluntarily decide to implement an MBM to the extent that such a measure is applicable to foreign aircraft operators.[166] The framework does not seek to impose on States the obligation to adopt MBMs. There have been different views of what the purpose of the framework should be. For instance, the United States suggested that the framework should be a "safety zone" that seeks to provide guidance to those States or group of States wishing to implement MBMs.[167] As another State brilliantly noted, the "safety zone" approach seeks to act as a "dispute settlement mechanism to resolve controversies amongst States. It seeks to set out the conditions of [the framework's] applicability and scope. Rather than enabling action to reduce GHG emissions from international aviation, this view is much more concerned with restricting the reach and the implementation of existing MBMs."[168] For others, the framework's main purpose is to "drive commonality and harmonization in order to avoid duplicative measures [and constitute] the building blocks for a future global scheme."[169] The Secretariat also suggested that the framework should "contribute to the achievement of [ICAO's global aspirational goals]."[170]

Although the Secretariat highlighted other areas where policy guidance was warranted,[171] States focused their discussions on the most controversial issue associated with the framework: the so-called geographical scope. In other words, States focused on whether a State implementing an MBM could include foreign aircraft operators and, if so, under what conditions and to what extent.[172] Depending on the approach adopted, the geographical scope of an MBM may provide the basis for allegations of extraterritorial application to be made. As will be discussed in Chapter 4, the initial design of the EU ETS covered incoming and outgoing flights of both EU and foreign aircraft operators. Moreover, emissions were computed for the whole duration of the flight. It suffices to note here that the central argument against the EU ETS was precisely that the extraterritorial elements present in the scheme were impermissible.[173]

---

166 See ICAO, C-WP/13861, *supra* note 164 at 3-4. See also ICAO, HGCC/3-WP/4, *Views of the UAE on a Framework for MBMs* (presented by the United Arab Emirates) (noting that the "framework should be a set of guiding principles that seek to guide States in the implementation of domestic and regional MBMs, when these measures are applied to foreign aircraft operators" at 3) [HGCC/3-WP/4].

167 ICAO, HGCC/2-WP 9, *Suggested Elements of the Framework for MBMs* (presented by the United States).

168 ICAO, HGCC/3-WP/4, *supra* note 166 at 3.

169 *Ibid.* at 2.

170 ICAO, HGCC/2-WP/4 *Outline of a Framework for MBMs* (presented by the Secretary of the HGCC) at Appendix A [ICAO, HGCC/2-WP/4].

171 These included issues such as whether the framework should incorporate provisions to accommodate the special needs of developing countries, whether the participants to an MBM scheme should be aircraft operators or the States themselves. See ICAO, HGCC/1-WP/5, *Means to Accommodate the Special Circumstances and Respective Capabilities (SCRC) of States* (presented by the Secretary of the HGCC).

172 As the State of registry, there is no question that the implementing State may apply an MBM to aircraft operators registered in such a State.

173 Here it is worth bearing in mind that the 36th Assembly had said that an emissions trading scheme should only cover foreign aircraft operators to the extent that their respective home States consented to. See ICAO, A36-22, *supra* note 17 (Chapter 2) at Appendix L.

The ICAO Secretariat identified three different approaches to the geographical scope. First, an MBM could be applied only to flights departing from airports situated in the implementing States.[174] The scheme would cover emissions for the whole duration of the departing flight.[175] A second alternative would capture all emissions produced by aircraft registered in the implementing State.[176] Finally, a third approach would be to allow the implementing State to cover all flights, both incoming and outgoing, of both national and foreign aircraft operators but only for the portion of emissions generated within its national airspace.[177]

At the time when States discussed potential options for the geographical scope for the framework, Europe had already announced the temporary suspension of its ETS. As a result, Europe was aware that it needed to make some concessions and show some flexibility. This explains why Europe supported the option of departing flights.[178] Most other States, however, supported the notion of national airspace.[179] Although this option would seem to be better aligned with principles such as the exclusive sovereignty over a State's territory, it is less attractive from an environmental perspective since it leaves out emissions produced over the high seas.[180] Some States suggested that as long as the implementing State is in conformity with the framework, such a State would not require the consent of other States to include foreign aircraft operators.[181] Other States, such as India, strongly advocated the need for mutual consent in all cases.[182]

During discussions held prior to the 38th Assembly, States were not able to reach consensus on the geographical scope of the framework. It was evident, however, that a majority of States favored the national airspace approach.[183] In fact, the text of the resolution that had been prepared for the 38th Assembly incorporated the notion of national airspace.[184] Industry, however, lobbied extensively against this provision, alleging that it could be interpreted as a license or encouragement for States to implement their own

174 See ICAO, HGCC/1-WP/3, *Designation of Coverage of a Framework for Market-Based Measures (MBMs)* (presented by the Secretary of the HGCC) at 2 [HGCC/1-WP/3]. See also ICAO, HGCC/2-WP/4, *supra* note 170 at 2.

175 See ICAO, HGCC/1-WP/3, *supra* note 174 at 2.

176 *Ibid.*

177 *Ibid.*

178 See ICAO, HGCC/3-WP/7, *CO_2 Emissions Coverage of the Geographic Scope Options for the Framework for* MBMs (presented by Belgium, France and the United Kingdom) [ICAO, HGCC/3-WP/7].

179 See, for instance, ICAO, HGCC/2-WP/9, *supra* note 167.

180 See ICAO, HGCC/3-WP/4, *supra* note 166 at 4.

181 See ICAO, HGCC/1-WP/2, *Role of the Framework for Market-Based Measures* (MBMs) (presented by the Secretary of the HGCC) at 1-2.

182 See ICAO, HGCC/3-WP/5, *Means to Accommodate the Special Circumstances and Respective Capabilities (SCRC) of States* (presented by the Secretary of the HGCC) at A-4.

183 See "Shuttle Diplomacy under way on Global Aviation Emissions Deal," *Reuters* (22 July 2013) online: <http://uk.reuters.com/article/2013/07/22/us-eu-aviation-emissions-idUKBRE96L0ZJ20130722>.

184 See ICAO, A38-WP/34 EX/29, *Consolidated Statement of Continuing ICAO Policies and Practices Related to Environmental Protection – Climate Change* (presented by the Council of ICAO).

schemes.[185] This was perceived as a real threat to the future of the global scheme.[186] In fact, Havel and Sanchez were the first authors to note that the very notion of a framework and the fact that ICAO had adopted some guiding principles for the implementation of MBMs presuppose that the organization is no longer the sole arbiter of climate change issues.[187] In other words, this recognizes that States may adopt measures outside the auspices of ICAO.

The incorporation of the national airspace approach as part of the Assembly resolution was also perceived as legitimizing a scaled-down version of the EU ETS. Consequently, with strong industry intervention, the 38th Assembly opted to remain silent on the issue of the geographical scope.[188] In fact, the Assembly resolution does not even mention the word framework.[189] However, according to the resolution, if a State wishes to implement a "new" or "existing" MBM, the State should "engage in constructive bilateral and/or multilateral consultations and negotiations with other States to reach an agreement."[190] This is a clear case of what Young calls "constructive ambiguity."[191] For some States, the requirement of an agreement equates to mutual consent in practice. In other words, the implementing State cannot include foreign aircraft operators, unless the affected States provide their consent. This would even be applicable to emissions generated over the implementing State's national airspace. For other States, on the other hand, this requirement only serves as a suggestion that they should use their best efforts to consult the other States concerned and attempt to reach an agreement. However, failure to reach an agreement does not stop the implementation of the scheme. In any case, this provision does not override the principle of exclusive sovereignty over a State's territory. In other words, consent is not required to include into a State's MBM scheme emissions produced over the implementing Sate's national airspace by foreign aircraft operators.

---

185 See "Aviation Industry Calls for Global Agreement and Climate Change Leadership by Governments Ahead of ICAO Assembly," *Green Air Online* (20 September 2013) online: <www.greenaironline:.com/news.php?viewStory=1745>.

186 *Ibid.*; "An NGO Message for the ICAO Assembly: Introduce a Global Market-Based Measure Now," *Green Air Online* (17 September 2013) online: <www.greenaironline:.com/news.php?viewStory=1754>; Letter from Nicolas E Calio, President and CEO A4A, to the Honorable Duane Woerth, Ambassador to the US Mission to ICAO; Ms. Julie Oettinger, Assistant Administrator, Policy, International Affairs and Environment, FAA; and Mr. Todd Stern, Special Envoy for Climate Change, US Department of State (29 August 2013).

187 See Havel & Sanchez, *supra* note 214 (Chapter 1) at 360.

188 See ICAO, A38-18, *supra* note 21 (Introduction).

189 *Ibid.*

190 *Ibid.* at para 16 (a).

191 Oran Young suggests that "[i]t is well known that 'constructive ambiguity' plays an important role in the realm of international governance and for that matter in governance more generally. It often makes sense to negotiate agreements containing ambiguities of this sort when the alternative is to admit defeat and to end up with no agreement at all." Oran R Young, "Deriving Insights from the Case of the WTO and the Cartagena Protocol" in Oran R Young et al, *Institutional Interplay* (New York: United Nations University Press, 2008) 131 at 145 [Young, "Deriving Insights from the Case of the WTO"].

The development of the framework has been a fascinating story in many ways. Firstly, it was conceived as a tool to correct the potential threat of uncoordinated approaches. Yet it also became its own worst foe as it was essentially abandoned in order not to encourage the surge of other schemes outside of the realm of ICAO. Secondly, for better or worse, the framework serves to demonstrate the enormous influence that the EU ETS has had on climate change discussions at ICAO.[192] Although these issues will be addressed in more detail in Chapters 4 and 7, it suffices to mention here that, from its initial conception to its actual abandonment by the 38th Assembly, the framework became more than a technical issue. It became a political battle that was seen by non-EU States as an appropriate avenue to demonize the EU ETS. For European States, however, this provided a means to legitimize the EU ETS. Thirdly, the framework is a classic example of a situation in which States get entangled into a political discussion with little or no legal or policy justification. From the get-go, it was self-evident that a non-binding framework will have no persuasive effect on the intended target. In other words, it is unrealistic to expect that European States would comply with the letter of a framework that would seem to contradict the very notion of the principle of exclusive sovereignty (e.g. the requirement to obtain the consent of other States over emissions produced within the airspace of the implementing State).

3.2.6 *A Global Scheme for MBMs*

Pursuant to the task of exploring the feasibility of a global MBM scheme assigned to the Council at the 37th Assembly, the President of the Council formed an Ad Hoc Working Group on MBMs (AHWG) on 20 January 2012 during the 195th Session of the Air Transport Committee.[193] The purpose of the AHWG was to identify potential measures for a global MBM scheme for international aviation with the assistance of a group of experts.[194] After extensive work, the experts assisting the AHWG along with the Secretariat identified the following four potential measures for a global scheme: (i) offsetting, (ii) offsetting with a revenue generation mechanism, (iii) emissions trading (cap & trade), and (iv) emissions trading (baseline & credit).[195]

Offsetting envisages a scheme where either States or aircraft operators (participants) are required to purchase emission credit units (ECUs) from other sectors.[196] The scheme's

192 See contra ICAO, C-MIN 196/7, *supra* note 124 (Chapter 1) at 9 (Statement of Cuba) (stating that ICAO was working under the mandate given by the Assembly and not as a result of the external pressure exerted by the EU ETS).

193 See ICAO, *Market-Based Measures (MBMs): Evaluation of Options for a Global MBM Scheme* (presented by the Secretary-General) at Appendix C [ICAO, C-WP/13894]. See also ICAO, C-WP/13799, *Study on Market-Based Measures (MBMs)* (presented by the Chairman of the Air Transport Committee) at 2.

194 See ICAO, HGCC/3-WP/4, *supra* note 166 at 2.

195 See ICAO, C-WP/13861, *supra* note 164 at 2.

196 *Ibid. at* Appendix A-1.

design involves the establishment of a set of criteria to determine which units will be eligible, as well as setting out a baseline.[197] Taking ICAO's CNG aspirational goal as an example, offsetting will mean that, in order to compensate for the growth in their emissions, participants will, as of 2021, be required to purchase ECUs taking into account their 2020 emission levels (baseline).[198] This was the least complicated alternative and the airline industry's preferred option.[199]

The second option, offsetting with a revenue generation mechanism, follows exactly the same principles except that it includes a revenue-generating mechanism, either through the imposition of a transaction fee or through the establishment of a price for emissions.[200] The rationale for adding the revenue-generating mechanism lies in the fact that such funds may be used for broader purposes such as climate change mitigation and adaptation in developing countries or to rechannel the proceeds within the aviation sector.[201] This, however, creates a fairly complex scheme as decisions will have to be made on matters such as the quantum of such levy, how and who will collect them, the purposes to which the revenues will be applied, and the beneficiaries thereof.[202] Given the well-known

197 *Ibid.*

198 *Ibid.*

199 See ICAO, A38-WP/68 EX/33, *Addressing CO_2 Emissions From Aviation* (presented by the Airports Council International (ACI), the Civil Air Navigation Services Organisation (CANSO), the International Air Transport Association (IATA), the International Business Aviation Council (IBAC), and the International Coordinating Council of Aerospace Industries Associations (ICCAIA)) (noting that "a simple carbon offsetting scheme would be the quickest to implement, the easiest to administer and the most cost-efficient" at 3) [ICAO, A38-WP/68 EX/33].

200 *Ibid.* at A-7.

201 *Ibid.*

202 Some institutions have actually suggested that ICAO has no experience in handling revenues. See WWF, Aviation Report, *supra* note 91 (Chapter 1) at 9. This criticism ignores that, since the 1950s, ICAO and its Member States have established cooperative arrangements for the provision of air navigation services over the North Atlantic. Two joint financing agreements have been adopted. Although the services per se are rendered by Denmark and Iceland, ICAO oversees the operation of the agreements and coordinates the establishment of air navigation charges paid by the users, as well as the assessments of the State parties. While users pay their air navigation charges to Denmark and Iceland through a UK service provider, States pay their annual assessment directly to ICAO. In a way, ICAO does have experience in handling revenues. See ICAO Doc 9585-JS/681, *Agreement on the Joint Financing of Certain Air Navigation Services in Greenland* (1956) as amended in 1982 and 2008, and ICAO Doc 9586-JS/682, *Agreement on the Joint Financing of Certain Air Navigation Services in Iceland* (1956) as amended in 1982 and 2008. In addition, in the context of modernizing the international liability regime for damages on the ground, in 2009, under the auspices of ICAO, States adopted *the Convention on Compensation for Damage to Third Parties, Resulting from Acts of Unlawful Interference Involving Aircraft*, 2 May 2009 [*Unlawful Interference Convention*]. Although not yet in force, under this instrument, passengers and cargo shippers will contribute to a fund that in turn will provide compensation to victims in cases of damages on ground caused as a result of an act of unlawful interference. To manage these revenues, the instrument sets up an international fund that, albeit a distinctive legal person, will function at ICAO headquarters. *Ibid.* What is interesting about both the joint agreements for the provision of air navigation services and the new liability regime for damages on the ground is that they have been established through international conventions. The *Chicago Convention* does not provide ICAO with specific powers to manage funds other than those that form part of its regular budget. See *Chicago*

industry rejection of fees and taxes as well as the position of some aviation-oriented States, it is unlikely that this option will ever see the light of day.[203]

The third option envisages an emissions trading scheme.[204] The system would cap the sector's overall international emissions.[205] It would then allocate allowances to the participants.[206] In essence, the allowance is a license to emit a specified quantity of CO_2.[207] The system would also establish compliance periods and would require participants to monitor their emissions.[208] If their emissions should exceed their allowances over a given compliance period, participants must purchase either ECUs or additional allowances.[209] The main difference between this option and offsetting is that under this option, allowances become a negotiable property right.[210] The presence of allowances, however, adds a layer of significant complexity to the scheme.

The fourth option also involves emissions trading but based on a baseline and credit system. Here, instead of allowances, the scheme establishes a baseline.[211] Should a participant emit more than the established baseline, such participant will be required to purchase ECUs.[212] If, on the other hand, at the end of the compliance period a participant's emissions are below the established baseline, the participant may either earn credits or bank them for future use.[213] In establishing the baseline, an efficiency metric may be taken into account.[214] Because of aviation's projected growth trends, it is expected that participants will have to purchase ECUs.[215] Therefore, this option does in fact closely resemble the offsetting scheme. In light of this, the Council ultimately decided to discard this option.[216]

All of these options will require the adoption of monitoring, reporting, and verification systems (MRVs); the establishment of a compliance period and enforcement mechanisms; a determination as to who the participants will be (in other words, whether States or aircraft

---

Convention, *supra* note 36 (Chapter 2) at Art 44. Therefore, if States would like to establish an offsetting scheme with a revenue generation mechanism, this will necessarily imply that a new international instrument will have to be adopted.

203 See ICAO, C-MIN 196/7, *supra* note 124 (Chapter 1) (Statement of UAE recommending abandoning consideration of offsetting with a revenue generation mechanism) at 8.

204 See ICAO, C-WP/13861, *supra* note 164 at 2.

205 *Ibid.* at Appendix A-8.

206 *Ibid.*

207 *Ibid.*

208 *Ibid.*

209 *Ibid.*

210 *Ibid.*

211 *Ibid. at* Appendix B.

212 *Ibid.*

213 *Ibid.*

214 *Ibid.*

215 *Ibid.*

216 See ICAO, C-DEC 196/7, *Council, 196th Session, Decision of the Seventh Meeting* at 2 [ICAO, C-DEC 196/7].

operators will bear compliance obligations); an assessment of the legal vehicles needed; and a means for distributing baselines and allocating allowances.[217]

The work of the experts assisting the AHWG sufficiently demonstrated that all of the options are "technically feasible and have the capacity to contribute to achieving [ICAO's aspirational goals]."[218] The studies have also revealed that the financial impact on the industry is expected to be minimal (at least for the time being).[219] On the basis of these findings, the 38th Assembly, working under intense pressure to demonstrate leadership and deliver results,[220] finally "[agreed] to develop a global MBM scheme for international aviation."[221] This decision attracted strong endorsement from the industry and no reservations were filed against it.[222] To develop such a scheme, the Assembly instructed the Council to consider a number of factors including proposals made by industry,[223] the special circumstances of States,[224] the need to explore a route-based and incremental approaches,[225] and the recognition of fast-growing carriers as well as those who have taken early action (e.g. early movers who have invested heavily in technology).[226] The Assembly also instructed the Council to present a concrete proposal to the 39th session scheduled for 2016.[227] At this time, the Assembly will be expected to adopt a global MBM scheme that will be implemented as of January 2020.

3.3 CONCLUSION

Environmental protection was not part of ICAO's original objectives as set out in its constitutional framework. Yet, through the adoption of various strategic objectives, the Secretary-General and Council have attempted to gradually incorporate environmental concerns as one of the organization's priorities. It is true, however, that the absence of a specific reference to the environment in the *Chicago Convention* is problematic, particularly when environment-related initiatives would seem to be in conflict with the convention's

217 See ICAO, C-WP/13861, *supra* note 164 at Appendices A1-9 and B.

218 See ICAO, C-WP/13894, *supra* note 193; ICAO, C-DEC 197/6, *Council, 197th Session, Decision of the Meeting.*

219 For instance, ICAO estimates that the cost of MBM could represent 0.8 to 1 percent of the industry's revenues. See ICAO, Assessment of the Impact of Market-Based Measures (MBMs), Council Informal Briefing (11 June 2013) [unpublished, archived on file with the author] at 11.

220 See ICAO, A38-18, *supra* note 21 (Introduction) at para 19.

221 See ICAO, A38-WP/68 EX/33, *supra* note 199.

222 See "Summary Listing of Reservations to Resolution A38-18," online: ICAO <www.icao.int/Meetings/a38/Documents/Resolutions/summary_en.pdf> [ICAO, Reservations to A38-18].

223 See ICAO, A38-18, *supra* note 21 (Introduction) at para 19 (a).

224 *Ibid.* at para 20.

225 *Ibid.* at para 21.

226 *Ibid.* at para 22.

227 *Ibid.* at para 19 (d).

overarching goals, such as fostering the development of the sector (i.e. growth). ICAO's governing structure seems to have strengthened the role of the Council. This poses some challenges when the need to engage a broader audience beyond those States that form part of the Council arises with the non-Council Member States having much less connection with the organization.

In theory, all of these deficiencies could be addressed through the introduction of substantial amendments to the international legal regime. This, however, is extremely unlikely to occur. The Council has already considered a proposal recommending amendments to the *Chicago Convention*, and the predominant view was that the septuagenarian instrument still provides an appropriate framework to address current and emerging challenges. For a number of reasons that extend beyond the realm of aviation and climate change issues, this view is not expected to change in the near future.

In this context then, ICAO should explore other alternatives to further engage its membership. Perhaps, this could be done through empowering its seven regional offices. These should be in much closer contact with its Member States. For instance, in an effort to address CAEP's imbalanced membership during the 38th Assembly, Argentina recommended that the regional offices should be much more involved in disseminating the committee's work within their respective regions.[228] Although unnoticed by the vast majority of States attending the Assembly, the Argentinean proposal was targeted at one of ICAO's most acute deficiencies: the disconnection between the organization and those Member States that are neither members of the Council nor participants in the technical work of the organization. I am not suggesting that this is a problem unique to ICAO. However, a proper resolution of this deficiency will go a long way to facilitate the design and implementation of the global MBM scheme for international aviation.

Ever since the *Kyoto Protocol* enjoined States to address the aviation sector's carbon footprint through ICAO, the organization has been extremely active in exploring alternatives, conducting studies, and presenting concrete proposals. The fact that GHG emissions continue to be unregulated is certainly not ICAO's fault, but it rather reflects the unwillingness of its Member States to undertake any serious commitments. Obviously, there are a number of exogenous and endogenous factors that explain the lack of political will on the part of Member States.

In spite of the enormous difficulties faced, what is clear is that ICAO has made concrete progress. The CO_2 standard is on track for adoption. This will not be a major regulatory breakthrough, but it is a significant step forward. Arguably, the State action plans are also a success story. They are in the process of changing the attitude of Member States on climate change issues. As a result of the State action plans, a large number of States are beginning

228 See ICAO, A38-WP/318 EX/110, *Environmental Protection, CAEP and the ICAO Regional Offices* (presented by Argentina).

to address issues involving aviation's climate change impact for the first time ever. In the near future, it will be desirable for the State action plans to become mandatory and included in ICAO audits.

The decision by the 38th Assembly to develop a global MBM scheme for international aviation has strengthened ICAO's leadership. Arguably, there is consensus that this is the best way forward to address GHG emissions from international aviation. However, forming consensus on the actual elements of the scheme will not be an easy task. Although the work of the group of experts has already provided a solid starting point for the development of the global scheme, a number of unresolved issues still remain. Tackling these issues will not only require technical and policy considerations but, above all, political decisions. As mentioned in Chapter 2, the success or failure of the global scheme will, to a large extent, be determined by how well it reconciles the principles of CBDR and non-discrimination,[229] the incentives provided to ensure broad participation and compliance, and the simplicity and cost-effectiveness of the scheme. These concerns must not only be articulated; they should also be operationalized.

229 See ICAO, C-MIN/196/7, *supra* note 124 (Chapter 1) (Statement of the Russian Federation suggesting the "need to address the political issues of CBDR and non-discrimination") at 10; *Ibid.* at 12 (Statement of South Africa).

4 The Inclusion of International Aviation in the EU ETS

This chapter seeks to understand the inclusion of international aviation in the EU ETS, the scheme's basic elements, the international opposition formed to defeat it, the political elements surrounding it, Europe's rationale to keep it alive, and its eventual suspension. This is central for the purpose of this project because the EU ETS has exerted an enormous influence in the aviation and climate change discourse. To a large extent, ICAO's work on climate change has been indirectly prompted by the EU ETS. For instance, as seen in Chapter 3, ICAO's framework for MBMs entirely geared toward legitimizing or constraining the EU ETS.

I begin by describing the historical context in which the scheme was adopted, as well as its elements, merits, and shortcomings. I then look into the scheme's specific characteristics as applied to international aviation. Significant attention is given to issues such as distribution of allowances, calculation of emissions, emission reduction units, the use of revenues, and its associated cost. Understanding the scheme's main features is relevant to demystify some, but definitely not all, of the criticism it has attracted. More importantly, it is not possible to examine the legal challenges against the EU ETS without a proper understanding of how it functions in practice.

I go on to examine efforts by the international community to derail the EU ETS. For a couple of years, non-EU States were more concerned in restricting the EU ETS than advancing ICAO's work on climate change issues. As will be explained, in many ways the inclusion of foreign aircraft operators into the scheme has attracted an unprecedented level of international opposition. This will bear profound consequences not only for the scheme's future geographical scope but also for the design of ICAO's global MBM. Similarly, the level of animosity will also bear significant influence on the scheme's legal analysis. Although the issue will be considered in more detail in Chapter 5, it suffices to say here that it is unquestionable that this serves as testimony that the international community does not in any way acquiesce the extraterritoriality present in the EU ETS.

Notwithstanding this international opposition, Europe was determined to keep the scheme in force for foreign aircraft operators. The final portion of this chapter attempts to explain why Europe took such a stance, given the high political and diplomatic cost.

4.1 The EU ETS

4.1.1 Background

For more than two decades, Europe has been exploring various regulatory alternatives to address the anthropogenic causes of global warming.[1] In the early 1990s, a number of European countries introduced carbon taxes,[2] and policy makers subsequently considered introducing a pan-European carbon tax.[3] This was perceived as an appropriate regulatory tool to ensure that firms incorporate environmental externalities as part of their production costs and as a means of ensuring the observance of the "polluter must pay" principle. However, a number of influential industries strongly opposed the proposal citing the potentially high costs associated therewith and concerns of a probable reduction of European competitiveness. In addition, some EU Member States felt that a tax dictated from Brussels would threaten their fiscal sovereignty.[4] Given that under community law, legislative proposals involving fiscal issues require unanimity, these concerns were sufficient to block the initiative. The pan-European carbon tax proposal was therefore never adopted.[5]

Having learned valuable lessons from this failed attempt, the Commission was determined to ensure that Europe would adopt adequate environmental measures in order to comply with its *Kyoto Protocol* obligations.[6] The Commission proposed borrowing from the United States the concept and know-how of emissions trading scheme (ETS).[7] Although

---

1 See Andrew Jordan, Dave Huitema, & Harro van Asselt, "Climate Change Policy in the European Union: An Introduction" in Andrew Jordan et al, eds, *Climate Change Policy in the European Union: Confronting the Dilemmas of Mitigation and Adaptation* (Cambridge: Cambridge University Press, 2010) 1 at 6.

2 See Per Kågeson, *Getting the Prices Right: A European Scheme for Making Transport Pay its True Costs*, (Stockholm: European Federation for Transport and Environment, 1993) at 41; Kathryn Harrison, "The United States as Outlier: Economic and Institutional Challenges to US Climate Policy" in Kathryn Harrison & Lisa McIntosh Sundstrom, *Global Commons, Domestic Decisions: The Comparative Politics of Climate Change* (Cambridge: MIT Press, 2010) 67 at 78 [Harrison, "The United States as Outlier"].

3 See Oberthür & Dupont, *supra* note 194 (Chapter 1) at 78.

4 The United Kingdom and Spain were among the strongest opponents to the pan-European carbon tax. See Brettny Hardy, "How Positive Environmental Politics Affected Europe's Decision to Oppose and then Adopt Emissions Trading" (2007) 17 Duke Envtl L & Pol'y F 297 at 302.

5 See Harro van Asselt, "Emissions Trading: The Enthusiastic Adoption of an "Alien" Instrument?" in Andrew Jordan et al, eds, *Climate Change Policy in the European Union: Confronting the Dilemmas of Mitigation and Adaptation* (Cambridge: Cambridge University Press, 2010) 125 at 127.

6 Brunnée and Toope point out that the *Kyoto Protocol* targets were not particularly onerous for the European Union mainly due to two reasons. First, before its ratification by the EU and its Member States, large European states had already achieved significant emission reductions. Second, the emission trajectories of Eastern European countries that had joined the EU through the second enlargement in 2004 did not present any significant level of threat. See Jutta Brunnée and Stephen J Toope, *Legitimacy and Legality in International Law* (Cambridge: Cambridge University Press, 2010) at 172.

7 See Vogler & Bretherton, *supra* note 194 (Chapter 1) at 9. See also Jonathan B Wiener, "Responding to the Global Warming Problem: Something Borrowed for Something Blue: Legal Transplants and the Evolution of Global Environmental Law" (2001) 21 Ecology LQ 1295 at 1297 (suggesting that the ETS was "vertical"

the Commission did not initially embrace ETS as a policy option,[8] it soon realized that it was much more acceptable to stakeholders and the public at large than taxes.[9] In particular, ETS was perceived as an attractive regulatory proposition for a variety of reasons. First, under community law, ETS qualifies as an environmental policy issue that only requires a qualified majority for its adoption (and not unanimity as with the carbon tax proposal).[10] Second, ETS provides much more flexibility to industry sectors that had hitherto been extremely preoccupied with the potential high costs of a carbon tax.[11] Third, ETS facilitates the transposition of global, country-based quantified reduction obligations into industry targets.[12] And fourth, a built-in cap facilitates the achievement of an environmental goal.[13]

In broad terms, the ETS establishes caps on certain types of emissions at specified levels (baseline or counterfactual) during a given compliance period.[14] The regulator in charge creates an artificial market by establishing European Union Allowances (EUAs) which are then distributed to entities (participants) subject to the regime.[15] In essence, an allowance is a license or a right to emit one tonne of carbon within a specified compliance period.[16]

---

or "trans-echelon" borrowing from US experience); Jos Delbeke, Written Testimony for the Senate Committee on Commerce, Science, and Transportation Hearing on the European Union's Emissions Trading System (saying that the EU ETS was "inspired in part by one of the most successful pieces of United States environmental legislation ever designed, the SO_2 allowance trading system under the Clean Air Act").

8 Pamela M Barnes, "The Role of the Commission of the European Union" in Rüdiger KW Wurzel & James Connelly, eds, *The European Union as a Leader in International Climate Change Politics* (London: Routledge, 2011) 41 at 52.

9 See Catherine Boemare & Philippe Quirion, "Implementing Greenhouse Gas Trading in Europe: Lessons from Economic Literature and International Experiences" (2002) 43 Ecological Econ 213 at 214.

10 See CE, *Treaty on the Functioning of the European Union* [2012] OJ, C 326/47 at Art 192 (e.g. Art 175 TEC) [*EU Treaty*]. See also Marcel Braun, "The Evolution of Emissions Trading in the European Union – The Role of Policy Networks, Knowledge and Policy Entrepreneurs" (2009) 34 Accounting, Organizations & Society 469 at 473; Barnes, *supra* note 8 at 49.

11 See Joe Kruger & Christian Egenhofer, "Confidence through Compliance in Emission Trading Markets" (2006) 6 Sustainable Dev L & Pol'y 2 at 8; Tsai & Petsonk, *supra* note 213 (Chapter 1) at 792.

12 See Kruger & Egenhofer, *supra* note 11 at 8.

13 *Ibid.*

14 See Michael T Hatch, "Assessing Environmental Policy Instruments: An Introduction" in Michael T Hatch, ed, *Environmental Policy Making: Assessing the Use of Alternative Policy Instruments* (Albany: State University of New York Press, 2005) at 7.

15 See Hans Opschoor, "Developments in the Use of Economic Instruments in OECD Countries" in Ger Klaassen & Finn R Førsund, eds, *Economic Instruments for Air Pollution Control* (Dordrecht: Kluwer Academic Publishers, 1994) 75 at 81.

16 See General EU ETS Directive, *supra* note 196 (Chapter 1) at Art 3 (a); Paul AU Ali & Kanako Yano, *Eco-Finance: The Legal Design and Regulation of Market-Based Environmental Instruments* (The Hague: Kluwer Law International, 2004) at 2; Matthieu Wemaëre, "Legal Nature of Kyoto Units" in W Th Douma, L Massai & M Montini, eds, *The Kyoto Protocol and Beyond: Legal and Policy Challenges of Climate Change* (The Hague: TMC Asser Press, 2007) 71 (describing emission reduction units as "tradable instruments which represent an entitlement to release a certain quantity of GHG emissions into the atmosphere" at 71); Perry S Goldschein, "Going Mobile: Emissions Trading Gets a Boost from Mobile Source Emission Reduction Credits" (1994) 13 UCLA J Envtl L & Pol'y 225 at 227.

Through these rights, which are measured in tonnes of CO_2, ETS[17] advances "market environmentalism."[18]

In order to meet their obligations, participants must surrender, at the end of each compliance period, enough allowances or other emission credit units (ECUs) to cover all the emissions they have emitted during such period.[19] The system therefore provides participants the flexibility to comply with their obligations through their own efforts (by undertaking measures to reduce their own emissions) or by purchasing allowances or ECUs from other participants in the market.[20] In other words, a participant for whom compliance is very expensive will reduce its emissions by purchasing allowances or credits from another participant whose cost of compliance with the system is lower.[21] Emission reductions then take place where is less expensive.[22] This maximizes the allocation of economic resources.[23] Even for CO_2-intensive sectors such as iron and steel production, which are highly sensitive to international trade, the financial impact of the ETS has been modest.[24] Similarly, participants in possession of excess allowances and ECUs may sell them.[25] To date, and in spite of its unquestionable difficulties, the EU ETS is the world's largest initiative to limit emissions and combat climate change.

The EU ETS only covers CO_2 emissions from large stationary power plants, iron and steel, mineral industry, and other sectors such as pulp from timber and paper.[26] Approxi-

17 ETS as a regulatory policy instrument also attracts detractors. Critics argue that ETS does not necessarily promote investment and innovation in new technology. Given that ETS revolves around the idea of abating where is less costly, participants may be encouraged to take advantage of this and delay investments for later. See David M Driesen, "Free Lunch of Cheap Fix?: The Emissions Trading Idea and the Climate Change Convention" (1998) 26 BC Envtl Aff L Rev 1 at 44 [Driesen, "The Emissions Trading Idea"]; David M Driesen, "Is Emissions Trading an Economic Incentive Program?: Replacing the Command and Control/Economic Dichotomy" (1998) 55 Wash & Lee L Rev 289 at 313 [Driesen, "Emissions Trading an Economic Incentive Program"].

18 See Daniel Cole, "What's Property Got to do With It? – A Review Essay by David M Driesen of Pollution & Property: Comparing Ownership Institutions for Environmental Protection" (2003) 30 Ecology LQ 1003 at 1003.

19 See Scott D Deatherage, *Carbon Trading: Law and Practice* (Oxford: Oxford University Press, 2011) at 19.

20 See Braun, *supra* note 10 at 470.

21 See Kerr, *supra* note 54 (Chapter 1) at 13. Kerr also notes that "[t]he overall environmental impact is zero if there are no concerns about the location of the emissions." *Ibid.*

22 See Slobodan Perdan & Adisa Azapagic, "Carbon Trading: Current Schemes and Future Developments" (2011) 39 Ener Pol'y 6040. A well-designed ETS may significantly reduce participants' per unit costs. See Thomas D Peterson & Adam Z Rose, "Reducing Conflicts between Climate Policy and Energy Policy in the US: The Important Role of the States" (2006) 34 Ener Pol'y 619 at 625.

23 See Hardy, *supra* note 4 at 298.

24 See Damien Demailly & Philippe Quirion "European Emission Trading Scheme and Competitiveness: A Case Study on the Iron and Steel Industry" (2009) 30 Energy Economics 2027.

25 See Holly Doremus & W Michael Hanemann, "Of Babies and Bathwater: Why the Clean Air Act's Cooperative Federalism Framework is Useful for Addressing Global Warming" (2008) 50 Ariz L Rev 799 at 807.

26 See General EU ETS Directive, *supra* note 196 (Chapter 1) at Annex 1.

mately 12,000 installations fall under the scheme.[27] EU Member States retain the prerogative to impose taxes on GHGs and installations presently not covered by the scheme.[28] The EU ETS follows a decentralized system that is run by EU Member States.[29] In this respect, each EU Member State designates a competent authority in charge of the implementation of the scheme,[30] develops a national allocation plan (NAP), allocates allowances,[31] monitors[32] and verifies[33] emissions, establishes a registry,[34] and imposes penalties whenever needed.[35]

In order to provide participants with less costly opportunities when complying with emission reduction obligations, the ETS was "linked" to the *Kyoto Protocol's* project mechanisms.[36] This allows participants to purchase and transfer ECUs as envisaged under the *Kyoto Protocol*, such as emission reduction units (ERUs)[37] generated under Joint Implementation (JI) project activities and certified emission reductions (CERs)[38] generated under *Clean Development Mechanism* (CDM) project activities.[39] When purchasing credits such as ERUs or CERs, participants benefit from the emission reductions that other firms have attained elsewhere.[40] The recognition or "linkage" of these credits is intended to supplement rather than replace participants' domestic actions.[41] In other words, participants are still expected to carry out most of their emission reduction obligations at home. Each EU Member State sets limits on the recognition of these credits.[42] Europeans, for instance, remain skeptical about the reliability and environmental credentials of some activities. As such, ECUs from projects involving nuclear installations, land use, and forestry are presently not recognized.[43]

---

27 EC, European Commission Press Release, IP/12/477, "Second year of emissions reporting from aircraft operators with very high level of compliance," *Europa* (15 May 2012) online: <http://europa.eu/rapid/pressReleasesAction.do?reference=IP/12/477&format=HTML&aged=0&language=EN&guiLanguage=en>.

28 See General EU ETS Directive, *supra* note 196 (Chapter 1) at 34 (recital 24).

29 *Ibid.* at 37 (Art 18).

30 *Ibid.*

31 *Ibid.*, at 35 (Art 6).

32 *Ibid.*, at 37 (Art 14).

33 *Ibid.*, at 37 (Art 15).

34 *Ibid.*, at 37 (Art 19).

35 *Ibid.*, at 37 (Art 16).

36 EC, *Parliament and Council Directive 2004/101 of 27 October 2004 Amending Directive 2003/87/EC Establishing a Scheme for Greenhouse Gas Emission Allowance Trading Within the Community, in respect of the Kyoto Protocol's Project Mechanisms* [2004] OJ, L 338/18 [Linking Directive].

37 ERUs stem from project activities that firms from developed countries carry out in other developed countries.

38 CERs arise from project activities that firms from developed countries carry out in developing countries. This is supposed to assist developing countries in achieving sustainable development.

39 See Linking Directive, *supra* note 36 at Art 1(m) and (n) at 20.

40 See Driesen, "The Emissions Trading Idea," *supra* note 17 at 43.

41 See General EU ETS Directive, *supra* note 196 (Chapter 1) at 41 (Art 30, para 3).

42 See Linking Directive, *supra* note 36, at 20 (Art 11 (a), para 1).

43 See *Ibid.*, Art 11 (a), para 3(a) & (b) at 21; Richard Benwell, "Linking as Leverage: Emissions Trading and the Politics of Climate Change (2008) 21:4 Cambridge Rev Int'l Affairs 545 at 557.

The EU ETS has also had opponents within Europe. In addition, it has been received with much less enthusiasm outside the continent.[44] In terms of its achievements, it is fair to say that emission reductions have thus far been "modest."[45] Its environmental integrity has also been called into question, given that it only covers CO_2 emissions and not other GHGs.[46]

4.1.2 EU ETS and Aviation

In November 2008, the European Parliament and the Council adopted Directive 2008/101/EC by virtue of which international aviation was included into the EU ETS.[47] The scheme applied to both EU and non-EU aircraft operators,[48] whether commercial or private, for flights to and from airports situated in the EU.[49] It also covered flights to non-EU States such as Norway, Iceland, and Liechtenstein, since these countries are members of the European Economic Area and the European Free Trade Association. The Commission especially established the so-called European Union Aviation Allowances (EUAAs). Airlines participating in the scheme could trade EUAAs among themselves but could not sell them to other sectors. In other words, airlines could surrender EUAs and EUAAs. However, entities from other sectors participating in the general EU ETS could only use EUAs.[50]

Initially, the cap, which is the total number of allowances to be allocated to aircraft operators, was set at 97 percent of the average of annual historical emissions from 2004

44 See Mar Campins Eritja, "Reviewing the Challenging Task Faced by Member States in Implementing the Emissions Trading Directive: Issues of Member State Liability" in Marjan Peeter & Kurt Deketelaere, eds, *EU Climate Change Policy: The Challenge of New Regulatory Initiatives* (Cheltenham: Edward Elgar Publishing Limited, 2006) 69 at 70.

45 See "The European Union Emission Trading Scheme (EU-ETS) Insights and Opportunities" *Pew Center on Global Climate Change* online: C2ES <www.c2es.org/docUploads/EU-ETS percent20Whitepercent20Paper.pdf>.

46 See Nick Farnsworth, "The EU Emissions Trading Directive: Time for Revision" in W Th Douma, L Massai & M Montini, eds, *The Kyoto Protocol and Beyond: Legal and Policy Challenges of Climate Change* (The Hague: TMC Asser Press, 2007) 29 at 29. Critics have also suggested that the scheme may hinder EU's competitiveness vis-à-vis other regions that are not subject to carbon regulation. See Tuerk et al, *supra* note 242 (Chapter 2) at 2. Defending the achievements of the scheme, the Commission reported a 2 percent drop in levels of GHG emissions in 2011 as compared to 2010 levels. See EC, European Commission Press Release, IP/12/477, "Second year of emissions reporting from aircraft operators with very high level of compliance," *Europa* (15 May 2012) online: <http://europa.eu/rapid/pressReleasesAction.do?reference=IP/12/477&format=HTML&aged=0&language=EN&guiLanguage=en>.

47 See Aviation Directive, *supra* note 17 (Introduction).

48 This includes passenger and freight airlines, business, and corporate jets.

49 See Mark Bisset & Georgina Crowhurst, "Is the EU's Application of its Emissions Trading Scheme to Aviation Illegal" (2011) 23:3 Air & Space L 1 at 5.

50 See ICAO, C-WP/13761 *European Emissions Trading Scheme ETS)* (presented by the Secretary-General) at 3.

through 2006.[51] The time frame against which the cap is compared is called the "baseline."[52] For each subsequent year, the cap was lowered to 95 percent of the participant's historical emissions. In February 2012, the Commission allocated free of charge to aircraft operators 85 percent of the total allowances.[53] Such allocation, which was proportional to the aircraft operator's reported share of activity in 2010, decreases to 82 percent for the trading period 2013-2020. Each year, 15 percent of the total allowances would be auctioned.[54] However, aircraft operators were not obliged to purchase allowances from the auctions.[55] Three percent of the allowances were set aside into a special reserve account for new entrants and air carriers growing at a rate higher than 18 percent annually (fast-growing carriers).[56]

4.1.3 *Calculation of Emissions*

The most controversial issue in its original design has been the way in which the EU ETS accounted for fuel consumption. In a simplistic fashion, the quantity of fuel burned is directly related to the amount of CO_2 emissions that an aircraft operator is deemed to have emitted. The quantity of fuel burned is one of the facts, although certainly not the only one, that determined the distribution of allowances. For the purpose of establishing the carbon footprint of an aircraft operator, the scheme takes into account the amount of fuel burned, and this is multiplied by an emission factor.[57] The emission factor is a fixed number that takes into account the "net calorific values" of each type of fuel used in aviation.[58] It measures the carbon content of these fuels.[59] In practical terms, fuel that was burned by aircraft while over the high seas or over non-EU airspace was also subject to

51 Gudo Borger explains that while international aviation was given a 2004-2006 baseline, other sectors have been subject to 1990 emission levels. This may, he argues, challenge the principle of equality. See Gudo Borger, "Al Things not Being Equal: Aviation in the EU ETS" (2012) 3 Climate L 265 at 280.

52 Establishing the baseline year is a very sensitive task. An inappropriate baseline may produce unjustified credits. See Tom Tietenberg, "The Tradable-Permits Approach to Protecting The Commons: Lessons for Climate Change" (2003) 19:3 Oxford Rev Econ Pol'y 400 at 419.

53 See Airport Watch, "EU Emissions Trading System" online: <www.airportwatch.org.uk/?page_id=8234>.

54 See Delbeke, *supra* note 7; Charlotte Burns & Neil Carter, "The European Parliament and Climate Change: From Symbolism to Heroism and Back Again" in Rüdiger KW Wurzel & James Connelly, eds, *The European Union as a Leader in International Climate Change Politics* (London: Routledge, 2011) 58 at 62; Peter Morrell, "An Evaluation of Possible EU Air Transport Emissions Trading Scheme Allocation Methods" (2007) 35 Ener Pol'y 5562 at 5566.

55 See Runge-Metzger, *supra* note 204 (Chapter 1) at 35.

56 See Aviation Directive, *supra* note 17 (Introduction), Art 3f. See also Isabelle Laborde, "EU Regulation of Aviation CO_2 Emissions" (2010) 24 WTR Nat Resources & Env't 54.

57 See Aviation Directive, *supra* note 17 (Introduction) at Annex IV; EC, *Commission Decision of 16 April 2009 Amending Decision 2007/589/EC as Regards to Monitoring and Reporting Guidelines for Emission and Tonne-Kilometre Data from Aviation Activities* [2009] OJ, L 103/10 at 103/19 (Annex xiv) [*EC, Monitoring and Reporting Decision*].

58 See *EC, Monitoring and Reporting Decision*, *supra* note 57 at 21.

59 *Ibid.*

the scheme when it was initially adopted.[60] Large quantities of fuel are consumed by the vast majority of flights outside European airspace.[61] Opponents of the EU ETS almost unanimously have asserted that this is an unequivocal expression of the scheme's (impermissible) extraterritorial reach. Michael Milde, for instance, claims that this is "in conflict with general international law" as it violates principles of exclusive sovereignty of States and the "the axiomatic concept that no State may claim sovereign rights over the high seas."[62] For the proponents of the scheme, this methodology is only a formula to calculate an aircraft's fuel consumption where some elements or "events" may take place "partly outside" of European airspace. Those in favor of the scheme argue that this does not necessarily imply a contradiction with international legal principles.[63]

4.1.4 *Distribution of Allowances: Surrendering of Allowances*

Aircraft operators must monitor their aviation activity, submit a benchmarking plan, and report data to their Administering Member State in order to be eligible to receive allowances free of charge.[64] The allowances are awarded on the basis of a benchmark allocation method that takes into account the aviation activity of each aircraft operator. Aviation activity is measured in tonne-kilometers (tkm).[65] In turn, tkm factors the shortest distance between two points that an aircraft operator flies on a given route multiplied by the flight's payload (mass of cargo plus mass of passengers).[66] To establish the number of allowances that an aircraft operator will obtain, the scheme takes into account the total amount of fuel burned during a given period (e.g. 2012 for allowances for the 2014 compliance period) and divides

60 See Brian F Havel & John Q Mulligan, "The Triumph of Politics: Reflections on the Judgment of the Court of Justice of the European Union Validating the Inclusion of Non-EU Airlines in the Emissions Trading Scheme" (2012) 37:1 Air & Space L 3 at 6.

61 As the US airline trade association, Airlines for America (A4A), has repeatedly noted, this formula subjects foreign aircraft operators to the EU ETS even though most of their emissions take place outside European jurisdiction. For instance, in a San Francisco to London flight, 29 percent of the emissions take place over the US airspace, 37 percent over Canadian airspace, 25 percent over the high seas, and yet only 9 percent over European airspace. See "Sovereignty the key issue as Europe, US airlines and environmental groups argue their EU ETS cases before the CJEU," *Green Air Online* (6 July 2012) online: <www.greenaironline.com/news.php?viewStory=1284>; "ATA Calls EU ETS Application to U.S. Airlines Illegal," *A4A* (5 July 2011) online: <www.airlines.org/Pages/ATA-Calls-EU-ETS-Application-to-U.S.-Airlines-Illegal.aspx>.

62 See Michael Milde, "The EU Emissions Trading Scheme – Confrontation or Compromise? – A Unilateral Action Outside the Framework of ICAO" (2012) ZLW 2012 at 9 [Milde, "EU Emissions Trading Scheme"]. Milde also describes the formula used to calculate emission as "almost arrogant." *Ibid.* at 7.

63 See Judgment of the Court (Grand Chamber) of 21 December 2011 in Case C-366/10 at paragraph 129 [*ATA Decision*].

64 See Aviation Directive, *supra* note 17 (Introduction) at Art 3e; See also Mark J Andrews et al, "International Transportation Law" (2010) 44 Int'l L 379 at 385.

65 See Aviation Directive, *supra* note 17 (Introduction) at Art 3e. See also Robert Malina et al, "The Impact of the European Union Emissions Trading Scheme on US Aviation" (2012) 19 J Air Transp Manag 36 at 37.

66 See *EC, Monitoring and Reporting Decision, supra* note 57 at 103/27.

it by the total number of TKs performed. The more TKs an aircraft operator performs with less fuel consumed, the better it is positioned to obtain more allowances. By using an efficiency metric (TKs), the benchmark allocation method rewards "early movers," that is to say those early adopters of new technology. In the aviation context, this concept covers those aircraft operators that have heavily invested in the most fuel-efficient aircraft.[67]

By the end of April of each year, aircraft operators ought to surrender to their Administering States allowances or ECUs corresponding to emissions generated throughout the preceding compliance period.[68] Given that the cost of in-sector abatement is extremely high, it is expected that aviation will be a net buyer of excess allowances from other industries.[69] International aviation will be in a position to buy allowances from other sectors but not to sell them.[70]

Administering Member States may impose "effective, proportionate, and dissuasive" penalties if aircraft operators fail to comply with the administrative requirements of the scheme such as the submission of fuel monitoring plans for instance.[71] Failure to surrender enough allowances or credits by 30 April of each year may subject aircraft operators to serious penalties.[72] In addition to publishing the names of the non-compliant carriers on a "wall of shame," Administering States may impose 100 euros per 1 tonne of CO_2 or allowance not surrendered.[73] Should these measures be insufficient to prompt the aircraft operator to comply with the requirements of the EU ETS, the Administering Member

67 Not all allowance allocation methods recognize these investments. For instance, the practice of "grandfathering" distributes allowances on the basis of the aircraft operator's own historical emissions. See ICAO, C-WP/13861, *supra* note 164 (Chapter 3) at A2. Those operators who emitted more will get more allowances. This system favors incumbent aircraft operators and acts as a disincentive to investments in more environmentally friendly technology. See Catherine Boemare & Philippe Quirion, "Implementing Greenhouse Gas Trading in Europe: Lessons from Economic Literature and International Experiences" (2002) 43 Ecological Economics 213 at 220.

68 See General EU ETS Directive, *supra* note 196 (Chapter 1) at 35 (Art 6, para 2 (c)). See also Braun, *supra* note 10 at 470. See also Martin Barlik, "The Extension of the European Union's Emission Trading Scheme to Aviation Activities" (2009) XXXIV Ann Air & Sp L 151 at 160.

69 For instance, retrofitting an aircraft's winglets to reduce fuel consumption is much more expensive than buying emission credit units from the market for the same amount of CO_2 reductions. See Annela Anger et al, "Air Transport in the European Union Emissions Trading Scheme, Final Report" *Aviation in a Sustainable World* (December 2008) online: <www.verifavia.com/bases/ressource_pdf/109/AU-OmegaStudy-17-finalreport-AAPMA-2-1-240209.pdf> at 8 [*Omega Report*].

70 See Anger & Köhler, *supra* note 205 (Chapter 1) at 39.

71 See General EU ETS Directive, *supra* note 196 (Chapter 1) at Art 16.

72 See Stephanie Switzer, "Aviation and Emissions Trading in the European Union: Pie in the Sky or Compatible with International Law" (2012) 39 Ecology L Currents at 2.

73 Non-compliance issues may arise where penalties are either too low or not clearly defined. See Jessica Coria & Thomas Sterner, "Tradable Permits in Developing Countries: Evidence from Air Pollution in Chile" (2010) 19:2 J Enviro & Develop 145 at 163.

State of such aircraft operator may request and the Commission may decide, as a measure of last resort, to impose operating bans upon the delinquent operators.[74]

4.1.5 Emission Credit Units (ECUs)

In addition to allowances distributed by the Commission free of charge and allowances purchased from other participants, aircraft operators may, in order to meet their compliance obligations, also surrender ECUs, such as ERUs and CERs. For the initial compliance period, aircraft operators were able to purchase such credits up to a maximum of 15 percent of the total amount of allowances that they are required to surrender.[75] Nevertheless, for the 2013-2020 compliance period, the limit substantially shrank to a level not exceeding 1.5 percent of the aircraft operators' verified emissions during those years.[76]

These limitations are intended to ensure that participants carry out domestic abatement in order to meet their compliance obligations.[77] They are designed to avoid a situation in which participants buy their way out by just purchasing ECUs from other carbon markets or by simply developing CDM[78] or JI projects elsewhere.[79] In theory, there is a risk that by increasing the levels available for ERUs and CERs, more emission reductions will occur outside rather than inside the EU. This is in line with the *Kyoto Protocol*'s supplementary principle.[80] ECUs should supplement and not replace domestic action.[81]

One can understand the rationale underlying this principle in the context of stationary sources where emissions are generated almost entirely within the territorial boundaries of EU Member States. The policy objective is to encourage European firms to carry out internal abatement within EU Member States. However, it does not automatically follow that these limitations should be applied mutatis mutandis to international aviation. A number of reasons may justify a different approach. First, in the case of aviation, the majority of emissions are produced outside Europe's airspace. Unlike emissions from

74 See Aviation Directive, *supra* note 17 (Introduction) at Art 16, paras 2, 3, and 5; Bartlik, *supra* note 68 at 161.

75 See Aviation Directive, *supra* note 17 (Introduction) at Art 11a.

76 See EC, *Parliament and Council Directive 2009/29 of 23 April 2009 Amending Directive 2003/87/EC so as to Improve and Extend the Greenhouse Gas Emission Allowance Trading Scheme of the Community* [2009] OJ, L 140/63.

77 See Perdan & Azapagic, *supra* note 22 at 6043.

78 CDM has been epitomized as an unmistakable example of "market liberalism." See Eva Lövbrand, Teresia Rindefjäll, & Joakim Nordqvist, "Closing the Legitimacy Gap in Global Environmental Governance? Lessons from the Emerging CDM Market" (2009) 9:2 Global Environmental Politics 74 at 75.

79 See Carlo Carraro & Alice Favero, "The Economic and Financial Determinants of Carbon Prices" (2009) 59:5 Czech J Econo & Finance 396 at 401.

80 See *Kyoto Protocol*, *supra* note 12 (Introduction) at Art 6, para 1 (d) and Art 17.

81 See Richard Benwell, "Linking as Leverage: Emissions Trading and the Politics of Climate Change (2008) 21:4 Cambridge Rev Int'l Affairs 545 at 557.

stationary sources, emissions from international aviation are not linked to national inventories under the *Kyoto Protocol.* Second, a large number of participants, if not the majority, are non-EU aircraft operators. Third, given the sector's notorious high abatement costs, aircraft operators will most likely either buy allowances or ECUs. At current low carbon prices, the sector is not expected to carry out "internal abatement" because it is simply too expensive to do so. After all, the whole concept of an ETS is to encourage participants to carry out emission reductions where it is least costly. Instead of carrying out more expensive internal abatement, any rational aircraft operator would either purchase allowances or ECUs. Fourth, unlike stationary sources, the inclusion of aviation in the EU ETS bears an undisputed cosmopolitan flavor.[82] There are just too many international elements present. To make things more complex, as will be discussed below, the EU ETS has been met with dire opposition from both industry stakeholders and non-EU States across the world. This level of resistance is not at all present in the case of stationary sources.

The limits of access to ERUs and CERs for the period 2013-2020 render aircraft operators' access to *Kyoto*-based projects such as JI and CDM almost meaningless. This implies a significant drop in financial flows for both developed and developing countries. Given that Europe is the largest buyer of CERs, the EU ETS' restrictions may play a significant role in the (negative) development of the CDM carbon market.[83] More participation from developing countries would have made perfect economic sense, for their lower abatement cost.[84] However, it is unlikely that developing countries will join any given climate change regime, unless there are significant economic incentives.[85] Limitations on ECUs are nothing but a mere reflection of the pervasive EU-centric approach that has framed the design of the scheme and alienated non-EU actors. If the purpose is to trigger collective action, providing some flexibility and incentive structure would have substantially contributed

82 The term "cosmopolitan" as applied to international aviation has been extracted from a very interesting article by Havel & Sanchez. See Brian Havel & Gabriel S Sanchez, "Restoring Global Aviation's Cosmopolitan Mentalité" (2010) 29 BU Int'l LJ 1 at 3 [Havel & Sanchez, "Cosmopolitan"].

83 See Fatemeth Nazifi, "The Price Impact of Linking the European Union Emissions Trading Scheme to the Clean Development Mechanism" (2010) 12 Enviro Econo & Pol'y Stud 164 at 184.

84 See Jared C Carbone, Carsten Helm, & Thomas F Rutherford, "The Case for International Emission Trading in the Absence of Cooperative Climate Policy" (2009) 58 J Enviro Econo & Management 266 at 279. Some authors have already noted the benefits of a sectoral CDM. See José Luis Samaniego & Christiana Figueres, "Evolving to a Sector-Based Clean Development Mechanism" in Kevin A Baumert, *Building on the Kyoto Protocol: Options for Protecting the Climate* (United States: World Resources Institute, 2002) at 92.

85 See Scott Barrett, "Climate Treaties and the Imperative of Enforcement" (2008) 24 Oxford Rev Econo Pol'y 239 at 253; Kristen Sheeran, "Environmental Consequences of Free Trade: Beyond Kyoto: North-South Implications of Emissions Trading and Taxes" (2007) 5 Seattle J Soc Just 697 (discussing that in order for developing countries to make any commitment to reducing GHGs, "alternative mechanisms must be used to ensured fairness" at 715).

to bring the parties closer.[86] This is something that should be taken into account when designing ICAO's global MBM.

4.1.6 Generation and Use of Revenues

The generation of revenues by public authorities often triggers significant political pressure with regard to their collection and use.[87] Under the EU ETS, EU Member States retain the discretionary prerogative on how to use revenues generated by the auctioning of allowances.[88] The legislation does not earmark but rather provides a number of non-mandatory "suggestions." These revenues may be used to tackle climate change in developing countries, to fund research and development for mitigation and adaptation purposes, and to cover the cost of administering the regime.[89] EU Member States must however report to the Commission how they will use these revenues.

The discretionary prerogative that EU Member States enjoy over these revenues attracted ferocious criticism. The US airline trade association called the ETS an "exorbitant money grab"[90] and a "new source of revenue for Europe."[91] For the Aerospace Industry Association, the scheme was "little more than a way for cash-strapped European governments to raise money."[92] EU airlines were concerned that proceeds will be used to "reduce [Europe's] budgetary deficit or to finance bailout plans."[93] There was a perception that the EU ETS is designed primarily as a revenue-raising mechanism to aid the general coffers of financially ailing European States.

Although the legislation acknowledges that the scheme should be viewed as one element of a comprehensive package of measures to address aviation's impact on climate change,[94]

86 See A Miola, M Marra, & B Ciuffo, "Designing a Climate Change Policy for the International Maritime Transport Sector: Market-Based Measures and Technological Options for Global and Regional Policy Actions" (2011) 39 Ener Pol'y 5490 (noting that "climate change policy needs to be able to promote collective actions while safeguarding flexibility and diversity" at 5495).

87 See Hahn, "Greenhouse Gas," *supra* note 197 (Chapter 1) at 176.

88 See Aviation Directive, *supra* note 17 (Introduction) at Art 3, para 4.

89 See Laborde, *supra* note 56 at 3.

90 "US Airlines Give Up on Legal Case Against Inclusion into the EU ETS But Call on their Government to Step Up Retaliatory Action," *Green Air Online* (29 March 2012) online: <www.greenaironline.com/news.php?viewStory=1444>.

91 Nancy Young, "Statement of Nancy Young Vice President of Environmental Affairs Airlines for America (A4A) before the Senate Committee on Commerce, Science and Transportation" (6 June 2012) online: A4A <www.airlines.org/Pages/A4A-Oral-Testimony-of-Nancy-Young,-VP-for-Environmental-Affairs.aspx> at 2.

92 Tim Devaney, "Leader of U.S. Aviation Group Slams 'Flawed' EU Emissions Scheme," *The Washington Times* (17 May 2012) online: <www.washingtontimes.com/news/2012/may/17/leader-of-us-aviation-group-slams-flawed-eu-emissi/>.

93 AEA, Position Paper, "Striving for an ETS that Supports a Sustainable Aviation Sector, Position," *AEA* (28 April 2011) online: <www.aea.be/assets/documents/positions/ETS%20Paper_April%2028.pdf> at 6.

94 See Aviation Directive, *supra* note 17 (Introduction) at recital 12.

the reality is that when combating aviation's carbon footprint, Europe has demonstrated very little progress on initiatives other than MBMs to combat aviation's carbon footprint. Operational and technological improvements, none of which raise revenues, have been extremely slow to materialize.[95] This has certainly alienated a number of actors. According to Fitzgerald, the failure to capitalize on non-MBM/non-revenue generating opportunities is one of the reasons why foreign aircraft operators and non-EU States have launched such a dire opposition against the scheme.[96]

Although Europe's notorious lack of progress on non-MBM measures reinforces this line of reasoning, it is difficult to characterize the EU ETS as a collection mechanism. It was not designed primarily to raise revenues. The adoption of the EU ETS preceded Europe's economic meltdown. Had the intention been to raise revenues, the scheme would have adopted a 100 percent auctioning of allowances.[97] The only way in which EU Member States could raise revenues is through the auctioning of allowances. It is not expected that the proceeds from these allowances would be significant. In addition, nothing in the legislation prevents airlines from completely foregoing the auctioning process. As Petsonk has brilliantly put it, "no airline participating in the ETS needs to send a dime into European government coffers if it doesn't want to."[98] In fact, airlines are given a number of alternatives to purchase allowances other than through auctions.[99] These may include accessing CERs from CDM projects or just buying allowances from other participants within the general EU ETS scheme.

Another aspect of the scheme's design that infuriates the airline industry relates to the fact that revenues may be used outside the sector.[100] This makes it tougher for airlines to make green investments to reduce GHG emissions, such as the acquisition of new fleet,

---

95 For instance, commentators have queried why no-regret options, such as continuous descent approach, are not more widely used within industry. See Anger & Köhler, *supra* note 205 (Chapter 1) at 42.

96 See Paul Fitzgerald, "Europe's Emissions Trading System: Questioning its *Raision d'Être*" (2010) 10:2 Issues Aviation L & Pol'y 189 at 211-12.

97 See Eckhard Pache, "On the Compatibility with International Legal Provisions of including Greenhouse Gas Emissions from International Aviation in the EU Emission Allowance Trading Scheme as a result of the proposed changes to the EU Emission Allowance Trading Directive, Legal Opinion Commissioned by the Federal Ministry for the Environment, Nature Conservation and Nuclear," online: <http://grist.files.wordpress.com/2011/06/aviation_emission_trading.pdf> at 14.

98 See Annie Petsonk, International Counsel Environmental Defense Fund (EDF), "Summary of the testimony of Annie Petsonk before the Committee on Commerce, Science, and Transportation United States Senate" (6 June 2012) online: EDF <www.edf.org/sites/default/files/EDF-Petsonk-Senate-Testimony-EU-ETS-Aviation-060612.pdf> at 25 [Petsonk, "Testimony"].

99 Given the fact that the proceeds from auctions are not earmarked for aviation-related projects such as the Single European Sky, the Association of European Airlines (AEA) characterizes the auctioning process as a tax. See AEA, Press Release, "AEA Challenges Five Emissions Trading Myths" (16 June 2011) online: AEA <http://files.aea.be/News/PR/Pr11-014.pdf>.

100 Devaney, *supra* note 92.

development of alternative fuels, and implementation of operational measures.[101] As some commentators point out, the use of "feebates" may be used to reward "good behavior" and to induce engagement of participants in any given scheme.[102] The earmarking or hypothecation of revenues in emissions trading is not at all new. In the United States, the *Clean Air Act* establishes that the proceeds from auctions must be rechanneled to participants of the scheme.[103] These are not considered revenues of the US government.[104]

From an aviation perspective, it may have been desirable to completely earmark these revenues to ensure that they would be used for mitigation purposes within the sector.[105] Yet earmarking revenues for aviation also ignores the fact that the sector's abatement cost is extremely high.[106] If the objective is to reduce CO_2 emissions, it makes much more sense to allocate revenues where mitigation could take place at the least cost. This is the very essence of an emissions trading. This will necessarily occur outside the sector.[107] Although from a purely environmental perspective, the desirable option is self-evident, the scheme's policy design should have taken into account the tenacious opposition it has created. The worldwide discontent is more than obvious, and it was certainly foreseeable. Industry stakeholders and some non-EU States have vehemently questioned the discretionary use of revenues from auctions.[108] Earmarking revenues within the sector may have helped to ease some of this tension.

---

101 "EU ETS Remains Bad News for Airlines," *A4A* (11 February 2013) online: <www.airlines.org/Pages/EU-ETS-Remains-Bad-News-For-U.S.-Airlines.aspx>. Dejong suggests that the fact that potential revenues from the EU ETS are not hypothecated contradicts Art 15 of the *Chicago Convention*, for such revenues are not tied to the provision of services. See Steven M Dejong, "Hot Air and Hot Heads: An Examination of the Legal Arguments Surrounding the Extension of the European Union's Emissions Trading Scheme to Aviation" (2013) 3:1 Asian J Int'l L 163 at 184.

102 See James Gustave Speth, *The Bridge at the Edge of the World: Capitalism, the Environment, and Crossing from Crisis to Sustainability* (New Heaven: Yale University Press, 2008) at 95.

103 *Clean Air Act*, 42 USC 7401-7626 at sec. 416.

104 *Ibid.*

105 See Jane A Leggett, Bart Elias, & Daniel T Shedd, "Aviation and the European Union's Emission Trading Scheme" *Congressional Research Service* (15 May 2012) online: FAS <www.fas.org/sgp/crs/row/R42392.pdf> at 14.

106 See Anger, "Including Aviation in the EU ETS," *supra* note 2 (Chapter 1) (noting that "marginal abatement costs for the aviation sector are set to be higher than those of other sectors covered by the EU ETS" at 102); ICAO, HGCC/3-IP/6, *Views of the Environmental NGO Community* (presented by the International Coalition for Sustainable Aviation, ICSA) at 2; WWF, Aviation Report, *supra* note 91 (Chapter 1) at 6.

107 See Inga J Smith & Craig J Rodger, "Carbon Emission Offsets for Aviation-Generated Emissions Due to International Travel to and from New Zealand" (2009) 37 Ener Pol'y 3438 (noting that abatement from international aviation will most likely take place outside the sector, at 3446).

108 Russia would seem to accept that revenues from MBMs may be used in sectors other than aviation but only in the "States over the territory of which [the] emissions [have] occurred." See Declaration of the Commitment of the Russian Aviation Authorities to the Protection of the Environment from the Effects of Aviation and to the Reduction of Greenhouse Gas Emissions [unpublished, archived on file with the author].

4.1.7 *Cost or Windfall Gains?*

Since the Commission first discussed the inclusion of aviation into the community trading scheme, one of the major concerns dealt with the potential high financial burden on the airline industry.[109] Some States viewed the scheme as a threat for the sustainable development of air transports.[110] Commentators warned that the EU ETS will "add considerable costs to civil aviation."[111] Passenger tickets and air freight rates would increase.[112] A4A's chief even suggested that the scheme will be responsible for the "the loss of US jobs"[113] and prevent (US) airlines from replacing their aging fleet.[114] Similarly, US Congressmen indicated that the increase will contribute to increasing the US unemployment rate.[115]

Initially, analysts predicted a yearly carbon bill in the order of 3-4 billion dollars (US). Most of these studies were carried out before the adoption of the EU ETS for international aviation on the basis of much higher prices of carbon.[116] Late in 2011, IATA suggested that the ETS would cost the airline industry around 1.2 billion dollars (US) in 2012. However, in light of the low carbon prices, later reports forecasted a cost in the range of 300 million

---

109 See generally Tate L Hemingson, "Why Airlines Should be Afraid: The Potential Impact of Cap and Trade and Other Carbon Emissions Reduction Proposals on the Airline Industry" (2010) 75 J Air L & Com 741 at 766; Hua Lan, "Comments on EU Aviation ETS Directive and EU – China Aviation Emission Dispute" (2011) 45 RJT 589 at 595.

110 *Joint Statement between the Civil Aviation Administration of the People's Republic of China and the Ministry of Transport of the Russian Federation on the European Union's Inclusion of Aviation into the European Union Emission Trading Scheme*, 27 July 2011 at para iv.

111 See Milde, "EU Emissions Trading Scheme," *supra* note 62 at 6.

112 See Pete Sepp, "EU Energy Tax Would Hike U.S. Airfares Even Higher", *US News* (6 June 2012).

113 "US Legislation to Prohibit Airlines from Joining EU ETS Moves a Step Closer as Senate Bill Receives Bipartisan Support," *Green Air Online* (8 March 2012), online: <www.greenaironline.com/news.php?viewStory=1436>; Nicholas E Calio, "EU's Emissions Tax Scheme Kills American Jobs," *Politico* (4 November 2012) online: <www.politico.com/news/stories/1112/83284.html>.

114 See "Airlines for America (A4A) Commends Senate Opposition of EU ETS," *A4A* (7 December 2011), online: <www.airlines.org/Pages/news_12-07-2011_2.aspx>. As strong supporters of economic laissez-faire, US airlines have long argued that environmental regulatory measures, such as the EU ETS, will only add a financial burden that, in turn, will make it very difficult to purchase new (aircraft) technology to improve fuel efficiency and reduce their carbon footprint. See Nicholas E Calio and Lee Moak, "Why U.S. Airlines Oppose EU's Emissions Tax Scheme" *USA Today* (27 March 2012) online: <www.usatoday.com/news/opinion/forum/story/2012-03-27/airlines-eu-planes-emissions-tax/53805612/1>. It is interesting to see that US airlines have been at the forefront of the EU ETS's criticism, in particular on the issue of cost. European commentators have said that the scheme is much more onerous for European aircraft operators, given that they must surrender allowances not only for long haul flights but also for their entire intra-European operations. See Petsonk, "Testimony," *supra* note 98 at 18; Ulrich Steppler & Angela Klingmüller, "EU Emissions Trading Scheme and Aviation: Quo Vadis?" (2009) 34:4 Air & Space L 253 at 257; AEA Position Paper, "Striving for an ETS that Supports a Sustainable Aviation Sector," *AEA* (28 April 2011) online: <www.aea.be/assets/documents/positions/ETS%20Paper_April%2028.pdf> at 3.

115 Pete Kasperowicz, "House Members Warn Carbon Tax would Increase Unemployment Rate," *The Hill* (3 December 2012), online: <http://thehill.com/blogs/floor-action/house/270579-house-members-warn-carbon-tax-would-increase-unemployment-rate>.

116 For instance, many reports assume a scenario where allowances would be allocated on the basis of a 100 percent auctioning. See Anger & Köhler, *supra* note 205 (Chapter 1) at 40.

dollars (US) for the whole aviation industry.[117] This is a drop in the ocean, if compared with the 39 billion dollar (US) fuel bill that IATA reported for the industry in 2012.[118] Therefore, in financial terms, the impact is rather minor.[119] It is also unlikely to threaten airline competitiveness or result in passengers changing airline choices.[120] Moreover, it has been suggested that the EU ETS was only expected to reach 1.25 percent of the industry cost.[121] Even if the price of carbon reached 40 euros, aviation demand would decline by just 1 percent.[122] For reference purposes, at the time of this writing, the EUA traded at approximately 4.7 euros.[123] The Commission estimated that the cost per passenger for a long-haul flight would be between 2 and 10 euros each way.[124] The airlines, on the other hand, argued that even if the increase in the price of tickets is low, this will be significant for an industry whose profit margins are particularly slim. One airline representative noted that "[t]his may be the difference between loss and profitability."[125]

---

117 See "Airlines face CO_2 bill of 300 Million Euros in 2012," *Reuters* (16 February 2011), online: <www.reuters.com/article/2012/02/16/carbon-aviation-barcap-idUSL5E8DG4O520120216>; Lorand Bartels, *The Inclusion of Aviation in the EU ETS: WTO Law Considerations*, Trade and Sustainable Energy Series, Issue Paper No 6 (Geneva: International Centre for Trade and Sustainable Development Switzerland, 2012) (suggesting that industry cost could range between 360-505 million per year); "China says EU carbon rule to cost $2.8 Billion by 2030," *Reuters* (5 March 2012), online: <www.reuters.com/article/2012/03/05/china-eu-emissions-idUSL4E8E51ME20120305> (indicating that the initial cost for Chinese airlines in 2012 would be US$ 127 million).

118 See IATA, "Jet Fuel Price Monitor," *IATA* online: <www.iata.org/whatwedo/economics/fuel_monitor/Pages/index.aspx>.

119 See Yue-Jun Zhang & Yi-Ming Wei, "An Overview of Current Research on EU ETS: Evidence from its Operating Mechanism and Economic Effect" (2010) 87 Applied Energy 1804 (arguing that the EU ETS will not have a significant financial impact for the airline industry, at 1811); Anger & Köhler, *supra* note 205 (Chapter 1) (suggesting that the EU ETS will not significantly affect the growth of demand, at 41); Laurel Pentelow & Daniel J Scott "Aviation's Inclusion in International Climate Policy Regimes: Implications for the Caribbean Tourism Industry" (2011) 17 J Air Transport Management 199 (underscoring that the EU ETS will not pose a financial threat to touristic destinations such as those in the Caribbean islands, at 204); Janina D Scheelhaase & Wolfgang G Grimme, "Emissions Trading for International Aviation – An Estimation of the Economic Impact on Selected European Airlines" (2007) 13 J Air Transport Management 253 (noting that the financial impact will be slightly higher for low-cost aircraft operators but it will not adversely affect profit margins, at 262); Anger, "Including Aviation in the EU ETS," *supra* note 2 (Chapter 1) (highlighting that the EU ETS is unlikely to be detrimental to EU's economic growth nor diminish its competitiveness, at 103); Malina et al, *supra* note 65 (indicating that the EU ETS will bear very little impact on US airlines operation and that, despite of its implementation, US traffic and its carbon footprint will continue to grow, at 39).

120 See Ascha Alberts, Jan-André Bühne, & Heiko Peters, "Will the EU ETS Instigate Airline Network Reconfiguration?" (2009) 15 J Air Transport Management 1 at 5.

121 See Jan Vesparmann & Andreas Wald, "Much Ado about nothing? – An Analysis of Economic Impacts and Ecologic Effects of the EU-Emission Trading Scheme in the Aviation Industry" (2011) 45 Transportation Research Part A 1066 at 1075 (also indicating that competition distortion will be minimum).

122 See *Omega Report*, *supra* note 69.

123 See Point Carbon, EAU Last 30 Days, online: Point Carbon <www.pointcarbon.com>.

124 See Runge-Metzger, *supra* note 204 (Chapter 1) at 29.

125 Elizabeth Rosenthal, "Your Biggest Carbon Sin May Be Air Travel," *The New York Times* (26 January 2013) online: <www.nytimes.com/2013/01/27/sunday-review/the-biggest-carbon-sin-air-travel.html>.

Notwithstanding complaints within the sector, aircraft operators may, to the surprise of many, profit from the EU ETS for international aviation. An interesting MIT study revealed that if US airlines pass on to passengers the cost of total allowances they are required to surrender, they stand to attain a windfall gain of 2.6 billion dollars (US) from 2012 to 2020.[126] This is because they would not only pass on the cost of allowance or emission reduction units they need to purchase but also those allocated free of charge.[127] The probability to pass on opportunity cost is not uncommon in competitive industries.[128] Allowances are "a significant new asset"[129] that may turn into "profit centers," at least for the most fuel-efficient aircraft operators.[130] Those defending the EU ETS point out that the scheme creates opportunities for aircraft operators to profit; they can charge for opportunity cost.[131] Ryanair capitalized on this opportunity.[132] Airline representatives, however, denied that the sector stands to profit from the scheme.[133]

4.2 INTERNATIONAL OPPOSITION

Notwithstanding the substantial support received from those in the environmental community, it is undeniable that the inclusion of foreign aircraft into the EU ETS has attracted an unprecedented level of opposition within the international aviation community. Never before has an international coalition of different players collectively strove to ruin a regulatory initiative dealing with international civil aviation.[134] In order to stop the scheme's

126 See Malina et al, *supra* note 65 at 39.

127 *Ibid.*

128 See "Briefing: The Billion Euro Aviation Bonanza," *Transport & Environment* (January 2013) online: <www.transportenvironment.org/sites/te/files/publications/Briefing_The_billion_Euro_Aviation_Bonanza.pdf> [Transport & Environment, "Briefing: The Billion Euro Aviation Bonanza"].

129 Petsonk, "Testimony," *supra* note 98 at 5; Wemaëre, *supra* note 16 (describing allowances as "an asset or (a quasi) property right for the owner if it is transferable" at 76).

130 See Petsonk, "Testimony," *supra* note 98 at 15.

131 See Written Observations of the Second Interveners in the Court of Justice of the European Union, reference from the High Court of Justice, London, Case C-366/10 at 21. Similarly, in the *ATA Decision*, the CJEU noted that it cannot be discarded that aircraft operators will profit from the scheme or will not have to surrender allowances at the end of the compliance period. See *ATA Decision*, *supra* note 63 at para 142.

132 See "Green Taxes – A Nice Little Earner for Some," *The Economist* (6 February 2012) online: <www.economist.com/blogs/gulliver/2012/02/airlines-and-emissions-permits> (estimating that in 2012, with the introduction of a € 0.25 passenger charge for each segment, Europe's largest low-cost carrier would approximately make € 10.8 million).

133 See Alessandro Torello, "Airlines Profit from CO_2 Plan?," *The Wall Street Journal* (22 January 2013) online: <http://blogs.wsj.com/brussels/2013/01/22/airlines-profit-from-emissions-plan/>.

134 See Goron, *supra* note 35 (Chapter 2) at 1; Elaine Fahey, "The EU Emissions Trading Scheme and the Court of Justice: The High Politics of Indirectly Promoting Global Standards" (2012) 13:11 German LJ 1247 (describing the dispute over the EU ETS as "one of the most controversial global regulatory disputes of modern times" at 1262); Sanja Bogojovic, "Legalizing Environmental Leadership: A Comment on the CJEU's Ruling in C-366/10 on the Inclusion of Aviation in the EU Emissions Trading Scheme" (2012) 24:2 J Enviro

implementation, myriad different States and industry stakeholders have resorted to multiple actions on both domestic and international fronts, including diplomatic, political, and legal measures. Both States and industry stakeholders have been firm and persistent objectors to the EU ETS.

This section examines some of the measures that non-EU States and industry stakeholders adopted in their quest to resist the EU ETS. Understanding these actions is important for two reasons. First, they not only exerted enormous pressure but also ultimately forced Europe to suspend the scheme as applied to foreign aircraft operators. Second, these actions were a clear testimony that, even if the scheme were to be found as being consistent with international law, the international community would not tolerate the inclusion of foreign aircraft operators.

4.2.1 Judicial Attacks: A4A's Legal Challenge

In December 2009, the trade association of US airlines, Airlines for America (A4A), as well as the four major US carriers brought before the High Court Justice of England and Wales (Queen's Bench Division) an application against the UK Secretary of State for Energy and Climate Change questioning the legality of including foreign aircraft operators into the European scheme.[135] A4A's legal challenge asserted that the EU ETS (i) is an extraterritorial and unilateral application of community legislation that violates the principle of exclusive sovereignty that each State is entitled to over its airspace under Article 1 of the *Chicago Convention*; (ii) is in conflict with Article 12 of the *Chicago Convention* and principles of customary international law, for no State may validly purport to subject any part of the high seas to its sovereignty, since aircraft overflying the high seas may only be subject to the jurisdiction of the country in which they are registered;[136] (iii) amounts to

L 345; Mayer, Case-366/10, *supra* note 133 (Chapter 1) at 1113; Hua Lan, *supra* note 109 at 589; Stephanie Koh, *supra* note 45 (Chapter 2) at 126.

135 See Andrea Gattini, "Between Splendid Isolation and Tentative Imperialism: The EU's Extension of its Emission Trading Scheme to International Aviation and the ECJ's Judgment in the ATA Case"(2012) 61 ICLQ 977 at 978; Markus W Gehring, "Case Note Air Transport Association of America v. Energy Secretary: Clarifying Direct Effect and Providing Guidance for Future Instrument Design for a Green Economy in the European Union" (2012) 2 RECIEL 149; Kulovesi, "Addressing Sectoral Emissions," *supra* note 214 (Chapter 1) at 199.

136 Although beyond of the scope of this book, opponents of the EU ETS also contend that the scheme's extraterritorial application of activity that takes places over the high seas violates principles of customary international law. The methodology used to calculate CO_2 emissions for the whole duration of the flight equates to a de facto claim of sovereignty over the high seas. See *United Nations Convention on the Law of the Sea*, at Art 89. Similarly, by regulating conduct that predominantly takes place over the high seas, the EU ETS contravenes Article 12 of the *Chicago Convention*. This provides that only ICAO is authorized to establish the rules of flight and maneuver of aircraft over the high seas. See *Chicago Convention*, *supra* note 36 (Chapter 2) at Art 12. Milde notes that ICAO carries out this function through the adoption of international standards in Annex 2 (Rules of the Air) to the Chicago Convention. See Milde, *International Air Law and*

an impermissible charge or fee levied solely for entering or exiting the territory of a State in breach of Article 15 of the *Chicago Convention*); (iv) contravenes provisions of the *Kyoto Protocol*, for ICAO is the only forum with jurisdiction over international aviation emissions; and (v) is contrary to various provisions of the EU-US open skies agreement.[137] In short, A4A contended that the scheme was unlawful in light of international treaties and well-established principles of customary international law.[138]

Pursuant to Article 267 of the *Treaty on the Functioning of the European Union* (TFEU),[139] the English Court asked the Court of Justice of the European Union (CJEU) for a preliminary ruling on whether a number of principles of customary international law and provisions of the *Chicago Convention* may constitute the basis to establish the scheme's validity.[140] On 21 December 2011, rejecting A4A's claim, the CJEU handed down its greatly anticipated decision.[141]

Although the court's decision will be examined in Chapters 5 and 6, at this stage it is sufficient to say that the CJEU expressly ruled that the principles of the *Chicago Convention* may not be used to assess the validity of the scheme, for the EU is not a party to this Convention. To the surprise of no one, the CJEU nonetheless found that the EU ETS is fully compatible with principles of customary international law.[142] As one could have expected,

ICAO, *supra* note 59 (Chapter 2) at 39. In this vein, Havel and Sanchez argue that EU ETS "disturbs the intent of [art] 12." The scheme "has a navigational impact" for aircraft operators, given the fact that "it adopts a unilateral navigational regime for EU airspace that conflicts with ICAO rules." See Havel & Sanchez, "Toward a Global Aviation Emissions Agreement," *supra* note 214 (Chapter 1) at 369. Only ICAO has jurisdiction over the high seas. *Ibid.* On the other hand, Europe has argued that Art 12 of the Chicago Convention only deals with issues relating to flight and maneuvers over the high seas. EU ETS does not require aircraft operators to "adhere to any particular flight path, specific speed limits, or limits on fuel consumption an exhaust gases." See Opinion of Advocate General Kokott delivered on 6 October 2011, Case C-366/10, online: Europa <http://ec.europa.eu/clima/news/docs/2011100601_case_c366_10_en.pdf at I-49> [Opinion of Advocate General Kokott]. See also Pietro Manzini & Anna Masutti, "The Application of the EU ETS System to the Aviation Sector: From Legal Disputes to International Relations?" (2012) 37 3-4 Air & Space L 307 at 320.

137 See Kati Kulovesi, "Make Your Own Special Song, Even if Nobody Else Sings Along": International Aviation Emissions and the EU Emissions Trading Scheme (2011) 2:4 Climate L 535 at 536 [Kulovesi, "Make your Own Special Song"]; Mayer, Case-366/10, *supra* note 133 (Chapter 1) at 1119.

138 See Written Observations of the Claimants in the Court of Justice of the European Union, reference from the High Court of Justice, London, Case C-366/10. See also Dejong, *supra* note 101 at 169; Fahey, *supra* note 134 at 1248; Kate Cook, "The Extension of the EU Emissions Trading Scheme to the Aviation Sector Does not Contravene International Law" *eutopia law* (1 November 2011) online: <http://eutopialaw.com/2011/11/01/the-extension-of-the-eu-emissions-trading-scheme-to-the-aviation-sector-does-not-contravene-international-law>; Robert "Bo" van Valkenburg, "Directive 2008/101: The EU Emissions Trading System," *International Relations* (13 January 2013) online: <http://internationalreports.wordpress.com/2013/01/13/directive-2008101>. See also Opinion of Advocate General Kokott, *supra* note 136 (noting that, in essence, the A4A claim argues that the EU ETS is an extraterritorial measure that impinges upon the sovereignty of non-EU States and it violates the customary principle of freedom of the high seas, at I-43).

139 See *EU Treaty*, *supra* note 10 at 164.

140 See *ATA Decision*, *supra* note 63 at para 1.

141 *Ibid.*

142 See Gattini, *supra* note 135 at 978.

the CJEU's judgment was subject to severe criticism outside the EU's boundaries.[143] Unfortunately, as Dejong correctly notes, this did not settle the controversy but rather alienated those who oppose the scheme.[144]

4.2.2 *Political Statements: From New Delhi to Moscow*

4.2.2.1 **The New Delhi Declaration**

During the summer of 2011, India and the United States intensively lobbied a group of like-minded States.[145] On 29-30 September 2011, this group, which is often referred to as the "Coalition of the Unwilling," met in New Delhi to discuss their official position with respect to the EU ETS.[146] As a result, 26 States adopted the Delhi Declaration.[147] Although not mentioning concrete actions, these States agreed to cooperate on non-MBM initiatives, such as operational and technical measures and alternative fuels.[148] The declaration further reaffirmed that these States "[o]ppose [Europe's] plan to include all flights by non-EU carriers [into the EU ETS]" and call upon "the EU and its Member States to refrain from including flights by non-EU carriers."[149]

The Coalition of the Unwilling constituted one of the most heterogeneous groups ever assembled in the annals of international civil aviation. It included developed, developing, and least developed countries, countries that adhered to CBDR, as well as those who abhor it. It encompassed all the major non-EU aviation powerhouses, emerging markets,

---

143 Havel & Mulligan argue that the ruling left aside long-standing aviation law principles. See Havel & Mulligan, *supra* note 60 (discussing that the judgment represents "questionable, environmentally motivated exceptions to the international law principles upon which the global operation of international air transport industry has depended for the past seven decades" at 5). In addition, IATA noted that the decision did not reflect the view of the majority within the international community. See IATA, Press Release, no 63, "IATA Disappointed with EU Court Decision on ETS" (21 December 2011) online: IATA <www.iata.org/pressroom/pr/pages/2011-12-21-01.aspx> (noting that the ECJ "may reflect European confidence in European plans. But that confidence is by no means shared by the outside world where opposition is growing"). Similarly, A4A noted that CJEU's decision "further isolated the EU from the rest of the world and will keep in place a unilateral scheme that is counterproductive to concerted global action on aviation and climate change." "A4A Comment on European Court of Justice Decision," *A4A* (21 December 2011), online: <www.airlines.org/Pages/news_12-21-2011.aspx>.

144 See Dejong, *supra* note 101 at 164; Goron, *supra* note 35 (Chapter 2) at 18.

145 See Doaa Abdel Motaal, "Curbing CO_2 Emissions from Aviation: Is the Airline Industry Headed for Defeat?" (2012) 3 Climate Change 1 at 19.

146 See Kulovesi, "Make Your Own Special Song," *supra* note 137 at 536.

147 Argentina, Brazil, Chile, China, Colombia, Cuba, Egypt, India, Japan, Republic of Korea, Malaysia, Mexico, Nigeria, Paraguay, Qatar, Russian Federation, Saudi Arabia, Singapore, South Africa, the United States of America, and the United Arab Emirates signed the New Delhi Declaration. However, Australia and Canada, also members of the ICAO Council, did not.

148 See New Delhi Declaration, *supra* note 132 (Chapter 1) at paras 2 and 3. It did not, however, establish a road map to achieve these objectives. This suggests that these objectives were merely reflected in the declaration as part of the political statements of its signatories, as opposed to a real and concrete will to achieve them.

149 See *Ibid.* at paras 5 and 6.

economies in transition, as well as a large share of the fastest-growing countries. Critics would say that the group never sought to present meaningful, alternative solutions to the European scheme, nor did it purport to genuinely advance the work in tackling international aviation's carbon footprint. In light of the diverse interests within its constituency, the main purpose of this group was predominately to halt the EU ETS.

4.2.2.2 ICAO Council Declaration of 2 November 2011

Shortly after the meeting in New Delhi, on 17 October 2011, 26 non-EU Member States (all of whom were members of the ICAO Council) tabled a proposal requiring the *in totum* adoption of the New Delhi Declaration by the ICAO Council.[150] This move sought to establish the opposition to the scheme as the official (political) position of the organization.[151] On 2 November 2011, at the second meeting of its 194th Session, the Council examined this proposal.[152] After a four-hour discussion and with 10 reservations, the Council finally passed the ICAO Declaration of 2 November 2011 against the EU ETS.[153] This made clear that the ICAO Council opposed the European scheme for being "inconsistent with applicable international law."[154] Although the declaration did not immediately prevent the implementation of the scheme, it was nonetheless indicative of the level of worldwide opposition that the scheme has attracted. Indirectly, it reaffirmed Europe's isolation on this issue.

Although it was perceived as a huge victory for the Coalition of the Unwilling, the process that led to the adoption of the declaration by the Council was unprecedented in a number of ways. First, as the United Kingdom noted, this marked the first time that the ICAO Council was called upon to pass a "declaration" which had essentially been negotiated and had already been adopted elsewhere.[155] Second, given that the 26 non-EU Member States clearly outnumbered European States, European Council members did not have a

150 See ICAO, C-WP/13790 *Inclusion of International Civil Aviation in the European Union Emissions Trading Scheme (EU ETS) and its Impact* (presented by Argentina, Brazil, Burkina Faso, Cameroon, China, Colombia, Cuba, Egypt, Guatemala, India, Japan, Malaysia, Mexico, Morocco, Nigeria, Paraguay, Peru, Republic of Korea, Russian Federation, Saudi Arabia, Singapore, South Africa, Swaziland, Uganda, the United Arab Emirates, and the United States) [Position of the Coalition of the Unwilling].

151 From a governance perspective, it may be questionable whether the ICAO Council – a group of only 36 Member States – is the competent body to articulate the organization's position over an issue with highly charged political elements. One could argue that this should be the sole purview of the ICAO Assembly.

152 See ICAO, C-MIN 194/2, *supra* note 86 (Chapter 1).

153 The eight European Council Delegations (Belgium, Denmark, France, Italy, Germany, Spain, Slovenia, and United Kingdom) filed an objection against the Council Declaration of 2 November 2011. Although expressly opposing the unilateral application of the EU ETS, Australia and Canada did not join the Position of the Coalition of the Unwilling. See ICAO, Declaration Adopted by the Council of the International Civil Aviation Organization (ICAO) at the Second Meeting of the 194th Session on 2 November 2011 [*ICAO Council Declaration of 2 November 2011*]. See also ICAO, C-MIN 194/2, *supra* note 86 (Chapter 1).

154 See ICAO, C-DEC 194/2, *Council, 194th Session, Decision of the Second Meeting* at Appendix, para 5 [ICAO, C-DEC 194/2].

155 See ICAO, C-MIN 194/2, *supra* note 86 (Chapter 1) at 12 and 20 (Statement of the United Kingdom).

chance to introduce a single change to the ICAO Council Declaration of 2 November 2011.[156] In fact, while presenting the Position of the Coalition of the Unwilling, India made very clear that this was a "non-negotiable" proposal.[157] Third, because the language was taken almost verbatim from the New Delhi Declaration, the text suffers from acute deficiencies and contradictions.[158] Fourth, in spite of the reservations filed, the adoption of the Declaration de facto constitutes a (political) pronouncement on the part of the Council on the illegality of the European scheme.

4.2.2.3 Moscow Declaration

Immediately after the ICAO Council meeting of 2 November 2011, the Russian Federation, in coordination with the United States, started the preparation of a follow-up meeting that would take place in Moscow from 20-21 February 2012.[159] Initially, as part of the agenda, the Russian Federation proposed to take concrete countermeasures to force the cancellation of the EU ETS. The idea was to move beyond a mere "declaratory" statement opposing the scheme. Although original drafts circulated by the Russian Federation prior to the meeting were much more aggressive,[160] on 20 February 2012, 23 States only agreed to "consider" a "basket of actions,"[161] such as triggering the dispute settlement mechanism of the *Chicago Convention* (Article 84), adopting blocking legislation to prohibit non-EU carriers from joining the EU ETS, ordering EU carriers to submit data, reviewing air service agreements, and levying charges[162] to EU carriers as a countermeasure.[163] The final wording

156 *Ibid.* at 4.

157 *Ibid.*

158 For instance, the Declaration states that the Council "[calls] on ICAO to continue to undertake efforts to reduce aviation's contribution to climate change." See ICAO, C-DEC 194/2, *supra* note 154 at Appendix 1, para 1. The Council is the governing body of ICAO. As such, it does not call on ICAO to do something, but rather either executes the mandate entrusted by the Assembly or decides what the policy of the organization should be on a set of issues. The Declaration also indicates that the Council "[i]ntends to accelerate the development and implementation of low-carbon aircraft technologies and sustainable alternative fuels." *Ibid.* at para 3. These would seem to be more germane to the actions of some of, but definitely not all, the States which form part of the Position of the Coalition of the Unwilling.

159 See Joint Declaration of the Moscow Meeting on the Inclusion of International Civil Aviation in the EU ETS of 20-21 February 2012.

160 See Dejong, *supra* note 101 at 180.

161 Motaal points that by extending the EU ETS to international aviation, the EC has de facto "turned itself into the global regulator." Motaal, *supra* note 145 at 21.

162 It has been noted that imposing charges against EU air carriers may be in conflict with Art 15 of the *Chicago Convention*, given that such a measure most likely will be discriminatory. See Manzini & Masutti, *supra* note 136 at 316.

163 See *Joint Declaration of the Moscow Meeting on the Inclusion of International Civil Aviation in the EU ETS*, 20-21 February 2012. Representatives of Armenia, Argentina, Azerbaijan, Republic of Belarus, Brazil, Cameroon, China, Cuba, Chile, India, Japan, Republic of Korea, Mexico, Malaysia, Nigeria, Paraguay, the Russian Federation, Saudi Arabia, Seychelles, Singapore, South Africa, Uganda, and the United States of America adopted the Moscow Declaration. See also Reuters, "Opponents of EU Airline CO_2 Scheme to Meet

of the Moscow Declaration left open for each signatory State to decide whether they will in fact take such measures.

4.2.3 *The Nail in the Coffin: The Anti-EU ETS US Bill*

In the 1970s, the continuous extraterritorial application of US antitrust law ignited ferocious responses and produced "retaliatory legislation by foreign governments." In fact, Lowe notes that during this time 16 countries passed legislation as a response to what they perceived was an unlawful extension of US antitrust law.[164] Blocking legislation involves a state passing a domestic statue where it deems that another State has unlawfully and extraterritorially applied its laws.[165] Thus, jurisdictional overreach prompts States to enact blocking legislation to protect areas or sectors that form part of their national sovereign interests.[166] For instance, the United States enacted blocking legislation regarding an EU noise regulation seeking to prevent the use of hushkitted aircraft.[167] Clearly, enacting

---

in Moscow" (6 February 2012), online: <http://uk.reuters.com/article/2012/02/06/uk-russia-aviation-idUKTRE8151QW20120206>.

164 See AV Lowe, ed, *Extraterritorial Jurisdiction: An Annotated Collection of Legal Materials* (1983). These include legislation in Australia, Belgium, Canada (the Uranium Information Security Regulations of 22 September 1976 and the Business Concerns Records Act of 1964 of the Province of Quebec), Denmark, Finland, France, Germany, Italy, the Netherlands, New Zealand, Norway, the Philippines, South Africa, Sweden, Switzerland (Art 47 of the Swiss Federal Bank Act), and the United Kingdom (the Shipping Documents and Commercial Documents Act of 1964). See also Maher M Dabbah, *The Internationalisation of Antitrust Policy* (Cambridge: Cambridge University Press, 2003) (suggesting that opposition against the extraterritorial application of US antitrust laws has increased,) at 187.

165 Mann conceives blocking legislation as "an act of self-defense" that seeks to remedy a perceived "international wrong inherent in the assumption of excessive jurisdiction." FA Mann, *Further Studies in International Law* (Oxford: Claredon Press, 1990) at 83. Similarly, Meessen suggests that blocking legislation "reflects a perceived impairment of foreign interests and may indicate the violation of an existing or an emerging rule of international law." Karl M Meessen, "Antitrust Jurisdiction under Customary International Law" (1984) 78 Am J Int'l L 783.

166 See Joseph P Griffin, "US Supreme Court Encourages Extraterritorial Application of US Antitrust Laws" (1993) 21 Int'l Bus L 389 at 391.

167 US, H Res 86, 106th Cong, 1999 (directing the US Secretary of Transportation to ban Concorde, as long as the EU adopts the hushkit regulation). By lodging a formal complaint at ICAO, the United States challenged the legality of the EU Regulation vis-à-vis the *Chicago Convention*. Therefore, as a result of extensive diplomatic negotiations, the EU was forced to repeal EU Regulation 925/99. This was done through EC, *Directive 2002/30/EC of the European Parliament and of the Council of 26 March 2002 on the Establishment of Rules and Procedures with regard to the Introduction of Noise-related Operating Restrictions at Community Airports* [2002] OJ, L 85/40. See also EC, *Councill Regulation (EC) No 925/1999 of 29 April 1999 on the Registration and Operation within the Community of Certain Types of Civil Subsonic Jet Aeroplanes which Have Been Modified and Recertificated as Meeting the Standards of Volume I, Part II, Chapter 3 of Annex 16 to the Convention on International Civil Aviation, Third Edition* (July 1993) [1999] OJ, L 115/1; EC, *Directive 2002/30/EC of the European Parliament and of the Council of 26 March 2002 on the Establishment of Rules and Procedures with regard to the Introduction of Noise-related Operating Restrictions at Community Airports* [2002] OJ, L 85/40.

blocking legislation has not been unusual in the international aviation setting.[168] In more recent times, the EU ETS once again epitomizes the tension created by blocking legislation.[169]

On 20 July 2011, the Chairman of the Transportation Committee, John Mica (R-FL),[170] introduced before the US House of Representatives a bill seeking to prohibit US airlines from adhering to the EU ETS.[171] This legislative proposal enjoyed strong support from industry trade associations such as A4A and IATA.[172] Although, on 24 October 2011, the House of Representatives passed this bill, proponents did not gather sufficient support in the US Senate. At that time, the bill was not approved. In light of this failure, later in December 2011, US Republican Senator John Thune (R-SD)[173] introduced another similar bill.[174] On 22 September 2012, the US Senate unanimously passed it.[175] Finally, and despite the pressure from a number of NGOs,[176] on 27 November 2012, US President Barack Obama signed into law the *European Union Emissions Trading Scheme Prohibition Act of 2011*, the first ever blocking legislation involving an initiative seeking to reduce GHGs.[177]

---

168 See Fahey, *supra* note 134 at 1249.

169 See Lea Brilmayer, “Extraterritorial Application of American Law: a Methodological and Constitutional Appraisal” (1987) 50 Law & Contemp Probs 11.

170 According to the Center for Responsive Politics (also known as Open Secrets.Org), a research group tracking money in US politics, for its 2012 congressional re-election, Mr. Mica received US$ 176,295 from donors related to the US air transport sector and US$ 149,595 from political action committees (PACs) related to the air transport sector. This included both donations from US airlines and its trade associations. See Open Secrets.org, John Micca, online: <www.opensecrets.org/politicians/summary.php?cid=N00002793&cycle=2012>.

171 See US, Bill HR 2594, *An Act to Prohibit Operators of Civil Aircraft of the United States from Participating in the European Union’s Emissions Trading Scheme, and for other Purposes*, 112th Cong, 2011.

172 See Amy Westervelt, “How Airlines Are Fighting Carbon Trading,” *Greenbiz* (8 August 2012) online: <www.greenbiz.com/news/2012/08/08/how-airlines-fighting-carbon-trading>; IATA, Press Release, “Statement on US Bill on the EU Emissions Trading” (20 July 2011) online: <www.iata.org/pressroom/pr/Pages/2011-07-20-01.aspx>; “A4A Commends Bipartisan Support of Senate EU ETS Bill,” *A4A* (7 March 2012) online: <www.airlines.org/Pages/news_3-7-2012.aspx>.

173 The Center for Responsive Politics also reports that, from 2007 through 2012, Senator John Thune received US$ 348,900 from both individual firms and PACs related to the air transport sector. See Open Secrets.Org, “Sen. John Thune 2007/2012,” online: <www.opensecrets.org/politicians/industries.php?cycle=2012&cid=N00004572&type=I&newmem=N>.

174 See US, S 1956, *An Act to Prohibit Operators of Civil Aircraft of the United States from Participating in the European Union’s Emissions Trading Scheme, and for other Purposes*, 112th Cong, 2012 [“EU ETS Blocking Legislation”]; “Following House Passage, Mirror Bill Introduced into US Senate to Block US airlines from participating in EU ETS,” *Green Air Online* (8 December 2011), online: <www.greenaironline.com/news.php?viewStory=1382>.

175 See “Senate Votes to Shield US Airlines from EU’s Carbon Scheme,” *Reuters* (24 September 2012) online: <www.reuters.com/article/2012/09/24/uk-usa-carbon-airlines-idUSLNE88N00K20120924>.

176 Keith Laing, “Greens Pressure Obama to Veto Airline Emissions Bill,” *The Hill* (14 November 2012) online: <http://thehill.com/blogs/transportation-report/aviation/267841-enviros-cast-airline-emissions-bill-as-obamas-first-post-sandy-climate-test>.

177 See “EU ETS Blocking Legislation,” *supra* note 174. See also Rachel Brewster, “US-Europe Fight over Airline Emissions Could Help Talks on Climate Change,” *The Christian Science Monitor* (15 January 2013) online: <www.csmonitor.com/layout/set/print/Commentary/Opinion/2013/0115/US-Europe-fight-over-airline-

Arguably, US airlines would have liked the blocking legislation to have been adopted long before the start of the EU ETS compliance period for 2012. However, the US presidential election proved to be an insurmountable obstacle. Owing to the fact that the Obama Administration pursued reelection in one of the most difficult and closest political contexts of recent times, the White House deliberately opted for a very cautious but firm diplomatic approach when handling the position of the US government with respect to the EU ETS.[178] Although the United States had engaged the Commission in diplomatic negotiations for a number of years, it led the aggressive attacks against the Commission to other countries such as China, India, and the Russian Federation.[179] A more belligerent approach by the US Government might have alienated President Obama's electoral base.[180] In fact, a number of NGOs and pro-environment personalities openly denounced the efforts of the US airlines and certain agencies within the US government to derail the EU ETS at all cost while at the same time blocking efforts in the international arena to reduce GHGs from international aviation.[181] Thus, it should not be a surprise that, although it did not openly express

---

emissions-could-help-talks-on-climate-change>. Petsonk comments that the EU ETS blocking legislation marked the third time that the United States has enacted blocking legislation. The two previous cases involved the South African apartheid and boycotts by the Arab League to Israel. See Annie Petsonk, "Aviation on the Flight Path to Success" *Carbon Finance*, (Spring 2013), online: EDF <www.edf.org/sites/default/files/CF Spring2013_On_the_flightpath_to_success_Petsonk.pdf> 22 [Petsonk, Aviation].

178 For instance, in 2011, in a joint letter US Secretary of State, Mrs. Hilary Clinton, and the US Secretary of Transportation, Mr. Raymond LaHood, advised the Commission that the United States would "take appropriate action," should the EU not stop the scheme's implementation. See letter from 19 NGOs addressed to Hilary Rodham Clinton, US Secretary of State, & Raymond H La Hood, Secretary of Transportation (30 July 2012) online: NBAA <www.nbaa.org/ops/environment/eu-ets/20120731-coalition-letter-us-hosted-aviation-climate-meeting.pdf>.

179 See "Obama Administration Pressed On EU ETS," *Airwise* (28 March 2012), online: <http://news.airwise.com/story/view/1332974498.html>.

180 The Keystone XL Pipeline Project (Keystone) represents another interesting episode that reveals how the domestic politics of the United States affect environmental policy decisions. Highly supported by Republicans but opposed by Democrats, Keystone involves the construction of a 2,700 kilometer pipeline that would transport oil from Canada's tar sands situated in the province of Alberta all the way down to US refineries in the Gulf of Mexico. The project, which was originally conceived as a major energy initiative to relieve the United States of its high dependency on Middle Eastern oil, is estimated to cost US$ 7 billion. In November 2011, by requesting further environmental impact assessment to the Keystone XL Pipeline project, the US State Department in fact significantly delayed the pipeline's construction. This created a considerable tension in the Canada-US relationship. The press reported that the White House opted for this move to avoid making a decision on the environmental suitability of the project before the 2012 Presidential Election. It was thought that the Obama Administration weighed in whether an eventual go-ahead to the project could have alienated its "green" constituents who had serious concerns about the project's environmental impact. See The Globe & Mail, "Video: US Delays Keystone XL Approval," *The Globe & Mail* (11 November 2011) online: <www.theglobeandmail.com/report-on-business/video/video-us-delays-keystone-xl-approval/article2233242/>; Konrad Yakabuski, "Keystone Pipeline Delay Throws a Bone to Obama's Democratic Base," *The Globe & Mail* (10 November 2011) online: The Globe & Mail <www.theglobeandmail.com/news/world/konrad-yakabuski/keystone-pipeline-delay-throws-a-bone-to-obamas-democrat-base/article2233033/>.

181 See letter from numerous Nobel Laureates and renowned economists addressed to President Barack Obama (14 March 2012) online: WWF <http://assets.wwf.org.uk/downloads/eu_ets_letter_from_economists_to_

opposition to the blocking legislation, the Obama administration did not expressly endorse it either.[182] It is fair to recognize that its passing by the US Congress is the result of the resilient and unyielding lobbying efforts of A4A.[183]

Unlike the previous proposal introduced in the House of Representatives, the EU ETS blocking legislation does not automatically prohibit, but rather authorizes, under certain conditions, the US Secretary of Transportation to "prohibit [US airlines] from participating in the [scheme]."[184] In theory, for the prohibition to become effective, the Secretary of Transportation must carry out an assessment. He or she ought to establish that the prohibition is in the US public interest.[185] In addition, factors such as "impacts on [US] consumers, [US airlines], and the impacts on the [US economy], energy and environmental security" should also be taken into account.[186] Although these factors may suggest that the US Secretary of Transportation must make a determination on the basis of an objective assessment, in practice, given their vagueness, it may well be a very subjective exercise ignited by the political pressure of the time.

The legislation encourages the FAA and the DOT to undertake international negotiations "to pursue a worldwide approach to address aircraft emissions, including the environmental impact of aircraft emissions."[187] Similarly, the Secretary of Transportation ought to ensure that US airlines be unharmed from the implementation of the EU ETS.[188] A priori, this provision may have been perceived as mere "wishful thinking" on the part of the US airline industry. It may be ludicrous to expect that the US Government would completely immunize the sector from any potential financial burden. Yet arguably, this congressional mandate bestowed upon the Secretary of Transportation is in fact one of A4A's major

---

obama.pdf> (asking the US President Barak Obama to "to support the European Union's innovative efforts to place a price on carbon from aviation through the emissions trading" and calling upon the US "not to oppose" European efforts). See also "Nobel Economists Urge Obama to Support EU Aviation Carbon Scheme," *Clean Technica* (15 March 2012) online: <http://cleantechnica.com/2012/03/15/nobel-economists-urge-obama-to-support-eu-aviation-carbon-scheme/>.

182 Perhaps the biggest disappointment for the US airlines on the lack of support from the Obama Administration in their crusade against the EU ETS occurred on 6 June 2012, when the US Senate held a Congressional Hearing on the then proposed bill to enact blocking legislation. At that time, US Secretary of Transportation, Mr. Raymond H. LaHood, openly said that, at that stage, the Administration was not in a position to endorse the bill. See Jen DiMascio, "LaHood Won't Endorse Senate Ban On EU ETS Compliance," *Aviation Week & Space Technology* (6 June 2012) online: <www.aviationweek.com/Article.aspx?id=/article-xml/awx_06_06_2012_p0-465479.xml>.

183 Environmental NGOs estimate that A4A spent US$ 4.3 million in lobbying efforts to get the blocking legislation passed by the US Congress. See Transport & Environment, "Briefing: The Billion Euro Aviation Bonanza," *supra* note 128.

184 See "EU ETS Blocking Legislation," *supra* note 174 at § 2 (a).

185 AEA criticized the blocking legislation, for it only contributes to "[increasing] the tension" between the two blocks. See "AEA Calls on the Commission to Act in Face of Mounting International Pressure on ETS," *AEA* (26 July 2011) online: <http://files.aea.be/News/PR/Pr11-015.pdf> [AEA].

186 See "EU ETS Blocking Legislation," *supra* note 174 at § 2 (1), (2), and (3).

187 *Ibid.* at § 3 (1).

188 *Ibid.* at § 3 (2).

achievements, an astute move with significant consequences. In practical terms, this frames the negotiating position of the United States regarding the EU ETS. This means that in its dealing with their European counterparts, the US negotiators should limit the scope of the scheme's application such as to reduce, to the extent possible, the potential financial impact on US carriers. Arguably, the blocking legislation has been designed as a negotiation tool to force the Commission to abandon the implementation of the scheme.[189] It has served as an unquestionable indication of the level of animosity that the scheme has generated.[190] It is no coincidence that President Obama's enactment of this legislation came just thirteen days after the Commission had announced that it would suspend the enforcement of the scheme for one year; its effectiveness in "persuading" the other side is undeniable. The blocking legislation has also encouraged other countries, such as China, Russia, and Saudi Arabia, to order their air carriers not to join the EU ETS.[191]

4.3 Why Did Europe Decide to Fly Solo?

The international community has vociferously condemned the inclusion of aviation into the community trading regime. Because of uncountable legal uncertainties, it was suggested that the EU should abandon the inclusion of international aviation.[192] Arguably, Europe has masterfully managed to upset the whole world. One cannot help but ask why Europe decided to pursue such dangerous and potentially costly (at least in terms of international diplomacy) regulatory approach to climate change. In this section, we attempt to find some explanations.

Contrary to the regulatory apathy that most countries have thus far professed, Europe has looked into different options to regulate CO_2 emissions from aviation since the early 1990s.[193] A contrario, albeit permissible under federal law,[194] the United States has been rather passive on the regulation of aircraft emissions.[195] A "cultural difference" may explain

189 Rep Micca said that hopefully the blocking legislation has "piled enough pressure on the EU for it to back down from applying its law." "US House OKs Bill To Shield Airlines from ETS," *Airwise* (14 November 2012), online: <http://news.airwise.com/story/view/1352872192.html>.

190 See AEA, *supra* note 185.

191 See Gattini, *supra* note 135 at 989; Goron, *supra* note 35 (Chapter 2) at 1.

192 See Daniel B Reagan, "Putting International Aviation into the European Union Emissions Trading Scheme: Can Europe Do it Flying Solo?" (2008) 35 BC Envtl Aff L Rev 349 at 384.

193 See Morrell, *supra* note 54 at 5564.

194 See, e.g., *Massachusetts v Environmental Protection Agency*, 549 US 497, 127 S Ct 1438 (2007). Chichilnisky & Sheeran point out that, since 1963, the US *Clear Air Act* "gives the federal government the authority to regulate and limit greenhouse gas emissions." Chichilnisky & Sheeran, *supra* note 7 (Chapter 2) at 119.

195 See Daniel Warner, "Commercial Aviation: An Unsustainable Technology" (2009) 74 J Air L & Com 553 at 567.

this difference in approach.[196] Environmental protection forms part of a set of values that most Europeans so dearly treasure.[197] When properly internalized, these values contribute to establishing norm formation which in turn triggers the adoption of policy imperatives, such as tackling aviation's carbon footprint.[198] While values are addressed in detail in Chapter 7, it suffices to say here that values place a significant role in setting environmental policy. This is not necessarily the case in most other regions where other pressing needs determine regulatory priorities.[199] Even in the general context of climate change, for a large number of developing countries, "economic and social development and the eradication of poverty" at the top of their policy agenda and emission reductions are merely an afterthought.[200] In Africa, for instance, basic aeronautical infrastructure and safety oversight are high priority areas of concern.[201]

In Europe, on the other hand, there is a perception that when it comes to internalizing aviation-generated environmental externalities, the sector is a notorious free rider. Mr. Andrew Haines, Chief Executive Officer of United Kingdom's Civil Aviation Authority, said that "unless the sector faces its environmental impact head-on, it will not be allowed to grow."[202]

196 See Vogler & Bretherton, *supra* note 194 (Chapter 1) at 19; Separate Opinion of Judge Weeramantry in *Gabcíkovo-Nagymaros Project, supra* note 228 (Chapter 2) (describing Europe as having "a deep-seated tradition of love for the environment, a prominent feature of European culture" at 108).

197 See Brunnée & Toope, *supra* note 6 (commenting that the "EU and its member states and, arguably, even its public, have actually internalized the goals and principles of the global climate regime to a significant degree" at 175).

198 See Vandenbergh, *supra* note 61 (Chapter 1) at 1102.

199 When most of the world is only concerned with aviation's CO_2 footprint, some commentators advocate that the policy instrument should also factor other impacts on climate change such as radiative forcing. See Piers M de F Forster, Keith P Shine, & Nicola Stuber, "It is Premature to Include non-CO_2 Effects of Aviation in Emission Trading Schemes" (2006) 40 Atmospheric Environment 1117.

200 See Haroldo Machado-Filho, "Financial Mechanisms under the Climate Regime" in Jutta Brunnée, Meinhard Doelle, & Lavanya Rajamani, *Promoting Compliance in an Evolving Climate Regime* (Cambridge: Cambridge University Press, 2012) 216 at 216.

201 Early in 2013, IATA indicated that, although the airline industry enjoyed its best safety year ever in 2012, Africa's accident rate was the worst in the world. See IATA, Press Release, "2012 Best in History of Continuous Safety Improvements" (28 February 2013) online: IATA <www.iata.org/pressroom/pr/Pages/2013-02-28-01.aspx>; Aaron Karp, "IATA: 2012 Western-built jet accident rate lowest in history," *Air Transport World* (23 February 2013) online: <http://atwonline.com/operations-maintenance/news/iata-2012-western-built-jet-accident-rate-lowest-history-0228>. On climate change, the African region is not so much focused on mitigation, but rather on adapting to the changes that the continent will experience as a result of climate change. See Paul Collier, Gordon Conway & Tony Venables, "Climate Change and Africa" (2008) 24:2 Oxford Rev Econo Pol'y 337.

202 "Aviation must tackle its environmental impact before it can be allowed to grow, says UK's aviation regulator," *Green Air Online* (27 January 2012) online: <www.greenaironline.com/news.php?viewStory=1421>.

As commentators note, Europe "has shown a considerably higher willingness to reduce its emissions than [other countries such as, for instance,] the United States and Japan."[203] Clearly, Europe wants to exercise global leadership on the issue of GHGs in order to prevent "dangerous anthropogenic interference with the climate system."[204] Since 1990, aviation's carbon footprint in Europe has grown by some 98 percent, while total emissions have declined.[205] At present, the sector's emissions account for approximately 3 percent of European CO_2 inventories.[206] These emissions are increasing at a rate of 6 to 7 percent per year.[207] To date, aviation's carbon footprint in the EU ranks second, right after electric power generation.[208] As Helm suggests, just the growth of aviation's emissions is sufficient to undermine the reductions achieved under the *Kyoto Protocol*.[209] Also, reports indicate that, in 2012, emissions from international passenger aircraft accounted for 360 million tonnes of CO_2. Europe was responsible for 36 percent of these emissions.[210] If aviation continues growing at this pace, emission reductions achieved by other sectors would become meaningless. This is particularly troublesome in the case of the United Kingdom where the government announced in 2003 a plan to reduce the country's CO_2 footprint by 60 percent by 2050.[211] In the UK, aviation emissions already doubled in 2010, as compared to 1990 levels.[212] As the fastest-growing source of CO_2 in the United Kingdom, the projected growth of aviation may destabilize the country's overall CO_2 reduction targets.[213] From a European public policy perspective, it is extremely difficult to justify that aviation CO_2 emissions continue to be unregulated *persecula seculorum*.[214] Clearly, Europe is the

---

203 Carbone et al, *supra* note 84 at 274; Harrison, "The United States as Outlier," *supra* note 2 (suggesting that the US government has adopted weak measures to tackle climate change. Policies have predominately relied upon voluntary measures, at 67).

204 General EU ETS Directive, *supra* note 196 (Chapter 1) at recital 2 at 3.

205 See Runge-Metzger, *supra* note 204 (Chapter 1). See also Daniel Calleja Crespo & Mike Crompton, "The European Approach to Aviation and Emissions Trading" (2007) 21:3 Air & Space L 1 (indicating that from 1990 to 2004 aviation emissions grew 87 percent).

206 See Anger, "Including Aviation in the EU ETS," *supra* note 2 (Chapter 1) at 110.

207 See Bows et al, *supra* note 94 (Chapter 1) at 106. Conversely, the road transport sector is accountable for a third of the total of Europe's CO_2 emissions. See Charlotte Streck & David Freestone, "The EU and Climate Change" in Richard Macrory, ed, *Reflections on 30 Years of EU Environmental Law: A High Level of Protection* (Groningen: Europa Law Publishing, 2006) 87 at 102.

208 Petsonk, "Testimony," *supra* note 98 at 5. See also Leggett et al, *supra* note 105 at 10.

209 See Helm, "Climate Change Policy," *supra* note 80 (Chapter 2) at 218.

210 Asia Pacific accounted for 29 percent, while North, Central America, and the Caribbean for 19 percent. See Southgate, *supra* note 97 (Chapter 1) at 13.

211 See Brand & Boardman, *supra* note 97 (Chapter 1) at 224.

212 See "UK Climate Advisers Recommend International Aviation Emissions be Included in National Carbon Budgets," *Green Air Online* (5 April 2012), online: <www.greenaironline.com/news.php?viewStory=1447>.

213 *Ibid.*

214 See Leonardo Massai, "Legal Challenges in European Climate Policy" in W Th Douma, L Massai, & M Montini, eds, *The Kyoto Protocol and Beyond: Legal and Policy Challenges of Climate Change* (The Hague: TMC Asser Press, 2007) 13 (suggesting that reducing aviation's carbon footprint is a key policy priority for the Commission, at 16).

region with the most (self-imposed) pressure to reduce emissions.[215] These are some of the reasons that may explain why Europe has decided to fly solo.

4.3.1 *Rationale for Including Non-EU Aircraft Operators*

The international community does not question Europe's authority to regulate the carbon footprint of EU aircraft operators.[216] What many States have challenged is the legal basis for the unilateral application of the EU ETS to non-EU aircraft operators, since most of their emissions occur outside European airspace. In fact, all of the legal arguments advanced against the EU ETS pivot on this point. Opponents of the extension of the EU ETS to international aviation contend that, in order for Europe to do so legitimately, the consent of the States concerned must first be obtained.[217] The Aviation Directive justifies the inclusion of foreign aircraft operators in the EU ETS by resorting to the provisions of Article 11 of the *Chicago Convention* (Application of Air Regulations),[218] under which a State is entitled to apply its laws to aircraft arriving in or departing from its territory, provided it does so in a non-discriminatory manner.[219] According to the European argument, this provision confers a prerogative and imposes an obligation. Thus, on the one hand, States can apply their national laws to foreign aircraft entering or leaving their territories. And on the other hand, in doing so, states are prohibited from discriminating among aircraft operators on the basis of nationality. The first element would authorize the EU to carry out its ETS. The second would require that foreign aircraft operators be subject to the scheme.

Mendes de Leon criticizes this rather superficial interpretation of the first element, on the ground that it does not take into account the fact that a State enjoys such prerogative inasmuch as the foreign aircraft subject to the regulation in question is, to use the language of the *Chicago Convention*, "within its territory."[220] Implicitly, the thrust of the argument lies in the fact that the purpose is not to regulate behavior that takes place beyond the territorial boundaries of the State imposing the measure. Be that as it may, the reality is that states often invoke it when they justify the adoption of measures that may have some extraterritorial flavor. The issue of extraterritoriality is addressed in Chapter 5.

215 See Morrell, *supra* note 54 at 5563.

216 See Havel & Sanchez, "Toward a Global Aviation Emissions Agreement," *supra* note 214 (Chapter 1).

217 On the basis of the principle of territorial sovereignty, Hardeman suggests that in order to include foreign operators, the EU requires the agreement of all non-EU States. See Hardeman, "Common Approach," *supra* note 113 (Chapter 1) at 15.

218 See *Chicago Convention*, *supra* note 36 (Chapter 2) at Art 11.

219 See Aviation Directive, *supra* note 17 (Introduction) at recital 9.

220 Pablo Mendes de Leon, "Enforcement of the EU ETS: The EU's Convulsive Efforts to Export its Environmental Values" (2012) 37:4 Air Sp. L. 287 [Mendes de Leon, "Enforcement of the EU ETS"]. See also *Chicago Convention*, *supra* note 36 (Chapter 2) at Art 11.

From Europe's policy perspective, the inclusion of non-EU aircraft operators within the scope of the regime is intended to achieve a twofold objective.[221] First, it aims to avoid, or at least limit, market distortions and carbon leakage.[222] Applying the regime exclusively to EU aircraft operators indirectly creates a competitive advantage in favor of non-EU aircraft operators, since they would not be subject to the financial, operational, and administrative burdens of the EU ETS. Second, by extending the scheme's reach to non-EU aircraft operators, the measure is more likely to attain the environmental goals pursued.

From a European environmental perspective, the inclusion of foreign aircraft operators is desirable. Excluding them would significantly undermine the scheme's environmental effectiveness.[223] Yet, the real question is whether the environmental gains outweigh the political cost that Europe has sustained in implementing the inclusion of international aviation into the EU ETS.[224]

4.3.2 *The "Temporary" Suspension of the EU ETS*

As a result of mounting international pressure, on 12 November 2012, the EU Commissioner on Climate Change, Miss Connie Hedegaard, announced that the Commission would put forward a proposal to the Parliament and the Council of Europe to temporarily suspend the enforcement of the EU ETS.[225] This is known as "stopping the clock."[226] In April 2013, the European Parliament approved a one-year suspension.[227] Although some would have preferred the complete cancellation of the scheme, the move was for the most part welcome.[228] The Commission indicated that the suspension was prompted by the

---

221 See Barton, Including Aviation in the EU ETS, *supra* note 2 (Chapter 1) at 185.

222 See Joshua Meltzer, "Climate Change and Trade: The EU Aviation Directive and the WTO" (2012) 15 J Intl'l Econ L 111 at 112.

223 See Aviation Directive, *supra* note 17 (Introduction) at recital 16 at 5. In justifying its legality, the CJEU reasoned that the inclusion of foreign aircraft operators was necessary to ensure a "high level of protection" in achieving the environmental policy objectives of the European Union. See *ATA Decision*, *supra* note 63 at para 128.

224 Kysar and Salzman note that in most environmental policy issues, there is a polarization of positions between those who support environmental measures and those who deny them outright. This confrontation leads to "environmental tribalism." The challenge for both sides is to find some middle ground where more eco-pragmatic solutions would be more likely to be implemented and accepted. Douglas A Kysar & James Salzman, "Environmental Tribalism" (2003) 87 Minn L Rev 1099 at 1103.

225 See EC, "Stopping the Clock of ETS and Aviation Emissions Following Last Week's International Civil Aviation Organization (ICAO) Council," MEMO/12/854 (Brussels, 12 November 2012).

226 *Ibid.*

227 "European Parliament Rapporteur Backs 'Stop the Clock' EU ETS Proposal but Calls for Clarity on EU Stance," *Green Air Online* (14 January 2013), online: <www.greenaironline.com/news.php?viewStory=1640> [Green Air, "Stop the Clock"].

228 Keith Laing, "Sen Thune Pleased with Airline Emission Fees Halt but Not Satisfied," *The Hill* (12 November 2012) online: <http://thehill.com/blogs/transportation-report/aviation/267345-sen-thune-pleased-with-airline-emission-fees-halt-but-not-satisfied->.

progress made by the ICAO Council at its meeting of 9 November 2012.[229] This was perceived as a necessary step to give an opportunity to ICAO to further elaborate its framework and global MBM scheme to be considered by the 38th Assembly in October 2013.[230] Yet it is hard to see any substantial progress achieved by ICAO at that meeting.[231] In fact, what emerged out of this meeting is a diluted outcome of what the Secretariat had proposed in the first place.[232] This suggests that other motives may have prompted Europe to stop the clock.

The days that preceded the meeting of 9 November 2012 saw an unusual number of aviation and climate change high-ranking officials, from both the Commission and the United States in attendance in Montreal. This suggests that extensive negotiations between the two blocks took place before the Council meeting. The fact that the decision to suspend the scheme was announced three days later may have two meanings. First, it was important for the Commission to give a vote of confidence to ICAO and to demonstrate goodwill with other States with whom the disagreement over the scheme had ensued. Second, the Commission probably knew that the US Congress would pass the EU ETS Blocking Legislation the following week. Given that the US elections were already over, it was not expected that President Obama would veto it. To avoid portraying a sign of weakness in the context of international relations, the Commission needed to make the announcement of the

229 See Valerie Volcovici & Barbara Lewis, "EU Sees Progress on UN airline Emissions Deal," *Reuters* (11 November 2012) online: <www.reuters.com/article/2012/11/11/uk-airlines-eu-us-co-idUSLNE8AA00I2012 1111>.

230 See "European Commission Backs Down on EU ETS and Agrees to 'Stop the Clock' on International Aviation Emissions," *Green Air Online* (12 November 2012), online: Green Air Online <www.greenaironline.com/news.php?viewStory=1620>.

231 At the sixth meeting of its 197th Session that took place on Friday, 9 November 2012, the Council only acknowledged the progress report of a group of experts that had been assembled to work on design features for a global MBM scheme and a framework that would enable States to implement their own MBMs. The group identified three measures (e.g. mandatory offsetting, mandatory offsetting with a revenue mechanism, and global emissions trading) as feasible MBMs for international aviation. The Council only noted the report presented by the experts. See ICAO, C-DEC 197/6, *supra* note 218 (Chapter 3) at 1. Due to the fact that there still were a number of policy issues that the group of experts could not resolve, at the suggestion of the Secretary-General, the Council decided to form a high-level group composed of high-ranking officials of 17 states with significant levels of aviation activity and sufficient geographical representation tasked with "[developing] policy recommendations" to further advance on the framework and the feasibility of a global scheme of MBM for international aviation. The hope was that the high-level group, which was to report back to Council for its 198th (March 2013) and 199th (June 2013) sessions, would be able to agree on a draft resolution that Council would present to the Assembly. This would include the recommendation for two draft Assembly resolutions: one addressing non-MBM issues (e.g. state action plans, sustainable alternative fuels, operational and technical measures to reduce GHG emissions) and one proposing a framework for regional and local MBMs. See ICAO, C-DEC 197/6, *supra* note 218 (Chapter 3) at 2. Arguably, the decision to establish a high-level group was the result of the lack of consensus within Council on MBM issues. It can hardly been seen as meaningful progress.

232 See ICAO, C-MIN 197/6 *supra* note 148 (Chapter 3) at 4.

scheme's suspension before the enactment of the EU ETS Blocking Legislation.[233] Arguably, the latter exerted an influential role.

Other economic factors may also help to explain why the Commission was forced to suspend the scheme. Some reports indicate that, in light of the poor conditions of the European economy at the time, major European players such as France might have exerted significant pressure on the Commission to abandon the scheme.[234] It is a well-known fact that through Airbus, the aviation sector had approached both the Commission and a number of European countries.[235] It is interesting to note that just a few days after the Commission had announced the temporary suspension of the EU ETS, Chinese airlines placed orders to purchase around 60 Airbus aircraft.[236] One must recall that, as part of their strategy to stop the EU ETS, the Chinese government had previously ordered its airlines to put these orders on hold.[237]

The initial suspension of the EU ETS was only applicable to the compliance year 2013. In the absence of new legislation, either repealing or extending the suspension, the scheme would automatically re-start in its initial form, that is to say covering emissions for the whole duration of the flight. In light of this, in October 2013, right after the 38th Assembly, and cognizant of the agreement that was reached at ICAO to develop the global MBM scheme, the Commission proposed to scale down the EU ETS' geographical scope by covering only emissions generated by (both foreign and European) aircraft operators within European airspace.[238] This would be applicable as of 1 January 2014. Concerned with the potential knock-on effect on the development of ICAO's global MBM scheme, the airline industry quickly expressed its opposition to the Commission's proposal to reduce the scheme's coverage to EU's airspace only.[239] In addition, major European countries such as France, Germany, and the United Kingdom made very clear that they did not support the Commission's proposal.[240] Instead, these countries suggested that the

233 See "Economics Turned EU Powers against ETS," *Airwise* (9 December 2012) online: <http://news.airwise.com/story/view/1355097725.html>.

234 *Ibid.*

235 See Mavis Toh, "Airbus and Partners Urge EU Leaders to Stop ETS Trade Conflict," *Flight Global* (12 March 2012) online: <www.flightglobal.com/news/articles/airbus-and-partners-urge-eu-leaders-to-stop-ets-trade-conflict-369357/>.

236 David Pearson, "Airbus Reaches Deal to Sell Jets to China," *The Wall Street Journal* (25 April 2013) online: <http://online.wsj.com/article/SB10001424127887324474004578444562817850562.html?mod=dist_smartbrief>. Later, in May 2013, Air China confirmed an order to buy 100 aircraft from Airbus. Johanne Chiu, "Air China's Board Approves Plan to Buy 100 Aircraft from Airbus," *The Wall Street Journal* (7 May 2013) online: <http://online.wsj.com/article/BT-CO-20130507-707347.html?mod=dist_smartbrief>.

237 See Airwise, *supra* note 233.

238 EC, Commission Proposes Applying EU ETS to European Regional Airspace from 1 January 2014 (16 November 2013) online: European Commission <http://ec.europa.eu/clima/news/articles/news_2013101601_en.htm>.

239 See IATA Press Release, "Remarks of Tony Tyler," *supra* note 141 (Chapter 1).

240 See "EU States Get Their Way as Deal Struck with Parliament Negotiators to Continue with Stop the Clock," *Green Air Online* (5 March 2014), online: <www.greenaironline.com/news.php?viewStory=1833>

EU ETS should be postponed until the end of 2016.[241] Mindful of the position of major actors within Europe, the European Parliament made a calculated decision. Acknowledging that "in order to sustain the momentum reached at the 38th Session of the ICAO Assembly in 2013 and facilitate progress at the upcoming 39th Session in 2016," at its first reading, the European Parliament decided to extend the suspension of the EU ETS as applied to foreign aircraft until 31 December 2016.[242] This suspension covers flights from third countries to the European Economic Area (EEA) of both foreign and EU aircraft operators.[243] The scheme is still in force for intra-European flights.[244] The proposed legislation by the European Parliament was sent to the Council of Ministers for its final approval.[245] Later, the Council of Ministers rubber-stamped the Parliament's text.[246] The decision to extend the "stopping the clock" is the result of the enormous political pressure that Europe has faced with regard to the EU ETS.[247]

4.4 CONCLUSION

There is no question that international aviation's climate change impact should be addressed. Yet the inclusion of foreign aircraft operators into the EU ETS has been one of the most hardly contested environmental policy decisions in recent times. Arguably, for industry stakeholders and non-EU States, the fact that the scheme initially intended to cover emissions generated for the flight's whole trajectory was extremely problematic. In spite of this, emissions trading as a policy instrument should not necessarily be rejected. In addition to affording significant flexibility, it allows participants to fulfill their compliance

241 *Ibid.*

242 See European Parliament, P7_TA-PROV (2014) 0278, Greenhouse Gas Emission Trading (International Aviation Emissions) online: <www.europarl.europa.eu/sides/getDoc.do?type=TA&reference=20140403&secondRef=TOC&language=en>.

243 *Ibid.*

244 *Ibid.*

245 See also European Parliament, MEPs back CO2 Permit Exemption for Long-Haul Flights, <www.europarl.europa.eu/pdfs/news/expert/infopress/20140331IPR41187/20140331IPR41187_en.pdf>.

246 The decision to extend the suspension of the EU ETS to foreign aircraft operators was met with strong opposition of climate change activists within the European Parliament. Satu Hassi (MEP Finland) said: "MEPs have today voted to let most of the aviation sector off the hook for its growing climate change impact in exchange for the vague hope of future global action. Excluding international aviation from the emissions trading scheme for 4 extra years will mean 4 more years of growth in airline emissions, undermining the emissions reductions from most other EU sectors."
The Greens: European Free Alliance in the European Parliament, "MEPs Vote to Let Aviation Off the Hook for Vague Hope of Future Global Action" (3 April 2014) online: <www.greens-efa.eu/airline-emissions-12161.html>.

247 "Europe Caves in to Foreign Pressure, Guts Aviation Emission-Reduction Law," *Transport & Environment* (3 April 2014) online: <www.transportenvironment.org/press/europe-caves-foreign-pressure-guts-aviation-emissions-reduction-law>.

obligations in a cost-effective manner, and it was not projected to represent a substantial cost for the industry.

By including international aviation, Europe sought compliance with its General ETS. In practice, this meant that the design of the scheme for international aviation had to take into account many of the overarching principles of the General ETS. From a European perspective, this was most unfortunate, because it ignored the particularities of international aviation and the notorious fact that the whole international aviation community opposed it. By standing firm on European environmental principles, the scheme lost the opportunity to gain acolytes. Arguably, Europe could have designed the scheme in manner more attractive for those opposing it. Lifting restrictions on ECUs and a different set of rules on the use of revenues could have eased some of the negative concerns.

What is undisputed is that by including foreign aircraft operators, Europe managed to upset the international aviation community in an unprecedented manner. Non-EU States and industry stakeholders have resorted to legal challenges, strong political statements, decisions of the ICAO Council, blocking legislation in various countries prohibiting their aircraft operators from joining the EU ETS, and considerations of potential countermeasures. The initiatives undertaken against the EU ETS demonstrate the animosity it has ignited and also that these actors will not tolerate the inclusion of foreign aircraft operators into the scheme.

Without completely capitulating, late in 2012, Europe was forced to temporarily suspend the scheme. Not doing so would have risked a trade war of unthinkable consequences for the aviation sector and the European economy. Right after the 38th ICAO Assembly, the Commission tabled a proposal to scale down the scheme's geographical scope to include only emission generated within European airspace. A number of major European countries, along with the airline industry, opposed such proposal. The European Parliament had no choice to suspend the scheme until 31 December 2016. At that time, it is expected that the ICAO global MBM scheme would already be adopted.

5 Legal Challenges against the EU ETS: Extraterritoriality

I examined the main design features of the EU ETS. I also described the multiple initiatives that the international coalition put together to defeat the scheme. As I discussed, critics of the EU ETS have strongly argued that the extension of the scheme to foreign aircraft operators and the manner in which CO_2 emissions are calculated for the whole duration of a flight – even for those parts of the flight that take place over the high seas and over the national airspace of other States – constitute an impermissible, extraterritorial application of EU law. By seeking to impose a set of environmental standards extraterritorially, the EU ETS de facto interferes with the rights of other States. As such, the EU ETS impinges upon the sovereignty of other States.[1] Effectively, these are the two main criticisms most frequently leveled against the scheme.

Opponents of the EU ETS place heavy emphasis on the unquestionable extraterritorial elements of the scheme and would seem to presuppose that the exercise of extraterritorial jurisdiction is unlawful. They ignore or chose to ignore the fact that, under certain circumstances, the exercise of extraterritorial jurisdiction may in fact be permitted.

As one of the most complex issues facing international law today,[2] extraterritorial jurisdiction has always been controversial since it involves dealing with foreign actors and infringing on national sovereignty, national interests, and potential threats to international

---

1 See ICAO, C-MIN 196/7, *supra* note 124 (Chapter 1) at 19 (Statement of Brazil noting that the EU ETS is unilateral and extraterritorial and it constitutes an infringement of States sovereignty); ICAO, C-MIN 194/2, *supra* note 86 (Chapter 1) at 11 (Statement of Mexico suggesting that the EU ETS imposes "a type of charge over international waters and in the airspace of third countries"); *Ibid.* at 4 (Statement of Argentina). A4A, the leading trade association of US airlines, described the EU ETS as an "unlawful tax [of] an egregious regulatory overreach, a violation of U.S. sovereignty and a clear cash grab to offset European debt." "A4A Commends U.S. House for Taking Further Action to Protect U.S. Airlines and Their Passengers from Unlawful EU Emissions Trading Scheme," *A4A* (28 June 2012) online: A4A <www.airlines.org/Pages/news_6-28-2012.aspx>; Hardeman, "Common Approach," *supra* note 113 (Chapter 1) (noting that the scheme interferes with the principle of State sovereignty, at 17); Nancy Young, *supra* note 131 (Chapter 1); Bartlik, *supra* note 68 (Chapter 4) (suggesting that the scheme violates the territoriality principle of jurisdiction, at 165); Dejong, *supra* note 101 (Chapter 4) (describing the EU ETS as a regulatory overreach, at 173).

2 See Vaughan Lowe, "US Extraterritorial Jurisdiction: The Helms-Burton and D'Amato Acts" (1997) 46:2 Int'l & Comp L Q 378 [Lowe, "US Extraterritorial Jurisdiction"]; Vaughan Lowe, "Jurisdiction" in Malcolm D Evans, *International Law*, 1st edn (Oxford: Oxford University Press, 2003) 329 (noting that unlawful extraterritoriality arises when one State interferes with the rights of other States to decide whether or not to regulate a particular conduct within its own national territory, at 329); Yumiko Tanaka, *The World Trade Organization and Dispute Over Extraterritorial Application: The Effectiveness and Function of the World Trade Organization Dispute Settlement Body in International Law* (LLM Thesis, McGill University Institute of Comparative Law, 2001) [unpublished] at 13.

peace and security.[3] Arnell notes that extraterritorial jurisdiction embodies a sense of "legal imperialism and chauvinism."[4] Frequently, as Dabbah contends, it disregards the interests of other countries.[5] In this chapter, I explore whether principles of international law and the "effects doctrine" and the theory of state responsibility justify Europe's exercise of extraterritorial jurisdiction. I identify the flaws of the Court of Justice of the European Union (CJEU) in its landmark *ATA Decision*.[6]

I begin by analyzing the principles of international law on jurisdiction, namely, territoriality and nationality as well as protective, passive, and universal jurisdiction. I outline examples provided both by international instruments and the internalization of those principles by States. I also examine some of the most famous international law cases in which these principles were considered. This chapter also explores the question of State responsibility with regard to obligations *erga omnes* and the circumstances that would allow a State other than the injured State to exercise an independent cause of action to defend the interests of the international community as a whole. I then look into the effects doctrine and the variations in its application in the United States and in Europe. I dissect the *ATA Decision* of the CJEU in which the extraterritoriality arguments leveled against the EU ETS were all dismissed as well as the approach the Commission has taken with regard to the extraterritorial allegations. My subsequent sections contrast the EU ETS scheme with the principles of international law on jurisdiction and other design alternatives available to Europe. I ask whether the scheme's jurisdictional overreach is unprecedented in international aviation and whether the doctrine of state responsibility may justify the EU ETS. My final section concludes that the extraterritorial elements present in the design of the EU ETS are not permissible under the principles of jurisdictions of international law.

---

3 See Bartram S Brown, "The Evolving Concept of Universal Jurisdiction" (2001) 35 New Eng L Rev 383 at 389.

4 See Paul Arnell, *Law across Borders: The Extraterritorial Application of United Kingdom Law* (London: Routledge, 2012) at 19.

5 See Dabbah, *supra* note 164 (Chapter 4) (noting that the exercise of extraterritorial jurisdiction needs to take into account the "legal, economic, and political interests of other countries" at 205); Mark P Gibney, "The Extraterritorial Application of U.S. Law: The Perversion of Democratic Governance, the Reversal of Institutional Roles, and the Imperative of Establishing Normative Principles" (1996) 19 BC Int'l & Comp L Rev 297 (suggesting that extraterritoriality requires "more sense of justice and fairness, not currently reflected" in States' practices, at 321).

6 See *ATA Decision*, *supra* note 63 (Chapter 4).

5.1 Principles of International Law on Jurisdiction

5.1.1 *Territorial Principle*

It is well established that a State may exercise jurisdiction over conduct that takes place within its territory.[7] In fact, this is an essential attribute of each State's sovereignty,[8] and it encompasses criminal, civil, and administrative jurisdiction. As a general norm, jurisdiction is first and foremost territorial. Territoriality is the least problematic aspect of jurisdiction. It only becomes controversial where foreign elements are present.[9] Arguably, an act carried out under one State's jurisdiction may present legitimate concerns for another State.[10]

International law provides copious examples where jurisdiction is exercised on the basis of territoriality.[11] For instance, the international air transport system has been

---

7 See *Laker Airways Limited v Sabena*, 731 F 2d 909 (DC Cir 1983) (discussing that "[t]he prerogative of a nation to control and regulate activities within its boundaries is an essential, definitional element of sovereignty. Every country has a right to dictate laws governing the conduct of its inhabitants. Consequently, the territoriality base of jurisdiction is universally recognized. It is the most pervasive and basic principle underlying the exercise by nations of prescriptive regulatory power. It is the customary basis of the application of law in virtually every country" at 918) [*Laker Airways*].

8 See Dong Xue, "A General Study of the Extraterritoriality of Criminal Forfeiture Law: Canada and China" (MA Thesis, Simon Fraser University School of Criminology, 1999) [unpublished] at 38; Eugene Kontorovich, "The Inefficiency of Universal Jurisdiction" (2008) U Ill L Rev 389 at 393 [Kontorovich, "Inefficiency"].

9 See FA Mann, "The Doctrine of Jurisdiction in International Law" (1964) 111 RCADI 1 at 14. See also R Higgins, "General Course on Public International Law" (1991) Rec des Cours 10 (discussing that all exercises of jurisdiction that are not based on the territoriality principle are exercises of extraterritorial jurisdiction, at 109); Austen Parrish, "The Effects Test: Extraterritoriality's Fifth Business" (2008) 61 Vand L Rev 1455 at 1464; Jeffrey A Meyer, "Dual Illegality and Geoambiguous Law: A New Rule for Extraterritorial Application of U.S. Law" (2010) 95:1 Minn Law Rev 110 at 121.

10 See Philippe Sands, "Turtles and Torturers: the Transformation of International Law" (2001) 33 NYU J Int'l L & Pol 527 at 535; Jennifer Zerk, *Extraterritorial Jurisdiction: Lessons for the Business and Human Rights Spheres from Six Regulatory Areas, Corporate Social Responsibility Initiative Working Paper No 59* (June 2010) (Cambridge; John F Kennedy School of Government, Harvard University) online: <www.hks.harvard.edu/m-rcbg/CSRI/publications/workingpaper_59_zerk.pdf> (noting that various "domestic measures" may pose "extraterritorial implications" at 86).

11 In the field of international criminal law, almost all treaties recognize the territorial jurisdiction of the State in which the offense took place. Among others, these include the *Convention for the Suppression of Unlawful Acts against the Safety of Civil Aviation*, 23 September 1971, 974 UNTS 14118 at Art 5, para 1 (a) [*Montréal Convention*]; *Convention on the Prevention and Punishment of Crimes against Internationally Protected Persons, including Diplomatic Agents*, 14 December 1973, 1037 UNTS 167 at Art 3, (a) [*International Protected Persons Convention*]; *International Convention against the Taking of Hostages*, 17 December 1973, 1316 UNTS 21931[*Hostages Convention*] at Art 5, para 1 (a); *Convention on the Physical Protection of Nuclear Material*, at Art 8, para 1 (a) [*Nuclear Material Convention*]; *Convention for the Suppression of Unlawful Acts against the Safety of Maritime Navigation*, 10 March 1988, I-29004 UNTS 1678 at Art 6, paral (1) (b) [*SUA Convention*]; *International Convention for the Suppression of Terrorist Bombings*, 15 December 1997, 2149 UNTS 37517 at Art 6, para 1 (a) [*Terrorist Bombings Convention*]; *International Convention for the Suppression of the Financing of Terrorism*, 9 December 1999, 2178 UNTS 38349 at Art 7, para 1 (a) [*Financing of Terrorism Convention*]; *International Convention for the Suppression of Acts of Nuclear Terrorism*,

structured on the basis of territorial sovereignty.[12] According to the *Chicago Convention*, each State enjoys "complete and exclusive sovereignty over the airspace above its territory."[13] As a result of the law's recognition of exclusive State sovereignty over territorial airspace, there is no complete freedom of the air to operate air services.[14] States engaged in international civil aviation have had to exchange commercial traffic rights (market access) with other States primarily through bilateral arrangements. Generally, there is a consensus that the principle of complete and exclusive State sovereignty over territorial airspace forms part of customary international law.[15] Therefore, it is binding on all States, not only those who are party to the *Chicago Convention*. As such, the principle also binds the European Union (EU), a non-party to the *Chicago Convention*.

A number of provisions of the *Chicago Convention* reflect the territorial principle. For instance, in full recognition of the principle of exclusive sovereignty over their territorial airspace, a State may (i) "require the landing at some designated airport of civil aircraft flying above its territory without authority";[16] (ii) refuse the right of cabotage to aircraft operators from other States;[17] (iii) deny or authorize operations of pilotless aircraft;[18] (iv) establish prohibited areas for military of public safety reasons where no operations can be

13 April 2005, 2445 UNTS 89 at Art 9, para 1 (a) [*Nuclear Terrorism Convention*]; *Convention on the Suppression of Unlawful Acts Relating to International Civil Aviation*, 10 September 2010 at Art 8, para 1 (a) [*Beijing Convention*], *Protocol Supplementary to the Convention for the Suppression of Unlawful Seizure of Aircraft*, 10 September 2010 at Art VII [*Beijing Protocol*]. Although the *Convention for the Suppression of Unlawful Seizure of Aircraft* does not expressly spell out the jurisdiction of the State where the offense was committed, this is implicitly recognized, given that the instrument "does not exclude any criminal jurisdiction exercised in accordance with national law." See *Convention for the Suppression of Unlawful Seizure of Aircraft*, 16 December 1970, 860 UNTS 12325 at Art 4, para 3 [*The Hague Convention*]. It is also common that international instruments impose obligations that States must implement within their territories. For instance, the *Convention on the Marking of Plastic Explosives for the Purpose of Detection* requires that its State parties "prohibit and prevent the manufacture in [their] territory of unmarked explosives." See *Convention on the Marking of Plastic Explosives for the Purpose of Detection*, 1 March 1991, 2122 UNTS 359, ICAO Doc 9571 at Art II [*Marking of Plastic Explosives Convention*].

12 See Milde, *International Air Law and ICAO*, *supra* note 59 (Chapter 2) at 35; Dempsey, *supra* note 5 (Chapter 2) at 43.

13 See *Chicago Convention*, *supra* note 36 (Chapter 2) at Art 1. This language replicates almost verbatim Art 1 of the *Paris Convention* of 1919. See *Paris Convention*, *supra* note 28 (Chapter 2) at Art 1.

14 In order to initiate a particular international flight, aircraft operators must first establish whether their home States enjoy traffic rights to carry out such operations. Should that not be the case, States need to negotiate either bilaterally or multilaterally such traffic rights with the concerned States. Only after these agreements are fully concluded can aircraft operators launch international flights to the sought destinations. See *Chicago Convention*, *supra* note 36 (Chapter 2) at Art 6.

15 See *Case Concerning Military and Paramilitary Activities in and Against Nicaragua (Nicaragua v United States of America)* [1986] ICJ Rep 14 at 392 (para 212) (recognizing the principle of State sovereignty). See also Bin Cheng, *The Law of International Air Transport* (London: Stevens & Sons Limited, 1962) at 120.

16 See *Chicago Convention*, *supra* note 36 (Chapter 2) at Art 3 para (b).

17 *Ibid.* at Art 7.

18 *Ibid.* at Art 8.

conducted;[19] (v) establish designated airports;[20] (vi) regulate the entrance of aircraft into its territory by applying its laws on issues such as customs, immigrations, and health;[21] (vii) adopt regulations to prevent the spread of diseases;[22] (viii) establish user charges;[23] and (ix) institute aircraft accident investigations.[24]

At the other end of the spectrum, international maritime law provides a number of concrete examples where States may exercise territorial jurisdiction within their own territorial sea to protect the environment. To this end, States bear the responsibility to prevent pollution of global marine environmental commons from land-based sources located in their territories.[25] The costal State is responsible for preventing maritime pollution in the seabed under its jurisdiction.[26] Similarly, while at its ports, a State may prevent the departure of an unseaworthy vessel threatening to damage the marine environment.[27]

Although the climate change regime has been established to mitigate the effect of anthropogenic emissions on the global climate, the calculation of emissions under the *Kyoto Protocol* strictly follows a territorial approach. Emissions are accounted for at the point of production, not consumption.[28] That is to say each country ought to report emissions generated within its territory.[29]

Two manifestations of the territorial principle are objective territoriality and subjective territoriality. Objective territoriality involves a State exercising jurisdiction over an act that is initiated in another State but impacts its territory.[30] The classic example is that of a man who stands at the border and shoots and kills another person in the territory of the

19 *Ibid.* at Art 9.
20 *Ibid.* at Art 10.
21 *Ibid.* at Arts 11 and 12.
22 *Ibid.* at Art 14.
23 *Ibid.* at Art 15.
24 *Ibid.* at Art 26.
25 See *United Nations Convention on the Law of the Sea*, 10 December 1982 1833 UNTS 3 at Art 207, para 1 (entered into force 16 November 1994) [*UNCLOS*].
26 *Ibid.* at Art 208.
27 *Ibid.* at Art 219.
28 See Alejo Etchart, Begum Sertyesilisik, & Greig Mill, "Environmental Effects of Shipping Imports from China and their Economic Valuation: The Case of Metallic Valve Components" (2012) 21 J Cleaner Production 51 at 52.
29 The *Kyoto Protocol* territorial methodology has also been subject to criticism. Analysts contend that the relevant factor should be the place where emissions are generated as a result of the consumption of products and services and not where these are manufactured. See Helm, "Climate Change Policy," *supra* note 80 (Chapter 2) at 220. If one were to adopt the place of consumption methodology, for instance, China's emissions produced in 2006 would drop from 5,500 to 3,840 mt. of CO_2. Similarly, China's emission growth rate would fall from 12.5 to 8.7 percent annually. See Jiahua Pan, Jonathan Phillips, & Ying Chen, "China's Balance of Emissions Embodied in Trade: Approaches to Measurement and Allocating International Responsibility" (2008) 24:2 Oxford Rev Econ Pol'y 354.
30 See RY Jennings, "Extraterritorial Jurisdiction and the United States Antitrust Laws" (1957) 33 Brit YB Int'l L 146 at 156; Zerk, *supra* note 10 at 27; Cedric Ryngaert, *Jurisdiction in International Law* (Oxford: Oxford University Press, 2009) at 76.

other State.[31] As Brownlie underscores, the "essential constituent element" is the consummation of the act in the territory of the State exercising jurisdiction.[32] According to Jennings, this is "where the offence takes effect or produces its effects."[33] As such, the link to the State exercising jurisdiction is substantial.[34] Conversely, subjective territoriality deals with an act whose constituent elements took place in the *forum* State but was consummated elsewhere.[35]

Traditionally, a State's territorial boundaries have established the limits of its jurisdiction. For instance, in the *Bering Sea Arbitration* concerning a fishery dispute between the United States and the United Kingdom, the panel found that, as the costal State, the United States was not entitled to apply its preservation laws on fur seals beyond its territory[36] and also to impose limitations on the access of other States to the high seas.[37] This rule would temporarily change. In 1927, at the Permanent Court of International Justice (PCIJ), France challenged Turkey's exercise of criminal jurisdiction over offenses committed on the high seas as a result of a collision between a French vessel (*S.S. Lotus*) and a Turkish vessel (*Boz-Kourt*) in which eight Turkish seamen perished.[38] Turkey instituted criminal proceedings against Lieutenant Demons, first officer of the S.S. Lotus.[39] The core legal question was whether, under international law, Turkey could exercise criminal jurisdiction.[40] While recognizing that in the absence of an express rule, a State cannot exercise its power in the territory of another State,[41] the PCIJ held that international law does not prevent a State from exercising its jurisdiction over persons, property, and acts outside of its territory.[42] The PCIJ suggested that international law gives States "a wide measure of discretion which is only limited in certain cases by prohibitive rules."[43] The Court's ruling appears to have

31 See Brownlie, *Principles of Public International Law*, 5th edn (Oxford: Oxford University Press, 1998) at 304.
32 *Ibid.*
33 Jennings, *supra* note 30 at 159.
34 Similarly, "subjective territoriality" takes place, for instance, where the act commences in the territory of the *forum* State but finishes elsewhere. *Ibid.* at 156.
35 See Christopher L Blakesley, "United States Jurisdiction over Extraterritorial Crimes" (1982) 73:3 J Crim L & Criminology 1109 at 1118-19.
36 See *Award between the United States and the United Kingdom Relating to the Rights of Jurisdiction of United States in the Bering's Sea and the Preservation of Fur Seal* (15 August 1893) vol XXVIII, p 263, online: UN <http://legal.un.org/riaa/cases/vol_XX>.
37 See Bodansky, *The Art and Craft of International Environmental Law*, *supra* note 62 (Chapter 1) at 22.
38 *The Case of S.S. "Lotus" (France v Turkey)* (1927), PCIJ (Ser A) No 70 [*Lotus*]. See also Brownlie, *supra* note 31 at 304.
39 See *Lotus*, *supra* note 38 at 11. See also Michael P Scharf, "Application of Treaty-Based Universal Jurisdiction to Nationals of Non-Party States" (2001) 35 New Eng L Rev 363 at 366 [Scharf, "Treaty-Based Universal Jurisdiction"].
40 *Ibid.* at 12.
41 *Ibid.* at 18.
42 *Ibid.* at 19.
43 *Ibid.* Scholars in the United States interpret this reasoning as laying down the foundational basis of the effects doctrine. See Parrish, *supra* note 9 (describing Lotus as a "precursor" of the effects doctrine, at 1471). However, this view is not shared outside the United States. For others, Lotus just contradicts the territoriality principle.

implied that, in the absence of an international rule to the contrary, States may legitimately exercise extraterritorial jurisdiction. This is certainly not the case. In his famous dissenting opinion in the case concerning *The Legality of the Threat or Use of Nuclear Weapons*, Judge Shahabuddeen explained that the *Lotus* Court failed to qualify the circumstances of such limits due to the less critical subject matter of that case.[44] Moreover, Judge Guillaume suggested that the *Lotus* rule was only intended to be applicable to the case's very specific nature.[45] He underlined the fact that the "exercise of that jurisdiction is not without its limits."[46]

Other decisions of the International Court of Justice (ICJ) have confirmed a clear departure from *Lotus*. In his dissenting opinion in the *Fisheries* case, Judge Alvarez indicated that States no longer enjoy absolute sovereignty.[47] This implies that the notion that States are entitled to do whatever they wish as long as such acts are not forbidden by international law can no longer be sustained.[48] In order to avoid situations in which States abuse their rights (*abus de droit*), the Charter of the United Nations and the interest of the international community as a whole limit the strict application of such principle.[49] The *Lotus* rationale is not illustrative of the status of extraterritorial application of law today under international law,[50] neither is it reflective of State practice.[51] In fact, it has been described as "obsolete"[52] and "not sound law."[53] Also, several commentators have suggested that it may have been replaced by the 1958 *High Seas Convention*.[54] Its strict application to other spheres of

---

See Ryngaert, *supra* note 30 at 4; Douglas E Rosenthal & William M Knighton, *National Law and International Commerce: The Problem of Extraterritoriality* (London: Routledge & Kegan Paul, 1982) at 15.

44 Dissenting Opinion of Judge Shahabuddeen in *The Legality of the Threat or Use of Nuclear Weapons, Advisory Opinion, [1996] ICJ Rep 226* at 395 [Shahabuddeen, *Legality of the Threat or Use of Nuclear Weapons*].

45 See Separate Opinion of President Guillaume in the *Case concerning the Arrest Warrant of 11 April 2000 (Democratic Republic of the Congo v. Belgium)*, [2002] ICJ Rep 3 at 37 [*Arrest Warrant*].

46 *Ibid.*

47 See Separate Opinion of Judge Alvarez in *Fisheries Case (United Kingdom v Norway)* [1951] ICJ Rep 116 at 152 [Separate Opinion of Judge Alvarez in *Fisheries*].

48 *Ibid.* As ably noted in the *Arrest Warrant Case*, "the [*Lotus*] *dictum* represents the high water-mark of laissez-faire in international relations, and an era that has been significantly overtaken by other tendencies." Joint Separate Opinion of Judges Higgins, Kooijmans, and Buergenthal in *Arrest Warrant*, *supra* note 45 at 78.

49 *Ibid.* at 152.

50 See Vaughan Lowe, "Blocking Extraterritorial Jurisdiction: The British Protection of Trading Interest Act, 1980" (1981) 75:2 Am J Int'l L 257 at 262 [Lowe, "Blocking Extraterritorial Jurisdiction"].

51 See Arnell, *supra* note 4 at 33. For instance, the UK practice reflects a clear rejection to Lotus, in particular with regard to the long-standing extraterritorial application of US antitrust laws. See Lowe, "Blocking Extraterritorial Jurisdiction," *supra* note 50 at 262; Michal Gondek, *The Reach of Human Rights in a Globalizing World: Extraterritorial Application of Human Rights Treaties*, (Oxford: Intersentia, 2009) at 50.

52 Ryngaert, *supra* note 30 at 26.

53 Jaye Ellis, "Extraterritorial Exercise of Jurisdiction for Environmental Protection: Addressing Fairness Concerns" (2012) 25:2 Leiden J Int'l L 397 at 402 [Ellis, "Fairness"]. Similarly, Brownlie describes *Lotus* as "characterized by vagueness and generality." Brownlie, *supra* note 31 at 305.

54 See Jose A Cabranes, "Our Imperial Criminal Procedure: Problems in the Extraterritorial Application of U.S. Constitutional Law" (2009) 118 Yale LJ 1660 at 1673.

international law may be problematic and prompted scholars to craft a modern interpretation of *Lotus*.[55]

Ryngaert, for instance, suggests that a restrictive approach has replaced *Lotus* and that extraterritoriality only takes place where a permissive rule of international law so allows.[56] This would include the nationality, protective jurisdiction principles, as well as universal jurisdiction. Similarly, some years later, in the *Fisheries Jurisdiction Case* (*UK v Iceland*), the ICJ found that Iceland's unilateral extension of its exclusive fishing rights to 50 nautical miles from the baseline was not permitted under international law. The ICJ also found that Iceland did not have the right to exclude UK-registered fishing vessels from areas lying between the 12-mile and 50-mile limits.[57]

More recently, in the famous *Tuna-Dolphin* trade controversy, the arbitral panel found an inherent territorial limit to the exception of GATT that permits unilateral trade-related measures.[58] Thus, the United States could only apply its conservation statute to protect dolphins from extinction within its territorial boundaries.[59] A second panel in this case may have implicitly accepted an "extraterritorial reach" of the exception.[60] Some years later, the Arbitral Body did not completely rule out extraterritoriality.[61] In *Shrimp-Turtle* – a case brought by India, Malaysia, Pakistan, and Thailand against the United States for an illicit ban on shrimp caught without turtle excluder devices (TEDs) – the Arbitral Body acknowledged a "sufficient nexus" between the jurisdiction that the United States sought to exercise and the location of the turtles, given the fact that these species "spent part of their migratory life cycle" in US waters,[62] just like foreign aircraft subject to the EU ETS.

55 See Shahabuddeen, *Legality of the Threat or Use of Nuclear Weapons supra* note 44 at 396.

56 Ryngaert, *supra* note 30 at 21.

57 *Fisheries Jurisdiction Cases (United Kingdom v Iceland), [1974] ICJ Rep 3.*

58 See *United States – Restrictions on Imports of Tuna* (3 September 1991), GATT DS21/R (Panel Report), unadopted, online: World Trade Law <www.worldtradelaw.net/reports/gattpanels/tunadolphinI.pdf>; Bruce Neuling, "The Shrimp-Turtle Case: Implications for Article XX of GATT and the Trade and Environment Debate" (1999) 22 Loy LA Int'l & Comp L Rev 1 at 21.

59 See Robert Howse, "The Appellate Body Rulings in the Shrimp/Turtle Case: A New Legal Baseline for the Trade and Environment Debate" (2002) 27:2 Colum. J Envtl L 489 at 490 [Howse, "Shrimp/Turtle Case"]; Zerk *supra* note 10 at 187.

60 See Ilona Cheyne, "Environmental Unilateralism and the WTO/GATT System" (1995) 24 Ga J Int'l & Comp L 433 at 453.

61 See *United States – Import Prohibition of Certain Shrimp and Ship Products* (12 October 1998), WT/DS58/AB/R Appellate Body Report), online: WTO <https://docs.wto.org/dol2fe/Pages/FE_Search/FE_S_S006.aspx?Query=(@Symbol=%20wt/ds58/ab/r*%20not%20rw*)&Language=ENGLISH&Context=FomerScriptedSearch&languageUIChanged=true#> at para 133. See also Peter Stevenson, "The World Trade Organization Rules: a Legal Analysis of their Adverse Impact on Animal Welfare" (2002) 8 Animal L 107 at 125.

62 Asif H Quereshi, "Extraterritorial Shrimps, NGOs and the WTO Appellate Body" (1999) 48 Int'l & Comp LQ 199 at 204; Bradly J Condon, "GATT Article XX and Proximity-Of-Interest: Determining the Subject Matter of Paragraphs B and G" (2004) 9 UCAL J Intl'l L & For Aff 137 at 159. Ultimately, the AB ruled against the United States because of the discriminatory manner in which it implemented the environmentally

The case did, however, recognize the US interest in protecting endangered species.[63] Unfortunately, none of these cases shed much light on the exact conditions and requirements for a State to engage in extraterritorial jurisdiction.

5.1.2 *Nationality Principle*

States have asserted jurisdiction over their citizens beyond their geographical boundaries for centuries.[64] Therefore, it is no surprise that, to date, robust consensus exists among international scholars that a State may legitimately exercise jurisdiction over its nationals for acts committed elsewhere.[65] However, as Brownlie notes, this is generally "confined to serious offences."[66] The exercise of jurisdiction on the basis of nationality also seeks to prevent the occurrence of jurisdiction-less situations in which a given State cannot extend its reach to govern behavior that takes places in areas beyond its sovereignty under the territorial principle.[67] Generally regarded as *res communis*, these include areas such as the high seas, outer space, or Antarctica.[68]

To date, more and more States have adopted laws that apply to the conduct of their nationals abroad on diverse issues such as terrorism, money laundering, corruption, sex tourism, child pornography, and other criminal activities.[69] For instance, the *US Foreign Corrupt Practices Act*[70] prohibits US citizens and US corporations from bribing foreign officials.[71] The *Logan Act* criminalizes the conduct of unauthorized negotiations with foreign

related trade measure. *Ibid.* at 159. See also Kulovesi, "Make Your Own Special Song," *supra* note 137 (Chapter 4).

63 See Sands, *supra* note 10 at 535. In light of these conflicting decisions, whether under GATS environmentally related trade measures may be applied extraterritorially "remains an open question." Condon, *supra* note 62 at 139.

64 See *Laker Airways*, *supra* note 7 at 926.

65 See Malcom N Shaw, *International Law*, 3rd edn (Cambridge: Cambridge University Press, 1991) at 403; DW Bowett, "Jurisdiction: Changing Patterns of Authority over Activities and Resources" (1982) 53 Brit YB Int'l L 1 at 7; Jennifer K Ranking, "U.S. Laws in the Rainforest: Can a U.S. Court Find Liability for Extraterritorial Pollution Caused by a U.S. Corporation? An Analysis of *Aguinada v Texaco, Inc.*" (1995) 18 BC Int'l & Comp L Rev 221 at 224; Brownlie, *supra* note 31 at 306; Harvard Research, Jurisdiction with Respect to Crime (1935) 29 Am J Int'l L 519.

66 See Brownlie, *supra* note 31 at 306.

67 See *American Banana Co v United Fruit Co*, 213 US 347 (1909) [*American Banana*] (holding that "in regions subject to no sovereign, like the high seas, or to no law that civilized countries would recognize as adequate, such countries my treat some relations between their citizens as governed by their own law, and keep, to some extent the old notion of personal sovereignty alive" at 355).

68 See Brownlie, *supra* note 31 at 306.

69 See Zerk, *supra* note 10 at 5. See also 18 USC § 2251; *Blackmer v United States*, 284 US 421 (1932) (where the US Supreme Court held that, as part of its jurisdiction in personam, a US Court may subpoena a US citizen residing abroad, at 437).

70 *US Foreign Corrupt Practices Act*, 15 USC § 78 dd-1, 2 (1977).

71 See Bruce Maloy, "Extraterritorial Application of United States Criminal Statutes to the Gaming Industry" (1997) 1 Gaming L Rev 491; Zerk, *supra* note 10 at 33.

governments by US citizens.[72] The *Age Discrimination in Employment Act of 1967* forbids discrimination on the basis of age against US citizens aged 40 and above employed by US corporations even outside the United States.[73] On various occasions, the US Supreme Court has also held that the United States may exercise jurisdiction in personam over US citizens for acts committed abroad.[74] Canada is known for extending its criminal jurisdiction over a wide range of offenses that its nationals might have committed abroad.[75] In the field of taxation, some States assess their nationals on the basis of their world (and not just domestically earned) income. This allows revenue authorities to tax nationals on the income earned abroad while residing outside of their country of nationality or permanent residence.[76] Nationality links the taxpayers to the taxing State. States also do not hesitate to regulate activities of their corporations and subsidiaries that are carried out abroad.[77]

Recognizing long-standing customary international law, international maritime law, air law, and space law has resorted extensively to the principle of nationality. Maritime vessels, aircraft, and space vehicles bear the nationality of the States in which they are registered.[78] Nationality creates a legal relationship between a person or an object and the State from which rights, duties, and liabilities may arise.[79] Accordingly, the *UN Convention on the Law of the Sea* (*UNCLOS*) recognizes that ships "have the nationality of the State whose flag they are entitled to fly" to the extent that "a genuine link between the State and

72 *Logan Act*, 18 USC § 953 (1799).

73 *Age Discrimination in Employment Act*, 29 USC § 623 (1967). The act did not initially protect the US citizens working abroad. However, as a result of extensive litigation, in 1984 it was amended to expressly cover those situations. See also Mary McKleen Madden, "Strengthening Protection of Employees at Home and Abroad: The Extraterritorial Application of Title VII of The Civil Rights Act of 1964 and the Age Discrimination in Employment Act" (1996) 20 Hamline L Rev 739.

74 See *Chafin v Chafin*, 133 SCt 1017 at 1025, 185 L Ed 2d 1 (2013); *Steele v Bulova Watch Co*, 344 US 280 at 289, 73 S Ct 252, 97 L Ed 319 (1952); *Tomoya Kawakita v United States*, 343 US 717, 72 S Ct 950 (holding that US "citizens [owe] allegiance to [their country] wherever they may reside" at 31).

75 This includes exercising jurisdiction over sexual offenses such as sexual interference, sexual touching, sexual exploitation, incest, anal intercourse, murder, manslaughter, assault, assault with a weapon causing bodily harm, aggravated assault, unlawfully causing bodily harm, torture, sexual assault, sexual assault with a weapon, aggravated sexual assault, kidnapping, trafficking in persons, threat against United Nations or associated personnel, attacks on premises, accommodation or transport of United Nations or associated personnel, punishment for theft, fraudulent concealment, fraud, robbery, hostage taking, offenses against internationally protected persons, and offenses involving explosive or other lethal device. See *Criminal Code*, RSC 1985, c C-46, ss 3.71, 4.11, 3.72.

76 See for instance, in the United States, 26 USC § 911. The US legislation taxes any foreign earned income in excess of 80,000 dollars (US). *Ibid.*, (B) (2) (d) (i).

77 See Zerk, *supra* note 10 at 12.

78 See Allyson Bennett, "That Sinking Feeling: Stateless Ships, Universal Jurisdiction, and the Drug Trafficking Vessel Interdiction Act" (2012) 37 Yale J Int'l L 433 at 438.

79 See Milde, *International Law and ICAO*, *supra* note 59 (Chapter 2) at 79. In a similar vein, Cheng comments that the nationality principle recognizes that these vehicles "possess legal personality under municipal law and are endowed with nationalities of their own. [This in turn] opens the way to other rights and liabilities being grafted directly onto aircraft, [vessels, and space vehicles], irrespective of their owners or operators." Cheng, *supra* note 15 at 128.

the ship" exists.[80] This implies that the ship is registered in a State. The flag State retains exclusive jurisdiction over its ships on the high seas.[81] In addition, *UNCLOS* lays out a number of specific "administrative, technical and social" obligations for the flag State relating to the exercise of jurisdiction and control of its ships.[82] Among others, these include duties such as (i) keeping a registry,[83] (ii) asserting jurisdiction,[84] (iii) exercising safety functions at sea,[85] (iv) preventing and punishing the carriage of slaves,[86] (v) requiring the master of its ships to render assistance to persons in distress,[87] (vi) adopting regulation to prevent maritime pollution,[88] (vii) enforcing regulations on pollution by dumping,[89] (viii) enforcing international rules and standards (on its ships),[90] and (ix) conducting investigations on alleged violation of these rules.[91] Where a collision occurs on the high seas, only the flag State and the State from which the master and the persons involved originate may exercise criminal jurisdiction.[92]

Likewise, the *Chicago Convention* provides that "aircraft have the nationality of the State in which they are registered."[93] The State of registry is responsible for providing the regulatory oversight of its aircraft.[94] With respect to aircraft registered therein, the obligations of the State of registry include, among others, ensuring that aircraft comply with the rules of flight and maneuver,[95] granting certificates of airworthiness,[96] licensing of personnel[97] and providing authorization to carry aircraft radio equipment,[98] establishing the code

80 See *UNCLOS*, *supra* note 25 at Art 91.
81 *Ibid.* at Art 92.
82 *Ibid.* at Art 94.
83 *Ibid.* at Art 94, para 2 (a).
84 *Ibid.* at Art 94, para 2 (b).
85 *Ibid.* at Art 94, para 3.
86 *Ibid.* at Art 99.
87 *Ibid.* at Art 98.
88 *Ibid.* at Art 211, para 2 and Art 211, para 1.
89 *Ibid.* at Art 216 (c).
90 *Ibid.* at Art 217, para 1.
91 *Ibid.* at Art 217, para 4.
92 *Ibid.* at Art 97.
93 See *Chicago Convention*, *supra* note 36 (Chapter 2) at Art 17. See also ICAO, Annex 7 to the *Chicago Convention*, *Aircraft Nationality and Registration Marks*, 5th edn (2003).
94 See ICAO Doc 9734 AN/959, *Safety Oversight Manual* (noting that "the State of Registry has the responsibility of ensuring that every aircraft on its register is maintained in an airworthy condition throughout its service life" at 3.6). See also ICAO Doc 8335, *Manual of Procedures for Operations Inspection, Certification and Continued Surveillance*, 5th edn (2010); Milde, *International Law and ICAO*, *supra* note 59 (Chapter 2) at 82.
95 See *Chicago Convention*, *supra* note 36 (Chapter 2) at Art 12. See also ICAO, Annex 2, *Rules of the Air*, 10th edn (2005) at 2.1.1.
96 *Ibid.*, Art 31. See also ICAO, Annex 8 to the *Chicago Convention*, *Airworthiness of Aircraft*, 11th edn (2010).
97 *Ibid.*, Art 32. See also ICAO, Annex 1 to the *Chicago Convention*, *Personnel Licensing*, 11th edn (2011) at 1.2.1.
98 *Ibid.*, Art 30.

of performance for aircraft,[99] and participating in the investigation of aircraft accidents.[100] According to the *Tokyo Convention 1963*, the State of registry retains jurisdiction over criminal offenses and other acts committed aboard an aircraft in-flight that may jeopardize its safety or good order and discipline.[101] Other international instruments also recognize the jurisdiction of the State of registry.[102]

Similarly, in the field of space law, the *Outer Space Treaty* follows the same trend, for it establishes the jurisdiction of the State of registry over objects and personnel "while in outer space or on a celestial body."[103] The States that launched the International Space Station agreed among themselves that criminal jurisdiction over acts occurring on board is vested in the State of nationality of the offender.[104] Canada has transposed this provision into its *Criminal Code*.[105]

5.1.3 Passive Personality Principle

International law also recognizes the assertion of jurisdiction over offenses that foreigners commit abroad against nationals of the *forum* State.[106] Although less widely accepted, State practice and numerous treaties suggest that this principle is increasingly gaining acceptance.[107] In this respect, more recently adopted treaties grant States the option of exercising

99 See ICAO, Annex 6 to the *Chicago Convention, Operations of Aircraft - International Commercial Air Transport - Aeroplanes*, 9th edn (2010) at 5.1.1.

100 See *Chicago Convention, supra* note 36 (Chapter 2), Art 26. See also ICAO, Annex 13, *Aircraft Accident Investigation*, 10th edn (2010) (noting that the State of registry may appoint observers to the aircraft accident investigations); Milde, *International Law and ICAO, supra* note 59 (Chapter 2) at 82.

101 See *Convention on Offences and Certain Other Acts Committed on Board Aircraft*, 14 September 1963, 704 UNTS 10106 [*Tokyo Convention*] at Art 3.

102 The following instruments recognize the non-exclusive jurisdiction of the State of registry: *The Hague Convention, supra* note 11 at Art 4, para 1 (a); *Montréal Convention, supra* note 11 at Art 5, para 1 (b); *International Protected Persons Convention, supra* note 11 at Art 3, para 1 (a) and (b); *Nuclear Material Convention, supra* note 11 at Art 8, para 1 (a) and (b); *Hostages Convention, supra* note 11 at Art 5, para 1 (a) and (b); *SUA Convention, supra* note 11 at Art 6, para1 (1) and (3); *Terrorist Bombings Convention, supra* note 11 at Art 6, para1 (b); *Financing of Terrorism Convention, supra* note 11 at Art 7, para 1 (b); *Nuclear Terrorism Convention, supra* note 11 at Art 9, para 1 (b); *Beijing Convention, supra* note 11 at Art 8; *Beijing Protocol, supra* note 11 at Art vii.

103 *Treaty on Principles Governing the Activities of States in the Exploration and Use of Outer Space, Including the Moon and Other Celestial Bodies*, 27 January 1967, 610 UNTS 205, 6 ILM 386 at Art viii (entered into force on 10 October 1967) [*Outer Space Treaty*].

104 See Julian Hermida, "Crimes in Space: A Legal and Criminological Approach to Criminal Acts in Outer Space" (2006) XXI Ann. Air & Sp L 405 at 410; Francis Lyall & Paul B Larsen, *Space Law: A Treatise* (Burlington: Ashgate Publishing Company, 2009) at 145.

105 See *Criminal Code*, RSC 1985, c C-46 at s 7 (2.3).

106 See Brownlie, *supra* note 31 at 306; Shaw, *supra* note 65 at 408; See R Higgins, "General Course on Public International Law" (1991) Rec des Cours 10 at 97.

107 See Joint Separate Opinion of Judges Higgins, Kooijmans, and Buergenthal in *Arrest Warrant, supra* note 45 at 76; Henry J Steiner, "Three Cheers for Universal Jurisdiction - or is it Only Two?" (2004) 5 Theoretical Inq L 199 at 202.

jurisdiction on the basis of the nationality of the victim.[108] Some courts have also embraced this principle. In *Yunis*, the US Court of Appeals for the DC Circuit affirmed US jurisdiction over a Lebanese citizen who, in 1985, hijacked and blew up a Royal Jordanian Airlines flight in Beirut.[109] The flight was not bound for the United States, and the only link to the *forum* was provided by the two US citizens on board. Relying on the provisions of the *Hostage Taking Act* – a domestic US statute internalizing the Taking of Hostages Convention – the Court held that the requirement that the offender be a US national was satisfied by the fact that two of the victims were US citizens, thereby intertwining the nationality and passive personality principles.[110]

5.1.4 Protective Principle

International law and State practice recognize the right of States to exercise jurisdiction over behavior that occurs beyond the State's territorial boundaries and which poses a threat to the State's national interests, national security, currency, immigration, and fiscal system.[111] Given that almost all States exercise this jurisdiction, it is hardly contested.[112] Here the exercise of jurisdiction is concerned with the "interests injured rather than the place of the act or the nationality of the offender."[113] In this vein, in the 1922 case of *Bowman*, the US Supreme Court held that the *locus* was immaterial in criminal offenses constituting fraud against the government.[114] Similarly, on various occasions, the President of the United States has authorized the Coast Guard to intercept illegal immigrants outside US territorial waters and force repatriation.[115] The US anti-money laundering legislation also expressly

108 See *Beijing Convention*, *supra* note 11 at Art 8; *Beijing Protocol*, *supra* note 11 at Art 4. The following international instruments also make optional the exercise of jurisdiction on the basis of the passive personality principle: *Hostages Convention*, *supra* note 11 at Art 5 (d); *SUA Convention*, *supra* note 11 at Art 6, para 2 (2); *Protocol for the Suppression of Unlawful Acts against the Safety of Fixed Platforms Located on the Continental Shelf*, 10 March 1988, 1678 UNTS 29004 at Art 3, para 2 (b); *Terrorist Bombings Convention*, *supra* note 11 at Art 6, para 2 (a); *Financing of Terrorism Convention*, *supra* note 11 at Art 7, para 2 (a); *Nuclear Terrorism Convention*, *supra* note 11 at Art 9, para 2 (a).

109 See *United States v Yunis*, 924 F 2d 1086 (DC Cir 1991).

110 *Ibid*. See also Abraham Abramovsky, "Extraterritorial Abductions: America's "Catch and Snatch" Policy Run Amok (1990) 31 Va J Int'l L 151 at 179.

111 See Vaughan A Lowe, "Extraterritorial Jurisdiction: The British Practice" (1988) 52:1/2 Rabels Zeitschrift für Ausländisches und Internationales Privatrecht 157 at 179 [Lowe, "Extraterritorial Jurisdiction"]; DW Bowett, "Jurisdiction: Changing Patterns of Authority over Activities and Resources" (1982) 53 Brit YB Int'l L 1 at 10.

112 See Brownlie, *supra* note 31 at 307.

113 See Harvard Research. "Part II - Jurisdiction with Respect to Crime" (1935) 29 Am J Int'l L Supp 435 at 543.

114 *United States v Bowman*, 260 US 94, 43 S Ct 39 (1922).

115 See 46 FR 48109, Exec Order No 12324, 1981 WL 404170 (Pres); 57 FR 23133, Exec Order no 12807, 1992 WL 12135821 (Pres).

provides for the exercise of jurisdiction over foreigners.[116] Likewise, the Paraguayan Penal Code asserts jurisdictions over offenses committed abroad against a Paraguayan legal interest (*bienes jurídicos paraguayos*).[117] This include offenses against the existence of the State,[118] preparation of war and aggression,[119] offenses against the constitutional order,[120] coercion against constitutional organs,[121] false testimony,[122] and exposure to common risk.[123] In addition, some international instruments provide the option for States to exercise jurisdiction relating to offenses against "government [facilities]" abroad, [124] as well as "[embassies] or other diplomatic or consular premises."[125]

5.1.5 *Universal Jurisdiction*

Under the principles of universal jurisdiction, a State may exercise extraterritorial jurisdiction over certain, specific types of offenses irrespective of the place of their occurrence and the nationality of the offenders and of the victims.[126] The conduct of these participants may bear no connection to the *forum* State.[127] The act taking place abroad over which jurisdiction is exercised does not threaten the *forum* State's essential interests, and neither does it pose a "discernible impact."[128] In fact, all the constituting elements of the offense

116 See 18 USC § 1956 (b) (2) (2006) (establishing US jurisdiction over foreign individuals and financial institutions for money laundry offenses taking place in whole or in part in the United States).

117 See Law No 1160/97, *Paraguayan Penal Code* at Art 7.

118 *Ibid.* at Art 269.

119 *Ibid.* at Art 271.

120 *Ibid.* at Art 273.

121 *Ibid.* at Art 286.

122 *Ibid.* at Art 242.

123 *Ibid.* at Art 203.

124 See *Financing of Terrorism Convention*, *supra* note 11 at Art 7, para 2 (B); *Terrorist Bombing Convention*, *supra* note 11 at Art 6, para 2 (b).

125 *Nuclear Terrorism Convention*, *supra* note 11 at Art 9, para 2 (b).

126 See Robert Cryer et al, *An Introduction to International Criminal Law and Procedure*, 2nd edn (Cambridge: Cambridge University Press, 2010) at 50; Michael P Scharf, "Universal Jurisdiction and the Crime of Aggression" (2012) 53 Harv Int'l LJ 357 at 365 [Scharf, "Aggression"]; Eugene Kontorovich, "The Piracy Analogy: Modern Universal Jurisdiction's Hollow Foundation" (2004) 45 Harv Int'l LJ 183 [Kontorovich, "Piracy Analogy"]; Zachary Mills, "Does the World Need Knights Errant to Combat Enemies of All Mankind? Universal Jurisdiction, Connecting Links, and Civil Liability" (2009) 66 Wash & Lee L Rev 1315 at 1316; Ariel Zemach, "Reconciling Universal Jurisdiction with Equality before the Law" (2011) 47 Tex Int'l LJ 143 at 145.

127 See Eugene Kontorovich, "Beyond the Article I Horizon: Congress Enumerated Powers and Universal Jurisdiction over Drug Crimes" (2009) 93 Min L Rev 1191 at 1192 [Kontorovich, "Beyond Article I Horizon"].

128 See Bruce Broomhall, "Towards the Development of an Effective System of Universal Jurisdiction for Crimes under International Law" (2001) 35 New Eng L Rev 399 at 400; Kendra Magraw, "Universally Liable? Corporate-Complicity Liability under the Principle of Universal Jurisdiction" (2009) 18 Minn J Int'l L 458 at 459.

may be foreign to the *forum* State, a form of "bystander justice."[129] Universal jurisdiction presents some unique additional characteristics. Given that it represents a significant departure from the territorial principle,[130] its application is restricted to offenses involving heinous, atrocious conduct[131] that affects the international community as a whole.[132] Often regarded as *hostis humani generis*,[133] these offenses are of universal concern.[134] As Colangelo points out, in principle, a single State by itself cannot establish the boundaries of heinousness.[135] Courts "[should] use the international legal definitions—as derived from customary law—of the universal crimes they adjudicate."[136] When exercising universal jurisdiction, a State acts on behalf of the international community to preserve commonly shared values.[137] Universal jurisdiction is usually restricted to criminal offenses although in some States it is also extended to civil matters.[138]

In theory, universal jurisdiction may be exercised in absentia: that is, when the defendant is not present in the *forum* State but is "either in hiding or openly living in [another] harboring [country]."[139] Although some commentators favor this type of unre-

129 See Diane F Orentlicher, "Whose Justice? Reconciling Universal Jurisdiction with Democratic Principles" (2004) 92 Geo LJ 1057 at 1133.

130 See Eric S Kobrick, "The Ex Post Facto Prohibition and Exercise of Universal Jurisdiction over International Crimes" (1987) 87 Colum L Rev 1515 at 1518.

131 See Brownlie, *supra* note 31 at 307; Shaw, *supra* note 65 at 411; Miriam Cohen, "The Analogy between Piracy and Human Trafficking: A Theoretical Framework for the Application of Universal Jurisdiction" (2010) 16 Buff Hum Rts L Rev 201 at 203; Jon B Jordan, "Universal Jurisdiction in a Dangerous World: A Weapon for all Nations against International Crime" (2009) 9 MSU-DCL J Int'l L 1 at 3; Meyer, *supra* note 9 at 145.

132 See Cryer et al, *supra* note 126 at 51; Brown, *supra* note 3.

133 Jennings, *supra* note 30 at 156.

134 See Scharf, "Aggression," *supra* note 126 at 365; Kontorovich, "Piracy Analogy," *supra* note 126 at 183; Mills, *supra* note 126 at 1316; Bennett, *supra* note 78 at 437; Zemach, *supra* note 126 at 145. Huang argues that, for instance, "hijacking, as well as other crimes against the safety of civil aviation under [ICAO's aviation security conventions] have been ranked by the international community as serious international crimes having similar status as the crimes against humanity, piracy. They constitute an affront to all humanity and are punishable by all." He does not, however, mention whether these offenses form part of customary international law. Jiefang Huang, *Aviation Safety and ICAO* (The Hague: Kluwer Law International, 2009) at 161.

135 See Anthony J Colangelo, "Universal Jurisdiction as an International False Conflict of Laws" (2009) 30 Mich J Int'l L 881 at 890 [Colangelo, "Conflict of Laws"].

136 Anthony J Colangelo, "The Legal Limits of Universal Jurisdiction" (2006) 47 Va J Int'l L 149 at 185 [Colangelo, "Legal Limits"].

137 Kai Ambos, "Accountability for the Torture Memo: Prosecuting Guantanamo in Europe: Can and Shall the Masterminds of the 'Torture Memos' Be Held Criminally Responsible on the Basis of Universal Jurisdiction" (2009) 42 Case W Res J Int'l L 405 at 443; Jennings, *supra* note 30 at 156 (suggesting that the "suppression of [the offence] is an interest common to all States and to all mankind"); Steiner, *supra* note 107 (noting that offenses subject to universal jurisdiction are "universally condemned" at 204); Zerk, *supra* note 10 at 20.

138 See Stephen, Macedo, Project Chair. *The Princeton Principles on Universal Jurisdiction Princeton Project on Universal Jurisdiction* (Princeton: Program in Law and Public Affairs, Princeton University, 2001) at 28 (principle 1); Scharf, "Aggression," *supra* note 126 at 366.

139 Anthony J Colangelo, "The New Universal Jurisdiction: in Absentia Signaling Over Clearly Defined Crimes" (2005) 36 Geo J Int'l L 537 at 543 [Colangelo, "Absentia"].

stricted, pure form of jurisdiction,[140] for the vast majority of others, it is exceedingly problematic.[141] To this end, universal jurisdiction is subject to the requirement of a link or a "factual connection with the *forum* State."[142] In practice, this means that, in addition to the heinous nature of the offense, the offender must be present in the *forum* State.

Universal jurisdiction is rooted in three main sources: (i) customary international law,[143] (ii) peremptory norms or *jus cogens*,[144] and (iii) international treaties.[145] For centuries, customary international law has recognized that States enjoy the right to exercise universal jurisdiction over acts of piracy on the high seas.[146] The atrocities of World War II prompted the international community to redirect the application of universal jurisdiction to crimes

140 *Ibid.* at 548; Mark A Summers, "The International Court of Justice's Decision in *Congo v Belgium*: How Has it Affected the Development of a Principle of Universal Jurisdiction that Would Obligate All States to Prosecute War Criminals?" (2003) 21 BU Int'l LJ 63 (discussing that not recognizing universal jurisdiction in absentia leads to a "limited form of impunity; the offender is safe so long as he does not leave the asylum state" at 99).

141 In 2002, the ICJ found that a warrant issued by a Belgium judge violated the immunity of Congo's Foreign Affairs Minister, Mr. Abdulaye Yerodia. See *Arrest Warrant*, *supra* note 45 at 208. In particular, the ICJ held that Belgium's arrest warrant "failed to respect the immunity from criminal jurisdiction and the inviolability which [Congo's incumbent foreign affairs Minister] enjoyed under international law." *Ibid.* Marks has criticized the rationale the ICJ followed in Arrest Warrant as leading to impunity. See Jonathan H Marks, "Mending the Web: Universal Jurisdiction, Humanitarian Intervention and the Abrogation of Immunity by the Security Council" (2004) 42 Colum J Transnat'l L 445 at 489. The ICJ was reluctant to recognize Belgium's assertion of universal jurisdiction to try alleged human rights violations of Congo's Minister for Foreign Affairs in absentia. *Ibid.* at 208.

142 Separate Opinion Judge Rezek in *Arrest Warrant*, *supra* note 45 at 92; Joint Separate Opinion of Judges Higgins, Kooijmans, and Buergenthal, *ibid.* at 76; Zerk *supra* note 10 at 173; Declaration of Judge Ranjeva in *Arrest Warrant*, *supra* note 45 (noting that a "requirement of a connection *ratione loci* [exists in order to apply] universal jurisdiction" at 56). Similarly, Zuppi recognizes that the current state of the law does not enable States to assert universal jurisdiction in absentia, but this rather is de lege ferenda. Alberto Luis Zuppi, "Immunity v. Universal Jurisdiction: The Yerodia Ndombasi Decision of the International Court of Justice" (2003) 63 La L Rev 309 at 338.

143 Customary international law recognizes piracy, war crimes, genocide, and crimes against humanity as being subject to universal jurisdiction. See Madeline H Morris, "Universal Jurisdiction in a Divided World" (2001) 35 New Eng L Rev 337 at 350; Douglass Cassel, "Empowering United States Courts to Hear Crimes within the Jurisdiction of the International Criminal Court" (2001) 35 New Eng L Rev 421 at 426.

144 The Vienna Convention on Law of Treaties defines *jus cogens* or peremptory norms as those "accepted and recognized by the international community of States as a whole... from which no derogation is permitted and which can be modified only by a subsequent norm of general international law having the same character." *VCLT*, *supra* note 131 (Chapter 2) at Art 53.

145 See Jon B Jordan, "Universal Jurisdiction in a Dangerous World: A Weapon for all Nations against International Crime" (2009) 9 MSU-DCL J Int'l L 1 at 6. International instruments have extended universal jurisdiction's reach to offenses involving torture, terrorism, apartheid, and slavery. See Kobrick, *supra* note 130 at 1520; Princeton Principles on Universal Jurisdiction, *supra* note 138 at 28 (principle 2); Cabranes, *supra* note 54 at 1673.

146 See Separate Opinion of President Guillaume in *Arrest Warrant*, *supra* note 45 at 38; Eugene Kontorovich & Steven Art, "Agora: Piracy Prosecution: An Empirical Examination of Universal Jurisdiction for Piracy" (2010) 104 AJIL 436 [Kontorovich & Art].

such as genocide, war crimes, and crimes against humanity.[147] In the 1990s, several European countries resorted to the principle of universal jurisdiction to prosecute various former heads of States as well as other high-ranking State officials for violations of human rights.[148]

Universal jurisdiction is important to promote accountability,[149] avoid impunity,[150] and defend the interests of the international community against outrageous offenses that violate some fundamental norms. Yet, in spite of recent surges,[151] the assertion of universal jurisdiction remains a "highly contentious" issue.[152] Bassiouni points out that the principle is "not as well established in conventional and customary international law."[153] It inherently infringes upon the sovereign equality of States.[154] It also is against democratic principles, for judges of the *forum* are essentially unaccountable.[155] Often times, there is a perception that universal jurisdiction is not fully aligned with rules of due process.[156] In spite of this criticism, some scholars have suggested that the application of universal jurisdiction should be expanded to include offenses dealing with environmental war crimes,[157] human traffick-

147 See generally Leila Nadya Sadat, "Crimes against Humanity in the Modern Age" (2013) 107 AJI.L 334 at 337.

148 See Orentlicher, *supra* note 129 at 1059; Wolfgang Kaleck, "From Pinochet to Rumsfeld: Universal Jurisdiction in Europe 1998-2008" (2009) 30 Mich J Int'l L 927 at 928.

149 Tanaz Moghadam, "Revitalizing Universal Jurisdiction: Lessons from Hybrid Tribunals Applied to the Case of Hissene Habre" (2008) 39 Colum Human Rights L Rev 471 at 527.

150 See Dissenting Opinion van den Wyngaert in *Arrest Warrant*, *supra* note 45 at 166. See also Naomi Roht-Arriaza, "The Pinochet Precedent and Universal Jurisdiction" (2001) 35 New Eng L Rev 311 (stressing that the case against former Chile's President Augusto Pinochet by Spanish courts have "helped revitalize the anti-impunity movement" at 316); Karinne Coombes, "Universal Jurisdiction: A Means to End Impunity or a Threat to Friendly International Relations?" (2011) 43 Geo Wash Int'l L Rev 419; M Cherif Bassiouni, "Universal Jurisdiction for International Crimes: Historical Perspectives and Contemporary Practice" (2001) 42 Va J Int'l L 81 at 82.

151 See Monica Hans, "Providing for Uniformity in the Exercise of Universal Jurisdiction: Can Either the Princeton Principles on Universal Jurisdiction or an International Criminal Court Accomplish this Goal" (2002) 15 Transnat'l L 357 at 368.

152 Scharf, "Aggression," *supra* note 126 at 358.

153 Bassiouni, *supra* note 150 at 83.

154 See *Arrest Warrant*, *supra* note 45 at 210 (argument raised by Congo); Ambos, *supra* note 137 (discussing that universal jurisdiction interferences with the "judicial affairs of the territorial state" at 444).

155 See Orentlicher, *supra* note 129 at 1063.

156 Morris, *supra* note 143 (describing serious due process flaws in the legal proceedings brought in Spain against Chile's former President, Augusto Pinochet, at 353).

157 Ryan Gilman, "Expanding Environmental Justice after War: The Need for Universal Jurisdiction over Environmental War Crimes" (2011) 22 Colo J Int'l Envtl L & Pol'y 447 at 470.

ing,[158] drug trafficking,[159] cyberterrorism,[160] terrorism,[161] and other human rights violations.[162] I contend that universal jurisdiction cannot serve as a gap filler to address those issues where conventional jurisdictional presents loopholes or seems inadequate. States should not loosely assert universal jurisdiction as "fundamentalist crusaders on behalf of humanitarianism"[163] or other legitimate causes. Experience has already shown that this only contributes to the creation and/or escalation of previously existing tensions among States, something that irritates other members of the international community. It should only be applied to the most serious criminal offenses and in a restrictive manner, provided there is a reasonable link to the *forum* State, such as the presence of the offender.[164]

5.2 STATE RESPONSIBILITY AND OBLIGATIONS *ERGA OMNES*

There is robust consensus that climate change is one of the issues of common concern to humankind.[165] There would also seem to be agreement that, in accordance with their respective capabilities and circumstances, each State bears an obligation *erga omnes* to minimize the impact that anthropogenic climate change causes to the environment.[166] The

158 See Cohen, *supra* note 131 at 221.

159 Anne H Geraghty, "Universal Jurisdiction and Drug Trafficking: A Tool for Fighting one of the World's Most Pervasive Problems" (2004) 16 Fla J Int'l L 371 (commenting that an extension of universal jurisdiction "over drug traffickers will provide a tool that will help states in ensuring that traffickers are face with a real threat of prosecution" at 403).

160 See Kelly A Gable, "Cyber-Apocalypse Now: Securing the Internet against Cyberterrorism and Using Universal Jurisdiction as a Deterrent" (2010) 43 Vand J Transnat'l L 57.

161 Sarah Mazzochi, "The Age of Impunity Using the Duty to Extradite or Prosecute and Universal Jurisdiction to End Impunity for Acts of Terrorism Once and for All" (2011) 32 N Ill UL Rev 75 at 100.

162 See Peter Weiss, "The Future of Universal Jurisdiction" in Wolfgang Kaleck et al, eds, *International Prosecution of Human Rights Crimes* (Berlin: Springer, 2007) 29 at 36. In light of the inherent problems in defining the crime, Nagle highlights that issues involving terrorism are "not suitable for universal jurisdiction." Luz E Nagle, "Terrorism and Universal Jurisdiction: Opening a Pandora's Box" (2011) 27 Ga St UL Rev 339 at 340.

163 Separate Opinion Judge Sayeman Bula-Bula in *Arrest Warrant*, *supra* note 45 at 136.

164 Even the International Criminal Court (ICC) has been established on the basis of very restricted jurisdictional rules, namely, the territorial and nationality principles. The ICC may also hear a case where the UN Security Council refers it or where the States consent its submission thereto. See Rome Statute of the Criminal Court, done at Rome on 17 July 1988, in force on 1 July 2002 UNTS Vo 2187, No 38544 at Art 13. See also Cedric Ryngaert, "The International Criminal Court and Universal Jurisdiction: A Fraught Relationship?" (2009) 12 New Crim LR 498 at 500.

165 See *UNFCCC*, *supra* note 11 (Introduction) at preambular clauses. See also Bodansky et al, *Oxford Handbook*, *supra* note 66 (Chapter 2) (noting that the international community has made substantial progress "in identifying issues that are of common concern" at 13); Shaw, *supra* note 65 at 559.

166 For instance, Mayer notes that the international community is responsible "towards populations affected by climate change, including setting up international resettlement programs for climate migrants." He offers various justifications, such as treaty-based obligations, humanitarian and fairness arguments, as well as the invocation of the polluter pays principle. See Benoit Mayer, "The International Legal Challenges of Climate-Induced Migration: Proposal for an International Legal Framework" (2011) 22 Colo J Int'l L Envtl L. & Pol'y 357 at 375. Similarly, the adoption of obligations *erga omnes* in treaties is not an uncommon feature in

question then is whether, in light of the reluctance of some States to take action to combat climate change in a meaningful way, States willing to protect global environmental commons may justify their actions on the basis of the theory of State responsibility and obligations *erga omnes*. Otherwise put, in light of the failure of the international community to address GHG emissions from international aviation, does the doctrine of State responsibility justify Europe's decision to include international aviation into its ETS?

In *Barcelona Traction*, the ICJ recognized that States bear some obligations "towards the international community as a whole."[167] Because of the high value nature of the rights embodied in these obligations, each State has a "legal interest in their protection."[168] Based on the principle of universality, these obligations are applicable to all States, and all States must observe them.[169] In this respect, the United Nations General Assembly has recognized that States must "safeguard and conserve nature in areas beyond national jurisdiction."[170] Although the ICJ introduced the notion of obligations *erga omnes*, it did not elaborate further on their scope and content.[171] Intentionally, the ICJ also did not provide an exhaustive list of which international obligations should fall under this concept.[172] Some commentators contend that the notion of obligations *erga omnes* may include issues such

---

international environmental law. See Antti Korkeakivi, "Consequences of 'Higher' International Law: Evaluating Crimes of State and *Erga Omnes*" (1996) 2 J Int'l Legal Stud 81 at 100. Gregory D Pendleton argues that "obligations to protect seamounts in the global commons should [also] be seen as an obligation *erga omnes*." Gregory D Pendleton, "State Responsibility and the High Seas Marine Environment: A Legal Theory for the Protection of Seamounts in the Global Commons" (2005) 14 Pac Rim L & Pol'y J 485 at 512.

167 *Case concerning The Barcelona Traction, Light and Power Company, Limited (Belgium v Spain) Second Phase*, [1970] ICJ Rep 3 [*Barcelona Traction*] at 32. For an interesting analysis of what is understood by the notion of the international community, see Kuan-Wei Chen, *In Search of the International Community (of States)* (LLM Thesis, Leiden University, 2008) [unpublished] [Chen, "International Community"].

168 *Barcelona Traction*, *supra* note 167 at 32.

169 See Pendleton, State Responsibility, *supra* note 166 at 511.

170 UNGA, World Charter for Nature, UN Doc A/RES/37/7 28 (October 1982), para 21 (e). See also Kuan-Wei Chen, *The Legality of the Use of Space Weapons: Perspectives from Environmental Law* (LLM Thesis, 2012) (McGill University Institute of Air and Space Law) [unpublished] [Chen, "Legality of the Use of Space Weapons"] at 47.

171 See Korkeakivi, *supra* note 166 at 97.

172 See Pendleton, State Responsibility, *supra* note 166 at 511. One could speculate that the ICJ left it unresolved to allow international law to further develop it overtime. Although the case did not deal with environmental issues per se, commentators do not hesitate to suggest that the notion of obligations *erga omnes* may well encompass norms dealing with environmental protection, including those relating to climate change. Jorge E Viñuales, "The Contribution of the International Court of Justice to the Development of International Environmental Law: a Contemporary Assessment" (2008) 32 Fordham Int'l LJ 232 at 235. Similarly, Huang advances the innovative idea that aviation safety obligations, such as regulatory oversight, imposing punitive measures, and preventing the use of weapons against international civil aviation, have acquired the rank of obligations *erga omnes*. Huang contends that, given the importance of these obligations for the orderly development of air transport, they are a common concern for the international community as a whole. See Huang, *supra* note 134 at 175. Although no one can contest the relevance of these obligations for the sector, the high level of non-compliance with ICAO standards in safety regulatory functions may be indicative that State practice does not necessarily conceptualize them as part of obligations *erga omnes*.

as the protection of global environmental commons and outer space.[173] This is consistent with the view that the ICJ held in *Gabčíkovo-Nagymaros* where it said that "great significance" must be given to "respect for the environment, not only for States but also for the whole of mankind."[174] Also, in *Furundžija*, the International Criminal Tribunal for Yugoslavia (ICTY) expressly recognized the right of the members of the international community to demand compliance with obligations *erga omnes* or the discontinuation of a breach of an international obligation such as torture.[175] Likewise, the ICJ noted that the prevention of genocide is also an obligation *erga omnes.*[176]

Drawing heavily from the ICJ's *dicta* in *Barcelona Traction* during the process of elaborating the Draft Articles on State Responsibility, the International Law Commission (ILC) suggested that States may act to remedy breaches of obligations *erga omnes.*[177] As a general principle, State responsibility necessarily involves the conduct (an act or omission) of a State that represents an internationally wrongful act constituting a breach of an international obligation.[178] This leads to a "loss or damage" that in turn gives rise to a right to reparation.[179] According to the new conceptualization of obligations *erga omnes* as reflected in the Draft Articles on State Responsibility, a State that has not directly suffered

173 See Chen, "Legality of the Use of Space Weapons," *supra* note 170 at 61; Maurizio Ragazzi, *The Concept of International Obligations Erga Omnes* (Oxford: Claredon Press, 1997) at 132.

174 *Gabčíkovo-Nagymaros Project, supra* note 228 (Chapter 2) at para 53. In this respect, Kaplan says that "[a] breach of an [*obligation erga omnes*] owed to the international community is a breach of an obligation to every member of that community. The State that committed the wrongful act owes to the international community secondary obligations of cessation, guarantees and assurances of no repetition, and appropriate reparations, and each State is in the position to demand those secondary obligations be fulfilled." Margo Kaplan, "Using Collective Interests to Ensure Human Rights: an Analysis of the Articles on State Responsibility" (2004) 79 NYU L Rev 1902 at 1910.

175 See *Prosecutor v Anto Furundžija, IT-95-17/1-T, Judgment (10 December 1998) (International Criminal Tribunal Yugoslavia, Trial Chamber)*, online: ICTY <www.icty.org/x/cases/furundzija/tjug/en/fur-tj981210e.pdf> at 260, para 151.

176 See *Case concerning East Timor (Portugal v Australia)* [1995] ICJ Rep 90; *Case Concerning Application of the Convention on the Prevention and Punishment of the Crime of Genocide (Bosnia and Herzegovina v Serbia and Montenegro)* [1996] ICJ Rep 595 at 616, para 31.

177 See "Draft Articles on Responsibility of States for Internationally Wrongful Acts," in *Report of the International Law Commission*, Fifty-third Session, 23 April-1 May and 2 July-10 August 2001, General Assembly Official Records, Fifty-sixth Session, Supplement No 10, UN Doc A/56/10 [*Draft Articles on State Responsibility*]. With the purpose of advancing the development international law and in particular its codification, in 1947, the United Nations General Assembly established the ILC. Since then, the ILC has been instrumental in the development of a number of draft articles that later formed the basis of influential international treaties. Shaw says that the ILC's drafts "may constitute evidence of customs as well as contribute to the corpus of usages which may create new law." Shaw, *supra* note 65 at 97. Although the Draft Articles on State Responsibility have not been codified in an international treaty, one must recognize that the ICJ has expressly referred to them in various occasions. See *Case Concerning the Application of the Convention on the Prevention and Punishment of the Crime of Genocide (Bosnia and Herzegovina v Serbia and Montenegro)*, Judgment of 26 February 2007 at paras 173 and 180 [*Bosnia Genocide Case 2007*]; *Legal Consequences of the Construction of a Wall in the Occupied Palestinian Territory*, Advisory Opinion of 9 July 2004 at paras 140, 150.

178 See *Draft Articles on State Responsibility, supra* note 177 at Art 2. See also Brownlie, *supra* note 31 at 387.

179 Shaw, *supra* note 65 at 482.

damage may invoke the responsibility of another State in cases involving the breach of two specific sets of obligations.[180]

The first obligation includes those that are owed to the "collective interest of [a] group [of States]" of which the invoking State is part.[181] The second encompasses those obligations that each State must observe merely because it is a member of the "international community as a whole."[182] Here, the link is more indirect, as the State invoking the responsibility may not have necessarily suffered the damage.[183] This State is normally referred to as the non-injured State.[184] The State is not acting to protect its own interests[185] but rather in defense of the interests of the international community as a whole. Thus, through a legal action to be presented before a court of law, such as the ICJ, the non-injured State may only seek restricted remedies such as the "cessation of the internationally wrongful act"[186] or reparation thereof.[187] The list of what a non-injured State can do is considered to be exhaustive.[188] A State other than the injured State cannot engage in countermeasures.[189] In addition to providing adequate notice of the claim, the State in question must clarify what specific actions it seeks from the responsible State (i.e. cessation of the wrongful act or reparation).[190] In principle, the action cannot proceed where the State has already acquiesced or waived its right to the claim.[191]

---

180 This implicitly seeks to reverse the position that the ICJ held in the Namibia case where it said that Ethiopia and Liberia did not enjoy legal standing in light of the fact that these States did not suffer damages. See Arnold N Pronto, "Human-Rightism and the Development of General International Law" (2007) 20 Leiden J Int'l L 753 at 755.

181 *Draft Articles on State Responsibility, supra* note 177 at Art 48, para 1 (a). See also ILC, Draft Articles on Responsibility of States for Internationally Wrongful Acts, with commentaries (2001) at 126 [ILC, *Commentaries on Draft Articles*].

182 *Draft Articles on State Responsibility, supra* note 177 at Art 48, para 1 (b). See also ILC, *Commentaries on Draft Articles, supra* note 181 at 217.

183 See Chen, "International Community," *supra* note 167 (suggesting that the Draft Art on Responsibility of States confer a "blank cheque for any State other than the injured State to take lawful measures against a perpetrator in the interest of the injured State or of the beneficiaries of the obligation breached" at 38).

184 Scholars have not unanimously accepted the notion of "non-injured States." Pierre-Maire Dupuy claims that "[i]t makes little sense to say that non-injured States have a right of action against another state. If they are not injured, what is the legal ground for them legitimately (and legally) to take remedial action? Rather more accurately, they are not affected in their individual and subjective interest, but in their objective interest in the adherence to those obligations that are of essential importance for the international community." Pierre-Marie Dupuy, "The Deficiencies of the Law of State Responsibility Relating to Breaches of 'Obligations Owed to the International Community as a Whole': Suggestions for Avoiding the Obsolescence of Aggravated Responsibility" in Antonio Cassese, ed., *Realizing Utopia: The Future of International Law* (Oxford: Oxford University Press, 2012) 210 at 216.

185 See Wolfrum, "International Environmental Law," *supra* note 67 (Chapter 2) at 56.

186 *Draft Articles on State Responsibility, supra* note 177 at Art 48, para 2 (a).

187 *Ibid.* at Art 48, para 2 (b).

188 See ILC, *Commentaries on Draft Articles, supra* note 181 at 127. See also Korkeakivi, *supra* note 166 at 113.

189 ILC, *Commentaries on Draft Articles, supra* note 181 at 139.

190 See *Draft Articles on State Responsibility, supra* note 177 at Art 43.

191 *Ibid.* at Art 45. See also, ILC, *Commentaries on Draft Articles on Responsibility, supra* note 181 at 128.

In essence, this *actio popularis*[192] is an independent cause of action that any State may exercise before a court in defense of the interests of the international community. Although some commentators would seem to suggest that there may be a growing trend to recognize these actions,[193] there is still significant disagreement over what States must actually do to respond to obligations *erga omnes*.[194] It still remains an abstract concept that has not been fully internalized by States, and so it must be exercised with caution.[195] States cannot act as mere "[policemen to protect those interests] of the international community."[196] State practice indicates that this is yet to be part of customary international law. A different issue arises when a State is confronted with a violation of *jus cogens* norms.[197] As part of its obligations *erga omnes*, a State may initiate an *actio popularis* to remedy these breaches[198] since they affect all States wherever they are committed.[199]

An interesting question is whether the principles of international law on State responsibility would allow a State to remedy a breach of an obligation *erga omnes* (such as the failure to take action on climate change to protect global environmental commons)

---

192 In 1966, long before the adoption of the Draft Articles on Responsibility of States, the ICJ expressly rejected the existence of an *actio popularis* in international law. In fact, in *West Africa (II)*, the ICJ found that "although a right of this kind may be known to certain municipal systems of law, it is not known to international law as it stands at present: nor is the Court able to regard it as imported by the 'general principles of law' referred to in Article 38, paragraph 1 (c), of its Statute." *South West Africa Cases (Ethiopia v South Africa; Liberia v South Africa) Second Phase* [1966] ICJ Rep 6 at 6, para 88. In his dissenting opinion, however, Judge Jessup noted that "[i]nternational law has accepted and established situations in which States are given a right of action without any showing of individual prejudice or individual substantive interest as distinguished from the general interest." See *South West Africa*, Second Phase (Dissenting Opinion Judge Jessup) at 387-88. Arguably, the ILC carefully avoided using the term *actio popularis*. Likewise in its Reparation for Injuries Advisory Opinion, the ICJ held that "only the party to whom an international obligation is due can bring a claim in respect of its breach." *Reparations for Injuries suffered in the Service of the United Nations*, Advisory Opinion, ICJ Reports 1949, paras 181-82. In spite of this, the recognition of the right to bring a legal action by a party other than the injured State in essence presents exactly the same characteristics as those of an *actio popularis*. See also Wolfrum, "International Environmental Law," *supra* note 67 (Chapter 2) at 55.

193 See Jennifer S Bales, "Transnational Responsibility and Recourse for Ozone Depletion" (1996) 19 BC Int'l & Comp L Rev 259 at 277. Kirgis, Jr., advances that each State has "universal standing" to defend and protect obligations *erga omnes*. See Frederic L Kirgis, Jr., "Standing to Challenge Human Endeavors that Could Change the Climate" (1990) 84 AJIL 525 at 528.

194 Bodansky et al, *Oxford Handbook*, *supra* note 66 (Chapter 2) at 11.

195 In this vein, Kaplan warns the dangers of empowering a State other than the injured State the prerogative to exercise an action to protect human rights obligations owned to the international community. See Kaplan, *supra* note 174 at 1915.

196 Korkeakivi, *supra* note 166 at 113.

197 The ILC has attempted to shed light on the relationship between obligations *erga omnes* and *jus cogens*. In this respect, the ILC has said that *jus cogens* are non-derogable norms that "override conflicting norms," whereas obligations *erga omnes* "designate the scope of application of the relevant law, and the procedural consequences that follow from this." Here is worth to clarify that not all obligations *erga omnes* may be classified as norms of *jus cogens*, but all *jus cogens* norms imply obligations *erga omnes*. See ILC, Fragmentation Report, *supra* note 27 (Chapter 2) at 193.

198 See Kenneth C Randall, "Universal Jurisdiction under International Law" (1998) 66 Tex L Rev 785 at 830.

199 *Ibid.* at 831.

through an *actio popularis*. Although heavily contested, for the sake of the discussion, I assume that combating climate change is an obligation *erga omnes*. The *actio popularis* claim – an independent cause of action that a non-injured State may invoke before a court – should be distinguished from the exercise of unilateral State action such as the EU ETS. The principles of international law on State responsibility and, more specifically, the invocation of responsibility by a State other than the injured State cannot serve as a justification carte blanche for unilateral measures. In addition, even though it is theoretically conceivable for a State to file a legal challenge against other States that have not taken measures to combat climate change, it is important to recall that the available remedies are limited to reparation and discontinuation of the internationally wrongful act. The applicant State may also be required to establish a causal link between the alleged international wrongful act and the supposed damage. In the context of a climate change dispute, this may prove to be extremely problematic. For instance, to invoke the doctrine of State responsibility, Europe would have to prove that the failure of non-EU States to adopt an MBM to tackle the carbon footprint of their aircraft operators has led to a concrete damage.

5.3 THE EFFECTS DOCTRINE

In this section, I examine the salient features of the effects doctrine. The doctrine has been largely used by some States to justify extraterritoriality and also because in the *ATA Decision* the CJEU and Advocate General Kokott implicitly alluded to it.

Although deeply rooted in the US legal tradition, the effects doctrine, just like the EU ETS saga, has been subject to a "lively [international] controversy."[200] In his seminal opinion in *Alcoa*, US Judge Learned Hand laid down the foundations of the doctrine as applied to antitrust law. This case involved cartel activities to monopolize the manufacturing of virgin aluminum ingot that occurred outside the United States.[201] Judge Learned Hand articulated the two main elements of this doctrine as follows: the behavior must be "intended" to have "some effect" in the United States.[202] Judge Learned Hand also (albeit incorrectly) suggested that "other States will ordinarily recognize" the exercise of jurisdiction in these types of situations.[203] His opinion in *Alcoa* came to be known as the "intended

200 *Timberlane Lumber Co v Bank of America*, 549 F (2d) 597 at 610 (9th Cir 1976) [*Timberlane*]; *Mannington Mills, Inc v Congoleum Corporation*, 595 F (2d) 1287 at 1291 (3rd Cir 1979) [*Mannington Mills*]. See also Ellis, Fairness, *supra* note 53 (noting that the effects doctrine is "solidly anchored in US law, [but it] is a source of controversy in international law" at 401).

201 *United States v Aluminum Co of America*, 148 F (2d) 416 at 421(2d Cir 1945) [*Alcoa*].

202 *Ibid.* at 443.

203 *Ibid.* (noting that "it is settled law... that any state may impose liabilities, even upon persons not within its alliance, for conduct outside its borders that has consequences within its borders which the state reprehends, and these liabilities other states will ordinarily recognize").

effect" test. Later, a number of courts followed this test[204] and formulated variations thereof.[205]

By the 1970s, the continuous extraterritorial application of US antitrust laws, through congressional statutes, judicial decisions, and enforcement from administrative agencies, prompted other States to adopt legislation to attenuate or "block" their effects on their nationals.[206] In 1976, in *Timberlane*, the US Court of Appeals for the Ninth Circuit applied a jurisdictional rule of reason to balance the interest of other States. In addition to reaffirming that the Sherman Act may govern extraterritorial behavior,[207] the Ninth Circuit held that factors other than the effects test needed to be considered in order to take into account the interest of other nations.[208] According to the Court, the analysis should include issues such as (i) whether there was some actual or substantial effect in the United States, (ii) whether such effect causes an injury to the plaintiff that constitutes a civil violation of antitrust laws, and (iii) whether the magnitude of such effect justifies an assertion of US jurisdiction.[209] A number of courts later followed the Timberlane test not only in the field of antitrust.[210] In *Hartford Fire Insurance*, a case that dealt with the alleged unlawful conspiracy of London underwriters, the US Supreme Court held that US antitrust law "applies

---

204 In *Continental Ore*, the Supreme Court held that "a conspiracy to monopolize or restrain the domestic or foreign commerce of the United State is not outside the reach of the *Sherman Act* just because part of the conduct complained of occurs in foreign countries." *Continental Ore Co v Union Carbide & Carbon Corp*, 370 US 690 at 704 (1962). Similarly, in *Pacific Seafarers, Inc v Pacific Far East Line*, the DC Circuit applied the *Sherman Act* to conspiracy activity of American shipping lines engaged in the transportation of cement and fertilizers between Taiwan and Vietnam. Although the Court noted that the shipping services did not include points in the United States, it nonetheless concluded that jurisdictional nexus constituted the fact that cargoes were financed through the US Agency for International Development (USAID). See *Pacific Seafarers, Inc v Pacific Far East Line, Inc*, 404 F 2d 804 at 807 (Ct App DC 1968). In *Matsushita*, the Supreme Court also looked into whether the alleged cartel activities abroad of Japanese electronics companies had some effect in the United States. See *Matsushita Electric Industrial Co v Zenith Radio Corp*, 475 US 574 at 583 (1985) [*Matsushita*].

205 See, e.g., *Hartford Fire Insurance Co v California*, 509 US 764 at 796 (1993) [*Hartford Fire Insurance*].

206 See Leigh Robin Lamendola, "The Continuing Transformation of International Antitrust Law and Policy: Criminal Extraterritorial Application of the Sherman Act in *United States* v. *Nippon Paper Industries*" (1998) 22:2 Suffolk Transnat'l L Rev 663 at 677; Knut Hammarskjöld, "About the Need to Bridge a Jurisdictional Chasm" (1983) VIII Ann Air & Sp L 97 at 105; AJC Paines, "Extraterritorial Aspects of Mergers and Joint Ventures" (1985) 13 Int'l Bus Law 344 (noting that the "UK has strongly and consistently opposed the application of the effects doctrine" at 344); Tony A Freyer, "Restrictive Trade Practices and Extraterritorial Application of Antitrust Legislation in Japanese-American Trade" (1999) 16 Ariz J Int'l & Comp L 159 (describing the Japanese opposition against the extraterritorial application of US and EU antitrust laws, at 160).

207 See *Timberlane, supra* note 200 at 608.

208 *Ibid.* at 611.

209 *Ibid.* at 614.

210 A number of courts later followed the Timberlane test. For example, in *Montréal Trading Ltd*, a case where a Canadian firm brought an antitrust action against various US potash producers alleging that the refusal to deal limited production and increased prices, the Tenth Circuit found that the predominantly foreign elements did not necessarily lead to sufficient effects on US commerce. *Montréal Trading Ltd v Amax Inc*, 661 F 2d 864 at 869 (10th Cir 1981) [*Montréal Trading*].

to foreign conduct that is meant to produce and does in fact produce some substantial effect in the [US]."[211]

The reach of the effects doctrine within the United States has not been static.[212] It has been expanded to other fields, including, but not limited to, securities and trademarks. Based on an analysis of copious judicial precedents, the authors of the *Restatement (Third) on Foreign Relations* concluded that US courts may assert prescriptive jurisdiction over conduct taking place abroad that "has or is intended to have substantial effect within [US] territory."[213] In principle, this exercise should not be unreasonable.[214] In practice, however, by resorting to extraterritorial jurisdiction through the effects doctrine, US courts seek to protect US economic interests.[215] Although some scholars in the United States have suggested that "modern international law" would justify the effects doctrine,[216] this is nowhere settled; the doctrine has been the subject of extensive criticism. In addition to its very broad reach,[217] scholars point out that it lacks a precise definition and that it leads to unpredictable results.[218] Others underscore the fact that the doctrine ignores rules of international law, such as the non-violability of sovereignty and extraterritoriality.[219] It is also problematic because the conduct is often lawful in the country where it takes place.[220]

Perhaps one of the most interesting aspects of the effects doctrine, and to some extent the extraterritorial application of US law in general, is the fact that courts tend to view it as an extension of the territorial principle that has evolved from objective territoriality.[221]

---

211 *Hartford Fire Insurance*, *supra* note 205 at 796.

212 See Won-Ki Kim, "The Extraterritorial Application of U.S. Antitrust Law and Its Adoption in Korea" (2003) 7 Singapore J Int'l & Comp L 386 (commenting that the US antitrust law has shifted from a strictly territorial conception in *American Banana* to the intended effects test in Alcoa, to the interest balancing and jurisdictional rule of reason in *Timberlane*, to return to the intended effects in *Hartford Fire Insurance* at 389).

213 Restatement (Third) of Foreign Relations Law of the United States § 402 (I) (c).

214 *Ibid.* § 403 (I). Similarly, Shaw points out that the effects doctrine presupposes that a State is entitled to assert jurisdiction where, despite of the fact that the conduct occurs predominantly abroad, it causes effects within its territory. See Shaw, *supra* note 65 at 423.

215 *Mannington Mills, Inc*, *supra* note 200 (holding that "[a]cts and agreements occurring outside the territorial boundaries of the [US] that adversely and materially affect American trade are not necessarily immune from [US] antitrust laws" at 1291).

216 Gerald F Hess, "The Trail Smelter, the Columbia River, and the Extraterritorial Application of CERCLA" (2005) 18 Geo Int'l Envtl L Rev 2 at 43.

217 See Parrish, *supra* note 9 (suggesting that it "gives license for near universal jurisdiction" at 1479).

218 See Chie Sato, "Extraterritorial Application of EU Competition Law: Is it Possible for Japanese Companies to Steer Clear of EU Competition Law?" 11 J Pol Sci & Soc 23 at 27; Parrish, *supra* note 9 (criticizing that the effects doctrine does not enjoy "doctrinal clarity" at 1480).

219 See Ellis, "Fairness," *supra* note 53 at 402; Dabbah, *supra* note 164 (Chapter 4) (discussing that some States regard the effects doctrine as an "intrusion of sovereignty" at 166).

220 See Vaughan Lowe, "The Problems of Extraterritorial Jurisdiction: Economic Sovereignty and the Search for a Solution" (1985) 34:4 Int'l & Comp L Q 724 at 725 [Lowe, "Problems of Extraterritorial Jurisdictions"].

221 See Sato, *supra* note 218 at 26; Patricia M Barlow, *Aviation Antitrust: The Extraterritorial Application of the United States Antitrust Laws and International Air Transportation* (Boston: Kluwer Law & Taxation Publishers, 1988) at 107 (justifying the application of the effects doctrine on the basis of the objective territorial principle).

Courts have resorted to creative ways to establish territorial justifications to the foreign elements that may be present in a given case. This may either provide a territorial nexus that is sufficient to justify the exercise of extraterritorial jurisdiction (e.g. intended effects in the United States), or it may simply turn the controversy into a purely domestic/territorial matter.[222] For instance, in the field of securities, although courts have acknowledged that the *Securities Exchange Act of 1934*[223] does not provide for extraterritorial application, they have ruled nonetheless that the statute governs transactions outside the United States through the adoption of the "conduct and effects" test.[224]

Although Europe has not strictly adopted the effects doctrine, it has followed an approach that is very similar to that of the United States. The leading European case is *Ahlström* where CJEU was asked to determine whether an agreement involving concerted practices of wood pulp producers, which took place outside of the EU but which effectively raised prices for customers situated in the EU, was contrary to EU competition laws.[225] The CJEU found that the "object and effect" of the agreement was to restrict competition in the EU.[226] In order to conclude that the effects in the territory of EU Member States were "not only substantial but intended,"[227] the CJEU drew a distinction between the formation and the "place of implementation" of the agreement.[228] The fact that the agreement was "implemented" in the EU provided sufficient legal grounds to assert territorial juris-

---

See also Ellis, Fairness, *supra* note 53 (noting that the objective territoriality is narrower in scope than the effects doctrine, at 401).

222 See, e.g., *Pakootas v Teck Cominco Metals, Ltd*, 452 F 3d 1066 at 1068 [*Pakootas*]. The controversial trail smelter case provides a good example of a court creatively interpreting one element as sufficient justification to regard the dispute as a domestic application of law when in fact it is simply full with extraterritoriality. See Ellis, "Fairness," *supra* note 53 at 397. Teck, a Canadian trail smelter, dumped effluents over the Columbia River in Canada that later were found to be located in the river's portion on the United States. Two individuals filed a legal suit seeking to enforce an administrative order of the US Environmental Protection Agency (EPA), which, on the basis of the Comprehensive Environmental Responsive Compensation and Liability Act (CERCLA), required Teck to conduct a feasibility study in order to clean-up the transboundary pollution. See also 52 USC § 9601. Teck moved for dismissal on the grounds that the US Congress did not intent CERCLA to apply beyond the US territory. *Ibid.* The Court of Appeals for the Ninth Circuit held that the case dealt with CERCLA's domestic as opposed to extraterritorial application. *Ibid* at 1074. The Court noted that, although the pollution was initiated in Canada, it has "come to be located" in the US. *Ibid.* Implicitly, the Court assimilated that pollution's ultimately location produced hazardous *effects* in the US. *Ibid* at 1079 (holding that "applying CERCLA [to Teck's activities] for the release of hazardous substances at the site is a domestic, rather than an extraterritorial application of CERCLA, even though the original source of the hazardous substances is located in a foreign country").

223 *Securities Exchange Act of 1934*, § 10 (b), 15 USCA § 78j (b); 28 US CA § 1331 [*Securities Exchange Act*].

224 See Ian L Schaffer, "An International Train Wreck Caused in Part by a Defective Whistle: When the Extraterritorial Application of Sox Conflicts with Foreign Laws" (2006) 75 Fordham L Rev 3 1829.

225 See *Ahlström v Commission*, Judgment of the Court of 27 September 1988, para 2 at 5240.

226 *Ibid.* at para 13.

227 *Ibid.*

228 *Ibid.* at 5243, para 18.

diction.[229] In practical terms, the place of implementation represents a variation of the effects doctrine as applied by US courts.[230]

5.4 THE EU ETS AND EXTRATERRITORIALITY

To examine the issue of extraterritoriality in connection with the EU ETS, I first look into the position taken by the Commission. This is substantially less complicated as the Commission has opted for an extremely basic approach. I then proceed to dissect the CJEU's decision in *ATA* and its creative defense of the exercise of extraterritorial jurisdiction.

5.4.1 *The Commission's Approach*

As stated several times throughout this book, it was evident that opponents of the EU ETS regarded the scheme as flagrantly extraterritorial. Arguably, the European Commission always knew about this. In this respect, one would have thought that the Commission would have elaborated a very carefully crafted response to the extraterritorial allegations. However, the Commission ignored these concerns almost in their entirety and responded with a rather simplistic and unarticulated approach. The Aviation Directive clearly states that "[i]n order to avoid distortions of competition and improve environmental effectiveness, emissions from all flights arriving at and departing from community aerodromes should be included [into the scheme]."[231] In essence, the Commission downplayed the relevance that opponents gave to the methodology used to calculate emissions for the whole duration of the flight. Foreign aircraft operators were included into the scheme to avoid market distortions. And as such, the scheme was made a condition to access the market. The Commission's views on extraterritoriality are best summarized in the statement delivered by Jos Delbeke, Director General of Climate Action, as part of his testimony before the US Senate. Delbeke contended that under the *Chicago Convention*:

> States have the sovereign right to determine the conditions for admission to or departure from their territory and require all airlines to comply. There is no extra-territorial effect because no obligations are imposed in the territory of another State. The EU fully recognizes that this is a fundamental principle of international aviation law, and should be fully respected. The requirement

229 The "implementation" approach has been criticized as being an omnibus "elusive concept" that can trigger EU jurisdiction to a wide variety of business transactions. Liad Whatstein, "Extraterritorial Application of EC Competition Law: Comments and Reflections" (1992) 26 Isr L Rev 195 at 237.

230 See Ryngaert, *supra* note 30 at 77.

231 See Aviation Directive, *supra* note 17 (Introduction) at preambular clauses, para 16.

> to report emissions and to surrender allowances under the EU ETS only arises when an aircraft enters or departs from an airport in an EU Member State.[232]

This statement is very illustrative of the position taken by the Commission for three reasons. First, although the statement indirectly makes reference to Article 11 of the *Chicago Convention*, it misinterprets it. The Commission conceives Article 11 as meaning that the laws of the State of arrival unrestrictedly govern the admission of foreign aircraft. Under the Commission's view, this would provide a sufficient territorial link to escape the extraterritorial trap. However, this is not *raison d'être* of Article 11. The provisions of Article 11 establish that the State of arrival may impose regulations with regard to the entrance and departure of foreign aircraft from its territory but "while within its territory."[233] In doing so, the measures cannot be applied in a discriminatory manner. The Commission also suffers from selective amnesia, for it forgets the geographical limitations embedded in the provision. Second, either consciously or unconsciously, the Commission mischaracterizes the scope of the scheme; arguing that it does not impose extraterritorial obligations is simply a fallacy. Aircraft operators must perform a number of tasks even before the aircraft departs from a foreign point to the EU. Third, the Commission's approach implies that the scheme is a mere condition to access the market.

5.4.2 *The* ATA Decision *and Extraterritoriality*

I now consider the CJEU's *ATA Decision*. In addition to examining the court's reasoning, I also look into the very influential opinion of Advocate General Kokott. I break down the court's ruling in six main parts, namely, (i) the symbolic statements on the scheme's territorial limits, (ii) the foreign aircraft operator's physical presence in the EU and the notion of unlimited jurisdiction of the State of arrival, (iii) the scheme's fuel consumption methodology as mere events taking place partly outside the EU, (iv) the scheme as a condition to access the EU air transport market, (v) the scheme as a vehicle to advance EU climate change policy following a rational actor's model, and (vi) the surreptitious embracement of the effects doctrine.

5.4.2.1 Drawing Symbolism

Following the reasoning of the Advocate General, the CJEU noted in its decision that the EU ETS scheme only applies to foreign aircraft operators to the extent that they operate

232 Delbeke, *supra* note 7 (Chapter 4) at 6-7.
233 See *Chicago Convention*, *supra* note 36 (Chapter 2) at Art 11.

flights to and from airports situated in the EU.[234] According to this rationale, the scheme is not concerned with behavior that takes place outside the jurisdictional boundaries of the EU,[235] neither does it cover operations over third countries.[236] These statements seek to downplay the extraterritorial elements present in the scheme. They are also very similar in nature to the symbolic statements on the limitation of jurisdiction made by US Courts. In *Microsoft Co. v AT&T*, while examining the extraterritorial application of US patent laws, the US Supreme Court famously said that US "law [only] governs domestically [as opposed to ruling the whole] world."[237] Similarly, in *Alcoa*, the landmark case of extraterritorial application of US antitrust rules, Judge Learned Hand recognized that jurisdiction is constrained by "the limitations customarily observed by Nations upon the exercise of their powers" and to this extent US law is only concerned with conduct that poses "consequences within the [country]."[238] In addition, in the field of securities, US courts have also symbolically recognized the geographical limitations of US law.[239] None of these hortatory statements, however, have prevented in any way the broad scope of extraterritoriality on either side of the Atlantic Ocean.[240]

5.4.2.2 Physical Presence Leads to Unlimited Jurisdiction

For the CJEU, the scheme's substantial connecting factor to the territorial principle is the physical presence of foreign aircraft when they either land or take off from European soil.[241] Rather than impinging upon the sovereignty of other States, according to the CJEU, the ubiquity of foreign aircraft in the EU constitutes irrefutable manifestation of the principle

234 See *ATA Decision*, *supra* note 63 (Chapter 4) at para 117; Opinion of the Advocate General Kokott, *supra* note 136 (Chapter 4) at paras 145 and 146.

235 See Petsonk, "Testimony," *supra* note 98 (Chapter 4) at 11.

236 See Opinion of Advocate General Kokott, *supra* note 136 (Chapter 4) at 144.

237 *Microsoft Co v AT&T Co*, 550 US 437 at 454 (2007). This statement, however, does not represent an abdication of extraterritoriality, but rather symbolic desiderata that, on a given set of facts, law is to be construed restrictively.

238 Alcoa, *supra* note 201 at 443.

239 See *Robinson v TCI/US West Communications Inc*, 117 F (3d) 900 at 906 (5th Cir 1997) [*Robinson*] (noting that the purpose of the US law is not to protect "the victims of any fraud that somehow touches the United States" at 906); *Leasco Data Processing Equipment Corporation v Maxwell Leasco Data*, 468 F (2d) 1326 (2nd Cir 1972) at 1334 [*Leasco*] (stating that US securities law "is much too inconclusive to lead [the Court] to believe that Congress meant to impose rules governing conduct through the world in every instance where an American company bought or sold a security" at 1335).

240 In the case of the United States, for instance, its own Constitution vested Congress with the authority "to [enact] all laws which shall be necessary." US Cons Art I § 8. Through a number of highly reported decisions, the US Supreme Court has read this mandate as not being circumscribed to any *locus* limitation. See *Blackmer v United States*, 284 US 421 at 437 (1932); *Foley Bros, Inc v Filardo*, 226 US 281 at 285 (1949); *EEOC v Arabian Am Oil Co*, 499 US 244 (1991). In other words, it is widely understood within the United States that Congress is empowered to enact laws with extraterritorial effect. See Gerald F Hess, "The Trail Smelter, the Columbia River, and the Extraterritorial Application of CERCLA" (2005) 18 Geo Int'l Envtl L Rev 2 at 25.

241 See Opinion of Advocate General Kokott, *supra* note 136 (Chapter 4) at para 149.

of territoriality.[242] It is hard to see how the inclusion of foreign aircraft into the EU ETS can be regarded as a manifestation of the territorial principle. The latter presupposes the regulation of activities and actors that take place within the territorial boundaries of a given State. The EU ETS deals with conduct that does not wholly take place within the EU territory. For the CJEU, however, the physical presence of foreign aircraft in EU territory provides sufficient nexus to justify the exercise of jurisdiction. As such, according to the CJEU, such exercise is not extraterritorial, but completely territorial. This is extremely similar to the approach and methodology that US courts resort to in order to validate the effects doctrine and the extraterritorial application of law in general.[243]

According to the CJEU, the physical presence of foreign aircraft in EU territory also grants the EU Member State "unlimited jurisdiction."[244] To support this assertion, the CJEU incorrectly relies on *Poulsen.*[245] In this case, the CJEU upheld the "unlimited jurisdiction" of a port State (Denmark) to confiscate salmon caught outside EU jurisdiction by a Panamanian-registered vessel.[246] The assertion of extraterritorial jurisdiction by Denmark, however, did not arise out of a will to exercise a role of the *Lord of the Seas* to protect fish stocks for the betterment of humankind, but rather, it was justified by community legislation that in turn internalized obligations contained in an international instrument to which both the EU and Denmark are parties.[247] As such, the authorization to regulate fisheries' behavior extraterritorially stems from a source of international law: a treaty.[248] In this sense, the EU ETS presents a striking difference. Neither the *UNFCCC* nor the *Kyoto Protocol*, directly or indirectly, authorizes the exercise of extraterritorial jurisdiction over international civil aviation to protect global environmental commons.[249]

The allegation that States have "unlimited jurisdiction" over foreign aircraft while they are in their territory reflects the extent to which the CJEU chooses to ignore basic principles of public international air law and how these have served the international community

---

242 See *ATA Decision*, *supra* note 63 (Chapter 4) at para 125. See also Manzini & Masutti, *supra* note 136 (Chapter 4) at 319.

243 US courts have examined whether extraterritorial activity falls under the basis of territorial jurisdiction. See Jonathan Turley, "When in Rome": Multinational Misconduct and the Presumption Against Extraterritoriality" (1990) 84 NW UL Rev 598 at 630.

244 *ATA Decision*, *supra* note 63 (Chapter 4) at para 125.

245 *Ibid.* at para 124.

246 *Ibid.* at para 128.

247 See EC, *Council Regulation No 3094/86 of 7 October 1986 Laying Down Certain Technical Measures for the Conservation of Fishery Resources* [1986] OJ, L 288/1 art 6; *Convention for the Conservation of Salmon in the North Atlantic*, 2 March 1982, 1338 UNTS 22433 (entered into force 1 October 1983).

248 See generally Statute of the International Court of Justice, 26 June 1945, 59 Stat 1055, 33 UNTS 993 Art 38 para 1.

249 As Master notes, one of the main challenges in designing policy to protect the global commons relates to the fact that it "means different things to different people." Julie B Master, "International Trade Trumps Domestic Environmental Protection: Dolphins and Sea Turtles are "Sacrificed on the Altar of Free Trade" (1998) 12 Temp Int'l & Comp LJ 423 at 426.

well for almost seventy years. It is certainly true that foreign aircraft must observe the laws of the State of arrival with respect to "the admission to or departure from [the] territory [of such State]"[250] and that such aircraft must also comply with regulations pertaining to customs, immigration, mail, and health (e.g. quarantine).[251] However, it is incorrect to assert that the State of landing enjoys "unlimited jurisdiction";[252] for instance, the State in question cannot require foreign operators to register all their aircraft in such State. All issues dealing with airworthiness certificates and licenses fall under the jurisdiction of the State of registry of the aircraft, not the State of arrival, even when a foreign aircraft is found in the territory of such a State.[253] Rules on substantial ownership and effective control are also subject to the State of registry or, alternatively in some cases, to the State of incorporation of the aircraft operator. The mere physical presence of foreign aircraft in the State of arrival does not change the nature of these long-standing rules.

In addition, the liability of an aircraft operator for damages that a passenger sustains in the context of an international flight after embarkation is subject to the rules of the Montreal Convention of 1999 or alternatively the Warsaw-Hague system, regardless of the fact that the aircraft on which the incident takes place is in the territory of the State of arrival. In this case, international rules will apply and not the rules of EU Member States.[254] The latter is a typical case of conflict of laws, as opposed to a conflict of jurisdiction per se. In spite of this, it helps to demonstrate that even where there are strong domestic elements, a State may still be bound by the international regime and not necessarily by the parochial approach of its own legal system. To this end, the fact that EU Member States may exercise jurisdiction on certain issues over foreign aircraft while they are in their territory cannot be assimilated into a notion of absolute jurisdiction as the CJEU would seem to suggest.

In addition, inferring that the EU has absolute jurisdiction over foreign aircraft from the sole fact that they are transitorily within EU territory may also suggest that there are grounds for the exercise of universal jurisdiction *in presentia* over any given subject matter. A number of scholars have said that the fact that flights depart and arrive in the territory of EU Member States cannot be interpreted as substantial link to the EU to justify the

250 *Chicago Convention, supra* note 36 (Chapter 2) at Art 11.

251 *Ibid.* at Art 13. In the United States, in *Laker Airways*, the DC Circuit reasoned that a foreign airline's landing rights tantamount to a license to undertake a commercial activity and as such it was required to subject itself to US antitrust laws. See *Laker Airways, supra* note 7. See also See Barbara A Bell, "The Extraterritorial Application of United States Antitrust Law and International Aviation: A Comity of Errors" (1988) 54 J Air L & Com 533 at 566.

252 See *Chicago Convention, supra* note 36 (Chapter 2) at Art 17.

253 *Ibid.* at Arts 31 and 32.

254 See *Convention for the Unification of Certain Rules for International Carriage by Air*, 28 May 1999, 2242 UNTS 309, ICAO Doc 9740 [*Hague Protocol*]; *Convention for the Unification of Certain Rules Relating to International Carriage by Air*, 10 October 1929, 3145 LON 137 13LoN-3145 [*Warsaw Convention*].

scheme's extraterritorial application to foreign aircraft.[255] Likewise, the absence of a substantial nexus is perceived by other States as an attempt by the EU to share externality costs that they otherwise should not.[256]

5.4.2.3 Events Partly Outside

The CJEU implicitly followed the Advocate General's rationale when it decided that computing fuel for the whole duration of the flight is only a "metric"[257] or calculation method that incorporates events that take place outside the territorial jurisdiction of the EU.[258] Unsurprisingly, the CJEU did not consider the fact that the majority of emissions covered by the scheme are generated outside EU jurisdiction to be problematic.[259] Drawing partly on the Advocate General's opinion, the CJEU was of the view that these are also "events" that take place "partly outside" of the territory of EU Member States.[260] The choice of words is very similar to that of US courts.[261] The reasoning also bears close resemblance with the *Shrimp-Turtle* trade controversy where, given the fact that these species "spent part of their migratory life cycle" in US waters, the Arbitral Body acknowledged a "sufficient nexus" between the jurisdiction that the United States sought to exercise and the location of the turtles.[262]

Without alluding to it, the CJEU implicitly approximated the EU ETS to the objective territoriality principle, given that flights initiate elsewhere but end in the EU. To justify

255 See Bartlik, *supra* note 68 (Chapter 4) at 164; Jason N Glennon, "Directive 2008/101 and Air Transport – A Regulatory Scheme Beyond the Limits of the Effects Doctrine" (2013) 78:3 J Air L & Com 479 (commenting that the "territorial linkage stemming from non-EU planes landing in the EU, which the Court claims is sufficient territorial linkage, is only sufficient for territorial emissions" at 495). See *contra* Gehring, *supra* note 135 (Chapter 4) (saying that, for instance, the extraterritorial application of US laws presents "much more tenuous connection to activities in foreign countries in the area of securities regulation or international taxation" at 152).

256 Bell argues that extraterritoriality in international aviation should only be applied when the effects of the extraterritorial conduct are substantial in the State exercising jurisdiction. See Bell, *supra* note 251 at 574.

257 Petsonk, "Testimony," *supra* note 98 (Chapter 4) at 11.

258 See Opinion of the Advocate General Kokott, *supra* note 136 (Chapter 4) at para 147.

259 See *ATA Decision*, *supra* note 63 (Chapter 4) at para 129.

260 *Ibid.* at para 129. See also Havel & Mulligan, *supra* note 60 (Chapter 4) at 17. The CJEU's reasoning in *ATA* to classify "events" as taking place "partly outside" EU territory resembles the language that the US Supreme Court used in early cases involving extraterritoriality in antitrust law. For instance, in 1913 the US Supreme Court said that the *Sherman Act* was applicable to an agreement entered into the United States that led to anticompetitive behavior that took place "partly within and partly [outside] the [country]." *United States v Pacific & Arctic Railway and Navigation Company*, 228 US 87 (1913) [*United States v Pacific & Arctic Railway*].

261 Similarly, in *United States v Sisal Sales Corporation*, the Supreme Court applied US antitrust law to a monopolistic agreement that "brought about forbidden results in the United States." See *United States v Sisal Sales Corporation*, 274 US 266 (1927).

262 See Quereshi, *supra* note 62 at 204; Condon, *supra* note 62 at 159. Ultimately, the AB ruled against the United States because of the discriminatory manner in which it implemented the environmentally related trade measure.

its rationale, the CJEU relied heavily on its previous decisions in *Commune de Mesquer*[263] and *Ahlström*.[264] None of these cases, however, provide a convincing analogy to the EU ETS.

Commune de Mesquer dealt with a shipwreck that occurred in a Member State's exclusive economic zone (EEZ). The CJEU was asked to consider whether the accidental oil spillage constituted waste under EU legislation.[265] However, in this case, although the pollution was originally generated outside the territorial waters of the Member State, damages were caused all along its coast.[266] Here, the link to the territorial jurisdiction of the Member State was not only direct but also substantial.[267] The same cannot be said about emissions occurring over the high seas or over the territory of other States.[268]

The facts of *Ahlström*, however, are substantially different from those presented by the EU ETS.[269] It is not clear how these cases support the position that the CJEU has taken.[270] Arguably, the CJEU's approach seeks to downplay the unmistakable extraterritorial elements present in the EU ETS.[271] The characterization of the fuel consumption methodology as a simple "event" that only "partly" occurs outside EU jurisdiction appears to be self-serving. The choice of words in the CJEU's decision is not only poor but has also contributed to infuriating opponents. Describing emissions as "partly" taking place outside of the territorial boundaries of the EU does not represent an accurate reflection of how things work in practice. For example, including in the calculations emissions generated by an auxiliary power unit (APU) while a foreign aircraft is taxiing in a foreign country is notorious evidence of the scheme's jurisdictional overreach. The vast majority of emissions that the scheme seeks to cover are produced elsewhere.

263 See *Commune de Mesquer v Total France SA and Total International Ltd*, C-188/07 [2008], ECR I-04501 [*Commune de Mesquer*].

264 See Ahlström, *supra* note 225.

265 See *Commune de Mesquer*, *supra* note 263 at paras 60 and 63. In this case, the CJEU was asked to shed light on the parties liable to provide compensation, as well as on whether the (unbreakable) limits of the International Oil Pollution Compensation Fund provided for some exceptions. *Ibid. at* para 89.

266 See Gattini, *supra* note 135 (Chapter 4) at 981.

267 See Mayer, Case-366/10, *supra* note 133 (Chapter 1) at 1130 (criticizing the analogy to *Commune de Mesquer*, where the oil spill directly affected France's territorial coast).

268 See Mendes de Leon, "Enforcement of the EU ETS," *supra* note 220 (Chapter 4) at 295.

269 See Gattini, *supra* note 135 (Chapter 4) at 981 (discussing that the fact that foreign aircraft operators serve the EU market represents a "technical necessity" that cannot be assimilated to the concept of "implementation").

270 See Dabbah, *supra* note 164 (Chapter 4) at 176.

271 See *contra* Kulovesi, "Make Your Own Special Song," *supra* note 137 (Chapter 4) (noting that the scheme does not "[regulate] the conduct of foreign aircraft outside the territory of its member states when requiring aircraft operating from airports within its jurisdiction to surrender emission allowances corresponding to the length of the entire flight" at 549).

5.4.2.4 A Condition to Access the Market

Zerk suggests that a State may invoke the territorial principle in those cases where the behavior regulated (extraterritorially) involves the terms and conditions under which a foreign firm may conduct business in that jurisdiction.[272] In the maritime context for instance, a State may adopt environmental regulation to prevent pollution as a precondition for foreign vessels to access its ports.[273] By rejecting the extraterritoriality claims, the CJEU's decision effectively implies that the scheme is simply a condition to access the European market, where through the exercise of its leverage, the EU compels foreign aircraft operators to comply with EU environmental policy that seeks to achieve the objectives of the *UNFCCC* and the *Kyoto Protocol.*[274] In other words, the EU uses its bargaining power to force environmental standards.[275] It is therefore pertinent to assess to what extent the scheme is a mere precondition to access the market or whether it presents significant extraterritorial elements that would render it unlawful.

Regulatory overreach on foreign firms as a precondition to access a given market is far from uncommon.[276] For example, in the United States, in the field of securities, the *Sarbanes-Oxley Act* imposes stringent internal control and financial reporting requirements for all publicly traded companies.[277] This represents an enormous administrative and financial burden for both US and foreign firms that must observe these requirements in order to be listed on the US Stock Exchange.[278] Ellis highlights emerging trends in the field of fisheries where ports take a number of innovative measures to combat the illegal, unreported, and unregulated (IUU) fishing of stocks on the high seas.[279] By denying access to ports to vessels with IUU stocks, these measures seek to induce compliance with internationally agreed conservation rules.[280] In the *Shrimp-Turtle* trade dispute, the WTO Appellate Body recognized the right of States to demand from other States, as a condition to access their markets, the adoption of environmental measures "comparable in effectiveness," provided this is not done in a discriminatory manner.[281]

---

272 See Zerk, *supra* note 10 at 14.

273 See *UNCLOS*, *supra* note 25 at Art 211, para 3.

274 See *ATA Decision*, *supra* note 63 (Chapter 4) at para 128.

275 Bogojovic suggests that the CJEU's decision sends a strong message that commercial activity in the EU must comply with EU's environmental policy imperatives. See Bogojovic, *supra* note 134 (Chapter 4) at 351.

276 See Zerk, *supra* note 10 (identifying the field of securities, at 61).

277 *Sarbanes-Oxley Act*, Pub L 107-204, 116 Stat 745 (2002).

278 See Clyde Stoltenberg et al, "A Comparative Analysis of Post-Sarbanes-Oxley Corporate Governance Developments in the US and European Union: The Impact of Tensions Created by Extraterritorial Application of Section 404" (2005) 53 Am J Comp L 457 at 468.

279 See Jaye Ellis, "Fisheries Conservation in an Anarchical System: A Comparison of Rational Choice and Constructivist Perspectives" (2007) 3 J Int'l L & Int'l Rel 1 at 43 [Ellis, "Fisheries"].

280 These measures may include "prohibitions on landings and transshipments of catch, port closures, denial of port services [as well as] trade-related measures." *Ibid.* at 43.

281 See WTO Appellate Body *Report on United States – Import Prohibitions of Certain Shrimp and Shrimp Products*, WT/DS58/AB/RW at para 144. See also Tracey P Varghese, "The WTO's Shrimp-Turtle Decisions:

Scott and Rajamani suggest that the question of whether the EU ETS is in fact extraterritorial will inevitably depend on the perspective from which the assessment is carried out. As such, the scheme "is extraterritorial when viewed through the lens of [a] production-based system boundary, [yet it is] merely differently territorial when viewed through a system boundary that posits market access as key."[282] Other scholars have suggested that the extraterritorial allegations become "less objectionable when [the scheme is] considered within a wider framework."[283] In this respect, Howse notes that the EU ETS may be assimilated to process production methods (PPMs)[284] which are not "an exercise of regulatory control over behavior outside the territory of the importing [S]tate, [but] they instead regulate the behavior of sellers and consumers on the territory of that State."[285] Other commentators recognize an "essential" extraterritorial component in trade measures.[286]

From an international aviation law perspective, Havel and Mulligan strongly criticize this position.[287] They point out that the rationale followed by the CJEU may help legitimize extending of EU emissions requirements to other sectors beyond the pure aviation domain as a precondition to access the EU market.[288] This is referred to as a border tax adjustment (BTA). In essence, a BTA is a levy imposed on imported products or services seeking to counterbalance higher production costs that domestic firms may have vis-à-vis foreign firms.[289] While seeking to correct different regulatory requirements, these taxes purport

---

The Extraterritorial Enforcement of U.S. Environmental Policy via Unilateral Trade Embargoes" (2001) 8 Envtl L 421 at 452.

282 See Joanne Scott & Lavanya Rajamani, "EU Climate Change Unilateralism: International Aviation in the European Emissions Trading Scheme" online: <www.indiaenvironmentportal.org.in/files/file/EU%20Climate%20Change%20Unilateralism.pdf> (arguing that the EU would seem to believe that for foreign registered aircraft the scheme simply represents a condition to access the EU market, at 10).

283 See Dejong, *supra* note 101 (Chapter 4) at 187.

284 Jason Potts describes PPMs as "any activity that is undertaken in the process of bringing a good to market." Jason Potts, *The Legality of PPMs under the GATT: Challenges and Opportunities for Sustainable Trade* (Winnipeg: International Institute for Sustainable Development, 2008). PPMs address how products are made or extracted (e.g. natural resources). See also Bernd G Janzen, "International Trade Law and the 'Carbon Leakage' Problem: Are Unilateral U.S. Import Restrictions the Solution?" (2008) 8 Sustainable Dev L & Pol'y 22 at 24.

285 Robert Howse, "EU Aviation Emissions Scheme Opinion" International Economic Law and Policy Blog *World Trade Law* online: World Trade Law <http://worldtradelaw.typepad.com/ielpblog/2011/10/eu-aviation-emissions-scheme-opinion.html> [Howse, EU ETS]. PPMs are not necessarily illegal under WTO rules. Steve Charnovitz, "The Law of Environmental PPMs in the WTO: Debunking the Myth of Illegality" (2002) 27 Yale J Int'l L 59 (noting that "the WTO legality of a PPM will depend both on its environmental rationale and on its implementation" at 110).

286 See Mark Edward Foster, "Making Room for Environmental Trade Measures within the GATT" (1998) 71 S. Cal L Rev 393 at 403.

287 See Havel & Mulligan, *supra* note 60 (Chapter 4) at 22.

288 *Ibid.*

289 Robert Ireland, "WCO Policy Research Brief – The EU Aviation Emissions Policy and Border Tax Adjustments" WCO Research Paper No 26, *World Custom Organization* (July 2012) online: World Custom Organization <www.wcoomd.org/en/topics/research/activities-and-programmes/~/media/5DE1056A53F4428EBD

to protect the competitiveness of local firms. Where BTAs target imported products or services subject to lower GHG regulations in their country of origin, they are called border carbon adjustments (BCAs).[290] These measures, which seek to reduce the risk of carbon leakage,[291] may lead to an increase in the price of goods and services at their point of origin.[292] BCAs may also pose significant tensions in the international relations setting.[293] For instance, in 2005, when Europe implemented its emissions trading scheme, mindful of the additional financial burden that the scheme may pose for European firms, and in order to maintain their competiveness, countries such as France and the United Kingdom entertained the idea of adopting BCAs.[294] In the United States, several proposals for domestic climate change legislation introduced before the US Congress also contemplated the adoption of BCAs to correct the imbalance of imports from environmentally unregulated countries.[295]

Karapinar and Holzer underscore the fact that the inclusion of foreign aircraft operators into the EU ETS acts as a de facto BCA.[296] Similarly, Ireland explains that both BCAs and the EU ETS seek to "equalize national climate policies by pressuring other countries to adopt comparable carbon pricing regimes."[297] Europe has always argued that the inclusion of foreign aircraft operators was done to avoid a discriminatory treatment of European carriers. Failure to do so would have led to considerable market distortions.[298] By participating in the EU ETS, European carriers would be exposed to more stringent requirements and higher abatement costs to reduce GHG emissions.[299] Indirectly, one may argue that

0908CBF80B6A9C.ashx> at 4.

290 See Benjamin Eichenberg, "Greenhouse Gas Regulation and Border Tax Adjustments: The Carrot and Stick" (2010) 3 Golden Gate U Envtl LJ 283 at 288.

291 See Charles E McLure, Jr., "The GATT-Legality of Border Adjustments for Carbon Taxes and the Cost of Emissions Permits: A Tiddle, Wrapped in a Mystery, Inside an Enigma" (2011) 11 Fla Tax Rev 221 at 224.

292 See Meltzer, *supra* note 222 (Chapter 4) at 117.

293 See Ireland, *supra* note 289 at 3.

294 See Elias Leake Quinn, "The Solitary Attempt: International Trade Law and the Insulation of Domestic Greenhouse Gas Trading Schemes from Foreign Emissions Credit Markets" (2009) 80 U Colo L Rev 201 at 228.

295 See Ryan van den Brink, "Competitiveness Boarder Adjustments in U.S. Climate Change Proposals Violate GATT: Suggestions to Utilize GATT's Environmental Exceptions" (2010) 21 Colo J Int'l Envtl L & Pol'y 85 (noting that proposed US climate change legislation violates GATT's national treatment and most favored nation principles and that it is unlikely that they could fit in one of the environmental exceptions of GATT's Art XX).

296 See Baris Karapinar & Kateryna Holzer, *Legal Implications of the Use of Export Taxes in Addressing Carbon Leakage: Competing Border Adjustment Measures*, NCCR Trade Regulation Working Paper No 2012/15, Swiss National Centre of Competence in Research (April 2012) online: World Trade Institute <http://goo.gl/ny1jgR>.

297 Ireland, *supra* note 289 at 4.

298 See Meltzer, *supra* note 222 (Chapter 4) (discussing that the inclusion of foreign aircraft operators into the scheme represents "a trade measure that seeks to respond to these competiveness and carbon leakage issues by equalizing the impact of its climate change policy on domestic and international airlines" at 154).

299 See Motaal, *supra* note 35 (Chapter 2) at 19.

the inclusion of foreign aircraft operators was intended to protect the competiveness of European carriers. This is the same argument that the United States applied when it decided to extend the bans on smoking and gambling on board aircraft, which it had originally imposed exclusively upon domestic airlines, to foreign aircraft operators. These cases are analyzed below.

Although one may certainly consider the inclusion of international aviation into the EU ETS as a condition to access the market and a PPM or a BCA to correct imbalanced environmental regulatory requirements, the fact remains that the behavior that is being regulated substantially takes place outside the territorial boundaries of the EU. In both PPMs and BCAs, the products are consumed, and the services are rendered entirely within the territory of the State that implements them. Although there might be some elements of extraterritoriality in these cases, the nexus to territorial jurisdiction is substantial. This is certainly not the case of the EU ETS, where the vast majority of emissions occur elsewhere. As such, international aviation cannot just be equated to an event that partly occurs outside Europe.[300] It is difficult to justify Europe's regulatory overreach even when it is considered from the perspective that it is only a condition to access its air transport market. It does not appear that there is "a substantial and bona fide connection" between the activities of foreign aircraft operators and the EU ETS.[301]

5.4.2.5 A Vehicle to Advance EU's Climate Change Policy: Where Extraterritorial Jurisdiction Meets Rational Choice

According to the Advocate General, the scheme's methodology for the calculation of fuel consumption over the entirety of the flight is intended to achieve three objectives: First, it seeks to shield the environmental integrity of the scheme.[302] From a European environmental perspective, this argument seems plausible, for other approaches may significantly weaken the scheme's environmental effectiveness. Adopting the concept of national airspace as the basis for establishing the scheme's geographical scope would basically mean that only a third of emissions generated by flights to and from Europe would have been covered.[303] Second, this formula attempts to embrace the "polluter pays" principle, which thus far has been elusive within aviation circles.[304] According to the EU, this is necessary to achieve "a fair pricing mechanism that recognizes the full environmental impact of each

300 See *contra* Pache, *supra* note 97 (Chapter 4) (noting that the scheme is a mild intervention, permissible under international law, at 82).
301 Brownlie, *supra* note 31 at 313.
302 See Opinion of the Advocate General Kokott, *supra* note 136 (Chapter 4) at para 153.
303 See "Scenario A: Airspace MBM and 1% De Minimis Exemption" (27 September 2013) [unpublished, archived on file with the author].
304 See Opinion of the Advocate General Kokott, *supra* note 136 (Chapter 4) at para 153.

flight."[305] It is certainly uncontroverted that aircraft operators ought to be accountable for the pollution they cause, but adopting this principle does not automatically empower Europe to exercise the role of the world's environmental watchdog[306] to correct the sector's global externalities.[307] Aside from the foregoing, opponents of the EU ETS do not necessarily agree with what the principle implies for international aviation. Third, the formula seeks to incorporate a sense of "proportionality" into the scheme.[308] This contention is most troublesome as it cannot be easily understood. A common reading of the term "proportional" would indicate that there has to be some relationship between the emissions an aircraft operator produces and the territorial jurisdiction of the EU in order for the latter to properly regulate the former. Thus, some analysts assert that, under the principle of proportionality, the EU would only be authorized to cover emissions that are generated over its national airspace.[309] It does not seem proportional to include emissions for the entirety of a flight, for instance, from Vancouver to London, where most of the pollution takes place over Canadian airspace and over the high seas.

Notwithstanding the foregoing, the stance Europe has taken by including international aviation into its emissions trading scheme is part of the very essence of EU environmental policy. Hans is of the view that EU law does not only seek to protect the European environment but also that outside its geographical boundaries.[310] The policy itself is embedded with extraterritorial objectives. This is exactly what the CJEU has done: advance EU's climate change policy through judicial interpretation. While laudable, this is neither justified under international law nor tolerated by other States.[311] As Parrish says, extraterritoriality "[create]s piecemeal solutions to global problems."[312]

305 See Kulovesi, "Make Your Own Special Song," *supra* note 137 (Chapter 4) at 549. See also Mariam Radetzki, "The Fallacies of Concurrent Climate Policy Efforts" (2010) 39 AMBIO 211 (discussing that cost efficiency may require that all emitters face the same carbon price, at 221).

306 Motaal points that by extending the EU ETS to international aviation, the EC has de facto "turned itself into the global regulator." Motaal, *supra* note 35 (Chapter 2) at 12.

307 See Gattini, *supra* note 135 (Chapter 4) (questioning Europe's role as a "legislator, fee collector, and exclusive beneficiary of revenues" at 983).

308 See Opinion of the Advocate General Kokott, *supra* note 136 (Chapter 4) at para 153.

309 See Gattini, *supra* note 135 (Chapter 4) at 983.

310 See Jan H Hans, *European Environmental Law* (Oxford: Europa Law Publishing, 2000) at 29.

311 Fahey comments that "[t]he choice by the EU to regulate such a high level of protection for the environment and to bring global climate change within the realm of its own legal order clearly represented a highly ambitious effort effectively to regulate the international aviation industry through EU law." Fahey, *supra* note 134 (Chapter 4) at 1257. Anderson advances the notion that adhering strictly to the principles of territorial jurisdiction and national sovereignty is detrimental to the protection of the environment. These principles must evolve to respond to current challenges. See Belina Anderson, "Unilateral Trade Measures and Environmental Protection Policy" (1993) 66 Temp L Rev 751 at 753.

312 See Parrish, *supra* note 9 at 1489. What has been recognized as part of customary international law, as the ICJ pointed out in the *Legality of the Threat or Use of Nuclear Weapons*, is the obligation of States to ensure that activities under their jurisdiction do not cause harm "in the environment of other States." *Legality of the Threat or Use of Nuclear Weapons, Advisory Opinion, supra* note 44 at 241-42. See also Rio Declaration, *supra* note 35 (Chapter 1) at Principle 2; Viñuales, *supra* note 172 at 257.

Some scholars suggest that extraterritoriality is intrinsically linked to the furtherance of the national interests of States.[313] Others are of the view that such extension of jurisdiction becomes desirable to attain a given set of policy objectives that, in light of the failure to trigger collective action, would otherwise not be achievable.[314] To this end, the extraterritorial application of law directly follows the theory of rational choice.[315]

Here I follow Snidal who argues that the "goal-seeking" attitude of actors should not be interpreted to include only "self-regarding or material interests" but to also encompass "normative or ideational goals."[316] States will assert extraterritorial jurisdiction to the extent that it helps them achieve their objectives and allows them to advance their policies.[317] The European approach is similar to the US trend under which Courts apply US laws extraterritorially whenever there is an American economic interest at stake, even where there are weak connecting elements to the *forum.*[318] The European approach seeks to advance EU policy objectives such as combating climate change through the inclusion of international aviation into its ETS.[319]

313 See Mark Gibney & R David Emerick, "The Extraterritorial Application of United States Law and the Protection of Human Rights: Holding Multinational Corporations to Domestic and International Standards" (1996) 10 Temp Int'l & Comp LJ 123 (noting that with extraterritoriality the United States pursues to "promote [its] national interest [and that of] its corporate entities" at 120); Mark P Gibney, "The Extraterritorial Application of U.S. Law: The Perversion of Democratic Governance, the Reversal of Institutional Roles, and the Imperative of Establishing Normative Principles" (1996) 19 BC Int'l & Comp L Rev 297 (suggesting that, although most of the US extraterritorial overreach involves the so-called market cases, they "serve [its] national interest [and protect] its corporate actors" at 304); Zerk, *supra* note 10 (commenting that after 9/11 events, States have shown a tendency to "extend criminal regulation extraterritorially [in order to protect] their own national interest" at 141); Gattini, *supra* note 135 (Chapter 4) (discussing that powerful States like the United States exercise extraterritorial to impose their views or interests, at 984).

314 See John M Connor & Darren Bush, "How to Block Cartel Formation and Price Fixing: Using Extraterritorial Application of the Antitrust Laws as a Deterrence Mechanism" (2008) 112 Penn St L Rev 813 (commenting that extraterritorial application of US antitrust laws is necessary to prevent cartel formation and price fixing abroad, at 856); Alan Toy, "Cross-Border and Extraterritorial Application of New Zealand Data Protection Laws to Online Activity" (2010) 24 New Zealand Universities L Rev 238 (underscoring that "[e]xtraterritorial application of data protection laws is necessary in order to facilitate" consumer confidence at 238); Brendan J Witherell "Trademark Law: The Extraterritorial Application of the Lanham Act: The First Circuit Cuts the Fat from the Vanity Fair Test" (2006) 29 Western New Eng L Rev 193 (noting that "establishing extraterritorial jurisdiction [on trademark infringement cases] in a global marketplace should be dependent solely on the effects of the defendant's conduct, and not upon the citizenship of the infringer or other comity considerations" at 196).

315 Snidal defines Rational Choice as a "a methodological approach that explains both individual and collective social) outcomes in terms of individual goal-seeking under constraints." Duncan Snidal, "Rational Choice and International Relations" in Walter Emmanuel Carlsnaes & Beth A Simmons, eds, *Handbook of International Relations* (London: Sage Publications Inc, 2013) 85 at 87.

316 *Ibid.* at 88.

317 See Gibney & Emerick, *supra* note 313 (noting that the United States applies its laws extraterritorially when it benefits its interests, at 141).

318 See *Alcoa, supra* note 201 at 443; *Montréal Trading, supra* note 210 at 869.

319 Cook, *supra* note 138 (Chapter 4).

5.4.2.6 The Surreptitious Effects Doctrine

In her opinion, the Advocate General Kokott described the extraterritorial allegations made by the US airlines before the CJEU as an "erroneous, and highly superficial reading" of the EU legislation.[320] What seems incorrect and simplistic is her justification that the borderless nature of GHGs causes an effect that concerns all States, including EU Member States.[321] This would suggest that, given that GHGs mix uniformly in the atmosphere and that pollution in one country affects the global commons, each country of the world is entitled to exercise jurisdiction where GHG emissions are concerned.[322] To this end, the EU is justified in extending its scheme to foreign aircraft operators for emissions produced outside the territorial jurisdiction of the EU.[323] Arguably, the Advocate General Kokott intuitively embraced the effects doctrine to protect global environmental commons.[324]

The CJEU did not expressly allude to the effects doctrine, but it is more than evident that the doctrine implicitly formed part of its decision. For the CJEU, as "events" that occur "partly" outside the EU, emissions from international aviation add to "the pollution of the air, sea or land territory of the Member States" and as such cannot invalidate the applicability of EU legislation.[325] The CJEU's innovative rationale presupposes that there is a direct link between the extraterritorial activity and the end result caused in the territory of Member States. In a way, it is similar to the objective territorial principle, under which a State may exercise jurisdiction over an act that takes place in the territory of another State but whose effects materialize in its territory.[326] The main flaw of the CJEU's reasoning lies in the fact that GHG emissions that result from international aviation activity do not only produce harm in the territory of EU Member States, but rather they affect the whole atmosphere. As a result, it does not seem plausible to claim that international aviation's carbon footprint leads to a direct, substantial, reasonable, and foreseeable effect in the territory of EU Member States.[327] Similarly, the harm caused by international aviation cannot be associated with the EU as its main place of implementation if one were to follow

---

320 See Opinion of the Advocate General Kokott, *supra* note 136 (Chapter 4) at para 144.

321 *Ibid.* at para 154. See also Mayer, Case-366/10, *supra* note 133 (Chapter 1) (suggesting that with its decision the CJEU "implies that emissions of GHGs by air carriers anywhere in the world in fact affect EU member States through climate change" at 1130).

322 See Robert A Van Valkenburg III, "Directive 2008/101: The EU Emission Trading System" (2013) online: <http://internationalreports.wordpress.com/2013/01/13/directive-2008101/>; Pache, *supra* note 97 (Chapter 4) (commenting that in light of the global nature of climate change, "each country has a justified interest in taking action to counter it, in the spirit of the effects doctrine" at 74).

323 See Gehring, *supra* note 135 (Chapter 4) at 152; Bogojovic, *supra* note 134 (Chapter 4) at 351.

324 See Opinion of the Advocate General Kokott, *supra* note 136 (Chapter 4) at para 155.

325 See *ATA Decision*, *supra* note 63 (Chapter 4) at para 129. See also Havel & Mulligan, *supra* note 60 (Chapter 4) at 17.

326 Jennings, *supra* note 30 at 159.

327 See Glennon, *supra* note 255 (suggesting that the EU ETS's extraterritoriality cannot be justified under the effects doctrine, at 497).

the effects doctrine as interpreted in Europe. Both the US and EU approaches to the effects doctrine do not justify the exercise of extraterritorial jurisdiction in this case.[328]

5.5 The EU ETS, Extraterritoriality, and Principles of International Law

Having discussed various principles of international law on jurisdiction and the justification of State responsibility, I turn to reconcile them with the implementation of the EU ETS as interpreted and justified by the CJEU and the Commission. The EU ETS will be contrasted with principles of international law that permit extraterritoriality. In particular, close consideration is given to the territorial and nationality principles as well as to universal jurisdiction. Principles such as protective jurisdiction and passive personality jurisdiction are not considered relevant to the study of extraterritoriality in connection with the EU ETS.

5.5.1 *The Nationality Principle*

The nationality principle is the least controversial. Under this principle, the EU ETS may certainly cover European-registered aircraft for emissions generated within European airspace, as well as over the high seas and over the airspace of other States. In fact, nobody questions Europe's authority to do so. We have seen many cases in international law where jurisdiction over a particular activity or conduct is vested on the State of registry. This is what ICAO calls the "State of Registry" approach to the geographical scope of MBMs.[329] In practice, the application of the EU ETS can only be extended to domestic and intra-European flights, thereby hampering the scheme's overall environmental integrity. Otherwise, it could lead to considerable market distortions since aircraft competing on the same extra-European routes may be subject to different emissions requirements.

5.5.2 *Territorial Principle: The National Airspace Approach*

An alternative basis for the design of the EU ETS would be to follow a national airspace approach and calculate aircraft fuel consumption only when the aircraft fly over European territory. Here, the notion of territory would also include the airspace over European ter-

328 See Havel & Mulligan, *supra* note 60 (Chapter 4) at 22. Conversely, Howse criticizes the Advocate General Kokott for having come out short. According to Howse, the Advocate General Kokott should have expressly recognized the desirability of exercising extraterritorial jurisdiction to protect global environmental comments. See Howse, EU ETS, *supra* note 285.

329 See ICAO, HGCC/1-WP/3, *supra* note 174 (Chapter 3) at 2.

ritorial waters.[330] Under this approach, both European-registered aircraft and foreign aircraft operators may be subject to the EU ETS but only for emissions produced within European airspace. Although more problematic, it may at least in theory include aircraft whose points of departure or arrival are not situated in Europe, but rather overfly European airspace.[331] Nevertheless, it would not permit the inclusion of emissions that occur over the high seas and over the national airspace of other States.

From a European perspective, the airspace approach presents some problems. It significantly weakens the environmental integrity of the EU ETS given that intra-European flights account for 40 percent of the continent's aviation carbon footprint, whereas flights to non-European countries contribute 60 percent.[332] The airspace approach would only cover one-third of the emissions that the scheme as originally adopted would have covered.[333] Clearly, most of these emissions are produced outside European airspace. There is also a fear that the system would encourage aircraft operators to opt for alternative routes to deviate from European airspace, without necessarily reducing their overall levels of emissions.[334] Yet, more importantly, adopting the airspace approach would leave most of aviation's CO_2 emissions outside the EU ETS, which, in practice, would render them simply "unregulated."[335] It has also been suggested that this approach poses considerable operational challenges to the calculation of emissions.[336] Although aircraft tend to fly on pre-assigned air navigational routes, a number of operational factors (i.e. weather condi-

330 See *Chicago Convention, supra* note 36 (Chapter 2) at Art 2.

331 See ICAO Doc 9885, *Guidance on the Use of Emissions Trading for Aviation, supra* note 89 (Chapter 3) at 3-5 (noting that the inclusion of overflights may pose administrative and enforcement difficulties).

332 See UK House of Lords, "Including the Aviation Sector in the European Union Emissions Trading Scheme" 21st Report of Session 2005-06.

333 See "The Clock Has Stopped: Where is ICAO Now," *Transport & Environment* (2 May 2013) online: Transport & Environment <www.transportenvironment.org/news/clock-has-stopped-where-icao-now>. Environmental NGO EDF estimates that the airspace approach would cover approximately 108 MMt CO_2 per year. See "Scenario A: Airspace MBM and 1% De Minimis Exemption," *supra* note 303.

334 See Written Observations of the Second Interveners in the Court of Justice of the European Union, reference from the High Court of Justice, London, Case C-366/10 at 13. The contention that aircraft operators would try to bypass or minimize flying over European airspace appears reasonable. However, this would only happen to the extent that the cost in doing so (diverting flights) is less than the cost of compliance with the scheme. At current low carbon prices and particularly high fuel prices, it will always be financially more attractive for aircraft operators to reduce fuel consumption by flying direct routes over European airspace.

335 See "Allocating Aviation CO_2 Emissions: The Airspace-Based Approach and its Alternatives," *Transport & Environment* online: Transport & Environment <www.transportenvironment.org/sites/te/files/publications/201301%20Airspace%20Version%209%20Final%20%281%29_0.pdf> (citing the Report of the Subsidiary Body for Scientific and Technological Advice of the *United Nations Framework Convention on Climate Change* on the work of its fourth session, Geneva 16-18 December 1996).

336 A study conducted by Manchester Metropolitan University, which European States presented to ICAO's High-Level Meeting on Aviation and Climate Change, concluded that if all States were to adopt an airspace approach in the context of a global scheme, only 22 percent of the total international aviation emissions would be covered. Arguably, this is far from optimal from an environmental perspective. See ICAO, HGCC/3-WP/7, *CO_2 Emissions Coverage of the Geographic Scope Options for the Framework for* MBMs (presented by Belgium, France, and the United Kingdom) at 1.

tions, route congestion) may cause alterations that would make it much more difficult to establish when exactly the aircraft enters or leaves European airspace.

The airspace approach neglects the fact that all GHG emissions, regardless of the location at which they are generated, do not remain within the territorial boundaries of the States from whose territory they were emitted. Thus, 1 tonne of CO_2 emissions over US airspace produces exactly the same effect as 1 tonne of CO_2 emissions over France's airspace.[337] In addition, with the airspace approach, nobody takes care of emissions generated over the high seas.[338]

From an operational and an environmental integrity standpoint, it is clear that the airspace approach is not the optimal solution.[339] However, unlike under the original geographical scope of the EU ETS, where the scheme would cover emissions generated outside the territorial jurisdiction of the implementing States, the airspace approach does not seem vulnerable to any significant legal barrier. Even though the airspace approach may not be regarded as politically and technically acceptable, international law does not in any way prohibit a State from implementing an MBM within its airspace.[340] The *Chicago Convention* recognizes "that every State has complete and exclusive sovereignty over the airspace above its territory."[341] This is a codification of customary internationally law. With the airspace approach, we are not dealing with issues of extraterritorial jurisdiction but rather with a strict manifestation of the territorial principle.

5.5.2.1 Flight Information Regions

To address Europe's concern about the lack of coverage under the national airspace approach, another option would be to extend the scheme not only to emissions produced

---

337 Kerr points out that "[aviation emissions] are uniformly distrusted (their effects are not localized), and the global climate system is one interrelated system, [where] the source does not matter." Kerr, *supra* note 54 (Chapter 1) at 9. As the US (then) Senator John Kerry noted "[t]he stuff that goes up there doesn't stay in any one's airspace. We get China's fumes. We get Indiana and Massachusetts." Keith Laing, "LaHood, Senate committee hammers 'lousy' EU Airline Emission Trading Rules," *The Hill* (6 June 2012) online: The Hill <http://thehill.com/policy/transportation/231295-lahood-senate-committee-hammer-lousy-eu-airline-emission-trading-rules>.

338 See Petsonk, "Testimony," *supra* note 98 (Chapter 4) at 5.

339 Lowe writes:

"whatever means are chosen to advance the policies, one must remember that the question, what can the law do? Is quite distinct from the question, what ought the law to do? It is easy to be tempted into thinking that, because laws can be drafted so as to close all the holes in the net, they should be so drafted. That is not so. The best is the enemy of the good. A law which is effective in relation to 70 percent of the persons at whom it is aimed, and which stays within the limits of international law and is acceptable to the neighbors of the [State exercising jurisdiction], is, in my view much preferable to one which catches 90 percent of the offenders but violates international law and sours relations between the United States and its allies."

Lowe, "US Extraterritorial Jurisdiction," *supra* note 2 at 389.

340 See Glennon, *supra* note 255 (suggesting that CJEU should have restricted the scheme's scope to the European airspace only, at 497).

341 *Chicago Convention*, *supra* note 36 (Chapter 2) at Art 1.

within the European territory but also to those generated within Flight Information Regions (FIRs) under the delegated responsibility of European States.[342] A number of European States are in charge of providing air navigation services over Oceanic FIRs that are situated in the Atlantic Ocean. In theory, the scheme may be extended to cover emissions produced over such FIRs. However, it is worth pointing out that FIRs over the high seas are established through regional air navigation agreements among the States affected and are ultimately approved by the ICAO Council.[343] The delegation of responsibility to carry out certain air navigation services within an FIR does not in any way imply a derogation of the sovereignty of the delegating States nor an attribution of sovereignty to the provider State to legislate over the high seas or to extend its territory.[344] It is simply a fictional, geographical creation that is used in international civil aviation to ensure the adequate provision of air navigation services that otherwise would not be available to the users. In the absence of an international agreement among concerned States, there is no basis under international law to justify the extension of the EU ETS to FIRs. Also, due to operational and technical reasons, FIRs often do not coincide with the territorial boundaries of States. Certainly, the territorial principle does not provide such justification. The same legal concerns that States have with regard to the original geographical scope of the EU ETS reappear in connection with the FIR option.

5.5.2.2 The 38th Assembly and the Non-Recognition of the Airspace Approach

In its original design, the EU ETS was intended to cover all emissions for the whole duration of the flight. However, in light of the significant international opposition, right after the "stopping the clock" was announced in November 2012, in the context of ICAO's High-Level Group on International Aviation and Climate Change (HGCC), the Commission proposed to restrict the scope of the scheme to only cover emissions for the whole duration of flights departing from European airports. Emissions produced by aircraft arriving in European airports (e.g. inbound flights) would be excluded. Unfortunately, the international aviation community outrightly rejected this proposal, which would have reduced the coverage of the EU ETS by 50 percent. This forced Europe to concede even more in the negotiations. During the discussions that took place in the summer of 2013 prior to the 38th Assembly (24 September to 4 October 2013),[345] Europe proposed to include in the

342 An FIR is an air traffic service concept that means "airspace of defined dimensions within which flight information service and alerting service are provided." ICAO, Annex 11 to the *Chicago Convention*, *Air Traffic Services*, 13th edn (2001) at I-5 [Annex 11].

343 *Ibid.*

344 See ICAO, A37-15, *Consolidated Statement of Continuing ICAO Policies and Associated Practices Related Specifically to Air Navigation* at Appendix M, Delimitation of Air Traffic Services Airspaces, para 6.

345 See Valerie Volcovici, "Shuttle Diplomacy under way on Global Aviation Emissions Deal," *Reuters* (22 July 2013) online: Reuters <http://uk.reuters.com/article/2013/07/22/us-eu-aviation-emissions-idUKBRE96L0ZJ20130722>.

draft Assembly Resolution on aviation climate change express language that would recognize the airspace approach as the geographical scope for the ICAO MBM framework.[346] This would have served as an indication of international approval for the EU to re-start the EU ETS in 2014 but only to cover emissions produced within the EU airspace by both intra-European flights (for their entire duration) and flights to or from third countries (for the portion that occurs within the EU airspace).[347] This proposal was part of the package that the Council put forward for the consideration of the Assembly.[348] Although the proposal fell "well short of the EU's starting position,"[349] as one commentator put it, "it [was] crucial for Europe to win ICAO's approval for the continuation of the bloc's carbon program in its own airspace."[350] Given that the global MBM to be developed under the auspices of ICAO is expected to be implemented only in 2020, Europe wanted to continue with its regional scheme in the interim.[351]

One would think that applying the EU ETS to cover emissions produced exclusively over EU's airspace would have attracted less criticism given that the approach is in accordance with the long-standing principles of territorial jurisdiction and State sovereignty. It certainly removes concerns of extraterritorial application and infringements on the sovereignty of other States.[352] However, this was not necessarily the view of States attending

346 See Frederic Tomesco & Ewa Krukowska, "EU to Defend Limited Carbon Market for Emissions from Airlines," *Bloomberg Business Week* (24 September 2013) online: Bloomberg Business Week <www.business-week.com/news/2013-09-24/eu-to-defend-limited-carbon-market-for-emissions-from-airlines>.

347 See "ICAO Moves Closer to Agreement on Limiting Growth of Aviation Emissions as EU Officials Justify Climb-Down," *Green Air Online* (9 September 2013), online: Green Air Online <www.greenairon-line.com/news.php?viewStory=1753> ["ICAO Moves Closer to Agreement on Limiting Growth of Aviation Emissions as EU Officials Justify Climb-Down"]. See also ICAO A38-WP/83 EX/38, *A Comprehensive Approach to Reducing the Climate Impacts of International Aviation* (presented by Lithuania on behalf of the European Union and its Member States and the other Member States of the European Civil Aviation Conference) (inviting the 38th Assembly to "work towards an enabling framework for MBMs implemented by States or groups of States, pending the entry into force of the global MBM").

348 See ICAO, A38-WP/34 EX/29, *supra* note 184 (Chapter 3) at para 17.

349 See "ICAO Moves Closer to Agreement on Limiting Growth of Aviation Emissions as EU Officials Justify Climb-Down," *supra* note 347.

350 See Ewa Krukowska, "Airlines Face Carbon Reduction Verdict on 708 Million Industry," *Bloomberg Business Week* (24 September 2013) online: Bloomberg Business Week <www.businessweek.com/news/2013-09-23/airlines-face-carbon-reduction-verdict-on-708-billion-industry>. See also Catalin Radu, "The European Contribution to the Positive Outcome of the 38th ICAO Assembly" (2013/2014) 50 The European Civil Aviation Conference Magazine at 4-5 (discussing that the recognition of the airspace approach "was very important for Europe as a means of solving the EU-ETS problem. Having not reached an agreement within ICAO, ETS will remain a difficult issue vis-à-vis the other ICAO Member States").

351 Valerie Volcovici & Joshua Schneyer, "EU Stands Firm on Aviation Emissions Position," *Reuters* (27 September 2013) online: Reuters <www.reuters.com/article/2013/09/27/us-aviation-climate-eu-idUS-BRE98Q17K20130927>.

352 Milde suggests that the EU should only apply its ETS on GHGs "that are produced in the sovereign air spaces of the EU countries." Milde, "EU Emissions Trading Scheme," *supra* note 62 (Chapter 4) at 14. See also ICAO, C-MIN 194/2, *supra* note 86 (Chapter 1) at 18 (Statement of Singapore).

the 38th Assembly.[353] With the tremendous lobbying efforts of a unified airline industry,[354] the recognition of the airspace approach was expressly rejected by an overwhelming majority of States.[355] The press described the events as a "humiliating defeat"[356] and "big blow to Europe's prestige."[357]

The rejection of the airspace approach is extremely interesting and in a way unprecedented in the annals of ICAO, for it contradicts its own long-standing practices, as well as customary international law that each State enjoys complete and exclusive sovereignty over its airspace.[358] In a far-reaching move, some States even proposed that the recognition of the airspace approach should be tied to the requirement of mutual consent.[359] A State wishing to implement an MBM within its airspace must first obtain the consent of the foreign home State of the aircraft.[360]

I am not aware of any other case, where, in the context of ICAO discussions, States have denied the sovereign right of States to take measures within their territory or order that the consent of other States must first be sought and obtained. It seems unreasonable that a State levying, for instance, an environmental tax upon passengers departing from its territory should require the consent of other States. Some States also feared that the recognition of the airspace approach in an ICAO Assembly resolution could be perceived as an invitation to introduce or maintain regional MBM measures that may be detrimental

---

353 Chad Trautvetter, "Aviation Groups Back ICAO Aircraft Emissions Framework" *AIN Online* (8 October 2013) online: Ain Online <www.ainonline.com/aviation-news/ainalerts/2013-10-08/aviation-groups-back-icao-aircraft-emissions-framework>.

354 See "Aviation Industry Calls for Global Agreement and Climate Change Leadership by Governments Ahead of ICAO Assembly," *Green Air Online* (20 September 2013) online: Green Air Online <www.greenaironline.com/news.php?viewStory=1745>; "An NGO Message for the ICAO Assembly: Introduce a Global Market-Based Measure Now," *Green Air Online* (17 September 2013) online: Green Air Online <www.greenaironline.com/news.php?viewStory=1754>; Letter from Nicolas E Calio, President and CEO A4A, to the Honorable Duane Woerth, Ambassador to the U.S. Mission to ICAO; Ms. Julie Oettinger, Assistant Administrator, Policy, International Affairs and Environment, FAA; and Mr. Todd Stern, Special Envoy for Climate Change, U.S. Department of State (29 August 2013)].

355 See Ewa Krukowska, "First Global Emissions Market for Airlines Wins Support," (3 October 2013), online: Bloomberg <www.bloomberg.com/news/2013-10-04/first-global-emissions-market-for-airlines-wins-support.html>.

356 See "ICAO Assembly Climate Change Outcome Hailed by Industry but Seen as a Missed Opportunity by Environmental NGOs," *Green Air Online* (6 October 2013), online: Green Air Online <www.greenaironline.com/news.php?viewStory=1763>.

357 See "ICAO States Reach Agreement on Roadmap Towards a Global MBM but Europe Suffers Defeat over EU ETS," *Green Air Online* (4 October 2013), online: Green Air Online <www.greenaironline.com/news.php?viewStory=1762>.

358 Annie Petsonk, "Aviation Emissions Deal: ICAO Takes One Step Forward, Half Step Back," *EDF* (3 October 2013) online: Environmental Defense Fund (EDF) <http://blogs.edf.org/climatetalks/2013/10/04/aviation-emissions-deal-icao-takes-one-step-forward-half-step-back/>.

359 See ICAO, C-MIN 199/13, *Council, 199th Session, Minutes of the Thirteenth Meeting* at 9 (Statement of India) [ICAO, C-MIN 199/13]; ICAO, A38-WP/425 EX/140 (working paper presented by Argentina, Brazil, China, Cuba, Guatemala, India, Iran, Pakistan, Peru, Russian Federation, Saudi Arabia, and South Africa).

360 See ICAO, A38-WP/176 EX/67, *supra* note 140 (Chapter 1) at 3.

for the future developing of the ICAO global scheme.[361] The surge of a patchwork of regional scheme may threaten the legitimacy of the multilateral process.[362] Some States incorrectly suggested that the recognition of the airspace in an ICAO Assembly resolution "may be interpreted as allowing the unilateral application of MBMs."[363] In addition, China, Brazil, and India questioned the sovereign right of both the European Union and European States to include foreign aircraft operators into the EU ETS even under the basis of an airspace approach.[364]

This criticism presupposes that the concept of exclusive sovereignty only applies to individual States, not to a group of States that collectively may decide to adopt a given measure. Moreover, given that the EU is not a State under international law, these States argue that it does not enjoy sovereignty. But the argument does not illustrate the practice of States at ICAO. For instance, to establish FIRs, a group of States may agree that State A will provide air navigation services over the airspace of two or more States. Here, the States in question do not delegate their sovereignty but rather on the basis of their sovereign rights agree that State A will perform such services.

Arguably, the refusal to recognize the airspace approach in the ICAO climate change resolution cannot be considered as a legal obstacle to the re-commencement of the EU ETS and to its inclusion of foreign aircraft operators. Sensu stricto, ICAO Assembly resolutions are not binding on its Member States. Yet they are indicative of the organization's policies, and in this particular case, it is extremely symbolic of the level of animosity that the EU ETS has generated. Both airlines and most non-EU States wanted to send a very clear message to Europe: the EU ETS should not be revitalized in any form.[365] This message outweighed any potential precedent that this incident may set for the broader international civil aviation context, namely, foregoing the recognition of state sovereignty.[366] As Koskenniemi notes, sometimes international law is "all political, too political."[367]

---

361 See ICAO, C-MIN 199/13, *supra* note 359 at 8.

362 *Ibid.* at 10-11 (Statements of the Representatives of Brazil and Cuba, respectively).

363 *Ibid.* at 12 (Statement of Paraguay).

364 *Ibid.* at 9, 10, and 11 (Statements of India, Brazil, and China, respectively).

365 Letter from Nicolas E Calio, President and CEO A4A, to the Honorable Duane Woerth, Ambassador to the U.S. Mission to ICAO; Ms. Julie Oettinger, Assistant Administrator, Policy, International Affairs and Environment, FAA; and Mr. Todd Stern, Special Envoy for Climate Change, U.S. Department of State (29 August 2013).

366 The words of France's Michel Wachenheim, President of the 38th Assembly, are very illustrative: "During the 25 bilateral meetings I have held, I have heard remarks about a lack of negotiation by Europe, or delays in engaging negotiations. It's at times like these that my role was the most difficult, as a European. To me, Europe must realize that it conveyed a negative image on [the issue of aviation and climate change], and take action to improve the situation, so that it does not happen again in other domains. It would be very risky to ignore this message and the political posture the [38th Session of the ICAO] Assembly expressed." Michel Wachenheim, "Interview with Michel Wachenheim – President of the 38th Session of the ICAO Assembly" in *The European Civil Aviation Conference Magazine, ECAC News* 50 (Winter 2013) 10 at 11.

367 See Koskenniemi, "The Fate of Public International Law: Between Techniques and Politics" (2007) 70:1 Modern L Rev 1 at 1.

5.5.2.3 Universal Jurisdiction

As Gehring puts it, the CJEU's decision "expands the EU's regulatory [overreach in order to] address environmental challenges of a transnational nature."[368] This notwithstanding, there is no rule of customary international law that would expressly authorize a State to exercise the role of a custodial climate change watchdog to protect the "general community interest" for whatever damages wherever they occur.[369] Accepting that there is such a right would be tantamount to extending the principle of universal jurisdiction to climate change issues.[370] This is not the state of the law.[371] This section has thoroughly explained the restrictive requirements that must be met in order to exercise universal jurisdiction. Failure to act to combat climate change cannot be categorized as a heinous or atrocious crime that would allow a State or group of States to exercise universal jurisdiction. The EU ETS does not fall under any of the sources of universal jurisdiction (e.g. customary international law, *jus cogens* norms, and international treaties). Also, the EU ETS does not meet the stringent conditions that States must observe to assert extraterritorial jurisdiction. The mere physical presence of foreign aircraft landing and departing from European airports should not be assimilated to extraterritorial jurisdiction *in persona*.

5.6 Is the Extraterritoriality of the EU ETS Unprecedented? Other Examples of Extraterritorial Jurisdiction in International Civil Aviation

Various examples reveal that exceptions to the principle of territoriality do exist in international civil aviation.[372] Many States do resort to the extraterritorial application of their own national law when the circumstances so demand.[373] As such, extraterritoriality is hardly the exception.[374] In fact, recognizing this tendency, a large number of international

368 Gehring, *supra* note 135 (Chapter 4) at 152. Similarly, for Bogojovic, the CJEU's decision legitimates Europe's entrepreneurial climate change role and "has a significant impact on shaping the global climate change regime." Bogojovic, *supra* note 134 (Chapter 4) at 347.

369 See Richard B Bilder, "The Role of Unilateral Action in Preventing International Environmental Injury" (1981) 14:1 Vand J Transnat'l L 51 at 73.

370 *Ibid.* at 74; Zerk, *supra* note 10 (suggesting that "it is doubtful whether universality would provide a legitimate basis for direct assertions of extraterritorial jurisdiction in the environmental field, except, perhaps in relation to deliberate and very serious environmental damage tantamount to war crimes or genocide" at 186).

371 See Gattini, *supra* note 135 (Chapter 4) (noting that "international law has [not] progressed so far as to impose on each and all an obligation to abate Greenhouse Gases (GHG) emissions" at 83).

372 See Dabbah, *supra* note 164 (Chapter 4) at 160; Marko Milanovic, *Extraterritorial Application of Human Rights Treaties: Law, Principles, and Policy* (Oxford: Oxford University Press, 2011) (commenting that the "impact of human rights treaties in an extraterritorial context is growing" at 1).

373 See Meyer, *supra* note 9 at 111; Parrish, *supra* note 9 at 1456.

374 *Ibid.* at 1457. Another area where States regularly engage in extraterritoriality is immigration control. Bernard Ryan describes these practices as "strategies of extraterritorialisation." Bernard Ryan, "Extraterritorial

agreements empower States to assert extraterritorial jurisdiction.[375] Yet the force of the extraterritorial criticism and the efforts assembled to defeat the EU ETS would seem to imply that this type of regulatory overreach is unprecedented in international civil aviation.

For instance, through statutes and regulatory enforcement, the US Congress and several administrative agencies have, on numerous occasions, exercised extraterritorial jurisdiction over foreign aircraft operators.[376] It is beyond controversy that the United States has the strongest and oldest extraterritorial tradition in the world.[377] Agencies such as the Civil Aeronautics Board (CAB), the Federal Aviation Administration (FAA), and most often the Department of Transportation have frequently engaged in this practice.[378] For instance, in the 1960s, the CAB proposed to regulate traffic of foreign aircraft between two foreign points (aka "blind sector traffic"), as long as one of the legs of the itinerary is initiated or terminated in the United States.[379] Years later, the US Congress enacted legislation prohibiting smoking[380] and gambling[381] on board aircraft that was applicable to foreign aircraft operators flying to and from the United States.[382] This measure was very similar to the original EU ETS, in terms of geographical scope, given that gambling and smoking aboard aircraft, for the most part, also took place beyond the territorial boundaries of the United States.[383] The US Congress also passed the *Air Carrier Access Act*[384] establishing requirements that both US and foreign aircraft operators must follow in the carriage of passengers with disabilities.[385]

Immigration Control: What Role For Legal Guarantees?" in B Ryan and V Mitsilegas, eds, *Extraterritorial Immigration Control: Legal Challenges* (The Hague: Martinus Nijhoff Publishers, 2010) 3 at 3.

375 See Zerk, *supra* note 10 at 9. See *UNCLOS*, *supra* note 25 at Art 218.

376 See Jol A Silversmith, "The Long Arm of the DOT: The Regulation of Foreign Air Carriers beyond US Borders" (2013) 38:3 Air & Space L 173 at 174.

377 See Gibney & Emerick, *supra* note 313 (describing that "it has been the [US] judiciary that has essentially created the law (or, more accurately, given the law an extraterritorial reading" at 132)); Adam I Muchmore, "Jurisdictional Standards (and Rules)" (2013) 46 Vand J Transnat'l L 171 (describing the United States as an "aggressive extraterritorial regulator" at 174).

378 *Ibid.* at 176.

379 *Ibid.* at 179.

380 See 14 CFR Part 252. On defense of the United States, it must be said that the banning of smoking in international flight followed an ICAO recommendation. In fact, in 1992, the 29th Assembly suggested that States "take necessary measures as soon as possible to restrict smoking progressively on all international passenger flights with the objective of implementing complete smoking bans by 1 July 1996." ICAO, A29-15, *Smoking Restrictions on International Passenger Flights.*

381 See 49 USC § 41311 (1994).

382 *Ibid.* at 188.

383 See Steven Grover, "Blackjack at Thirty Thousand Feet: American's Attempt to Enforce its Ban on In-flight Gambling Extraterritorially" (1999) 4 Gaming Research & Rev J 2 (discussing that the United States "bans in-flight gambling devices on flights to or from the [US], even while a foreign aircraft is outside of U.S. airspace" at 13).

384 *Air Carrier Access Act*, 49 USC § 41705.

385 See Silversmith, *supra* note 376 at 201.

Since the 1990s and before ICAO adopted international standards on Advance Passenger Information (API), US authorities ordered all aircraft operators flying to/from the United States to transmit passenger data prior to flight departure.[386] Yet perhaps the most controversial initiative prior to 11 September 2001 (9/11) was the proposed amendments that the FAA sought to carry out to implement provisions of the *Antiterrorism and Effective Death Penalty Act of 1996*.[387] In November 1998, the FAA announced a Notice of Proposed Rulemaking (NPRM) that would require foreign aircraft operators and foreign airports to adopt "identical security measures" to those imposed on US airlines and US airports.[388] These measures were more stringent than the standards that ICAO had in force at the time. The proposal, which came to be known as the "Hatch Amendment," triggered dire opposition from the international community.[389] In addition to being perceived as a unilateral and unlawful imposition of US security measures, opponents contended that the United States did not have authority to regulate behavior taking place at foreign airports and aboard foreign airlines abroad.[390]

On 5 February 1999, the ICAO Council declared that such amendments "infringe basic principles of the *Chicago Convention*, and run counter to the spirit of multilateralism."[391] In order to avoid a diplomatic conflict, the FAA was forced to abandon the amendment. This case presents a striking resemblance to the EU ETS, namely, strong international opposition, unquestionable extraterritorial components, and a threat to ICAO's multilateralism.[392] Having said this, one must bear in mind that the Hatch Amendment predates the 9/11 events. It would have been interesting to see whether, in a post 9/11 landscape, the level of international opposition would have been the same. It would also have been interesting to observe whether the United States would have decided to repeal the proposal in the face of such opposition. I am inclined to believe that neither of these situations would have arisen.

After 9/11, the United States tightened its standards on aviation security. US authorities ordered that all aircraft operators should install reinforced cockpit doors.[393] In addition

386 US authorities not only requested the passenger's full name but also, in some cases, sensitive information such as religious orientation. This conflicted with data protection legislation of European countries.

387 *Antiterrorism and Effective Death Penalty Act of 1996* Pub L No 104-132, 110 Stat 1214 (1996).

388 *Security Program of Foreign Air Carriers*, 63 Fed Reg 64764 (1998).

389 Foreign airlines, foreign airports, airline trade associations, as well as a numerous governments urged the FAA to abandon the proposed amendments. See FAA-98-4758-4 (comments by IATA), 24 May 1999 [FAA-98-4758-4]; FAA-98-4758-5 (comments by Vancouver International Airport); FAA-98-4758-13 (comments by the British Embassy in Washington); FAA-98-4758-14 (comments by the French Embassy in Washington).

390 See FAA-98-4758-4, *supra* note 389.

391 ICAO, *Council Resolution of 5 February 1999* [ICAO *Hatch Resolution*].

392 The ICAO *Hatch Resolution* expressly declared that the proposed amendment was extraterritorial. *Ibid.*

393 See Michael B Jennison, "Regulating Greenhouse Gas Emissions: Legal Aspects of Levies and Trading Systems" IATA Legal Symposium (February 2007) at 5 [unpublished, archived on file with the author] [Jennison, "EU ETS"].

to Advance Passenger Information (API), the newly created Transportation Security Administration (TSA) requested the transfer of the passenger flight bookings – PNR in the airline jargon – as part of its Secure Flight Program.[394] This was later extended to flights overflying US territory.[395] Similarly, the United States compelled all foreign aircraft operators flying to/from Washington, D.C., to have on board in-flight security officers (IFSOs).[396] As US Homeland Security Secretary Tom Ridge stated, "any sovereign government retains the right to revoke the privilege of flying to and from a country or even over their airspace. So ultimately a denial of access is the leverage that you have."[397]

Arguably, at the time, foreign aircraft operators were forced to deploy IFSOs as a condition to serve certain routes to the United States. This was clearly an imposition. Some years later, US pressure led ICAO to adopt new international standards on the deployment of IFSOs.[398] FAA's Michael Jennison justifies these measures as "[governing] conduct with substantial impact on U.S. citizens or U.S. territory."[399] Jennison also points out that, in most cases, ICAO later followed with multilateral measures.[400] For US authorities, these requirements are just conditions to access the US market. Unfortunately, for foreign aircraft operators, it involves regulating behavior outside US territory.

Arguably, the United States has extensively resorted to extraterritoriality in dealing with security issues in the post 9/11 era. To defend the US position, one could argue that

---

394 See TSA, "Secure Flight Program," *TSA* online: Transportation Security Administration <//www.tsa.gov/stakeholders/secure-flight-program>.

395 See Pablo Mendes de Leon, "The Fight against Terrorism through Aviation: Data Protection Versus Data Production" (2006) 31:4-5 Air & Space L 320.

396 According to Annex 17 to the *Chicago Convention*, IFSOs are law enforcement officers, duly authorized by both the State of the operator and the State of registry of the aircraft, deployed in certain flights with the purpose of protecting the aircraft, passengers, and crew against potential acts of unlawful interference. See Annex 17 to the *Chicago Convention, Security: Safeguarding International Civil Aviation Against Acts of Unlawful Interference*, 9th edn (March 2011) at Chapter 1 [Annex 17]. See also Alejandro Piera, "Report of the Rapporteur of the Special Sub-Committee on the Preparation of an Instrument to Modernize the Convention on Offences and Certain Other Acts Committed on Board Aircraft of 1963" (2012) XXXVII Ann Air & Sp L 485 at 523.

397 Fox News, "U.S. Bolstering Foreign-Flight Security," *Fox News* (3 December 2003) online: Fox News <www.foxnews.com/story/0,2933,106941,00.html>. Those defending the EU ETS argue that the scheme is far less intrusive than security-related measures that the United States adopted after 9/11, such as reinforced aircraft cockpit doors, screening of liquids and gels, and spot check at foreign airports. See Petsonk, "Testimony," *supra* note 98 (Chapter 4) at 12.

398 Right after 9/11, in December 2011, ICAO introduced swift amendments on aviation security in Annex 17. At the time, the concept of "armed personnel" was used. Later, in 2006, the concept of IFSOs was adopted. Clearly, the United States was the driving force behind these normative changes. The spirit of the ICAO rules is to subject the deployment of IFSOs, not to one State's will, but rather to the consent of all the States concerned. Thus, in theory at least, States are required to consider, but definitely not obliged to accept, requests from other States wishing to deploy IFSOs on international flights. See Annex 17, *supra* note 396 at Chapter 4, Standard 4.7.5 at 4-4.

399 Jennison, "EU ETS," *supra* note 393 at 5.

400 *Ibid.*

in these cases, there was a "substantial and bona fide connection"[401] between the measures implemented and the State exercising extraterritorial jurisdiction. These extraterritorial measures were designed to protect US soil, the travelling public, and the airline industry at large. Had these measures not been taken, the effects could have been catastrophic. However, this level of immediate threat is not necessarily present in the EU ETS. Having said this, extraterritorial measures on in-flight gambling and smoking bans do not seem to belong to the same category. It is certainly much more difficult to justify the permissibility of these measures under international law.

Moreover, some extraterritorial measures engender a strong level of opposition. In spite of the fact that they represented a similar level of extraterritorial overreach, unlike the EU ETS, the patently extraterritorial security measures adopted by the United States in the post 9/11 era did not prompt a unified international coalition attempting to defy them. Lowe links the acceptance of extraterritorial jurisdiction to values. When a State resorts to extraterritorial jurisdiction to "[uphold] generally-held values" such as combating terrorism or enhancing aviation security measures, other States will most likely tolerate it.[402] However, this is not the case where extraterritoriality is used to advance what is perceived as "parochial policies" such as antitrust laws or even climate change mitigation.[403] As Silversmith puts it, "aviation safety matters have seemed to engender greater international cooperation than economic matters."[404] Therefore, whenever a State exercises extraterritorial jurisdiction, different issues will prompt different reactions from the international community.[405] The degree of tolerance will vary in accordance with the particularities of the case and the nature of the subject matter. This explains why States have grudgingly acquiesced to the extraterritorial overreach of the United States on aviation security measures but, on the other hand, have vehemently reacted against the EU ETS.

5.7 Can the Doctrine of State Responsibility Exonerate the EU ETS?

In light of the reluctance of some States to take action to combat climate change in a meaningful way, it must be determined whether States willing to protect global environmental commons may justify their actions on the basis of the theory of State responsibility. Can the invocation of responsibility by a State, other than the injured State, provide the EU Member States sufficient legal grounds to justify the inclusion of foreign aircraft operators into its ETS? I argue that the particularities of the EU ETS do not.

401 Brownlie, *supra* note 31 at 313.

402 Lowe, "Problems of Extraterritorial Jurisdiction," *supra* note 220 at 734.

403 *Ibid.*

404 Silversmith, *supra* note 376 at 211.

405 As Meyer notes, "despite its own track record of extraterritorial adventurism, the [US] bristles when foreign States seek to extend their laws to the territorial [US]." See Meyer, *supra* note 9 at 112.

As mentioned above, the invocation of responsibility by a State other than an injured State in order to defend obligations *erga omnes* is, in practice, subject to very restricted application. In fact, it cannot be interpreted as providing a carte blanche to States to perform the role of *comunis* watchdog. The issue must necessarily involve an internationally wrongful act placing the responsible States in breach of an international obligation. This is perhaps the biggest obstacle for the EU ETS. Non-EU States, who have not yet taken steps toward regulating GHGs from international aviation, have not breached any international obligation. In fact, the only obligation that may be applicable is Article 2.2 of the *Kyoto Protocol*.[406] The provision does not require that developed States take the "bull by the horn" and adopt measures to mitigate GHG emissions from international aviation. Rather, it directs them to work through ICAO.[407] Albeit without much success, all these States have long been engaged in the ICAO process. In addition, given international aviation's current GHG contribution, the damage or loss that the international community has suffered as the consequence of the sector's inaction is not clear and cannot be realistically quantified.

Even if, for the sake of discussion, one assumes that non-EU States breached an international obligation and that there was damage to the environment, EU Member States could only be entitled to demand the remedies identified above. Lack of tangible action to combat international aviation's impact on climate change on the part of non-EU States does not serve as a valid justification for the EU Member States to act unilaterally. Also, EU Member States are not entitled to institute a collective countermeasure with the object of protecting the international community as a whole.[408] The EU ETS cannot be regarded as a countermeasure but rather as a unilateral measure. The invocation of State responsibility by a State other than the injured State is an independent cause of action to bring a legal challenge before a court such as the ICJ with formalistic and procedural requirements, such as the consent of the respondent.[409]

A different question is whether this novel notion should evolve and apply more broadly in a way that it would allow true universal standing. That is to say that those States willing to take action in defending certain obligations *erga omnes* in the interest of the international community may do so without having to observe prescriptive requirements that, as present to date in the theory of State responsibility (e.g. presence of internationally wrongful act), thwart its value in practice. Some commentators have already criticized this restrictive

406 See *Kyoto Protocol*, *supra* note 12 (Introduction) at Art 2.2.

407 *Ibid.*

408 See generally James Crawford, "Responsibility to the International Community as a Whole" in James Crawford, ed, *International Law as an Open System* (London: Cameron May Ltd, 2002) 341.

409 See Hugh Thirlway, "Injured and non-Injured States before the International Court of Justice" in Maurizio Ragazzi, ed, *International Responsibility Today: Essays in Memory of Oscar Schachter* (Leiden: Martinus Nijhoff Publishers, 2005) 311 at 317.

approach as making international law "ineffective."[410] After all, the ILC has already recognized that obligations *erga omnes* may evolve over time.[411] This could facilitate their internalization by States. However, relaxing existing rules on invoking State responsibility could lead to chaos since decisions to act upon these obligations necessarily involve subjective assessments which are heavily influenced by different sets of values, which may differ among countries.[412] As Dupuy writes, this rule should not be "used in an anarchic way by individual states or by a group of States acting collectively to take measures outside any international control by turning the alleged defense of a community interest into a convenient alibi for the realization of specific political strategies."[413]

More internationalization of obligations *erga omnes* may also prompt an unwarranted increase of unilateral actions. This in turn may jeopardize the well-being of international relations between States. It would, therefore, seem that State practice will evolve in such a way that more domains will eventually fall under the notion of obligations *erga omnes*. While this does not mean that rules will be relaxed, it seems unlikely that States will be given universal standing to protect international public goods.

5.8 CONCLUSION

This chapter has examined principles of international law dealing with jurisdictional questions as a means to assess the legality of the EU ETS. The initial design of the EU ETS had clear extraterritorial elements.[414] Fuel consumption was computed for the whole duration of the flight, although for flights originating from or destined to foreign countries, most of the emissions are produced outside the territorial limits of the EU. Although these factors are extremely relevant in understanding how the system works, they cannot constitute the basis for determining whether the EU ETS is permissible. In fact, this has been

410 Wolfrum, "International Environmental Law," *supra* note 67 (Chapter 2) at 55.

411 See ILC, *Commentaries on Draft Articles*, *supra* note 181 at 217.

412 See Chen, "International Community," *supra* note 167 (discussing that in many cases, with the pretext to defend the international community, some States champion unilateral actions without observing the procedural requirements of the UN Security Council, at 39).

413 See Dupuy, *supra* note 184 at 210.

414 See Manzini & Masutti, *supra* note 136 (Chapter 4) (noting that the EU ETS does not contravene the principles enshrined in the *Chicago Convention*, at 323); Gareth Price, *The EU ETS and Unilateralism within International Air Transport* (LLM Thesis, McGill University) [unpublished] at 10; Meltzer, *supra* note 222 (Chapter 4) at 152 (saying that the EU ETS does not represent an infringement on State sovereignty); Pache, *supra* note 97 (Chapter 4) (arguing that it is not an intrusion on State sovereignty given that the measure is "merely a temporary and not necessarily desirable interim step" and because the equivalent measures provision de facto recognizes the principle of comity, at 69); Benoit Mayer, "A Defense of the EU Emission Trading Scheme in Aviation Activities" (November 7, 2011) online: <http://dx.doi.org/10.2139/ssrn.1955817> [Mayer, "Defense of the EU ETS"] (noting that the scheme does not "interfere with other Sates policies through forcing honest public debates on issues that politicians do not want to discuss" at 26).

one of the main flaws in the criticism advanced against the scheme. The mere presence of extraterritorial or foreign elements does not automatically render the measure impermissible.

The question of its permissibility must be analyzed through the principles of international law, an analysis that the CJEU consciously opted to disregard. In addition to being self-serving, the CJEU's examination of the extraterritoriality aspects of the EU ETS in its *ATA Decision* essentially followed the typical US methodology to validate the EU's extraterritorial exercise of jurisdiction. This approach sought to interpret that at least one element, such as the physical presence of foreign aircraft landing and departing from EU airports, provides sufficient nexus between the operation and European territory, thereby justifying the assertion of jurisdiction. In the CJEU's view, such exercise of jurisdiction is not extraterritorial, but rather a manifestation of territoriality. The ferocious opposition that almost the entire world has launched against the EU ETS certainly suggests that, outside Europe, the international community is not ready to accept what is perceived as an unlawful, extraterritorial assertion of jurisdiction.[415]

None of the principles of international law on jurisdiction justifies the scheme's original geographical scope (e.g. emissions for the whole duration of the flight). Likewise, it is very clear that the EU and its Member States cannot avail themselves of the principles of State responsibility. Again, the purpose of the invocation of responsibility by a State other than the injured State to remedy the breach of obligations *erga omnes* is to defend the interest of the international community as a whole. It is not a vehicle to immunize the exercise of unilateral measures. Thus, I contend that, in its original form, the EU ETS is not permissible under international law.

These legal concerns are however not present at all in the proposal to limit the geographical scope of the scheme to European airspace. Under the *Chicago Convention* and principles of customary international law, European States do have the sovereign right to establish a market-based measure (MBM) such as an ETS, and they may choose to include foreign aircraft operators in the scheme, to the extent that the computation of their emissions is restricted to that which is generated within the European airspace. Challenges may be based on political and operational considerations, but certainly not on plausible legal grounds.

We must also recognize that the extraterritoriality present in the original scope of the EU ETS was not at all unprecedented. On countless occasions, the United States has engaged in similar practices. It is worth recalling the bans on gambling and smoking aboard aircraft. Both of these initiatives were extremely similar in scope to the EU ETS. The United States

415 See Jacques Hartmant, "The European Emissions Trading System and Extraterritorial Jurisdiction" (23 April 2012), online: Blog of the European Journal of International Law <www.ejiltalk.org/the-european-emissions-trading-system-and-extraterritorial-jurisdiction/>.

has also expanded its jurisdiction beyond its borders on numerous initiatives relating to aviation security. However, as suggested earlier, different issues trigger different responses. The acceptance or rejection of extraterritorial jurisdiction is inexorably linked to the values that actors associate to the behavior in question. There is not much appetite to tolerate extraterritorial jurisdiction involving climate change and international civil aviation.

Furthermore, in pursuing extraterritorial jurisdiction, the *leitmotiv* of both the United States and Europe is very similar. US courts and regulatory agencies do not hesitate to assert extraterritorial jurisdiction where there is a US interest at stake. Similarly, in Europe, extraterritoriality is intrinsically linked to advancing European policy objectives, including but not limited to climate change issues. Therefore, this will likely not be the last chapter on extraterritorial jurisdiction involving international civil aviation. It is certainly likely that the issue will reappear.

Although extraterritoriality remains an extremely controversial subject, it is fair to say that, to the extent that there are reasonable links to the State exercising jurisdiction, international law does not prohibit it.[416] It is generally accepted that States may extend their jurisdiction beyond their territorial boundaries.[417] However, this exercise is not without limits.[418] The question is not whether extraterritoriality is lawful, but rather in under what circumstances a State may legitimately exercise extraterritorial jurisdiction.[419]

Way back in 1957, just twelve years after Judge Learned Handed delivered his opinion in *Alcoa*, Jennings wisely articulated some limitations on extraterritoriality.[420] For Jennings, international law recognizes the right of a State to engage in extraterritoriality where a legitimate interest so requires.[421] However, such interest must be shared by the common practices of the States concerned.[422] Drawing from the theory of abuse of rights, Jennings clarified that extraterritoriality becomes unacceptable where it "[interferes] with the exercise of the local territorial jurisdiction."[423] One can only hope that policy makers will take Jennings' wise words into accounts when designing schemes with possible extraterritorial elements.

416 See Sigrun I Skogly, *Beyond National Borders: States' Human Rights Obligations in International Cooperation*, (Oxford: Intersentia, 2006) at 23.
417 See Barlow, *supra* note 221 at 106.
418 See Separate Opinion of President Guillaume in *Arrest Warrantsupra* note 45 at 37.
419 See William R Slomanson, *Fundamental Perspectives on International Law*, 4th edn (United States, Thomson West: 2003) at 213.
420 Jennings *supra* note 30 at 153.
421 *Ibid.*
422 *Ibid.*
423 *Ibid.*

6 Additional Legal Issues Involving the EU ETS

Although extraterritoriality has been the main legal argument against the EU ETS, non-EU States and industry stakeholders have also contended that the scheme contravenes a number of other legal principles. These actors suggest that the EU ETS constitutes an unequivocal manifestation of unilateralism, an unlawful tax that flagrantly violates provisions of the *Chicago Convention* and clauses of myriad air services agreements (ASAs), and WTO rules. Moreover, developing countries have also asserted that the design of the EU ETS ignores CBDR.

These other legal concerns are extremely relevant in analyzing the scheme's legality, particularly if one takes into account that in October 2013 Europe announced its intention to scale down the scheme's geographical reach to cover only emissions produced by both foreign and EU aircraft operators within European airspace. Although in April 2014 the European Parliament ultimately decided to suspend the application of the EU ETS to foreign aircraft operators until the end of 2016, the airspace approach would have weakened the extraterritoriality argument. Also, it is pertinent to consider these concerns because it may well be the case that in 2016 or early in 2017, the EU decides to re-start the EU ETS as originally proposed, should ICAO not reach a global agreement.

This chapter assesses whether these additional legal concerns can invalidate the EU ETS. I examine the arguments put forward by the main actors as well as the justifications provided by both the CJEU and Advocate General Kokott in the *ATA Decision*. I explore the potential for unilateralism in the EU ETS, examining the scheme's unilateral elements through Bodansky's balanced, liberal approach. I attempt to establish whether such unilateral action may be permissible under international law or whether it is beneficial or detrimental in light of the failure to regulate GHG emissions from international aviation through ICAO.

Careful consideration is given to the issue of whether the EU ETS may be regarded as an impermissible tax. In order to assess the nature of emissions trading as a policy option, it tracks its genesis. The analysis reveals that emissions trading was created precisely to provide an alternative to taxes in order to internalize environmental externalities. The section also considers whether the EU ETS may constitute an impermissible levy under certain provisions of the *Chicago Convention*. It explains what makes the EU ETS different from taxes.

In addition to examining whether the scheme contravenes WTO and CBDR, close attention is given to the CJEU's argument that in the EU ETS saga the European Union is not bound by the provisions of the *Chicago Convention*. This proposition has significant consequences for international civil aviation and explains why States whose aircraft operators did not have a single flight to and from Europe opposed the EU ETS.

6.1 Unilateral Action

Unilateral State action is not at all uncommon in international law.[1] In fact, some notable examples contributed to the formation of principles of customary international law and their later codification in international instruments. Unilateral State action is an unquestionable reality, more and more present in the environmental field.[2] It is the legitimate manifestation of States' sovereignty.[3] In this respect, the right of States to take domestic measures is incontestable.[4] Yet this becomes problematic when such action affects the "rights, claims and interests" of other States.[5] It is also challenging when it is regarded as a mechanism to advance the hegemonic will of stronger over weaker States,[6] or as a vehicle through which to impose a set of norms or values that not all States necessarily share. In addition, unilateral State action raises issues of legitimacy, for it is essentially unidirectional, non-inclusive, and, as such, antidemocratic.[7] As will be explained in this section, not all unilateral acts constitute violations of international law.[8] In fact, a number of them are either permissible or have received the tacit acquiescence of the international community.

In its original form, the EU ETS presented a conclusive extraterritorial component impermissible under international law. Concerns over its extraterritorial reach are directly linked to its unilateral imposition. From a European perspective, the scheme is the (unilateral) regulatory response to address an environmental problem, given the reluctance to do so through multilateralism. On the other hand, the scheme's detractors contend that

1 See Cedric Ryngaert, *Jurisdiction in International Law* (Oxford: Oxford University Press, 2009) at 9 (claiming that the United States always resorts to unilateralism in world politics) [Ryngaert, *Jurisdiction in International Law*].

2 See Richard B Bilder, "The Role of Unilateral Action in Preventing International Environmental Injury" (1981) 14:1 Vand J Transnat'l L 51 at 52; Laurence Boisson de Chazournes, "Unilateralism and Environmental Protection: Issues of Perception and Reality of Issues" (2000) 11:2 EJIL 315 at 319 (commenting that "States have, and will continue, to resort to [unilateral] measures for various environment-related objectives") [Boisson de Chazournes].

3 See Boisson de Chazournes, *supra* note 2 at 316; Daniel Bodansky, "What is so Bad about Unilateral Action to Protect the Environment" (2000) 11:2 EJIL 339 at 340 ["Bodansky, Unilateralism"].

4 Bodansky, "Unilateralism," *supra* note 3 at 341.

5 Bilder, *supra* note 2 at 54; *ibid.* at 341.

6 See Boisson de Chazournes, *supra* note 2 at 318.

7 *Ibid.* at 320.

8 See Bodansky, "Unilateralism", *supra* note 3 at 341; Kulovesi, "Addressing Sectoral Emissions", *supra* note 214 (Chapter 1) (saying that unilateralism bears a "devoid legal meaning" at 194); Bilder, *supra* note 2 (commenting that the notion of unilateral State action carries a pejorative sense) at 52; Boisson de Chazournes, *supra* note 2 (commenting that unilateralism is not well defined) at 315.

international aviation's emissions should be handled by ICAO through a multilateral process.[9] They maintain that the scheme's unilateral imposition on foreign aircraft represents an unlawful and impermissible action that impinges upon the sovereignty of other States.[10]

6.1.1 *The* Kyoto Protocol, *ICAO, and Unilateral State Action*

States opposing the EU ETS are of the view that Europe's unilateral approach infringes the implicit mandate that the *Kyoto Protocol* bestowed upon ICAO. For these States, the mandate implies a sense of exclusivity in favor of ICAO.[11] In other words, no State or group of States can regulate GHGs arising from international civil aviation. ICAO is the (sole) forum to address it. The mandate inherently comes with exclusive jurisdiction on the subject matter.

Assuming that the *Kyoto Protocol's* mandate does in fact imply the exclusive jurisdiction of ICAO, the question is then what States should do in light of the fact that ICAO has not been able to establish, at least not yet, a system to regulate GHGs from international aviation. Do States refrain from taking any measures until ICAO finally finds a solution? What if this is simply unfeasible? What if there is no political will? Given that the mandate does not impose any timeline or deadline on ICAO,[12] the impasse may continue for years to come. Similarly, interpreting the mandate as conferring exclusive jurisdiction, in the absence of ICAO's action, prevents States from exercising leadership in this field and, at the same time, makes things easier for those abhorring any type of regulation. All they

9 See New Delhi Declaration, *supra* note 132 (Chapter 1), preambular clauses; ICAO Council Declaration of 2 November 2011, *supra* note 132 (Chapter 1), preambular clauses; Moscow Declaration, *supra* note 132 (Chapter 1), preambular clauses. Although not a binding source of international law, the Rio Declaration states that environmental mitigation measures to address problems of a global commons nature should be addressed internationally on the basis of consensus. See Rio Declaration, *supra* note 35 (Chapter 1) at Principle 12.

10 See ICAO, C-MIN 196/7, *supra* note 124 (Chapter 1) (Statement of Brazil, describing the EU ETS as an "illegal, unilateral and extraterritorial [measure violating] State sovereignty" at 19). In practical terms, some States and analysts fear that the EU ETS may constitute a precedent that would spur the unilateral adoption of similar regional or national measures elsewhere. ICAO, C-MIN 194/2, *supra* note 86 (Chapter 1) at 18 (Statement of Singapore); Dejong, *supra* note 101 (Chapter 4) (commenting that the *ATA Decision* "set a worrisome precedent, for the regulatory regime behind international aviation will quickly collapse if states look to impose unilateral solutions on multilateral issues" at 187); Goron, *supra* note 35 (Chapter 2) at 4; Position of the Coalition of the Unwilling, *supra* note 150 (Chapter 4) (underscoring the "likelihood of similar competing schemes being introduced by other States (some as a retaliatory measure), bringing about a chaotic situation adversely affecting the sustainability of air transport" at 3).

11 See Heather L Miller, "Civil Aircraft Emissions and International Treaty Law" (1997) 63 J Air L Com 697 at 722; Goron, *supra* note 35 (Chapter 2) (arguing that Art 2.2 of the *Kyoto Protocol* is not "a green light to unilateral measures" at 18).

12 See Warren B Fitzgerald, Oliver JA Howitt & Inga J Smith, "Greenhouse Gas Emissions from the International Maritime Transport of New Zealand's Import and Exports" (2011) 39 Ener Pol'y 1521 at 1523.

should worry about is to stall discussions at ICAO. Such interpretation is then the perfect recipe to justify the do-nothing approach. Pache adds that ICAO's inaction "strengthens the [right of EU Member States] to take unilateral action."[13] Although, like most Europeans, Pache recognizes that the "multilateralism" is the best way forward, it should not constitute the basis on which to assess the permissibility under international law of a unilateral action such as the EU ETS.[14]

An exclusive mandate also means that even bilateral agreements between States are precluded. Moreover, it runs counter to the *UNFCCC's* objective, which is precisely to achieve the "stabilization of [GHG] concentrations in the atmosphere at a level that would prevent dangerous anthropogenic interference with the climate system."[15] The *Kyoto Protocol* is an international instrument to achieve this goal.[16] As Advocate General Kokott noted in the *ATA Decision*, the meaning of the actual scope of the mandate given to ICAO should not be construed in such a way as to defy the very purpose underlying the constituting instrument.[17]

Not surprisingly, while the United States is not a party to the *Kyoto Protocol*, in *ATA Decision* US airlines argued that the *Kyoto* mandate conferred exclusive jurisdiction upon ICAO, and therefore, the EU had no authority to extend its ETS to international aviation.[18] Preferring a broad interpretation of the mandate, the CJEU ruled that the *Kyoto Protocol* does not prohibit the implementation of a regional environmental measure such as the EU ETS. The CJEU noted that in the context of the *UNFCCC*, national and regional policies are permitted.[19] The last ICAO resolutions on climate change may have also implicitly accepted that certain measures may be taken outside of ICAO, and by doing so, it has de facto accepted the non-exclusive mandate given by *Kyoto*.[20] However, this cannot be interpreted in any way as a license to exercise unilateral measures outside the boundaries of international law.

---

13 Pache, *supra* note 97 (Chapter 4) at 6.

14 *Ibid.*

15 *UNFCCC, supra* note 11 (Introduction) at Art 2.

16 See Pache, *supra* note 97 (Chapter 4) at 5.

17 See Opinion of Advocate General Kokott, *supra* note 136 (Chapter 4) at I-51.

18 Markus W Gehring, "Case Note Air Transport Association of America v. Energy Secretary: Clarifying Direct Effect and Providing Guidance for Future Instrument Design for a Green Economy in the European Union" (2012) 2 RECIEL 149 at 151.

19 See Opinion of Advocate General Kokott, *supra* note 136 (Chapter 4) at I-51.

20 See Havel & Sanchez, "Toward a Global Aviation Emissions Agreement," *supra* note 214 (Chapter 1) at 360. In 2010, the 37th Session of the ICAO Assembly instructed the Council to "develop a framework for [MBMs] in international aviation." See ICAO A37-19, *supra* note 207 (Chapter 1) at para 13. To this end, the Assembly elaborated a set of "guiding principles for the design and implementation of [MBMs]." *Ibid.* at Annex. It also encouraged States to respect these principles "when designing new and implementing existing MBMs." *Ibid.* at para 14. Similarly, the Assembly recommended that States revise their existing MBMs to ensure that they are aligned with ICAO's guiding principles. *Ibid.* at para 17. In 2013, the 38th Assembly also recognized the existence of "new and existing MBMs for international aviation." See A38-18, *supra* note 21 (Introduction) at para 16.

6.1.2 *Types of Unilateral State Actions*

International law scholarship conceives at least four types of unilateral State actions. The first category addresses the adoption of a unilateral measure by a State within the context of an (international) legal framework.[21] For instance, under Article XX of the General Agreement on Tariffs and Trade (GATT), States may take non-discriminatory, environmentally related unilateral trade measures that are necessary to protect, inter alia, "human, animal or plant life or health"[22] or "exhaustible natural resources."[23] In the *Shrimp-Turtle* case, WTO Appellate Body (AB) expressly recognized that States can resort to unilateral trade measures as long as the execution of such measures is not carried out in a discriminatory matter.[24] Similarly, both the *Agreement on Trade-Related Aspects of Intellectual Property* and GATS also allow State parties to take any measure "which [they consider] necessary for the protection of its essential security interest."[25]

In the environmental field, the Montreal Protocol authorizes State parties to unilaterally adopt import ban restrictions on ozone-depleting substances from non-State parties.[26] Under the Wellington Convention, a State party may unilaterally ban "the transshipment of driftnet catches"[27] as well as "restrict port access and port servicing facilities for driftnet fishing vessels."[28] In addition, in the *Tuna-Dolphin* dispute, the United States argued that some international instruments, such as the *Basel Convention on the Control of Transboundary Movements of Hazardous Wastes and Their Disposal* and the *Convention on the International Trade in Endangered Species of Wild Fauna and Flora*, expressly allow the use of unilateral environmentally related trade measures to protect the global commons.[29] All these instruments do authorize the exercise of unilateral measures. To this end, they are

21 See also Steve Charnovitz, "The Law of Environmental PPMs in the WTO: Debunking the Myth of Illegality" (2002) 27 Yale J Int'l L 59 (commenting that a number of international instruments empower States to adopt unilateral measures to preserve their integrity and to ensure that their objectives are achieved) at 105.

22 See *General Agreement on Tariffs and Trade*, 1947, 55 UNTS 194 at Art XX (b) [GATT]. Similar provisions exist in GATS, *supra* note 3 (Chapter 2) at Art XIV, Annex 1B. See also Howse, "Shrimp/Turtle Case," *supra* note 59 (Chapter 5).

23 GATT, *supra* note 22 at Art XX (g).

24 See Tracey P Varghese, "The WTO's Shrimp-Turtle Decisions: The Extraterritorial Enforcement of U.S. Environmental Policy via Unilateral Trade Embargoes" (2001) 8 Envtl L 421 at 432; Slayde Hawkings, "Skirting Protectionism: A GHG-Based Trade Restriction under the WTO" (2008) 20 Geo Int'l Envtl L Rev 427 at 439. States also resort to unilateral trade measures to advance environmental policy objectives. See Anderson, *supra* note 311 (Chapter 5) at 751.

25 See *Trade-Related Aspects of Intellectual Property Rights*, Annex 1 C to the Marrakesh Agreement Establishing the World Trade Organization at Art 73 (b) [*TRIPS*]; *GATS*, *supra* note 3 (Chapter 2) at Art XIV bis (b).

26 See *Montreal Protocol on Substances that Deplete the Ozone Layer*, 16 September 1987 522 UNTS 3, 26 ILM 1550 (1987) at Art 4 (1) [Montreal Protocol].

27 See *Convention for the Prohibition of Fishing with Long Driftnets in the South Pacific*, adopted at Wellington on 24 November 1989, Art 3 (2) (b) (entered into force 17 May 1991) [*Wellington Convention*].

28 *Ibid.* at Art 3 (2) (d).

29 See Ilona Cheyne, "Environmental Unilateralism and the WTO/GATT System" (1995) 24 Ga J Int'l & Comp L 433 at 451-54.

permissible as between the parties to the international instrument in question.[30] The *Poulsen* case best exemplifies the legality of certain unilateral measures. The CJEU found lawful the confiscation (unilateral act) by Danish authorities of salmon caught outside the jurisdiction of EU Member States by a Panamanian-flagged vessel.[31] The justification of this unilateral exercise was rooted in community legislation internalizing obligations contained in an international agreement.[32]

In the aviation context, another good example is the International Aviation Safety Assessment (IASA) that the US FAA launched in 1992.[33] Through this program, the FAA evaluated whether States were in compliance with ICAO standards. Although it targeted States, in practice it meant that aircraft operators whose States were found not to be ICAO-compliant by the FAA could not fly to the United States.[34] The United States justified its unilateral, unpopular measure in Article 33 of the *Chicago Convention*, where a signatory State is not obliged to recognize the airworthiness certificates and personnel licenses granted by another State that were below ICAO's minimum (safety) standards.[35] Some years later, Europe followed a similar trend with the implementation of its Safety Assessment of Foreign Aircraft (SAFAEC SAFA).[36] In practice, and despite much criticism,[37] these unilateral programs act as indirect enforcement mechanisms of the international safety regime.[38] They not only paved the way for ICAO's Universal Safety Oversight Audit Program (USOAP),[39] but they have also served as exogenous enforcers of its standards.[40]

---

30 In theory, in light of the fact that the instrument in question could not be opposed to a non-party State, such a State could still argue that the measure in question is unlawful.

31 See Judgment of the Court of 24 November 1992 (*Anklagemyndigheden v Peter Michael Poulsen and Diva Navigation Corp*, Case C-286/90 at para 11 I-6053 [*Poulsen Case*]

32 See EC, Council Regulation (EC) 3094/86 of 7 October 1986 Laying Down Certain Technical Measures for the Conservation of Fishery Resources, [1986] OJ, L 288/1 at Art 6; *Convention for the Conservation of Salmon in the North Atlantic*, 2 March 1982, 1338 UNTS 22433.

33 See Bodansky, "Unilateralism", *supra* note 3 at 343.

34 See Michael B Jennison, "The *Chicago Convention* and Safety after 50 Years" (1995) XXI Ann Air & Sp L 283.

35 See *Chicago Convention*, *supra* note 36 (Chapter 2) at Art 33.

36 See Jimena Blumenkron, "Implications of Transparency in the International Civil Aviation Organization's Universal Safety Oversight Audit Program" (2009) XXXIV Ann Air Sp L 31 at 57.

37 See Michael Milde, "Aviation Safety & Security: Legal Management" (2004) XXIX Ann Air & Sp L 1 at 4.

38 See Michael Milde, "Aviation Safety Oversight: Audits and the Law" (2001) XXVI Ann Air & Sp L 165 at 170; Paul Dempsey, Compliance & Enforcement in International Law: Achieving Global Uniformity in Aviation Safety" (2004) 30 NC J Int'l L & Com Reg 1; Paul Dempsey, *supra* note 5 (Chapter 2) at 85. See also Bodansky, "Unilateralism", *supra* note 3 (noting that a unilateral State action may facilitate the enforcement of the international regime where international organizations lack effective enforcement tools or where they are simply nonexistent) at 344.

39 See Silversmith, *supra* note 376 (Chapter 5) at 216.

40 This issue is revisited in Chapter 8 where potential enforcement mechanisms for ICAO's global MBM are discussed in detail.

A second category deals with issues of self-defense and state of necessity.[41] In these situations, a State carries out a unilateral act to avoid an imminent threat or correct an existing violation.[42] Here it is understood that the State need not "resort to or observe the procedures set out in international law."[43] To this end, the unilateral act is permissible. The 1967 bombing by the United Kingdom's Royal Air Force of Liberian-flagged supertanker Torrey Canyon in the high seas to limit the damage caused by an oil spillage over the British and French coasts provides an example where other States acquiesced to a unilateral action resulting from a looming environmental threat.[44]

A third category includes actions where States attempt to tweak the international legal regime in order to advance their own policy agenda or economic interest. This is often referred to as "policy-forging."[45] In principle, these unilateral acts are not permissible since they seek to influence legal regimes. However, their permissibility and lawfulness can only be determined on a case-by-case basis. A classic example of this is the 1945 Proclamation of President Harry S Truman through which he asserted the US jurisdiction on its continental shelf contiguous to its territory.[46] Similarly, in the 1960s, Iceland sought to unilaterally extend its exclusive fishing rights to 50 nautical miles from its coastline. The ICJ ultimately

41 In the *Gabcíkovo-Nagymaros Project*, the ICJ had a chance to address the principle of state of necessity and the conditions that ought to be met in order to exclude the responsibility of a State for the commission of a wrongful act. See *Gabcíkovo-Nagymaros Project, supra* note 228 (Chapter 2) at 39. Relying heavily on the principles articulated by the International Law Commission (ILC) on the *Draft Articles on International Responsibility of States*, the ICJ expressly recognized that the state of necessity is part of customary international law. *Ibid.* at 41; see *Draft Articles on State Responsibility, supra* note 177 (Chapter 5). As a principle of an exceptional nature, the ICJ noted that a State may avail itself of these grounds to exonerate its international responsibility to the extent that its actions are the only way to protect its "essential interest" with respect to a real, "grave and imminent peril" that such State is confronted with. *Gabcíkovo-Nagymaros Project, supra* note 228 (Chapter 2) at 39. Drawing from the works of the ILC, the ICJ shed light on what it should be understood by the notion of a "peril." In this respect, the ICJ noted that such notion may be assimilated to a "risk" which for the State in question must be concrete and proximate. With this, the ICJ excluded those "possible" risks as justified basis for exoneration. The ICJ also recognized the possibility that such risks may take place not immediately but rather "far off" in time. However, in these cases, the ICJ clarified that the risk, although not immediate, ought to be "certain and inevitable." *Ibid.* at 42. For a very interesting commentary on the evolution of the theory of State Necessity see Robert D Sloane, "On the Use and Abuse of Necessity in the Law of State Responsibility" (2012) 106 AJIL 447.

42 See Bilder, *supra* note 2 at 54.

43 Boisson de Chazournes, *supra* note 2 at 68.

44 See Bilder, *supra* note 2 at 61; Bodansky, *The Art and Craft of International Environmental Law, supra* note 62 (Chapter 1) at 27.

45 Boisson de Chazournes, *supra* note 2 at 317. In this respect, Anderson defines unilateral action as a "nation state's use of its administrative and enforcement agencies to secure a policy goal set by its domestic political process." Anderson, *supra* note 311 (Chapter 5) at 754.

46 See Proclamation 2667 of 28 September 1945 of President Harry S Truman on the Continental Shelf of the United States. Steinberg argues that this unilateral declaration was done out of self-interest to protect the Alaskan salmon. See Richard H Steinberg, "Power and Cooperation in International Environmental Law" in Andrew T Guzman & Alan O Sykes, eds, *Research Handbook in International Economic Law* (Northampton: Edward Elgar, 2007) 485 at 496.

ruled against Iceland.[47] Although at the time of their adoption they were not necessarily fully in conformity with international law, these two precedents were instrumental in the codification of the limits of the territorial sea and concept of the exclusive economic zone (EEZ) which some years later were incorporated in the *United Nations Convention on the Law of the Sea.*[48] More recently, the French government explored the possibility of introducing an emission charge to products imported from countries that were not parties to the *Kyoto Protocol.* Although the proposal was never adopted, it sought to induce non-*Kyoto* parties to accede to the instrument.[49]

A fourth type involves actions by a State that expressly contravene international law or impinge on the rights of other States.[50] In this case, the unilateral action is impermissible. For instance, the shooting down of Korean Airlines flight 007 by the Soviet Union air force in 1983, which resulted in the loss of 269 lives, was found contrary to "the norms governing international behavior and elementary considerations of humanity."[51] Similarly, it is generally acknowledged that the 2003 US-led Iraq invasion was not in conformity with the *Charter of the United Nations.*[52]

Without constituting a separate category, a distinctive feature of unilateral State action comprises those situations in which these acts either incite or promote the adoption of international standards.[53] Although it may be present in any of the four types described above, this feature generally occurs in any of the following three settings: (i) when there is a lacuna and the issue is not expressly addressed by the international regime; (ii) when multilateralism fails to provide an adequate response; and (iii) when a grave international catastrophe or disaster has taken place.[54] In the words of Bodansky, "[w]hen unilateral

47 See *United Kingdom v Iceland,* Judgment of 25 July 1974.

48 See *UNCLOS, supra* note 25 (Chapter 5), Arts 3 and 55. See also Boisson de Chazournes, *supra* note 2 at 320.

49 See Benjamin Eichenberg, "Greenhouse Gas Regulation and Border Tax Adjustments: The Carrot and Stick" (2010) 3 Golden Gate U Envtl LJ 283 at 362.

50 See Boisson de Chazournes, *supra* note 2 at 316.

51 ICAO, *Council Resolution of 16 September 1983*; ICAO, C-WP/7696, *Draft Resolution* (presented by Australia, Canada, Denmark, France, Federal Republic of Germany, Italy, Japan, Kingdom of the Netherlands, Spain, United Kingdom, United States); ICAO, C-WP/7697, *Draft Resolution* (presented by the Soviet Union); ICAO, C-WP/7698, *Draft Resolution Interception of Civil Aircraft* (presented by France); ICAO, News Release, PIO 15/83, New Release, "ICAO Council Takes Action on Korean Air Lines Incident" (19 September 1983) online: ICAO <http://legacy.icao.int/icao/en/nr/1983/pio198315_e.pdf>. As a result of this incident, the international aviation community introduced an amendment to the *Chicago Convention.* Thus, on 10 May 1984, the ICAO Assembly adopted a protocol introducing Art 3 *bis* to the Convention. This obliges Member States "to refrain from resorting to the use of weapons against civil aircraft in flight." See *Chicago Convention, supra* note 36 (Chapter 2) at Art 3 *bis.*

52 See Irene Gendzier, "Just a War: Reflections on the U.S. Invasion of Iraq: Evidence, International Law, and Past Policy" (2004) 9 Nexus J Op 101 at 106; Max Hilaire, "International Law and the United States Invasion of Iraq" (2005) 44 Mil L. & L War Rev 125 (finding that the US "invasion of Iraq was a flagrant violation of international law and the Charter of the United Nations" at 125).See *contra* James D Fray, "Remaining Valid: Security Council Resolutions, Textualism, and the Invasion of Iraq" (2007) 15 Tul J Int'l & Comp L 609.

53 See Bodansky, "Unilateralism", *supra* note 3 at 344.

54 *Ibid.* at 345.

action is aimed at developing multilateral standards that are impartial and advances shared objectives, rather than parochial national interests, then it can play a beneficial role in the international standard-setting process."[55] For instance, in the aftermath of the Exxon Valdez disaster, in 1990, the United States enacted the *Oil Pollution Act*, which phased-out single hull tank vessels.[56] As a result of the US normative push, IMO later incorporated double hull standards for oil tankers by introducing amendments to the *International Convention for the Prevention of Pollution from Ships* (MARPOL Convention).[57] Similarly, the renowned Torrey Canyon oil disaster and the resulting unilateral action taken by the United Kingdom led to the codification in 1969 of the *International Convention Relating to Intervention on the High Seas in Cases of Oil Pollution Casualties.*[58] In addition, both IASA and EC SAFA have proved to be driving forces for ICAO to adopt a comprehensive safety audit program (USOAP, ICAO's Universal Safety Oversight Audit Program) for international civil aviation that eventually required the adoption of copious international standards within the 19 Annexes to the *Chicago Convention.*[59]

6.1.3 *Where Does the EU ETS Stand?*

Placing the EU ETS in one of the four categories of unilateral State action is a difficult exercise, given that the scheme bears elements and characteristics present in all of them.

---

55 *Ibid.*

56 See *Oil Pollution Act* 33 USC ch 40 § 2701 (1990).

57 See *International Convention for the Prevention of Pollution from Ships*, adopted on 2 November 1973, online: IMO <http://goo.gl/Akl2NV>; *Protocol of 1978 relating to the International Convention for the Prevention of Pollution from Ships*, 1973, 17 February 1978, 10 February 1983, 1340 UNTS 61. Later in 2002, the EU passed community legislation to accelerate the phase-in of double hull vessels. This applied to both EU-flagged and non-EU-flagged vessels accessing ports or offshore terminals within the territory of an EU Member State. See EC, *Regulation No 417/2002 of the European Parliament and of the Council of 18 February 2002 on the Accelerated Phasing-in of Double Hull or Equivalent Design Requirements for Single Hull Oil Tankers and Repealing Council Regulation (EC) No 2978/94* [2002] OJ, L64/1. Later, the Prestige oil tanker disaster that took place on 21 November 2002 off the coast of Galicia prompted Europe to adopt even more stringent measures. See *EC, Regulation (EC) No 1726/2003 of the European Parliament and of the Council of 22 July 2002 amending Regulation (EC) No 417/2002 of the Accelerated Phasing-in of Double-Hull or Equivalent Design Requirements for Single-Hull Oil Tankers* [2003] OJ, L 249/1. See Gattini, *supra* note 135 (Chapter 4) at 982.

58 See *International Convention relating to Intervention on the High Seas in Cases of Oil Pollution Casualties*, 29 November 1969, 970 UNTS 211 [*Intervention on the High Seas Convention*]. The *Intervention on the High Seas Convention* recognizes the right of States to take "proportional" and "reasonable" (extraterritorial) measures on the high seas to address an imminent threat or danger to their coastline as a result to pollution from oil spillage. In addition, it clarifies that such measure should not be interpreted as constituting a violation to the customary international law principle of freedom of the high seas. *Ibid.* at preambular clauses, Arts 1, 5. See also Boisson de Chazournes, *supra* note 2 at 320; Barrett, *Why Cooperate*, *supra* note 57 (Chapter 1) at 161.

59 See Jiefang Huang, *Aviation Safety and ICAO* (The Netherlands: Kluwer Law, 2009) at 219 [Huang, *Aviation Safety*].

The distinctions are not necessarily self-evident and vary depending on how one views the scheme. Those in favor argue that Europe has implemented the scheme to comply with its obligations as a signatory of the *Kyoto Protocol*. If this was the only consideration, one could say that the EU ETS is a unilateral act within the context of an international legal framework that so permits it. However, this must not be analyzed in isolation. In the case of emissions from international civil aviation, the *Kyoto Protocol* bestows a non-exclusive, implicit mandate to ICAO. This may suggest that the existing international legal regime does not necessarily grant an express license to the EU to exercise a unilateral act. It does not therefore seem appropriate to include the EU ETS in the first category.

Arguably, the EU ETS does not constitute a regulatory response to a state of necessity. Using the criteria that the ICJ articulated in the *Gabcíkovo-Nagymaros Project*, it is evident that the scheme's unilateral application cannot be justified on the basis that the measure was part of an "essential interest" of the EU or that it was necessary to protect it from a "grave and imminent peril."[60] Climate change issues clearly fall outside of these considerations.[61]

It is unquestionable that the EU ETS resembles in part the category of policy-forging. As discussed in Chapter 5, the measure has been introduced to advance community policy interests. European constituencies are very much in favor of tackling anthropogenic climate change. It is, therefore, inconceivable that international aviation emissions continue to be unregulated as other sectors' carbon footprints are regulated. That said, the scheme also belongs in the fourth category. Whether the EU's unilateralism expressly contravenes an existing rule of international law or directly impinges on the rights of other States is a question of interpretation. While Europe contends that the scheme does neither, there is a compelling argument that it, in fact, does. In particular, as has been discussed previously, under its original geographical scope, the scheme constitutes an impermissible extraterritorial act that contravenes a number of well-established principles of international law. In addition, although it would seem de jure that the EU ETS does not affect the rights of other States, as it only applies to private firms (aircraft operators) flying to or from Europe, the scheme de facto does. With the scheme's implementation, Europe presupposes that emissions from international aviation ought to be regulated by and subject to MBMs. Most States outside Europe are not necessarily willing to regulate international aviation emissions through the use of MBMs. Clearly, the focus has thus far only been on (voluntary) operational and technological measures. What happens, therefore, if a State decides not to reg-

60 Philip Bender points out that "[t]he ILC Commentaries to Article 25 confirm that "peril" must be objectively established and it cannot be something that is a mere possibility. There must also be sufficient temporal proximity between the action taken and the peril." Philip Bender, "A State of Necessity: IUU Fishing in the CCAMLR Zone" (2008) 13 Ocean & Costal LJ 233 at 257.

61 Bartlik, *supra* note 68 (Chapter 4) (suggesting that there is not sufficient evidence of serious consequences for global climate change and the direct link to aviation) at 162.

ulate emissions from its aircraft operators engaged in international operations? The scheme does not provide a way out other than deciding not to fly to and from Europe. In this respect, the EU ETS directly impinges upon the rights of other States that may decide for themselves whether to take any action.[62] As such, the unilateral State action becomes impermissible under international law.[63]

6.1.4 *The EU ETS under the Lenses of Bodansky's Balanced, Liberal Approach*

Bodansky recognizes that, although multilateralism is desirable, "unilateralism can still serve important functions."[64] Unilateralism has a role to play when multilateralism fails, for the alternative may simply be "inaction."[65] Similarly, Kulovesi notes that, in the case of international aviation, unilateralism should be explored as a way to fulfill, for instance, the purpose and objectives of the *UNFCCC*.[66] Rather than intuitively rejecting unilateral State action on the basis of its perceived illegality, Bodansky proposes a utilitarian assessment to establish whether they are in fact more beneficial or detrimental to achieve a given set of environmental goals.[67] Each particular unilateral State action should be carefully examined. In order to establish its permissibility, Bodansky advances five critical elements through which to assess unilateral State action. These include whether the measure in question: (i) is necessary to prevent an environmental threat; (ii) arises as a result of the inability of the multilateral system to produce the desirable environmental outcome;[68] (iii) impinges upon the rights and interests of other States;[69] (iv) constitutes a response of a

62 See Havel & Mulligan, *supra* note 60 (Chapter 4).

63 Wolfrum points out that "[e]ffective protection of the environment cannot be undertaken unilaterally, but rather requires international cooperation and international regulation." Wolfrum, "International Environmental Law", *supra* note 67 (Chapter 2) at 3.

64 See Bodansky, "Unilateralism", *supra* note 3 at 344. In this respect, Marc Rietvelt contends that unilateralism may be beneficial "to protect a specie's already delicate status"). See Marc Rietvelt, "Multilateral Failure: A Comprehensive Analysis of the Shrimp/Turtle Decision" (2005) 15 Ind Int'l & Comp L Rev 473 at 495; Anderson, *supra* note 311 (Chapter 5) at 753. Julie B. Master, "International Trade Trumps Domestic Environmental Protection: Dolphins and Sea Turtles are Sacrificed on the Altar of Free Trade" (1998) 12 Temp Int'l & Comp LJ 423 (noting that unilateral trade embargoes "are a potentially useful means of encouraging international cooperation in developing environmental standards" at 454); Ilona Cheyne, "Environmental Unilateralism and the WTO/GATT System" (1995) 24 Ga J Int'l & Comp L 433 (saying that WTO/GATT should not discard unilateral measures aimed at protecting "non-trade values" at 465).

65 See Bilder, *supra* note 2 at 91.

66 See Kulovesi, "Addressing Sectoral Emissions", *supra* note 214 (Chapter 1) at 195. See also Tracey P Varghese, "The WTO's Shrimp-Turtle Decisions: The Extraterritorial Enforcement of U.S. Environmental Policy via Unilateral Trade Embargoes" (2001) 8 Envtl L 421 (noting that environmental trade embargoes may be, in some cases, beneficial) at 427.

67 See Bodansky, "Unilateralism", *supra* note 3 at 346.

68 *Ibid.* at 345.

69 *Ibid.* at 346.

given State in a state of necessity;[70] and (v) forms part of a "universal rule" that other States may also follow.[71]

Although his views have been strongly criticized for being biased toward legitimating unilateralism, Bodansky's liberal approach provides an interesting platform from which to assess the permissibility of unilateral State action, such as the EU ETS, and the extent to which the international community will acquiesce to such a measure.[72] Arguably, more conservative approaches would outright dismiss the EU ETS.

It is uncontroverted that tackling aviation's CO_2 emissions is desirable. It should be part of international aviation's policy agenda. However, it is difficult to argue that the sector's carbon footprint constitutes an "environmental threat" to humankind. There are other sectors with much larger contributions to climate change. To this end, as a unilateral measure, the EU ETS does not pass Bodansky's necessity element. But, thus far, the scheme represents Europe's (unilateral) response to the failure by the international community to address the issue.[73] Both Advocate General Kokott and the CJEU acknowledged that the adoption of the EU ETS was prompted by the lack of progress at ICAO.[74] Here, the EU ETS is in compliance with Bodansky's second requirement. Although the scheme is not as intrusive as the extraterritorial abduction of persons,[75] as it has been explained above, it does impinge upon the rights of other States.

In addition, the EU ETS was not enacted to respond to a state of necessity. The international community will most likely tolerate unilateral actions that, although extraterritorial and unilateral, constitute "reasonable" and "proportionate" responses to environmental threats or harms.[76] For instance, as the *Torrey Canyon* case demonstrated, and as it was later codified in *UNCLOS*, a coastal State will be justified in "tak[ing] or enforc[ing] measures beyond [its] territorial sea proportionate to the actual or threatened damage to protect [its] coastline or related interests, including fishing, from pollution or threat of pollution following upon a maritime casualty or acts relating to such a casualty, which may reasonably be expected to result in major harmful consequences."[77] This is hardly the

---

70 *Ibid.* at 346.

71 *Ibid.* at 347.

72 Instead of finding innovative ways to recognize the permissibility of unilateral State action, other commentators advocate the need to establish a special forum, such as WTO in order to restrict their application. See Boisson de Chazournes, *supra* note 2 at 338.

73 Fahey, *supra* note 134 (Chapter 4) (saying that the EU ETS constitutes a "major success on the part of the EU to regulate where other global governance mechanisms have failed" at 1260).

74 *Ibid.* at 1247.

75 See generally Abraham Abramovsky, "Extraterritorial Abductions: American's Catch and Snatch Policy Run Amok" (1991) 31 Va J Int'l L 151 (noting that the United States foregoes international extradition to extraterritorially abduct alleged offenders at 152. These acts, which "officially sponsored lawlessness [,] offends not merely the [US] constitution, but also the broader realm of what is understood to be customary international law" at 209).

76 Bilder, *supra* note 2 at 72.

77 *UNCLOS*, *supra* note 25 (Chapter 5) at Art 221.

case of the EU ETS, which fails to meet the third and fourth elements. The greatest difficulty is encountered with regard to the fifth element. As mentioned above, at present, there is no rule of international law that authorizes States to unilaterally exercise extraterritorial jurisdiction to protect global environmental commons. Principle 12 of the Rio Declaration recognizes that unilateral State action to deal with environmental commons should be "avoided."[78]

Notwithstanding the foregoing, it is necessary to move beyond a simplistic and formalistic legal analysis. As Bodansky suggests, the quest is to determine whether on balance the unilateral State action is more beneficial or detrimental for the international community,[79] despite its possible impermissibility under international law. Here, one must recognize that the EU ETS has had an enormous influence upon the aviation and climate change discourse at ICAO. For Kulovesi, even if unilateral, the EU ETS is better than nothing at all.[80] Arguably, adopting unilateral action may be instrumental in advancing international environmental norms. They may in fact serve as a catalyst for change. That said, it is also unquestionable that they may lead to serious and long-lasting international conflicts.[81] As Schoenbaum suggests, "the goal of raising international environmental standards should be pursued, but there are more effective ways than licensing unlimited unilateral action."[82] Unfortunately, EU's normative push has not been translated into the setting of international standards. In addition, the unprecedented level of opposition against the scheme is very indicative that other States are not willing to tolerate it. As Kennedy notes, unilateralism is often repudiated.[83] Likewise, it is worth noting that ICAO certainly does not favor unilateral action.[84] As Legault writes, unilateral State action, "even where it may not be contrary to international law, will be ineffective over the long term unless it is supported or at least

78 Rio Declaration, *supra* note 35 (Chapter 1) at Principle 12.
79 See Bodansky, "Unilateralism", *supra* note 3 at 347.
80 See Kulovesi, "Make Your Own Special Song", *supra* note 137 (Chapter 4) at 557.
81 See Mark Edward Foster, "Making Room for Environmental Trade Measures within the GATT" (1998) 71 S Cal L Rev 393 at 415.
82 See Thomas J Schoenbaum, "International Trade and Protection of the Environment: The Continuing Search for Reconciliation" (1997) 91 AJIL 268 at 312.
83 Kevin C Kennedy, "The Illegality of Unilateral Trade Measures to Resolve Trade-Environment Disputes" (1998) 22 Wm. & Mary Envtl L & Pol'y Rev 375 at 498.
84 See ICAO, A37-20, *Consolidated Statement of Continuing ICAO Policies in the Air Transport Field*, Appendix A Economic Regulation of Air Transport at III-5 (stressing the need to "to avoid adopting unilateral measures that may affect the orderly and harmonious development of international air transport and to ensure that domestic policies and legislation are not applied to international air transport without taking due account of its special characteristics"); ICAO, *A28-7, Aeronautical Consequences of the Iraqi Invasion of Kuwait* (declaring invalid "the unilateral registration of aircraft of Kuwait Airways" by Iraq); ICAO, A36-6, *State Recognition of the Air Operator Certificate of Foreign Operators and Surveillance of their Operations* (requesting Member States "to refrain from unilateral implementation of specific operational requirements"); ICAO, A36-22, *supra* note 17 (Chapter 2) at Appendix L (calling upon States "to refrain from unilateral implementation of greenhouse gas emissions charges"); ICAO, A35-5, *supra* note 17 (Chapter 2) (urging States "to refrain from unilateral implementation of greenhouse gas emissions charges" at Appendix I).

not actively opposed by major trading [actors]."[85] The unilateralism present in the EU ETS, as originally designed, is not permissible under international law. However, this may not necessarily be the case if the scheme is downsized to cover only emissions generated within European airspace.

6.2 The Tax Controversy

I now turn to considering whether the EU ETS may be conceptualized as a tax. States opposed to the EU ETS, industry stakeholders, and a number of commentators contend that the scheme is a tax or a charge that is not permissible under international law. They claim that the scheme would violate various provisions of the *Chicago Convention* and international air services agreements (bilaterals).[86] In this regard, opponents have often used epithets such as "unilateral taxation scheme"[87] and "exorbitant tax" to describe the EU ETS.[88] Arguably, the tax argument has been one of the most forceful criticisms against the EU ETS.[89]

Establishing whether the EU ETS is a tax is relevant because such assessment is not at all linked with the scheme's geographical scope. In other words, if the EU ETS is deemed to be a tax, and therefore contrary to various provisions of the *Chicago Convention*, this would invalidate it, even when applied to foreign aircraft operators flying to Europe for emissions produced over European airspace or for aircraft operators flying between two points within Europe (e.g. intra-European flights).

In ICAO parlance, a "tax" is a levy that is not connected to civil aviation in its entirety or on a cost-specific basis, but is rather intended to increase revenues of a given State,

---

85 See Leonard H Legault, "U.S. Assertions of Extraterritorial Jurisdiction: A Canadian Perspective" in John R Lacey, ed, *Act of State and Extraterritorial Reach: Problems of Law and Policy* (Chicago: American Bar Association, 1983) 129 at 132.

86 See ICAO, C-MIN 194/2, *supra* note 86 (Chapter 1) at 4 (Statement of Argentina) (discussing that the EU ETS is contrary to Arts 15 and 24 of the *Chicago Convention*). See also Havel & Mulligan, *supra* note 60 (Chapter 4) at 31; Katherine B Andrus, "Beyond Aircraft Emissions: The European Court of Justice's Decision May Have Far-Reaching Implications" (2012) 24:4 Air & Space L 1 (characterizing the EU ETS as a "tax on fuel" at 16).

87 Green Air Online, "US Legislation to Prohibit Airlines from Joining EU ETS Moves a Step Closer as Senate Bill Receives Bipartisan Support" (8 March 2012), online: Green Air Online <www.greenaironline.com/news.php?viewStory=1436>.

88 "The Carbon Caper", *Travel Weekly* (9 January 2012), online: <www.travelweekly.com/Editorials/The-carbon-caper/>.

89 In their dire attacks against the EU ETS, critics have often used labels such as a "thinly tax on commercial aviation" and a "tax scheme." See Statement of Captain Sean Cassidy, First Vice President Air Line Pilots Association Before the Committee on Commerce, Science and Transportation U.S. Senate on The European Union Emissions Trading System (6 June 2012); Statement of Nancy N Young Vice President of Environmental Affairs Airlines for America (A4A) before the Senate Committee on Commerce, Science and Transportation (6 June 2012).

either at the national or local level.[90] In contrast, a "charge" is a levy designed and destined to recover the cost associated with the provision of services and facilities related to civil aviation.[91] Stemming primarily from ICAO's non-binding guidance materials,[92] this classification is insufficient and does not provide much guidance on where the EU ETS falls.[93] In the section that follows, I examine the genesis of the scheme and discuss whether it constitutes an impermissible tax.

6.2.1 *Understanding the Genesis of ETS: An Alternative to Taxes*

In the 1920s, renowned British economist Arthur Cecil Pigou conceived the idea that taxes should be imposed on activities that generate negative externalities such as pollution.[94] By placing a price, taxes internalize the external cost that otherwise polluters would not bear.[95] Although initially Pigou's proposal did not gain much traction, it later became the cornerstone of the "polluter pays" principle.[96] Thus, much of the scholarship on internalizing the costs of environmental externalities has centered on Pigouvian taxes. A carbon tax, for instance, falls under the notion of a Pigouvian tax.[97] Those in favor of taxes highlight simplicity, cost certainty, and ability to send a price signal to correct market deficiencies.[98]

---

90 See also Xu Yan, "Green Taxation in China: A Possible Consolidated Transport Fuel Tax to Promote Clean Air?" (2010) 21 Fordham Envtl. L Rev 295 at 300.

91 See ICAO Doc 9082/7, *ICAO's Policies on Charges for Airports and Air Navigation Service*, 7th ed (2004).

92 See ICAO Doc 8632, *ICAO's Policies on Taxation in the Field of Air Transport* 3rd edn (2000); ICAO Doc 9082, *ICAO's Policies on Charges for Airports and Air Navigation Services* 9th edn (2012).

93 Some European countries have already adopted "charges" to address local aircraft emission issues. See Janina D Scheelhaase, "Local Emission Charges – A New Economic Instrument at German Airports" (2010) 16 J Air Transport Manag 94 at 98. Likewise, some experts favored the use of "effluent charges" to tackle aircraft noise abatement problems. See Jerold B Muskin, "An Effluent Charge Approach to Aircraft Noise Abatement" (1978) 5 J Env Eco & Manag 333.

94 See Arthur Cecil Pigou, *The Economics of Welfare*, 4th edn (London, Macmillan, 1932).

95 See Inho Choi, "Global Climate Change and the Use of Economic Approaches: The Ideal Design Features of Domestic Greenhouse Gas Emissions Trading with an Analysis of the European Union's CO_2 Emissions Trading Directive and the Climate Stewardship Act" (2005) 45 Nat Resources J 865 at 879; Paul Ekins et al, "Increasing Carbon and Material Productivity through Environmental Tax Reform" (2012) 42 Ener Pol'y 365 at 366.

96 Jeff Pope & Anthony D Owen, "Emission Trading Schemes: Potential Revenue Effects, Compliance Costs and Overall Tax Policy Issues" (2009) 37 Ener Pol'y 4595 at 4596. In the 1970s, the Organization for Economic Co-operation and Development (OECD) had already embraced this principle. See Par Kågeson, *Getting the Prices Right: A European Scheme for Making Transport Pay its True Costs* (Stockholm: European Federation for Transport and Environment, 1993) at 41. The OECD noted that the principle seeks to allocate the "costs of pollution prevention and control measures [to the person producing the pollution in order to] encourage rational use of scarce environmental resources." OECD, *The Polluter Pays Principle: Definition Analysis Implementation* (Paris: OECD, 1975) at 12.

97 See Boqiang Lin & Xuehui Li, "The Effect of Carbon Tax on Per Capita CO_2 Emissions" (2011) 39 Ener Pol'y 5137 at 5138.

98 See Reuven S Avi-Yonah & David M Uhlmann, "Combating Global Climate Change: Why a Carbon Tax is a Better Response to Global Warming than Cap and Trade" (2009) 28 Stan Envtl LJ 3 at 43.

Critics, on the other hand, identify the likelihood of higher production costs, potential weakening of competiveness, and political resistance as some of the main drawbacks.[99] There is little doubt that the ETS is much more complex in its design than a (simple) carbon tax.[100]

Later, in the 1960s, Nobel Laureate economist Ronald Coase noted that Pigouvian taxes did not necessarily lead to an efficient allocation of resources.[101] Unlike Pigou, Coase argued that the purpose of (environmental) policy should not only be to "restrain the activity which produces the externality" but also to maximize the value of production.[102] Opposing the notion of governmental regulation through the imposition of taxes to correct negative externalities such as pollution, Coase thought that, if allowed to bargain, parties can best allocate externalities through private arrangements to obtain a more optimal allocation of resources.[103] However, this would only take place to the extent that the cost of maximizing the value of production is greater than the cost of carrying out the market transaction (i.e. the transaction cost).[104] In Coase's words, "what has to be decided is whether the gain from preventing the harm is greater than the loss which would be suffered elsewhere as a result of stopping the action which produces the harm."[105]

One of the underlying assumptions in Coase's theory is that private parties enjoy property rights with which they may bargain. Coase, however, did not address a situation in which parties with conflicting interests would interact in an unrestricted *common* environment where no one enjoys property rights. Thus, in 1968, building on the foundations of Coase, University of Toronto Professor John H Dales developed the notion of "transferable property rights" to dispose waste in the province of Ontario.[106] Like Coase, Dales recognized that, for a better allocation of resources to occur, parties must not only be allowed to bargain but also need to be granted property rights for the use of some scarce

99 Lin & Li, *supra* note 97 at 5138.

100 See Pope & Owen, *supra* note 96 at 4597.

101 See Ronald Coase, "The Problem of Social Cost" (1960) 3 JL & Econ 1 at 2.

102 *Ibid.* at 27.

103 *Ibid.* at 4. See also Bodansky, *The Art and Craft of International Environmental Law*, *supra* note 62 (Chapter 1) at 49.

104 Coase, *supra* note 101.

105 *Ibid.* at 27. To demonstrate that Pigouvian taxes do not necessarily lead to an optimal solution, Coase provided the following example: Let us think of a factory that while emitting smoke causes damages to those in adjacent properties valued at 100 US dollars. As a result of this action, the government decides to enact a Pigouvian tax to all factories producing similar smokes. The amount of the tax is set at 100 US dollars. In order not to pay the tax, the factory's management finds a smoke-preventing device for 90 US dollars. By installing such device, the factory would be better off by 10 US dollars. However, let us assume that the cost of moving elsewhere for those in adjacent properties is only 40 US dollars. An offer of 50 US dollars to reallocate to another place will mean that those in adjacent properties will be better off by 10 US dollars and the factory by 50 US dollars. *Ibid.* at 41.

106 JH Dales, *Pollution, Property & Prices* (Toronto: University of Toronto Press, 1968) at 71. See also Isabel Rauch, "Developing a German and an International Emissions Trading System: Lessons from U.S. Experiences with the Acid Rain Program" (2000) 11 Fordham Envtl LJ 307 at 308.

resources (such as air and water) for the market transaction to take place; otherwise, "they have nothing to negotiate with."[107] The allocation of property rights on scarce resources facilitates the internalization of pollution cost.[108] Under Dales' model, the government would set levels of pollution and allocate rights to individuals to discharge waste. Later, in 1972, Montgomery advanced the concept by advocating the creation of a market in emissions licenses to control pollution.[109] As is the case under the EU ETS, through property rights, governments create artificial markets.[110] Those capable of reducing their own waste consumption could sell their excess rights to other interested parties. Bargaining between parties allows for a better allocation of resources at "the smallest possible total cost to society."[111] The works of Coase, Dales, and Montgomery helped to build the foundations of ETS.[112]

Initially, environmental policy makers were skeptical about the convenience of introducing MBMs such as tradable property rights as a means for controlling pollution. Aside from academic studies, there was not much familiarity with economic instruments.[113] In fact, policy makers perceived them as granting a "right to pollute" with no certainty that environmental goals would be achieved.[114] Regulators preferred command-and-control regulation,[115] which sought to encourage firms to use the best available technology to control pollution.[116] However, as pollution costs increased, so did dissatisfaction with such policy option.[117] The requirement to use a given technology to meet a predetermined environmental goal proved extremely inflexible and expensive.[118] For instance, in the

107 *Ibid.*

108 See Peter H Pearse, "Developing Property Rights as Instruments of Natural Resources Policy: The Case of the Fisheries" in OECD, *Climate Change: Designing a Tradable Permit System* (Paris: OECD, 1992) 110.

109 See W David Montgomery, "Markets in Licenses and Efficient Pollution Control Programs" (1972) 5 J Econ Theory 395.

110 See Kirk W Junker, "Ethical Emissions Trading and the Law" (2005) 13 U Balt J Envtl L 149 at 174.

111 Dales, *supra* note 106 at 107.

112 See Harro van Asselt, "Emissions Trading: The Enthusiastic Adoption of an "Alien" Instrument?" in Jordan et al, *supra* note 1 (Chapter 4) 125 at 126; Johannes Stripple and Eva Lövbrand, "Carbon Market Governance Beyond the Public-Private Divide" in Frank Biermann, Philipp Pattberg & Fariborz Zelli, eds, *Global Climate Governance Beyond 2012: Architecture, Agency and Adaptation* (Cambridge: Cambridge University Press, 2010) 165 at 169; Ralf Antes, Bernd Hansjürgens & Peter Letmathe, "Introduction" in Ralf Antes Bernd Hansjürgens & Peter Letmathe, eds, *Emissions Trading and Business* (Heidelberg: Physica-Verlag, 2006) 1 at 2.

113 See Robert W Hahn, "Marketable Permits: What's all the Fuss About?" (1982) 2:4 J Pub Pol'y 395. [Hahn, "Marketable Permits"].

114 See Speth, *supra* note 102 (Chapter 4) at 94.

115 See Carol M Rose, "Hot Spots in the Legislative Climate Change Proposals" (2008) 102 Nw UL Rev Colloquy 189 at 189; Scott D Deatherage, *Carbon Trading: Law and Practice* (Oxford: Oxford University Press, 2011) at 16.

116 See Speth, *supra* note 114 at 93.

117 See Hahn, "Marketable Permits", *supra* note 113 at 399.

118 See Susan J Kurkowski, "Distributing the Right to Pollute in the European Union Efficiency, Equity, and the Environment" (2006) 14 NYU Envtl LJ 698 at 704; Tim Jackson, *Efficiency Without Tears: No-Regrets*

United States, emissions performance standards (EPS), a type of command-and-control regulation, were extremely helpful to address local air pollution issues.[119] However, EPS impose emission reduction levels upon polluting firms without considering the cost implications thereof.[120] It was clear that other alternatives were warranted.[121]

Regulators distributed tradable permits over fisheries that previously were subject to open exploitation.[122] Likewise, in the United States, the Environmental Protection Agency (EPA) started to gradually incorporate economic instruments into environmental policy, in particular for its Acid Rain Program.[123] Thus, the US *Clean Air Act* amendments of 1990[124] introduced ETS to cap sulfur emissions from power plants in order to tackle acid rain pollution.[125] A number of economists underscored the theoretical efficiency advantages of ETS.[126] Similarly, regulators and politicians preferred ETS not only because it provided more flexibility to participating firms[127] and was a cost-effective means of addressing pollution but mainly because it was also a more politically palatable policy instrument as compared to taxes which are considered abhorrent.[128] Stakeholders demonstrated a keenness

---

Energy Policy to Combat Climate Change (London: Friends of the Earth, 1992) (noting that command-and-control regulation is not cost effective because "it ignores the efficiency of the market" at 30).

119 See Peter Radgen, Jane Butterfield, & Jürgen Rosenow, "EPS, ETS, Renewable Obligations and Feed in Tariffs: Critical Reflections on the Compatibility of Different Instruments to Combat Climate Change" (2011) 4 Ener Procedia 5814 at 5816.

120 *Ibid.*

121 See Robert W Hahn & Robert N Stavins, "Incentive-Based Environmental Regulation: A New Era from an Old Idea" (1991) 18 Ecology LQ 1 at 27 [Hahn & Stavins, "Incentive-Based Environmental Regulation"].

122 See Pearse, *supra* note 108 at 109.

123 See Evan Goldenberg, "The Design of an Emissions Permit Market for RECLAIM: A Holistic Approach" (1993) 11 UCLA J Envtl L & Pol'y 297 at 297; Nancy Kete, "Air Pollution Control in the United States: A Mixed Portfolio Approach" in Ger Klaassen & Finn R Førsund, *Economic Instruments for Air Pollution Control* (Dordrecht: Kluwer Academic Publishers, 1994) (discussing that, in 1976, EPA established "four-prong emissions trading" that included offsetting, netting, bubbles, and banking) at 131.

124 In addition to laying out EPA's responsibilities, the US Clean Air Act tackles issues of air quality and stratospheric ozone layer. See 42 USC § 7401.

125 See Kathryn C Wilson, "The International Air Quality Management District: Is Emissions Trading the Innovative Solution to the Transboundary Pollution Problem" (1995) 30 Tex Int'l LJ 369 at 379; Nancy Kete, "Air Pollution Control in the United States: A Mixed Portfolio Approach" in Ger Klaassen & Finn R Førsund, *Economic Instruments for Air Pollution Control* (Dordrecht: Kluwer Academic Publishers, 1994) at 139.

126 See Holly Doremus & W Michael Hanemann, "Of Babies and Bathwater: Why the Clean Air Act's Cooperative Federalism Framework is Useful for Addressing Global Warming" (2008) 50 Ariz L Rev 799 at 801.

127 See Robert W Hahn, "Book Review: Regulatory Reform at EPA: Separating Fact from Illusion Reforming Air Pollution Regulation: The Toil and Trouble of EPA's Bubble by Richard A Liroff" (1986) 4 Yale J on Reg. 173 at 174.

128 See Heller, *supra* note 191 (Chapter 2) at 118; Soltau, *supra* note 192 (Chapter 1) (discussing that there is little appetite for taxes for environmental purposes) at 73. Although recognizing that the system gives participants the option to reduce emissions where it is least costly, ETS pundits argue that such abatement may not necessarily take place where it is most beneficial to a given society. It may well be that an abatement project represents the maximization of resources from an economic point of view but, for the affected community, constitutes the least desirable outcome. See Lily N Chinn, "Can the Market Be Fair and Efficient? An Environmental Justice Critique of Emissions Trading" (1999) 26 Ecology LQ 80 at 116.

for ETS over taxes.[129] This is one of the main reasons why, in 1989, US President George HW Bush supported ETS over command-and-control regulation and proposed its inclusion into the 1990 amendments to the *Clean Air Act.*[130] A Republican-led administration introduced ETS in the US environmental policy landscape precisely as an alternative to command-and-control regulation and Pigouvian taxes. It was the experience of implementing ETS in the Acid Rain Program that prompted the United States to push for its inclusion in the *Kyoto Protocol.* After extensive US lobbying, Europe grudgingly accepted ETS at the eleventh hour.[131]

There is significant consensus that the origin of ETS may be traced back to Coase's treatment of negotiated solutions to (environmental) externalities.[132] Coase articulated his whole theory as a rebuttal to Pigou's ideas. Dales and Montgomery significantly advanced Coase's initial proposal by providing additional elements to address externalities generated in an unrestricted *common.* Arguably, their overriding intention was to offer policy alternatives other than taxes to internalize externalities through market transactions in a more efficient manner.

6.2.2 *Does Article 15 of the* Chicago Convention *Prohibit the EU ETS?*

Under Article 15 of the *Chicago Convention,* States cannot impose fees, dues, or other charges upon aircraft from other States for "the sole right of transit over or entry into or exit from [their] territory."[133] There are at least three conflicting interpretations on the scope of this prohibition.[134] A literal interpretation would imply that no levies of any kind may be levied when an aircraft flies into, out of, or over a given State. However, through guidance materials, ICAO has clarified that States do have the right to recover the costs of services rendered to such aircraft.[135] What is forbidden is the imposition of a levy on

129 See Hahn & Stavins, "Incentive-Based Environmental Regulation," *supra* note 121 (commenting that "interest group attitudes towards proposals for sharp cuts in sulfur dioxide emissions can vary dramatically depending on the mechanism that is selected to reach that goal" at 4).

130 See Environmental Protection Agency, "The Clean Air Act Amendments of 1990", *Environmental Protection Agency* online: <www.epa.gov/air/caa/caaa_overview.html>; see President George HW Bush "Statement on Signing the Bill Amending the Clean Air Act" *The American Presidency Project* (15 November 1990) online: UC Santa Barbara <www.presidency.ucsb.edu/ws/index.php?pid=19039>.

131 See Soltau, *supra* note 192 (Chapter 1) at 84.

132 See Hahn & Stavins, "Incentive-Based Environmental Regulation", *supra* note 121 at 8; Larry Lohmann, "Toward a Different Debate in Environmental Accounting: The Cases of Carbon and Cost-Benefit" (2009) 34 Accounting, Organizations and Society 499 at 504.

133 *Chicago Convention, supra* note 36 (Chapter 2) at Art 15 (3).

134 See ICAO Doc 9562, *Airport Economics Manual* 2nd edn (2006) at 1-1.

135 *Ibid.* at 1-2.

account of the authorization granted to the aircraft to fly into, out of, or over the territory of a State.[136]

The second (broadly construed) interpretation would suggest that levies are permitted only to the extent that a service is rendered or a facility is provided to the levied user in connection with the operation of an aircraft entering into, exiting from, or overflying the territory of the levying State. If aircraft operators' obligation to surrender sufficient allowances for GHGs emitted within a given compliance period was categorized as a "levy," the EU ETS would contravene Article 15. This would be so because the allowances or the requirement to purchase additional emission credit units is not associated with a service or facility.[137]

A third (simplistic) interpretation would posit that the subject matter of Article 15 essentially deals with charges for the use of airports and air navigation facilities and, as well, lays down some additional rules (e.g. that charges must be imposed in a non-discriminatory manner).[138] The EU ETS does not relate in any way to airport or air navigation charges.[139] Thus, the scheme falls outside of its scope.[140] As Advocate General Kokott noted, Article 15 of the *Chicago Convention* addresses airport access and it is therefore not applicable to the EU ETS.[141]

6.2.3 *Jennison's Functional Equivalency Theory*

Advocating the broadly construed interpretation, Jennison notes that the EU ETS imposes costs upon and limits the operations of aircraft operators.[142] Aircraft operators would incur costs by having to purchase additional allowances or emission credit units in order to

136 See Michael Milde, "Can Airlines Get Any Relief" IATA Legal Symposium, Bangkok, Thailand, 10 February 2009; Robert Lawson "UK Air Passenger Duty Held to be Consistent with the *Chicago Convention*" (2008) 33:1 Air & Space L 1; Pache, *supra* note 97 (Chapter 4) (indicating that Art 15 of the *Chicago Convention* "prohibits only such charges as are levied solely for the right to transit, enter or exit" at 23).

137 See Macintosh, *supra* note 100 (Chapter 1) (noting that the EU ETS generates revenues that are not to "defray the cost of facilities and services" at 417); Gabriel S Sanchez, "In Defense of Incrementalism for International Aviation Emissions Regulation" (2012) 53:1 Va J Int'l L 1 (emphasizing that the EU ETS may be "construed as an illicit charging mechanism that is also inconsistent with the *Chicago Convention*" at 4); Bartlik, *supra* note 68 (Chapter 4) (saying that the requirement to purchase allowances are "*de-facto* payments made to obtain the right to enter into or exit from the airspace of the EU when arriving or departing from an aerodrome situated in the territory of an EU Member State" at 167).

138 See PPC Haanappel, "The Impact of Changing Air Transport Economics on Air Law and Policy: A Short Commentary (2009) XXXIV Ann Air & Sp L 519 at 523.

139 See Pache, *supra* note 97 (Chapter 4) (suggesting that the EU ETS does not fall under the scope of Art 15) at 21.

140 See Manzini & Masutti, *supra* note 136 (Chapter 4) at 324.

141 See Opinion of Advocate General Kokott, *supra* 136 (Chapter 4) at I-56.

142 See Jennison, EU ETS, *supra* note 393 (Chapter 5).

comply with their obligation at the end of each compliance period.[143] Such costs are not related to providing aeronautical facilities or services.[144] Therefore, in Jennison's view, the EU ETS is the "functional equivalent of [an impermissible] levy" under Article 15.[145] This analysis suggests that under the *Chicago Convention*, taxes and levies are forbidden unless they are linked to services for the provision of facilities. State practice, however, does not support this interpretation. For instance, countries such as Argentina and Chile levy upon arrival a reciprocity fee to passengers whose countries of origin require a visa to their nationals to obtain a visa.[146] One could argue that this is in fact an admission fee for the sole right to enter the country. Costa Rica and Paraguay levy international embarkation taxes and/or earmark part of the revenues to finance the promotion and development of their tourism ministries.[147] Australia's Customs and Border Protection Services levies a 55-dollar passenger movement charge for every passenger departing from Australia by air.[148] There is no indication that these levies are used to defray the cost of providing services or facilities.

When deciding to impose a given tax or a levy associated with international aviation, states do not seem to be constrained in any way by the limitations prescribed in Article 15 of the *Chicago Convention*. Even if one were to assume that this provision only permits the imposition of levies that are linked to the provision of services or facilities, it does not in any way provide guidance concerning what the level of such levies should be. The *Chicago Convention* does not establish any "test of reasonableness" to determine if the amount of such levy is "commensurate with the costs of the services provided."[149] If the sole requirement is to tie the levy to services or facilities, in theory at least, a State could impose an exorbitant tax where 90 percent of the revenues are allocated to purposes other than aviation-related activities (e.g. to finance climate change adaptation in developing countries) and only 10 percent are allocated to defray the costs of aviation. Strictly speaking, this would keep the levying State in compliance with the requirement that the tax be (partially) linked to services and facilities.

However, it does not seem reasonable to conclude that this is the real intention underlying the provisions of Article 15. What this prohibition purports to avoid is a situa-

143 *Ibid.*

144 *Ibid.*

145 *Ibid.* See also Mayer, Case -366/10, *supra* note 133 (Chapter 1) at 1135; Stephanie Koh, *supra* note 45 (Chapter 2) at 134.

146 See Charles E Smith, "Air Transportation Taxation: The Case for Reform" (2010) 75 J Air L & Com 915 at 939 [Charles Smith].

147 See *Law No 1331/88, Which Modifies Law No 85*, 16 December 1991 at Art 1 (Paraguay); *Presidential Decree No 793/2008, Which Establishes the International Embarkation Tax by Air for Each Passenger* (Paraguay).

148 See *Passenger Movement Charge Act of 1978*, Act No 118, online: Australian Government <www.comlaw.gov.au/Details/C2012C00605> at Art 6. See also, *Passenger Movement Charge* online: Australian Custom and Border Protection Service <www.customs.gov.au/site/page6068.asp>.

149 See Charles Smith, *supra* note 146 at 940.

tion in which a State would impose a levy as a "right of easement" or "right of way" for simply granting access to its territory (e.g. transit, entry, or exit). It should not be construed as prohibiting the levying of any type of taxes not associated with the provision of a service or facility.[150] As Advocate General Kokott indicated, it is unlikely that, with Article 15, the drafters of the *Chicago Convention* intended to preclude the possibility of levying all other taxes, such as those for environmental purposes, not directly related to international aviation activities.[151] In light of the foregoing, it is reasonable to assert that even if the EU ETS were to be considered a tax, Article 15 would not forbid it.

In addition, Jennison's functional equivalence theory incorrectly associates "cost" with a tax. Supporting this line of reasoning, Bartlik contends that the EU ETS "introduces a system that results in financial burdens equal to a direct imposition of a fee, duty or other charge."[152] In my view, this is an oversimplification. The fact that the scheme may potentially (and most definitely not in all cases) generate a cost to aircraft operators does not automatically turn such cost into a levy or a tax.[153] If this were the case, complying with complex and stringent regulatory requirements, such as those of the *Sarbanes-Oxley Act* of 2002[154] for companies listed in the New York Stock Exchange, would be tantamount to a "tax" on the sole basis that these requirements impose costs on regulated agents. Under this line of reasoning, almost every regulatory requirement would be the "functional equivalent" of a tax. This assimilation also presupposes that the cost of the scheme will, as a matter of necessity, be assumed entirely by aircraft operators and that they will not be able to pass it on to passengers and freight forwarders.

6.2.4 *Why the EU ETS Is Not a Tax*

Most commentators outside aviation circles contend that the EU ETS is not a tax.[155] Similarly, it cannot be regarded as a charge either, given that it is not linked to the use of airports

150 *Ibid.* (commenting that the "*Chicago Convention* fails to prohibit aviation taxes and user fees" at 939).

151 See Brian Havel & Niels van Antwerpen, "Dutch Ticket Tax and Article 15 of the *Chicago Convention* (continued)" (2009) 34:6 Air and Space L at 449.

152 Bartlik, *supra* note 68 (Chapter 4) at 166.

153 See Annie Petsonk & Adam Peltz, "Why Europe's Climate Program for Airlines is not a Tax", *EDF* (13 February 2012) online: <http://blogs.edf.org/climatetalks/2012/02/13/why-europe percent_E2_per cent80 per cent99s-climate-program-for-airlines-is-not-a-tax> [Petsonk, "EU ETS is not a Tax"].

154 See *Sarbanes-Oxley Act*, Pub L 107-204, 116 Stat 745 (2002).

155 See Macintosh, *supra* note 100 (Chapter 1) (advocating that the EU ETS is neither a tax nor a levy, but rather a "regulatory requirement concerning transferable property rights" at 417); Petsonk, "EU ETS is not a Tax", *supra* note 153 (suggesting that the EU ETS is a "market-based cap on pollution that lets [aircraft operators] find the best and cheapest way to reduce emissions."); Bartels, *supra* note 117 (Chapter 4) (noting that the EU ETS is not a fiscal measure nor a tax either); Pache, *supra* note 97 (Chapter 4) (arguing that EU ETS is not a tax) at 15; Written Observations of the Second Interveners in the Court of Justice of the European Union, reference from the High Court of Justice, London, Case C-366/10 (underscoring that the EU ETS is not a tax but rather a market-based measure) at 3; Written Observations of the United Kingdom in the Court

or air navigation facilities but rather to the "output of emissions."[156] Also, the fact that ICAO distinguishes between emissions-related charges and emissions trading clearly suggest that the latter constitutes something different.[157] Yet if the EU ETS is not an operational standard, a tax, or a charge, the question then arises as to its intrinsic legal nature. Many commentators, particularly in the field of economics, regard ETS as an economic instrument or an MBM designed to achieve a predetermined environmental goal. Although this situates ETS as an environmental regulatory instrument option,[158] it does not provide much clarification about its most salient legal characteristics. Likewise, this does not help to distinguish the scheme from taxes, which also form part of that genre.

In this vein, Macintosh's conceptualization of the EU ETS is most enlightening. For Macintosh, the EU ETS is a "regulatory requirement concerning transferable property rights."[159] This would seem more in line with the categorization used by the founding fathers of ETS (Coase, Dales, and Montgomery). However, on its face, the notion of "transferable property rights" is not self-explanatory. To understand this, one must resort to an examination of one of the pivotal elements of the scheme: the "allowances."[160] This is a foreign concept to air law.[161] Here, it is worth recalling some of its basic features: Based on past emissions, the regulator sets a cap and distributes allowances free of charge to participants; the allowances are exchangeable among participants; in most cases, these allowances cover a percentage but not all of the participants' emissions within a given compliance period; for their remaining emissions, participants must purchase additional allowances either from fellow participants or from government-run auctions; in order to comply with their obligations, participants may also purchase ECUs from the market.

---

of Justice of the European Union, reference from the High Court of Justice, London, Case C-366/10 (suggesting that the EU ETS "is not a charge for the use of airports or a means of raising revenues from fuel" at 7); Delbeke, *supra* note 7 (Chapter 4) (underscoring that the EU ETS is neither a tax nor a charge).

156 Pache, *supra* note 97 (Chapter 4) at 14; Hardeman, "Common Approach", *supra* note 113 (Chapter 1) (noting that the EU ETS cannot be a charge since it is a "commercial transaction which is entirely unrelated to the provision or use of airport or air navigation facilities" at 12).

157 See ICAO, A36-22, *supra* note 17 (Chapter 2) at Appendix L.

158 According to Stewart, environmental regulatory instrument choices include the following: (i) command-and-control regulation, (ii) economic instruments, (iii) information-based approaches, and (iv) hybrid regulatory approaches. Among others, economic instruments may encompass taxes, tradable pollution rights (such as ETS), subsidies, property rights in natural resources, and allocation of liability rules. See Richard B Stewart "Instrument Choice" in Daniel Bodansky, Jutta Brunnée & Ellen Hey, eds, *The Oxford Handbook of International Environmental Law* (Oxford: Oxford University Press, 2006) 147 at 152.

159 See Macintosh, *supra* note 100 (Chapter 1) at 417.

160 For ICAO, an "allowance" or an "emission allowance" is a "tradable emission permit that can be used for compliance purpose in an emissions trading system. An allowance grants the holder the right to emit a specific quantity of pollution once." See ICAO Doc 9885, *Guidance on the Use of Emissions Trading for Aviation*, *supra* note 89 (Chapter 3) at ix.

161 Pache rightly points out that the bilaterals do not address the distribution of allowances. See Pache, *supra* note 97 (Chapter 4) at 35.

In this context, the "allowance" bears two distinctive legal features. First, it is a license that the regulator grants to participants to emit certain GHGs up to a specified level (e.g. 1 allowance = 1 tonne of CO_2).[162] Second, the ability to transfer or exchange the allowances transforms such licenses into de facto property rights that participants enjoy.[163] Participants may sell or purchase allowances almost without restrictions. In economic terms, the allowances become a tradable asset that may be quantified in the participant's accounting books.[164] In fact, in accordance with standard no 38 (intangible assets) of the International Accounting Standards (IAS), allowances should be recorded as "intangible assets."[165] They may also serve as collateral for credit purposes. As Soltau notes, a tax on the other hand "extracts revenue."[166]

For the most part, the price of allowances is subject to market forces and it is determined by supply and demand.[167] Although this is not a characteristic feature of the EU ETS, in theory, the regulator may set floors and ceilings for the prices of allowances that are available through auctions.[168] In any event, regulatory intervention would not affect those allowances that are exchanged outside these auctions. This is one of the main distinctions from a tax scheme where the government sets the tax rate. Andrus has criticized those drawing this distinction on the ground that it presupposes that a tax has always a fixed base.[169] According to Andrus, there are taxes, such as real estate tax, where market conditions may cause fluctuations in the amount to be paid to the tax authority. As in the case of the EU ETS, the market intervenes to determine the outcome of the tax.[170] This criticism confuses the tax rate with the final amount that the tax payer ought to pay. With respect to almost all taxes, market forces do play a significant role. For example, supply and demand may reduce or increase tax exposure (e.g. income tax, sales tax, luxury tax). However, the government sets the tax rate and not the final amount owed. In the EU ETS, the regulator plays a much lesser role in the establishment of the price for allowances or ECUs.

Another major distinction between the EU ETS and taxes is that, under the ETS, revenues raised from the exchange of allowances do not necessarily go to the general coffers

162 See Daniel Cole, "What's Property Got to do With It? A Review Essay by David M Driesen of Pollution & Property: Comparing Ownership Institutions for Environmental Protection" (2003) 30 Ecology LQ 1003 (advocating that "the term allowance itself suggests that a government license to pollute is at issue, rather than ownership right" at 1008).

163 See Robert Howse, "The Political and Legal Underpinnings of Including Aviation in the EU ETS" in Bartels, *supra* note 117 (Chapter 4) [Howse, EU ETS].

164 See Deatherage, *supra* note 115 at 21.

165 Edeltraud Günther, "Accounting for Emission Rights" in Ralf Antes Bernd Hansjürgens & Peter Letmathe, eds, *Emissions Trading and Business* (Heidelberg: Physica-Verlag, 2006) 219 at 230.

166 See Soltau, *supra* note 192 (Chapter 1) at 73.

167 See Howse, EU ETS, *supra* note 163.

168 See Hahn, "Greenhouse Gas," *supra* note 197 (Chapter 1) at 172.

169 See Andrus, *supra* note 86 at 16.

170 *Ibid.*

of EU Member States, whereas taxes are their main source of funding.[171] Revenues are only kept by the governments when participants purchase allowances from government-run auctions. Here, the government raises revenues for issuing licenses to emit a certain amount of GHGs.[172] Just like governments issue a number of other licenses or permits (e.g. driving licenses). However, aircraft operators can trade allowances between themselves without any gains being made thereon by the governments of EU Member States.[173] As mentioned above, there is no obligation to purchase allowances from these auctions.[174] Airlines can entirely comply with their obligations without allocating a penny to EU Member States. They may completely bypass the auctions. A tax scheme does not allow tax payers the choice of who to credit. The revenue is always received by the government. As economic instruments, both taxes and ETS may provide price signals to correct environmental externalities.[175] From a purely economic viewpoint, where 100 percent of all allowances are auctioned to participants, an ETS may initially assimilate to a tax.[176] In this case, all revenues will go to the government. No allowances are allocated free of charge. This would only affect the initial allocation of allowances. It would not prevent participants from engaging in secondary trading of allowances.

One must also bear in mind the likelihood that aircraft operators may charge for the opportunity cost of obtaining allowances free of charge. In other words, the EU ETS may create an opportunity for windfall profits. If the scheme were in fact a tax, and aircraft operators decided to charge for opportunity cost, in most jurisdictions such action would probably constitute a criminal offense in the form of fraud, false pretense, or exaction. It is simply inconceivable that, while recouping from or passing the cost of a tax through to final consumers, a given tax payer would also dare to charge for opportunity cost and seek to make a profit. Taxes do not offer this alternative. In fact, one of the main advantages of

---

171 See Macintosh, *supra* note 100 (Chapter 1) at 417.

172 *Ibid.*

173 See Written Observations of the Second Interveners in the Court of Justice of the European Union, Reference from the High Court of Justice, London, Case C-366/10 at 3.

174 *Ibid.*

175 See Jane Andrew, Mary A Kaidonis & Brian Andrew, "Carbon Tax: Challenging Neoliberal Solutions to Climate Change" (2010) 21 Critical Perspectives on Accounting 611 at 613.

176 See Hardeman, "Common Approach", *supra* note 113 (Chapter 1) at 12; Stefan Fölster & Johan Nyström "Climate Policy to Defeat the Green Paradox" (2010) 39 AMBIO 223 at 224 (noting that "if emission rights are auctioned off, [an ETS] is much the same as a tax"); Koushik Ghosh & Peter Gray, "Rushing to Copenhagen? Is Cap-and-Trade the Answer? (2010) Challenge (January/February) 5 (commenting that an ETS and a tax may have the "same distributional [effect, as long as the allowances are auctioned off] by the government, [and that they] generate [equal] revenues" at 17); Inho Choi, "Global Climate Change and the Use of Economic Approaches: The Ideal Design Features of Domestic Greenhouse Gas Emissions Trading with an Analysis of the European Union's CO_2 Emissions Trading Directive and the Climate Stewardship Act" (2005) 45 Nat Resources J 865 (recognizing that in essence an ETS and a tax are different. However, they can be assimilated depending on design features such as auctioning) at 900.

the design of an ETS is precisely the array of different options provided to participants to attain a given environmental objective at the least cost.[177]

6.2.5 *Does the EU ETS Contravene Article 24 of the* Chicago Convention*?*

Those who view the EU ETS as a tax also argue that the taxable subject matter is fuel, given that emissions are computed on the basis of fuel consumption.[178] These critics point out that Article 24 of the *Chicago Convention* exempts from taxation any fuel carried on board an aircraft upon arrival and retained on board upon leaving the territory of another ICAO Member State.[179] Similarly, critics note that in their various clauses that address customs duties and charges, more than 4,000 bilaterals exempt fuel uplifted for international flights from taxation. To this end, the scheme would be impermissible under international law and existing bilaterals.

In order to interpret Article 24, one must understand where it falls within the *Chicago Convention*. Along with clauses on formalities,[180] customs and immigration procedures,[181] and exemption from seizure on patent claims,[182] Article 24 forms part of a chapter of the *Chicago Convention* on "measures to facilitate air navigation."[183] The provision seeks to avoid the levying of customs duties on goods on board the aircraft once that aircraft arrives in the territory of another State. The exemption also covers "inspection fees" and "similar national or local duties and charges."[184] Its underlying rationale is to prevent a situation in which the State of arrival would apply its customs levies to goods on board an aircraft that are only temporarily subject to its jurisdiction. Thus, the exemption was conceived as a necessary measure to ensure that the aircraft could continue with its operations without having to be assessed for customs duties or similar charges on goods on board (including fuel) every time it reached a foreign destination. This interpretation is in line with the protection that the *Chicago Convention* also affords to aircraft and their operators against potential patent infringement claims when such aircraft enters or transits through the territory of another State.[185] Clearly, the drafters of the *Chicago Convention* intended to facilitate the air navigation of aircraft with these provisions.

---

177 Tol, for instance, underscores that a potential tax on aviation fuel would be much costly to generate any meaningful carbon footprint reduction from international aviation. See Richard SK Tol, "The Impact of a Carbon Tax on International Tourism" (2007) 12 Transportation Research Part D 129 at 138.

178 See Written Observations of the Claimants in the Court of Justice of the European Union, Reference from the High Court of Justice, London, Case C-366/10 at 46.

179 See *Chicago Convention, supra* note 36 (Chapter 2) at Art 24.

180 *Ibid.* at Art 22.

181 *Ibid.* at Art 23.

182 *Ibid.* at Art 27.

183 *Ibid.* at Chapter iv.

184 *Ibid.* at Art 24.

185 *Ibid.* at Art 27.

On 24 February 1999, the ICAO Council adopted a consolidated resolution on taxation issues involving air transport in which it stated that "customs and other duties" include "import, export, excise, sales, consumption and internal duties and taxes of all kinds levied upon the fuel, lubricants and other consumable technical supplies."[186] This interpretation only confirms the intrinsic nature of the exemption: taxes and levies imposed on fuel of whatever kind. It does not address the whole universe of regulatory requirements that may demand; for instance, for safety reasons, an aircraft operator must conduct its operation in such a manner that it is likely to consume more fuel. Although the objective is generally to optimize operations and reduce fuel consumption to the greatest extent possible, there might be situations in which safety regulatory requirements impose additional fuel burdens on aircraft operators. For instance, ICAO Annex 6 requires that aircraft operators "carry a sufficient amount of usable fuel, to complete the planned flight safely."[187] To this end, aircraft operators must conduct a preflight calculation of such fuel. The regulator, however, always retains the prerogative to introduce operational variations to such calculations.[188] For instance, the regulator may include new critical fuel scenarios to account for a sudden engine failure or loss of pressurization.[189] The regulator may also impose additional requirements for extended diversion time operations (EDTO). Both of these safety-related operational requirements compel the aircraft operator to uplift additional fuel at the point of departure.[190] This entails additional costs for aircraft operators. Yet it is inconceivable to label these measures as "taxes." Again, not all costs or regulatory requirements associated with fuel consumption are taxes or levies. Therefore, given that their subject matters are acutely different, it does not seem reasonable to infer that the exemptions provided in Articles 15 and 24 of the *Chicago Convention* preclude the application of an emissions trading scheme such as the EU ETS. Just like fuel-related operational requirements are not "taxes," an environmental regulatory requirement that deals with tradable permits cannot be categorized as a customs duty, a charge, a levy, an inspection fee, or other similar charges.

Aircraft operators in the United States have criticized this interpretation as an "exercise in linguistic evasion."[191] This could hardly be the case. Even though taxes and ETS may, in some restricted circumstances, possess similar characteristics, they are in essence diametrically opposed legal and economic concepts.

---

186 See ICAO, *Consolidated Council Resolution on Taxation of International Air Transport, adopted on 24 February 1999* [*Council Resolution on Taxation*]. See also ICAO, C-DEC 156/3; ICAO, C-MIN 156/3.

187 See Annex 6 to the *Convention on International Civil Aviation* (Operations), Part I at 4.3.6.1 [Annex 6].

188 *Ibid.* at 4.3.6.6.

189 See generally ICAO Doc 9976, *Flight Planning and Fuel Management Manual* 1st ed (2012).

190 See Annex 6, *supra* note 187 at 4.7.

191 See Written Observations of the Claimants in the Court of Justice of the European Union, Reference from the High Court of Justice, London, Case C-366/10 at 50.

6.2.6 *Is* Braathens *Relevant?*

Proponents of the tax on fuel argument often cite the *Braathens* decision that the CJEU handed down in 1999.[192] In this case, Braathens Sverig AB, a Swedish aviation conglomerate, brought a challenge against the Swedish Tax Authority with regard to a national environmental protection tax levied on domestic aviation.[193] Calculated on the basis of fuel consumption, the tax was levied for each flight at a rate of 1 Swedish krona (SEK) per kilogram of fuel burned.[194] At the time, Council Directive 92/81 harmonized excise duties on mineral oils.[195] The community legislation in question, however, exempted "mineral oils supplied for use as fuels for the purpose of air navigation."[196] The core legal issue in the case was whether the taxable event fell within the exemption provided by the community legislation.[197] The Swedish Tax Authority argued that the taxable event was not fuel, but rather "the polluting emissions of carbon dioxide" generated from domestic commercial aviation activities."[198] Siding with Braathens, the CJEU observed a "direct and inseverable [sic] link between fuel consumption and the polluting substances emitted in the course of such consumption."[199] Fuel consumption was therefore held to be a taxable event.[200] As such, a tax on emissions was deemed as a tax on fuel. The CJEU ruled that community legislation in place at the time prevented the application of the tax.[201]

The CJEU had another chance to consider the elements of Braathens in the most recent *ATA Decision.*[202] US airlines claimed that the EU ETS violates Article 24 of the *Chicago*

---

192 See *Braathens Sverige AB v Riksskatteverket*, Case 346/97, [1999] ECR I-3433 [*Braathens*].

193 *Ibid.* at I-3435.

194 *Ibid.* at I-3438.

195 See EC, *Directive 92/81 of 19 October 1992 on the Harmonization of the Structures of Excise Duties on Mineral Oils* [1992] OJL 316/12.

196 *Ibid.*

197 See Braathens, *supra* note 192 at I-3440.

198 *Ibid.* at I-3441.

199 *Ibid.*

200 The CJEU noted that the tax "must be regarded as levied on consumption of the fuel itself." Braathens, *supra* note 192 at I-3443.

201 *Ibid.* at I-3446. As part of the new community tax framework for energy products and electricity, in 2003, through EC Directive 2003/96, the European Council repealed EC Directives 92/81 and 92/82. The new tax scheme still exempted fuel used for air navigation purposes. It did, however, provide EU Member States the prerogative to limit the tax exemption to international and intra-community flights. This de facto cleared the way for EU Member States to impose taxes on fuel on entirely domestic flights. It also suggested that the tax exemption could be waived in cases where there is a bilateral agreement between EU Member States. See EC, *Council Directive 2003/96/EC of 27 October 2003 Restructuring the Community Framework for the Taxation of Energy Products and Electricity* [2003] OJL 283/51, Art 14. It may be argued that by providing more flexibility to tax aviation fuel under limited circumstances, the legislator acknowledged the pressure from EU (environmental) constituencies, thereby implicitly recognizing the need to gradually incorporate the "polluter pays principle" into the sector.

202 See *ATA Decision, supra* note 63 (Chapter 4) at para 144 at I-13898.

Convention as well as the customs duties provisions of the US-EU Open Skies agreement.[203] In light of the fact that, in the claimant's view, the EU ETS is an impermissible tax, the core holding in *Braathens* that a Swedish GHG tax was a tax on fuel seemed directly relevant.[204] Unfortunately for US airlines, neither the Advocate General nor the CJEU were sympathetic to the arguments they had advanced. Both viewed *Braathens* as not fully applicable to the case.[205] In particular, they noted that the EU ETS as such is not a tax.[206] Conversely, Braathens dealt with a mandatory tax imposed by the Swedish government.[207] Under the EU ETS, operators may purchase allowances from a variety of sources.[208] In addition, the Advocate General was of the view that fuel consumption in and on itself does not lead to a final estimation of the total amount of GHGs that an aircraft operator emits. For calculating this, the EU ETS takes into account an emission factor that also forms part of the equation.[209] Similarly, the CJEU failed to conclude that there was a "direct and inseverable link between the quantity of fuel held or consumed by an aircraft and the pecuniary burden on the aircraft's operator in the context of the allowance trading scheme's operation."[210] In addition, the CJEU found that other factors, such as the market price of allowances, will contribute to determine the cost for aircraft operators.[211] I concur with the position of the CJEU: fuel consumption is one but definitely not the only factor which the schemes take into account in order to distribute allowances among participants.

6.3 A Scheme Violating WTO Rules?

The EU ETS has also been attacked as being inconsistent with rules of the World Trade Organization (WTO).[212] Most of these allegations focus on the effect that the scheme will have on exporters of goods. In general terms, under the WTO regime, a State must provide equal treatment to exports from other States (Most-Favored Nation Treatment principle).[213]

203 See Written Observations of the Claimants in the Court of Justice of the European Union, Reference from the High Court of Justice, London, Case C-366/10 at 51.

204 *Ibid.* at 50.

205 See *ATA Decision*, *supra* note 63 (Chapter 4) at para 144 at I-13898. Bisset & Crowhurst recognize that Braathens is not "directly relevant." Mark Bisset & Georgina Crowhurst, "Is the EU's Application of its Emissions Trading Scheme to Aviation Illegal" (31 March 2011), online: Clyde & Co <www.clydeco.com/news/articles/is-the-eus-application-of-its-emissions-trading-scheme-to-aviation-illegal>.

206 See Opinion of Advocate General Kokott, *supra* note 136 (Chapter 4) at I-62; *ATA Decision*, *supra* note 63 (Chapter 4) at para 142.

207 See *ATA Decision*, *supra* note 63 (Chapter 4) at para 145.

208 *Ibid.* at para 146.

209 See Opinion of Advocate General Kokott, *supra* note 136 (Chapter 4) at I-62.

210 *ATA Decision*, *supra* note 63 (Chapter 4) at para 142.

211 *Ibid.*

212 See Meltzer, *supra* note 222 (Chapter 4) at 118.

213 See GATT, *supra* note 22 at Art I; *GATS*, *supra* note 3 (Chapter 2) at Annex 1B to the Uruguay Round.

States cannot grant a preferential treatment to its local products (National Treatment principle).[214] Again, this analysis is important because these allegations may invalidate the scheme, even if it follows a different geographical scope (e.g. covering only emissions generated with the EU).

Albeit minor, the EU ETS will impose a cost on aircraft operators. This will then be passed to shippers, thereby raising the cost of exporting goods. These goods will be placed at a disadvantage vis-à-vis those produced within the European market which are not carried by air.[215] Following this line of reasoning, the scheme would impose a trade restriction on exported goods, thereby violating the national treatment principle.[216] Bartels disagrees with this position on the basis that the scheme would also be applicable to goods carried on intra-EU flights.[217] Others are of the view that the scheme is only discriminatory with respect to EU aircraft operators.[218] Their exposure and financial implication are much greater than those of non-EU carriers. In addition, Howse suggests that the EU ETS may be regarded as a "quantitative restriction contrary to [GATT's] Article XI."[219] Even assuming that it contradicts GATT's Articles III and XI, one must also consider whether the scheme could fall under one of the environmental safeguards.[220] Article XX allows a State to introduce trade-restricted measures that are "necessary to protect human, animal or plant life or health" or "relating to the conservation of exhaustible natural resources."[221] For these measures to be considered lawful, they cannot constitute "a means of arbitrary or unjustifiable discrimination [amongst States]"or a "disguised restriction on international trade."[222]

In *US Gasoline*, the panel was of the view that "clean air" was an "exhaustible natural resource."[223] A measure, such as the EU ETS, designed to reduce GHGs from a particular sector in order to protect the atmosphere most likely will fall under such exception.[224] Other WTO-related instruments, such as the Agreement on the Application of Sanitary and Phytosanitary Measures (SPS) and the Agreement on Technical Barriers to Trade

---

214 See GATT, *supra* note 22 at Art III.

215 See Meltzer, *supra* note 212 (Chapter 4) at 128. See *contra* Bartels, *supra* note 117 (Chapter 4).

216 See GATT, *supra* note 22 at Art III.

217 See Bartels, *supra* note 117 (Chapter 4).

218 See Gehring, *supra* note 18 at 153; Gudo Borger, "All Things not Being Equal: Aviation in the EU ETS" (2012) 3 Climate Law 265 at 280 (explaining that non-EU carriers will be less affected); Mayer, Case -366/10, *supra* note 133 (Chapter 1) at 1133; Dejong, *supra* note 101 (Chapter 4) at 175. See *contra* Henri J Nkuepo, "EU ETS Aviation Discriminates against Developing Countries" (2012) 7 African Trade L at 1.

219 Howse, EU ETS, *supra* note 163.

220 See Pache, *supra* note 97 (Chapter 4) at 60.

221 See GATT, *supra* note 22 at Art XX (a) & (g).

222 *Ibid.* See also Sydnes, *supra* note 152 (Chapter 2) at 77.

223 See *United States - Standards for Reformulated and Conventional Gasoline* (20 May 1996), WTO Doc WT/DS2/9 (Appellate Body Report), online: WTO <www.wto.org/english/tratop_e/dispu_e/2-9.pdf> at para 6.40.

224 See Bartels, *supra* note 117 (Chapter 4).

(TBT), also provide for similar exceptions to the general rules of the international trade regime.[225] Under the SPS, States may adopt measures "necessary to protect human, animal or plant life or health," as long as they are not done in a discriminatory or arbitrary manner.[226] Under the TBT, a technical regulation cannot be "more trade-restricted than necessary to fulfill a legitimate objective," such the protection of the environment.[227]

In theory, the EU ETS may also be regarded as constituting an unfair trade restriction on the provision of (air) services (e.g. the price of carbon is so high that passengers are preventing from flying anymore). In this respect, the WTO regime also provides for an Agreement on Trade in Services (GATS).[228] In turn, GATS contains a specific Annex on Air Transport Services (AATS).[229] AATS has however very limited application to international aviation. It only applies to services relating to aircraft repair and maintenance, computer reservation systems, and marketing. Traffic rights and "services directly related to the exercise of traffic rights" are excluded from GATS.[230] As Melzer notes, the EU ETS cannot not be categorized as a traffic right.[231] As such, the scheme does not fall under GATS scope. Having said this, GATS also provides for almost identical environmental safeguards as those of GATT, which I alluded to above.[232] Therefore, even if the GATS is applicable to the scheme, the EU may have reasonable grounds to invoke one of its environmental safeguards.[233] In sum, the EU ETS does not appear to be inconsistent with WTO rules.[234]

6.4 AN ATTACK UPON CBDR?

A large number of developing countries perceive the EU ETS as a burden-sharing mechanism that Europe is using to transfer to them mitigation obligations and the costs

225 See *Agreement on the Application of Sanitary and Phytosanitary Measures* [SPS], 15 April 1994, 1867 UNTS 493; *Agreement on Technical Barriers to Trade*, 15 April 1994, 1868 UNTS 120 [TBT]. See also Young, "Deriving Insights from the Case of the WTO", *supra* note 191 (Chapter 3) at 143.

226 SPS, *supra* note 225 at Art 2, para 2.

227 See TBT, *supra* note 225 at Art 2, para 2.2. See also Richard H Steinberg, "Power and Cooperation in International Environmental Law" in Guzman & Sykes, *supra* note 46 at 520; David Hunter, James E Salzman & Durwood Zaelke, "International Trade and Investment Law" in David Hunter, James E Salzman & Durwood Zaelke, eds, *International Law and Policy*, 3d ed (New York: Foundation Press, 2007) at 1259.

228 See GATS, *supra* note 3 (Chapter 2).

229 *Ibid.*

230 *Ibid.* at Annex on *Air Transport Services* at para 2 (a) (b). AATS only applies to: (a) aircraft repair and maintenance services, (b) the selling and marketing of air transport services, and (c) computer reservation system (CRS) services. *Ibid.* at para 3.

231 See Meltzer, *supra* note 212 (Chapter 4) at 128.

232 See GATS, *supra* note 3 (Chapter 2) at Art XIV.

233 See Bartels, *supra* note 117 (Chapter 4).

234 See Pache, *supra* note 97 (Chapter 4) at 60; Meltzer, *supra* note 212 (Chapter 4) at 157.

associated with such actions that they otherwise do not have to bear.[235] As the Coalition of the Unwilling put it, the scheme disregards the "different social and economic circumstances"[236] of developing countries. This may "curb the sustainable growth of international aviation."[237] Given that under the EU ETS all aircraft operators flying to and from airports situated in Europe must surrender allowances at the end of the commitment period, regardless of whether they are from developing or developed countries, some developing countries regard this non-differentiated scheme as directly contravening CBDR.[238] This may be the case even under a different geographical scope, for instance, if an airline from a developing country exercises traffic rights between two European cities (e.g. London and Madrid).

Although academic scholarship analyzing the legality of the EU ETS has primarily focused on other issues, few commentators have noted that the regional scheme does not pay much respect to CBDR.[239] The Commission, on the other hand, responded to critics by saying that the EU ETS is not in any way in conflict with CBDR. The obligation to surrender allowances is imposed on aircraft operators, not States.[240] The special recognition of CBDR in the international climate change regime applies only to developing countries. These States are not required to undertake quantified emission reductions. However, the exclusion does not extend to firms from developing countries established and conducting businesses in developed countries. If this were the case, a power-generating company from a developing country that sets up a branch in Europe would be excluded from the general ETS solely on the basis that its principal place of business is situated in, or that its main

---

235 See ICAO, A36-WP/130 EX/50, *Viewpoint of the Latin American Civil Aviation Commission on the Aviation Emissions Trading Scheme* (presented by the 22 Member States of the Latin American Civil Aviation Commission); ICAO, A36-WP/285, A36-WP/285 EX/92, *Chile's Position on the Inclusion of Civil Aviation in Emissions Trading* (presented by Chile to the 36th Assembly); ICAO, A36-WP/251 EX/82, *Environment and Emission Trading Charges* (presented by Nigeria on behalf of African States) (arguing that the EU ETS would impose emission reduction obligations "through the back door" at 3); ICAO, C-MIN 194/2, *supra* note 86 (Chapter 1) (Statement of Cuba arguing that the EU ETS transfers mitigation costs to developing countries at 13); UNFCCC, *Proposals by India for Inclusion of Additional Agenda Items in the Provisional Agenda of the Seventeenth Session of the Conference of the Parties*, FCCC/CP/2011/INF.2/Add 1, online: UNFCCC <http://unfccc.int/resource/docs/2011/cop17/eng/inf02a01.pdf> (stressing that the EU ETS "stands in violation of the *UNFCCC* as it does not respect the principle of CBDR"). See also Hua Lan, "Comments on EU Aviation ETS Directive and EU: China Aviation Emission Dispute" (2011) 45 RJT 589 at 604.

236 Position of the Coalition of the Unwilling, *supra* note 150 (Chapter 4) at 3.

237 *Ibid.*

238 See ICAO, A36-WP/235 EX/76, *Addressing Aviation Emissions Based on the Principle "Common But Differentiated Responsibilities"* (presented by China) (opposing a non-differentiated ETS); see Lan, *supra* note 235 (arguing the EU ETS infringes the CBDR principle and that it will be detrimental to the development of the Chinese aviation industry, since its growth rate is significantly higher than those of other developed countries) at 593.

239 See Joanne Scott & Lavanya Rajamani, "EU Climate Change Unilateralism: International Aviation in the European Emissions Trading Scheme" *India Environmental Portal* online: <www.indiaenvironmentportal.org.in/files/file/EUper_cent20Climateper_cent20Change_per_cent20Unilateralism.pdf>.

240 See Runge-Metzger, *supra* note 204 (Chapter 1) at 40; Pache, *supra* note 97 (Chapter 4) at 11.

shareholders are from, a developing country. It is simply unrealistic to expect that European countries would exempt these entities from the application of community legislation. Although Scott and Rajamani contend that there seems to be no reason to exclude the application of CBDR to business enterprises,[241] robust State practice indicates otherwise. The whole international climate change regime addresses the obligation of States, not business enterprises. It is therefore reasonable to conclude that CBDR only applies to States.

Notwithstanding the foregoing, some academics remain unconvinced that the EU ETS only addresses aircraft operators. In their view, the EU ETS very clearly seeks to influence State behavior.[242] Nowhere is this better exemplified than in the equivalent measures provision. In the scheme's original geographical scope, an aircraft operator's incoming flights to Europe would be excluded from the scheme, provided that the home State of such aircraft operator adopts measures equivalent to those of the EU ETS.[243] In fact, ever since the EU ETS was adopted, the Commission has lobbied extensively around the world to attract support from other States on equivalent measures. It is worth noting that this lobbying campaign has primarily targeted States and not aircraft operators. Although it is undisputed that the EU ETS indirectly seeks to affect State behavior, this in and of itself is not enough to support the conclusion that the scheme contradicts the CBDR principle. The EU ETS does not impose quantified emission reductions on developing countries. The original design of the scheme provided some incentives for all countries that decide to regulate GHG emissions from international aviation (i.e. by exempting incoming flights from the EU ETS).[244] Yet it cannot be construed in any way as imposing an obligation. Developing countries reserve the right to decide whether or not to take action. Similarly, developing countries wishing to shield themselves behind CBDR may certainly continue doing so. The EU ETS will certainly not prevent them from doing so. In designing the EU ETS, Europe opted not to recognize a differentiated scheme for aircraft operators from developing countries. Its administrative de minimis provision only excluded aircraft operators either performing minimum activity to/from European airports or emitting an insignificant amount of CO_2 per year. Although international law does not require Europe to adopt such a differentiated scheme, given the tremendous level of opposition that the EU ETS has been met with, perhaps it would have been advisable to make every effort to accommodate to some extent the concerns of developing countries. This might have facilitated the overall acceptance of the scheme.

241 See Scott & Rajamani, *supra* note 239 at 15.

242 *Ibid.*; See Mariam Radetzki, "The Fallacies of Concurrent Climate Policy Efforts" (2010) 39 AMBIO 211 at 221.

243 See Aviation Directive, *supra* note 17 (Introduction) at Art 25.

244 *Ibid.*

6.5 Does the *Chicago Convention* Matter?

Another highly controversial part of the *ATA Decision* dealt with the fact that the CJEU expressly ruled that the EU was not bound by the provisions of the *Chicago Convention*.[245] Although the CJEU did however acknowledge that the EU was bound by principles of customary international law,[246] its rejection to the *Chicago Convention*, as one could expect, enraged the international aviation community.[247] The examination of this issue is important because it explains why States with no aircraft operators flying to and from Europe nonetheless fiercely opposed the scheme. It is therefore relevant to analyze the arguments put forward by the claimants, the CJEU's reasoning, as well as the potential fallout.

6.5.1 *Background*

The *Chicago Convention* entered into force on 4 April 1947,[248] whereas the Treaty of Rome establishing the European Economic Community (EEC) did so on 1 January 1958.[249] Clearly, the *Chicago Convention* predates the EU. To date, all EU Member States are parties to the *Chicago Convention*. In spite of its continuous interaction with ICAO, the EU is not, in the strict sense, a party to the *Chicago Convention*.[250] Similarly, it is noteworthy that, from the point of view of the European legal framework, both the *Treaty on the Functioning of the European Union (TFEU)* and the *Treaty of the EU* cannot, theoretically, affect Member States' rights and obligations derived from international agreements that were concluded before the EEC came into existence or before the dates of accession for States that joined the EU at a later stage.[251] Also, the EU ETS was adopted through community legislation. That said, as explained at length in Chapter 4, Member States bear a number of obligations with regard to the implementation of the scheme.

---

245 See *ATA Decision, supra* 63 (Chapter 4) at para 101

246 *Ibid.* See also Opinion of Advocate General Kokott, *supra* note 136 (Chapter 4) at para I-20.

247 See Mayer, Case -366/10, *supra* note 133 (Chapter 1) at 1127; Andrus, *supra* note 86 at 16; Havel & Mulligan, *supra* note 60 (Chapter 4) at 10; van Valkenburg III, *supra* note 322 (Chapter 5).

248 See *Chicago Convention, supra* note 36 (Chapter 2).

249 See CE, *Treaty Establishing the European Economic Community*, 25 March 1957, online: Eur-Lex <http://eur-lex.europa.eu/LexUriServ/LexUriServ.do?uri=CELEX:11957E/TXT:EN:NOT>.

250 Unlike other more recent international instruments adopted under the auspices of ICAO, the *Chicago Convention* does not contemplate the possibility that regional international organizations could become parties. See *Chicago Convention, supra* note 36 (Chapter 2) at Art 92.

251 See *Treaty on the Functioning of the European* Union, [2012] OJ, C 326/47 at Art 351.

6.5.2 The Theory of Functional Succession

In *ATA Decision*, the claimants acknowledged that, in the strict sense, the EU was not a party to the *Chicago Convention*. They nonetheless contended that the EU was bound by its provisions and that international aviation's *Magna Carta* could be used to assess the validity of a community act (e.g. the directive that implements the EU ETS).[252] In order to bypass the lack of accession to the *Chicago Convention*, the claimants invoked the theory of functional succession.[253] To understand what this theory entails to, it is necessary to consider its origin, as well as the way in which the CEJU had interpreted it in previous cases.

Many years ago, drawing heavily from principles of the German Civil Code, and given the fact that, at the time, principles of State succession did not clearly address the transfer of delictual obligations to the successor State,[254] the German Supreme Court had skillfully constructed the theory *Funktionsnachfolge*.[255] This presupposes that, where a successor entity (State) carried out "essentially the same function" as those of its predecessor,[256] the former should bear obligations incurred by the latter. The German Supreme Court used this rationale in a number of cases to establish the liability of the Federal Republic for delictual obligations that the *Reich* had caused.[257] Arguably, because of the reconstruction process that the Federal Republic went through after World War II, it may be reasonably inferred that courts resorted to this theory to provide a remedy on the basis of fairness and equity that otherwise would have not been available to claimants.

Later, in the 1960s, the theory of functional succession tacitly landed in the European context. In *International Fruit Company*, the CJEU found that GATT bound the EC – despite the fact that the EC had not acceded to it at the time, all its Member States were parties to it.[258] Advocate General Mayras noted that, since its establishment, the EC had

252 See Gehring, *supra* note 18 at 150.

253 It is undisputed that the EU has taken up certain functions in the air transport field that its Member States previously used to perform. Yet both the EU and its Member States continue to coexist. Given that there is no "definite replacement of one State by another," this is not a case governed by the international law principles of State succession. See Brownlie, *supra* note 31 (Chapter 5) at 649.

254 See *ST v The Land N* (1952) Bundesgerischtshof (Supreme Court of the Federal Republic of Germany) Entscheidungen des Bundesgerichtshofes in Zivilsachen, 8: 22 169 in "Succession of States and Governments", *Yearbook of the International Law Commission 1963*, vol 2 (New York: UN, 1964) (A/CN.4/SER.A/1963/Add 1) 147; Brownlie, *supra* note 31 (Chapter 5) at 654; Shaw, *supra* note 65 (Chapter 5) at 604.

255 See Wladyslaw Czaplinsky, "State Succession and State Responsibility" (1990) 28 Can YB Int'l L 339 at 347; Noëlle Quénivet, "Binding the United Nations to Human Rights Norms by Way of the Laws of Treaties" (2010) 42 Geo Wash Int'l L Rev 587 at 606; FA Mann, "Germany's Present Legal Status Revisited" (1967) 16 Intl'l & Comp LQ 760 at 779.

256 See Michael John Volkovitsch, "Righting Wrongs: Towards a New Theory of State Succession to Responsibility for International Delicts" (1992) 92 Colum L Rev 2162 at 2194.

257 See *ST v The Land N*, *supra* note 254 at 148. See also *ibid.* at 2194.

258 See *International Fruit Company NV and others v Produktschap voor Groenten en Fruit*, C-21 to 24/72, [1972] ECR I-01219 at 1224 [*International Fruit Company*].

already participated in a number of GATT trade negotiations.[259] Similarly, the CJEU observed that the Member States had always voted following the Commission's position.[260] In other words, the views of both Member States and the Commissions had been aligned. Below, I illustrate the similarities between the EU and Member States' approaches to climate change issues at ICAO.

To establish that GATT was binding, the CJEU reasoned that the EC "[had not only] assumed the powers previously exercised by [its] Member States in the area governed by [GATT, but it had also taken the] functions inherent in the tariff and trade policy."[261] Unfortunately, the CJEU failed to provide further guidance on the specific conditions that ought to be met to consider that a transfer of functions in favor of the EC has taken place. In this case, however, the CJEU clearly emphasized on the assumption of powers, but it did not indicate whether this requires a full or partial assumption. The CJEU also indicated that, given the fact that GATT does not confer a direct cause of actions to persons other than States,[262] a fruit importer cannot avail itself of this instrument as the basis to invalidate community legislation that established the requirement to obtain a license prior to importing apples.[263]

International Fruit Company constitutes the only reported case where the CJEU implicitly embraced the theory of functional succession. Perhaps the court's willingness to recognize, albeit timidly, this theory had to do with the fact that doing so did not in any way invalidate the community acts that were the subject of the legal challenge. After all, the CJEU did not attribute a direct effect to GATT. Its recognition, however, did make it clear that trade policy was a matter of (exclusive) community competence. These considerations may have not necessarily been present in the cases that will follow below, where the CJEU expressly ruled out the application of the functional succession theory.

More recently, in *Intertanko* the CJEU concluded that, given the fact that there has not been a "transfer of powers" from Member States, the *International Convention for the Prevention of Pollution from Ships* (MARPOL 73/78) did not bind the EU.[264] What is most daunting about this ruling is the fact that, although the EU is not a party, the EU Directive 2005/35 expressly internalizes some (but unfortunately not all) of the MARPOL 73/78 obligations.[265] For the CJEU, this factor was insufficient to resort to the international

---

259 See Opinion of Mr. Advocate General Mayras Delivered on 25 October 1972 in Joined Cases 21 to 24/72 at 1237.

260 See *International Fruit Company, supra* note 258 at 1225.

261 *Ibid.* at 1227. See also Robert Schütze, "EC Law and International Agreements of the Member States: An Ambivalent Relationship?" (2006) 9 Cambridge YB Eur Leg Stud 387 at 398.

262 *Ibid.* at 1228.

263 *Ibid.*

264 Judgment of the Court (Grand Chamber) of 3 June 2008 in Case C-308/06 at paras 47 & 49 [*Intertanko*].

265 See EU, *Directive 2005/35/EC of the European Parliament and of the Council of 7 September 2005 on Ship-Source Pollution and on the Introduction of Penalties for Infringements* [2005] OJ, L 255/11. A completely different approach would have been to rely on the so-called mortgage theory advanced by some commentators

instrument as the basis "to review the directive's legality."[266] In *Bogiatzi*, the CJEU raised the standard even higher by declaring that the EU was not bound by the 1929 Warsaw Convention because no "full transfer of powers" has taken place.[267] Similarly, a year later, in *TNT Express*, the CJEU found that it had no jurisdiction to interpret provisions of the *Convention on the Contract for the International Carriage of Goods by Road* (*CMR*).[268] Again, in light of the fact that the EU was not a party to *CMR* and that the EU has not fully assumed the powers that Member States previously exercised, the CJEU declined to apply the theory of functional succession.[269] To this end, provisions of these international instruments could not be used to challenge the legality of a community act.

6.5.3 *When* Strasbourg *Negated* Chicago

In light of these precedents, it should then not be a surprise that in *ATA Decision*, the CJEU followed exactly the same approach noting that "only if...the European Union has assumed the powers previously exercised by its Member States in the field, the [*Chicago Convention's*] provisions would have the effect of binding the [EU]."[270] Moreover, the CJEU said that the EU "must have assumed... all the powers previously exercised by the Member States that fall within the [scope of the *Chicago Convention*]."[271] Although acknowledging that there is vast community legislation dealing with the issues addressed, and internalizing obligations contained, in the *Chicago Convention*,[272] the CJEU opined that this was not a controlling factor.

For the CJEU, the central issue was that the EU does not enjoy "exclusive competence in the entire field of international civil aviation."[273] The EU has not assumed entirely all the powers that were previously exercised by Member States.[274] As Advocate General Kokott noted, in the air transport field, not all functions have passed from Member States to the EU.[275] To this end, Member States still enjoy prerogatives in this field.[276] Moreover,

in the 1970s. This presupposes that Member States cannot explicitly transfer any powers to the EC. The latter is bound by international agreements that Member States concluded before joining the Treaty of Rome. Given that the theory does not recognize the autonomy of the community legal regime, it has been strongly criticized and seldom used. See Schütze, *supra* note 261 at 395.

266 See *Intertanko*, *supra* note 264 at para 50.

267 See Judgment of the Court (Fourth Chamber) of 22 October 2009 in Case C-301/08, paras 32 & 33.

268 Judgment of the Court (Grand Chamber) of 4 May 2010 in case 533/98 at para 61 [*TNT Express*].

269 See Opinion of Advocate General Kokott, *supra* note 136 (Chapter 4) delivered in Case C-533/08 of 28 January 2010 at paras 61 at I-4126.

270 See *ATA Decision*, *supra* note 63 (Chapter 4) at para 62.

271 *Ibid.* at para 63.

272 *Ibid.* at para 65.

273 *Ibid.* at para 69.

274 *Ibid.* at paras 71, 72, & 90.

275 See Opinion of Advocate General Kokott, *supra* note 136 (Chapter 4) at I-23.

276 See *ATA Decision*, *supra* note 63 (Chapter 4) at para 71.

Advocate General Kokott also pointed out that the EU does not act on behalf of its Member States at ICAO, but it rather only serves as a coordinator acting as a liaising agent among them.[277] Therefore, the CJEU concluded that the functional succession theory cannot be applied.[278] As such, given that the EU is not bound by the *Chicago Convention*,[279] the latter cannot form the basis to assess the validity of the EU Directive implementing the EU ETS.[280]

The CJEU's interpretation of the functional succession theory, as applied consistently in *Bogiatzi*, *TNT Express*, and *ATA Decision*, is much more restrictive than what the German Supreme Court had originally envisioned when it had developed the concept.[281] In these three transport-related cases, the CJEU has required a complete transfer of prerogatives and competences to the EU, whereas for the German Supreme Court, it was only necessary that the successor entity assumes or exercises "concrete functions" from the predecessor entity.[282] This is a substantial difference. Arguably, the German approach did not purport to capture the whole universe of activities that the predecessor entity may perform. On the contrary, the CJEU's approach considerably restricts the theory's application to very limited circumstances.[283]

From an extremely formalistic viewpoint, it is correct to say that the EU is not bound by the *Chicago Convention*.[284] It is simply unquestionable that it has never acceded to this international instrument. It is also undeniable that Member States have retained a number of prerogatives in the air transport field. Among others, these include issues such as aviation security and the exchange of traffic rights (with the exception of a handful of multilateral open skies agreement that the EU and its Member States have jointly entered into with third parties, such as the United States, Canada, and Brazil). There is a strong case that the CJEU's position is correct: the transfer of powers in favor of the EU has not been complete if one looks at the whole spectrum of activities dealing with international civil aviation.

Notwithstanding the foregoing, the perspective is remarkably different if one narrows down the focus to concentrate only on aviation and climate change issues. Here, it is more than evident that the EU has taken exclusive jurisdiction on the subject matter. This is an

277 See Opinion of Advocate General Kokott, *supra* note 136 (Chapter 4) at I-23.

278 *Ibid.* at I-22. See also van Valkenburg III, *supra* note 322 (Chapter 5); Gattini, *supra* note 135 (Chapter 4) at 985; Dejong, *supra* note 101 (Chapter 4) at 169.

279 See Opinion of Advocate General Kokott, *supra* note 136 (Chapter 4) at I-22. See also Andrus, *supra* note 86 at 14; Bartlik, *supra* note 68 (Chapter 4) at 162; Havel & Mulligan, *supra* note 60 (Chapter 4) at 9.

280 See *ATA Decision*, *supra* note 63 (Chapter 4) at para 72. See also Dejong, *supra* note 101 (Chapter 4) at 168.

281 See Kulovesi, "Make Your Own Special Song", *supra* note 137 (Chapter 4).

282 See KV Schleswig-Hoslstein (1951) Bundesgerichtshof, Entscheidungen des Bundesgerichtshofes in Zivilsanchen, 4:30, p. 266 in International Law Commission, "Succession of States and Governments" (1963) UNYB Int'l L Comm'n 95 at 147.

283 See Schütze, *supra* note 261 at 396.

284 See Bogojovic, *supra* note 134 (Chapter 4) at 350; Mayer, Case -366/10, *supra* note 133 (Chapter 1) at 1121.

area where one can reasonably say that a complete transfer of functions has taken place. Member States have almost no leverage and lack autonomy. Law-making and policy setting occur at the pan-European level. On behalf of the Commission, the Directorate-General for Climate Action (DG Clima) clearly exercises the leading role. Although the views of Member States are taken into account, it is unquestionable that Brussels dictates the rules of the game. In the international area, DG Clima's head serves as the chief negotiator and counterpart to deal with those opposing the scheme. In spite of the fact that both the EU and its Member States are parties to the *UNFCCC*, the former's influence is indisputable.

Strictly speaking, the EU only enjoys observer States at ICAO. However, its experts take part in all of the organization's working groups, panels, task forces, and committees. It is true that the EU does not participate as a State party in the ICAO Assembly, meetings of Council, HGCC, GIACC, high-level meetings, and the like. Instead, its Member States do. However, this does not prevent the EU from determining what the common European position should be. Although on its face it would appear as if Member States are representing none other than themselves, in practice their positions are completely circumscribed by the instructions they receive from Brussels. This has been patently evidenced by the fact that Europe's chief climate change negotiator, Mr. Jos Delbeke,[285] served as Belgium's chief delegate to ICAO's HGCC.[286]

By requiring a universal transfer of functions, the CJEU intentionally or unintentionally overlooked how things work in practice. At least at ICAO, the role of the EU is not that of a mere "coordinator," as Advocate General Kokott suggested. It is rather a conclusive example of an entity setting and negotiating policy. It is therefore difficult not to conclude that this is tantamount to an implicit transfer of functions, at least with regard to aviation and climate change issues.

6.5.4 *Repercussions*

The CJEU's view on the *Chicago Convention* places its Member States in a very delicate position. As one commentator notes, by exploiting a formalistic "loophole," the CJEU may

285 Since February 2010, Mr. Jos Delbeke has served as Director General for Climate Action. See EC, European Commission, "Our Director General" online: European Commission <http://ec.europa.eu/clima/about-us/director/index_en.htm>.

286 See "Composition of the High-level Group on International Aviation and Climate Change: HGCC" online: ICAO <https://portal.icao.int/HGCC/Membership/HGCC.Composition%20-%20Members%20and%20Advisors.%20%2025%20March%202013.Revised.pdf>. India vehemently protested to the President of the ICAO Council for allowing the participation of Mr. Jos Delbeke as head of the Belgium delegation to HGCC. With the support of countries such as China, Brazil, and the Russian Federation, India regarded "the participation of an EU official in the ICAO deliberations [as] a travesty to [the *Chicago Convention* and ICAO]" given that the CJEU had expressly declared that the EU was not bound by this instrument. See Letter from Prashant Sukul, Representative of India on the Council of ICAO, to Mr. Roberto Kobeh Gonzalez, President of the ICAO Council, untitled (14 March 2013) [unpublished, archived on file with the author].

put its Member States in a situation to either "circumvent [or violate their obligations] to an international agreement" to which they have been parties to before joining the EU.[287] Arguably, the *ATA Decision* defeats the spirit of Article 351 of the TFEU: membership to the EU should not invalidate prior international obligations of its Member States. One cannot but wonder why the CJEU opted for rejecting the theory of functional succession, thereby completely discarding the *Chicago Convention*'s applicability. At the end, most of the *Chicago Convention*'s provisions that the claimants relied on already form part of the well-established principles of customary international law (e.g. exclusive sovereignty, freedom to fly over the high seas, no extension of sovereignty over the high seas).[288] The CJEU nonetheless examined each and every one of these principles in its decision. The substantive outcome would have probably been the same, with the sole exception of perhaps Article 12 of the *Chicago Convention* (e.g. rules over the high seas are those of ICAO).[289] Even accepting the applicability of the *Chicago Convention*, the CJEU could have easily blessed the scheme's legality, thereby finding that it was not extraterritorial nor did it purport to represent an infringement on the sovereignty of third States.

Perhaps the CJEU consciously decided to reject the *Chicago Convention* in order to reaffirm the autonomy of the European legal system vis-à-vis international law.[290] More explicitly, the *ATA Decision* confirms that the CJEU is reluctant to apply foreign instruments to determine whether a given community act is valid or not. It was re-emphasized that this would only be permitted in very limited circumstances and only to the extent that when doing so would not be detrimental to advance European interests. As some pundits have suggested, perhaps it is not advisable to recognize the preeminence of international law when in fact the European legal regime would appear to be more progressive.[291] This Euro-centric view implies that it is up to Europe to determine what is to be considered "progressive" as well as when and how such notion is to be interpreted. For the rest, on the other hand, this represents a disguised attempt to export and colonize European (environmental) values around the world.

---

287 See van Valkenburg III, *supra* note 322 (Chapter 5).

288 See Manzini & Masutti, *supra* note 136 (Chapter 4) at 331. See also *ATA Decision*, *supra* note 63 (Chapter 4) at para 45.

289 See Mayer, Case -366/10, *supra* note 133 (Chapter 1) at 1128; Gattini, *supra* note 135 (Chapter 4) at 985.

290 See Mayer, Case -366/10, *supra* note 133 (Chapter 1) (noting that there is a "love-hate relationship between EU and international law" at 1114); Schütze, *supra* note 261 (discussing that the relationship between community law and international law had remained a neglected problem during the foundational period of the community legal order) at 388.

291 See Mayer, Case -366/10, *supra* note 133 (Chapter 1) at 1114.

6.6 Conclusion

In addition to the extraterritorial allegations, opponents have advanced other grounds for the impermissibility of the EU ETS. This chapter has outlined some of them. The state of affairs in international law reveals that Europe's unilateralism is simply not permissible under the original geographical scope of the EU ETS.[292] States do not unilaterally engage in actions to correct global environmental externalities.[293] In fact, for the most part, States have shown a tendency to prefer "cooperative multilateral approaches."[294] Therefore, and in the absence of an express rule of international law, a State cannot unilaterally and extraterritorially implement a policy measure in order to protect diffuse environmental commons.[295]

It also should be said that emissions trading as a policy instrument is a type of economic measure different than a tax. The tax characterization disregards the genesis of ETS. Outside aviation circles, there is little support for the argument that the EU ETS is a tax. Most commentators regard the EU ETS as an economic instrument or, using the ICAO terminology, an MBM but definitely not as a tax or a charge. Arguably, in light of the fact that the Commission did a poor job in explaining the scheme's main features and how it differs from the intrinsic characteristics of a tax, detractors saw a magnificent opportunity to attack its legality. After all, in most countries, taxes are toxic. This argument resonates with the general public. In fact, in the United States, A4A championed this crusade.[296]

In this study, I did not find that the EU ETS contravenes WTO rules or the CBDR principle. Although not an attack on the EU ETS itself, the rationale of the CJEU in *ATA Decision* to disregard the applicability of the *Chicago Convention* is extremely problematic, for it undermines the whole integrity of the system of international civil aviation, which – under ICAO's patronage – has strived to achieve harmonization and commonality.[297] It may also be a negative precedent that may be exported to other areas beyond the purely aviation domain. As one critic points out, the *ATA Decision* allows the EU to choose and

292 See Pablo Mendes de Leon, "ATA and others v. the UK Secretary of State for Energy and Climate Change (2009)" (2010) 35:2 Air and Space L 199 (commenting that the claimants argued that the EU ETS represents a unilateral action "infringing international law" at 200).

293 See Dejong, *supra* note 101 (Chapter 4) (noting that the *Chicago Convention* and international law do not support "a unilateral scheme which subjects foreign airlines to a charge for flight in foreign and international airspace" at 171).

294 Zerk, *supra* note 10 (Chapter 5) at 187.

295 *Ibid.* (commenting that "States do not, by and large, act unilaterally to protect global environmental resources" at 208).

296 A4A's President and CEO, Nicholas E Calio has relentlessly attacked the EU ETS. As part of A4A's campaign against the scheme, Mr. Calio has been on record by calling it an "exorbitant money grab" and a "job killer." "A4A EU ETS Lawsuit Defines Clear Path for Government Action", *A4A* (27 March 2012) online: <www.airlines.org/Pages/news_3-27-2012.aspx>.

297 See Havel & Mulligan, *supra* note 60 (Chapter 4) at 10.

pick when to completely disregard the *Chicago Convention.*[298] In addition to being extremely self-serving, it has been regarded as an insult for those outside Europe. As Havel and Mulligan put it, the CJEU's decision represents a "diplomatic destabilizing view."[299] This also may explain why 98 ICAO Member States, whose air carriers did not have commercial operations to and from Europe, in one way or another, have joined the opposition choir against the EU ETS. Many of them even formed part of the Coalition of the Unwilling. For these countries, the fight against the EU ETS was not to protect their airlines, oppose a 3-US dollar airfare increase, or to challenge the scheme's alleged extraterritorial application, but rather a quest to preserve the integrity of the *Chicago Convention.*[300] Arguably, the political costs of leaving behind the *Chicago Convention* do not seem to have justified potential gains (or the gains are not evident from an international perspective). Europe may opt for foregoing the *Chicago Convention*, but for the world at large, adherence to this instrument and the system that it entails matters.

298 See Andrus, *supra* note 86 at 16.
299 See Havel & Mulligan, *supra* note 60 (Chapter 4) at 12.
300 See Mayer, Case -366/10, *supra* note 133 (Chapter 1) at 1127.

7 We Are All Ahead of the Curve

A number of international aviation actors claim to be ahead of the curve with respect to mitigating aviation industry's contribution to climate change. Having established the first global emissions trading system (ETS), Europe asserts that the inclusion of international aviation lays down the foundational grounds for a sectoral and global market-based measure (MBM).[1] IATA claims that the airline industry is the only global industry with ambitious targets to mitigate its carbon footprint.[2] ICAO has publicized the facts that it was the first UN agency to negotiate and adopt universal goals.[3] Similarly, the United States, the world's largest aviation market, reports that its emissions have been cut by almost 14 percent since 2000.[4] While these actors have invested considerable efforts to make changes in order to improve their environmental performance, the ultimate goal of reducing GHG emissions from international aviation is yet to be achieved. The sector's emissions continue to grow.

Through the lenses of norm entrepreneurship and the theory of norms in general, this chapter examines the roles, achievements, and shortcomings of some of the key players

---

1 See EC, European Commission "Reducing Emissions from Aviation" online: European Commission <http://ec.europa.eu/clima/policies/transport/aviation/>.

2 See IATA, "Remarks of Tony Tyler at the Media Round Table, Singapore", *IATA* (16 October 2013) online: <www.iata.org/pressroom/speeches/Pages/2013-10-16-01.aspx>.

3 See ICAO, 14/10, "ICAO Member States Agree to Historic Agreement on Aviation and Climate Change" (8 October 2010) online: ICAO <www.icao.int/Newsroom/Pages/icao-member-states-agree-to-historic-agreement-on-aviation-and-climate-change.aspx>.

4 According to the US Bureau of Transportation Statistics, in 2000, the total fuel consumption of US airlines, for both scheduled and non-scheduled services, domestic and international, was approximately 1.6 billion gallons. At the time, fuel costs were 0.81 US dollars/gallon. In 2012, US airlines consumed roughly 1.39 billion gallons of jet fuel. By that time, fuel cost rose to 2.88 US dollars per gallon. See Bureau of Transportation Statistics, "Airline Fuel Cost and Consumption", *US Department of Transportation* online: <www.transtats.bts.gov/fuel.asp?pn=1>. Given that fuel consumption is directly linked to CO_2 emissions, this statistic shows that US airlines emitted approximately 14 percent less emissions in 2012 as compared to their levels of activity in 2000. Both the US government and the US airlines are quick to point out this reduction whenever possible. See ICAO, C-MIN 194/2, *supra* note 86 (Chapter 1) (Statement of the United States stressing that CO_2 emissions from US airlines dropped by 15 percent in the last decade at para 100); "Oral Testimony of Nancy Young, VP for Environmental Affairs Before the Senate Committee on Commerce, Sciences and Transportation", *A4A* (6 June 2012) online: <www.airlines.org/Pages/A4A-Oral-Testimony-of-Nancy-Young,-VP-for-Environmental-Affairs.aspx> (highlighting that US airlines "burned 11 percent less fuel in 2011 than they did in 2000, resulting in an 11 percent reduction in CO_2 emissions, even though they carried almost 16 percent more passengers and cargo on a revenue-ton-mile basis"). Although higher fuel prices are a built-in incentive mechanism to reduce fuel consumption, other factors such as industry consolidation, fleet rationalization, increasing load factors, and negative growth may also explain the overall decrease in CO_2 emissions. Airbus points out that in the last 20 years, US airlines have improved their load factors from 61 to 82 percent. In practical terms, this means performing fewer flights while transporting more passengers. Needless to say, fewer flights directly correlate to less CO_2 emissions. See Airbus, "Global Market Forecast: 2013-2032," *supra* note 1 (Introduction).

involved in aviation and climate change issues. Norm entrepreneurship is relevant to this project because it identifies actors that promote change to modify or instill a new social norm. It is a valuable tool for detecting which actors seek to position themselves ahead of the curve. The theory also sheds light on the conditions under which these norm changes may occur and the motivations behind agents' decisions that promote change. I utilize normative change to refer to a market-based measure (MBM) regulating GHG emissions from international aviation, such as ICAO's Global MBM or the EU ETS. This chapter also demystifies actors' rhetorical statements about proposals to curb GHG emissions from aviation, which has been a tendency by States who seek to avoid seriously addressing proposals. I illustrate the predicaments about the behavior of actors and identify the constraints under which they operate and the conditions that are necessary for norm internalization. I argue that the process of norm internalization is extremely relevant because a large number of actors involved in the aviation and climate change discourse contend that MBMs are not warranted. These considerations should be taken into account when designing ICAO's Global MBM for international aviation.

Section 7.1 examines the theory of norms in general and norm entrepreneurship. Particular attention is given to the definition of norms both from the rational choice and constructivist perspectives. I examine who norm entrepreneurs are, their characteristics, the formation of transactional networks, the motives behind norm entrepreneurship, as well as the conditions that may induce it and influences that may be detrimental to its development. I also look into the life cycle of norms and the process of norm internalization. These are central to understanding not only why the EU ETS was rejected but most importantly why it did not engender action by those outside Europe.

Section 7.2 reflects on the norm entrepreneurship roles of the EU, IATA, ICAO, and the United States in light of the international aviation emissions debate. Although climate change issues in aviation involve a number of actors, these have been chosen for the following reasons: (i) The EU has been the first actor to suggest a regulatory model to control GHG emissions from international aviation, having a profound impact on the behavior of all other actors, which for the most part have reacted to its influence; (ii) as the world's largest and most well-established airline trade association, IATA had redefined the industry's role with regard to climate change; (iii) ICAO is the institutional platform where most of the international discussions on aviation and climate change take place. In addition, most actors would seem to agree that ICAO, not another international organization, should be the forum to handle GHG emissions from international aviation; (iv) the United States is still the largest domestic and international aviation market. In fact, inaction on the part of the United States with respect to addressing climate change profoundly influences aviation and climate change discourse. There cannot be a meaningful deal on this topic without US participation.

7.1 The Theory of Norms and Norm Entrepreneurship

7.1.1 What Are Norms?

Rational choice theory and constructivism have dominated conceptualizations of norms. Snidal explains that rational choice is "a methodological approach that explains both individual and collective (social) outcomes in terms of individual goal-seeking under constraints."[5] Norms constrain behavior.[6] Economists and rational choice theorists have long studied how norms affect the behavior of actors.[7] They do so by placing heavy reliance on interests, sanctions, and incentive structures.[8] Bodansky argues that "to change [the actors'] behavior, we need to change their incentive structure by giving them an interest in engaging in environmentally sound behavior."[9] Rational choice theory may also be useful in explaining the emergence of an increasing importance of non-State actors in the international context.[10] The application of the rational choice theory to the study of norms has not been without extensive criticism. Björkdahl points out that the rational choice theory does not "capture adequately the influence of ideas, values and norms on the identities and interests of actors."[11] Abbott and Snidal acknowledge the value of rational

---

5 Snidal, *supra* note 315 (Chapter 5) at 87.

6 Oran R Young argues that "the role of regimes is to alter incentives in such a way as to prevent individualistic behavior likely to lead to collective-action problems in situations involving strategic interactions." Oran R Young, "Regime Effectiveness: Taking Stock" in Oran R Young, ed, *The Effectiveness of International Environmental Regimes: Causal Connections and Behavioral Mechanism* (Cambridge: MIT Press, 1999) 249 at 269. Similarly, while discussing the level of success of the *Montréal Protocol* and the *Kyoto Protocol*, Barrett points out that the key issue is to ensure that the regime provides a better incentive structure so states will be better off by joining than they would be otherwise. According to Barrett, the *Montréal Protocol* achieved this but the *Kyoto Protocol* failed in doing so. See Barrett, *Why Cooperate*, *supra* note 57 (Chapter 1) at 93.

7 See Richard H McAdams, "The Origin, Development, and Regulation of Norms" (1997) 96 Mich L Rev 338 at 339. Conversely, Posner writes that "social norms are usefully understood as mere behavioral regularities with little independent explanatory power and little exogenous power to influence behavior. They are the labels that we attach to the behavioral regularities that emerge and persist in the absence of organized conscious directions by individuals. These behavioral regularities result from the interactions of individuals acting in their rational self-interest broadly understood (to include altruism and other forms of interdependent utility)." Eric A Posner, *Law and Social Norms* (Cambridge: Harvard University Press, 2000) at 8.

8 See Kenneth W Abbott & Duncan Snidal, "Values and Interests: International Legalization in the Fight against Corruption" (2002) 31 J Legal Stud 141 at 141-42 [Abbott & Snidal, "Values and Interests"]; Jeffrey T Checkel, "International Norms and Domestic Politics: Bridging the Rationalist-Constructivist Divide" (1997) 3 Eur J Int'l Relations 473.

9 See Daniel Bodansky, *The Art and Craft of International Environmental Law*, *supra* note 62 (Chapter 1) at 46.

10 See Snidal, *supra* note 315 (Chapter 5) at 101. For instance, Glensy defines norms as "forces that explain how society shapes individual behavior." Rex D Glensy, "Quasi-Global Social Norms" (2005) 38 Conn L Rev 79.

11 Annika Björkdahl, "Norms in International Relations: Some Conceptual and Methodological Reflections" (2002) 15:1 Cambridge Rev Int'l Affairs 1:9.

choice's methodology, but criticize the fact that values cannot be reduced "to another category of interests."[12] Druzin notes that actors are not always guided by self-interest.[13]

Adler explains that the constructivist approach to international relations conceives norms as "social identities [that] give national interests their content and meaning."[14] Björkdahl writes that norms are characterized by "shared knowledge and intersubjective understandings."[15] Placing strong emphasis on the process of their transmission or "inheritance," Florini conceives norms as a "set of beliefs, attitudes and values."[16] For Finnemore and Sikkink, norms are "standards of appropriate behavior for actions with a given identity."[17] This is often referred to as the "logic of appropriateness" or what ought to be done.[18] One of the criticisms of the constructivist's conception of norms is that it underplays the role of interests in the process of norm construction.[19]

As Abbott and Snidal note, both rational choice and constructivism provide valuable contributions to the study of norms.[20] However, in real life, values and interest are not isolated, but rather they interact.[21] In fact, international issues, legal institutions, and a number of different actors "depend on the deeply intertwined interactions of 'values' and 'interests'."[22] Kratochwil points out that "values inform the attitudes of actors."[23] Moreover, some international regimes are built on the basis of the "interplay [between] values and

12 See Abbott & Snidal, "Values and Interests," *supra* note 8 at 142.
13 See Bryan Druzin, "Law, Selfishness, and Signals: An Expansion of Posner's Signaling Theory of Social Norms" (2011) 24:5 Can JL & Juris 1 at 5.
14 Emanuel Adler, "Constructivism in International Relations: Sources, Contributions, and Debates" in Walter Emmanuel Carlsnaes & Beth A, Simmons, eds, *Handbook of International Relations* (London: Sage Publications, 2013) at 126. See also Jaye Ellis, "Fisheries," *supra* note 279 (Chapter 5) (commenting that "[c]onstructivists consider legal rules and systems to be constituted by shared understandings developed through iterative processes of interaction" at 2).
15 See Björkdahl, *supra* note 11 at 20. See also Checkel, *supra* note 8 (discussing that norms "constitute" shared understandings that [form part of actors'] "identities and interests" at 473).
16 See Ann Florini, "The Evolution of International Norms" (1996) 40 Int'l Stud Q 363 at 367.
17 See Martha Finnemore & Kathryn Sikkink, "International Norm Dynamics and Political Change" (1998) 52:4 International Organization 887 at 891.
18 *Ibid.* Although much of his work would seem to be based on rational choice with significant emphasis on interests and incentive structures, Cass Sunstein's definition would seem to follow the logic of appropriateness. Sunstein defines norms as "attitudes of approval or disapproval, specifying what ought to be done and what ought not to be done." Cass R Sunstein, "Social Norms and Social Roles" (1996) 96 Colum L Rev 4: 903 at 914. For Andrew Hurrel & Terry MacDonald, norms suggest "patterns of behavior and give rise to expectations as to what will, in fact, be done in a particular situation." Andrew Hurrel and Terry MacDonald, "Ethics and Norms in International Relations" in Walter Emmanuel Carlsnaes & Beth A Simmons, eds, *Handbook of International Relations* (London: Sage Publications, 2013) 57 at 69.
19 See Abbott & Snidal, "Values and Interests," *supra* note 8 at 142-43.
20 *Ibid.* at 141-42.
21 *Ibid.* at 144.
22 *Ibid.* at 142.
23 Friedrich V Kratochwil, *Rules, Norms, and Decisions: On the Conditions of Practical and Legal Reasoning in International Relations and Domestic Affairs* (Cambridge: Cambridge University Press, 1989) at 64.

interests."[24] The European normative push to regulate GHG emissions from international aviation is a clear example that some actors, as Abott and Snidal discuss, may be "simultaneously motivated by both values and interests."[25] Strahilevitz advocates a neutral definition of norms as "patterns of behavior that are widely adhered to by [a group of actors]."[26] Although I share this broad definition of norms as common patterns of behavior, legal scholarship emphasizes their coercive nature. Hart notes that "[t]he most prominent general feature of law at all times and places is that its existence means that certain kinds of human conduct are no longer optional, but in some sense obligatory."[27] Actors are expected to comply with the norm.[28]

For the purposes of this project, I am primarily concerned with legal norms: those adopted through explicit legal vehicles such as statutes, resolutions of international organizations, and international treaties. I deploy "norm" to mean the regulation of GHG emissions from international aviation through an MBM scheme. I am not so much concerned with other legal norms such as general principles of international law or norms of customary international law. Given that IATA has put forward some industry proposals (e.g. industry targets, proposal for a global offsetting scheme), I will also examine extra-legal social norms and their influences upon legal norms. My understanding of norm entrepreneurship is, therefore, intended to closely analyze a potential global MBM scheme for international aviation and the numerous exogenous and endogenous influences affecting it. I adopt this approach in order to identify elements that may be considered for the design of ICAO's global MBM.

7.1.2 *The Emergence of Non-State Actors*

Although it is recognized that the legal system may affect individual and social behavior,[29] changes in social behavior may also be achieved through channels other than law.[30] In fact, Ellickson claims that "[i]n many contexts, law [may] not be central to the maintenance of social order."[31] Boyd notes that the idea that the State is the sole actor in the international

24 See Abbott & Snidal, "Values and Interests," *supra* note 8 at 144.

25 *Ibid.* at 154.

26 Lior Jacob Strahilevitz, "Charismatic Code, Social Norms, and the Emergence of Cooperation on the File-Swapping Networks" (2003) 89 Va L Rev 505 at 537. See also Björkdahl, *supra* note 11 (describing international norms as "standards of behavior" at 13).

27 See HLA Hart, *The Concept of Law* (Oxford: The Clarendon Press, 1967) at 6.

28 *Ibid.* at 21.

29 See Robert E Scott, "The Limits of Behavioral Theories of Law and Social Norms" (2000) 86 Va L Rev 8 1603.

30 See Clayton P Gillette, "Lock-in Effects in Law and Norms" (1998) 78 BIL Rev 813.

31 Robert C Ellickson, *Order without Law: How Neighbors Settle Disputes* (Cambridge: Harvard University Press, 1991) at 280 [Ellickson, *Order without Law*].

arena is no longer valid.[32] Arguably, other agents may also lead to a change of norms. The role of non-state actors[33] in international relations has significantly increased over the last few decades.[34] As some commentators have noted, "[g]one are the days when [S]tates were the only subjects of international law."[35] Non-State actors frequently participate in the design of new structures in international regimes.[36] In addition to promoting a balance between economic and environmental objectives,[37] and exercising pressure over domestic agents to engage in normative change,[38] non-governmental organizations (NGOs)[39] have been active in the negotiation of numerous international environmental regimes.[40] They have also supported ethical reforms in other fields, such as anti-bribery international regulations.[41] More specifically, in the context of aviation and climate change, as we will see

32 See William Boyd, "Climate Change, Fragmentation, and the Challenges of Global Environmental Law: Elements of Post-Copenhagen Assemblage" (2010) 32 U Pa J Int'l L 457 at 466.

33 For the purposes of this book, non-state actors refer both to environmental NGOs and industry trade associations, such as IATA and A4A.

34 See Eric Dannenmaier, "The Role of Non-State Actors in Climate Compliance" in Jutta Brunnée, Meinhard Doelle & Lavanya Rajamani, *Promoting Compliance in an Evolving Climate Regime* (Cambridge: Cambridge University Press, 2012) at 175; Kenneth W Abbott, "Toward a Richer Institutionalism for International Law and Policy" (2005) 1 J Int'l L & Intl'l Rel 9 at 46; Alhaji BM Marong, "From Rio to Johannesburg: Reflections on the Role of International Legal Norms in Sustainable Development" (2003) 16 Geo Int'l Envtl L Rev 21 (discussing that, although States continue to have a predominant role in international law, the emergence of a "broad range of non-state actors" in the legal process is undisputed) at 49; Anne-Marie Slaughter, Andrew S Tulumello & Stephan Wood, "International Law and International Relations Theory: A New Generation of Interdisciplinary Scholarship" (1998) 92 Am J Int'l L 367 at 370 [Slaughter et al].

35 Suzannah Linton & Firew Kebede Tiba, "The International Judge in an Age of Multiple International Courts and Tribunal" (2009) 9 Chi J Int'l L 407 at 408.

36 See Koremenos et al, *supra* note 4 (Chapter 2) at 763.

37 See Anne Gerlach, "Sustainability Entrepreneurship in the Context of Emissions Trading" in Ralf Antes et al, eds, *Emissions Trading and Business* (Heidelberg: Physica-Verlag, 2006) at 74.

38 See Slaughter et al, *supra* note 34 at 371.

39 The United Nations Charter recognizes the role of non-governmental organizations in international relations. In this respect, the Charter provides that "[t]he Economic and Social Council may make suitable arrangements for consultation with [NGOs] which are concerned with matters within its competence." *Charter of the United Nations*, 26 June 1945, Can TS 1945 No 7 at Art 71.

40 Michele M Betsill and Elisabeth Corell, "NGO Influence in International Environmental Negotiations: A Framework for Analysis" in Peter M Haas, ed, *International Environmental Governance* (Burlington: Ashgate, 2008) 453 at 469. Charnovitz points out that "by boldly advocating new forms of cooperation, NGOs helped to make international law more responsive to the needs of the international community." Steve Charnovitz, "Nongovernmental Organizations and International Law" (2006) 100 Am J Int'l L 348 at 9; Wolfrum, "International Environmental Law", *supra* note 67 (Chapter 2) (describing the significant role of non-state actors in the development of international environmental law) at 5.

41 Zerk, *supra* note 10 (Chapter 5) at 57. See also Hildy Teegen, "International NGOs as Global Institutions: Using Social Capital to Impact Multinational Enterprises and Governments" (2003) 9 J Int'l Manag 271 (discussing that "international nongovernmental organizations (INGOs) have emerged as informal institutions operating globally to significantly change the context within which governments and [multinational enterprises] interact") at 271. See generally Pierre-Marie Dupuy & Luisa Vierucci, eds, *NGOs in International Law: Efficiency in Flexibility?* (Cheltenham: Edward Elgar, 2008).

in the following sections, IATA has played a predominant role in shaping ICAO's work and advancing concrete proposals for the global MBM scheme.

7.1.3 Who Are Norm Entrepreneurs?

Sunstein advances the idea that norm entrepreneurs[42] may prompt changes in social norms.[43] They may also invite reconsideration of existing policies.[44] In this vein, Adut notes that norm entrepreneurs may also work "towards the solidification of existing yet under-enforced norms."[45] Abbott and Snidal distinguish between "interest entrepreneurs" and "value actors." The former "seeks to advance their interests through engagement in politics, including transnational politics,"[46] whereas the behavior of the latter is guided by beliefs.[47]

In essence, a norm entrepreneur is an agent that advocates change[48] – an actor that promotes "a change of norms" to influence the behavior of others by targeting, local, national, or transnational issues.[49] In addition to setting new trends,[50] norm entrepreneurs are active in creating new social norms, social meanings for those norms, and sanction

42 Legal scholarship attributes Cass Sunstein for having coined in 1995 the term "norm entrepreneurship." However, Ann Florini claims that John Mueller first used this in May 1993 at a conference held at the University of California Los Angeles (UCLA) on "The Emergence of New Norms in Personal and International Behavior." See Florini, *supra* note 16 at 375. In addition, it is worth pointing out that, as far back as 1963, to refer to those actors promoting and enforcing norms, Howard Becker coined the term "moral entrepreneurs." Howard Becker, *Outsiders: Studies in the Sociology of Deviance* (New York: The Free Press of Glencoe, 1963).

43 Sunstein, *supra* note 18 at 909.

44 See Catherine Powell, "The Role of Transnational Norm Entrepreneurs in the U.S. 'War on Terrorism'" (2004) 5 Theoretical Inq L 47 (citing the case where norm entrepreneurs forced President George W Bush to re-think the US detention policy) at 71.

45 See Ari Adut, "Scandal as Norm Entrepreneurship Strategy: Corruption and the French Investigating Magistrates" (2004) 33 Theory and Society 5 529 (discussing the use of scandals by French magistrates as part of a "norm entrepreneurship strategy" to combat cases of corruption) at 530.

46 Abbott & Snidal, "Values and Interests", *supra* note 8 at 145.

47 *Ibid.*

48 See David Pozen, "We Are all Entrepreneurs Now" (2008) 43 Wake Forest L Rev 283 at 327; Florini, *supra* note 16 (describing norm entrepreneurs as "an individual or organization that sets out to change the behavior of others") at 375.

49 Robert C Ellickson, "The Market for Social Norms" (2001) 3 Am L Econ Rev 1 at 10 [Ellickson, "Market"]. See also Sandeep Gopalan, "Changing Social Norms and CEO Pay: The Role of Norms Entrepreneurs" 39 Rut LJ 1 [Gopalan, "Changing Norms"] (discussing that a "norm entrepreneur is an agent responsible for the invention or evolution of new social norms") at 30; Daniel Gilman, "Of Fruitcakes and Patriot Games" (2002) 90 Geo L J 2387 at 2395; Kristin Madison, "Government, Signaling, and Social Norms" (2001) U Ill L Rev 867 (commenting that norm entrepreneurship may play a decisive role in establishing new social norms) at 869. See also Annegret Flohr et al, *The Role of Business in Global Governance: Corporations as Norm-Entrepreneurs* (New York: Palgrave Macmillan, 2010) (discussing that norm entrepreneurs possess a strong sense of ownership and are highly committed to achieving their goals) at 96.

50 See Gilman, *supra* note 49 at 2395; Madison, *supra* note 49 (commenting that norm entrepreneurship may play a decisive role in establishing new social norms) at 869.

regimes to enforce them.[51] Focusing on their impact upon transnational corporations, Segerlund claims that norm entrepreneurs are actors "engaged in international norm construction aimed at changing the standards by which these enterprises are judged."[52] However, a norm entrepreneur may be an individual, a firm, a State, a government agency, or an international organization. I provide a few examples for better understanding.

In addition to being the founder of the International Committee of the Red Cross, Nobel Peace laureate Jean Henry Dunant was instrumental in the adoption of the 1864 Geneva Convention on international humanitarian law.[53] Harold Koh points out that Henry H Shelton, then Chair of the Joint Chief of Staff, and Madeleine Albright, then Secretary of State, acted as norm entrepreneurs in promoting the extension of the 12-mile territorial sea limit through presidential proclamation, given that the United States has yet to ratify *UNCLOS*.[54] Abbott and Snidal comment that the role of General Electric (GE) and international non-governmental organization Transparency International (TI) were instrumental in the adoption of the 1997 Organisation for Economic Co-operation and Development (OECD) *Anti-Bribery Convention*.[55] Although private norm entrepreneurs may exert significant influence, to be effective, they also need a government to sponsor or to "buy into" the proposal they seek to achieve.[56] In particular, this seems to be the case when dealing with legal norms. For instance, as will be seen below, IATA has advanced a number of concrete proposals to address GHG emissions from international aviation. IATA has however sought the endorsement of States gathered at ICAO.

Although it has been argued that private norm entrepreneurs may be more successful than States in creating and advancing changes in norms,[57] various commentators have indicated that governments can also play an active role as norm entrepreneurs, both at the

---

51 See Gopalan, "Changing Norms," *supra* note 49 at 3. Similarly, Berman notes that "individuals or groups who try to influence popular opinion in order to inculcate a social norm, may consciously try to mobilize social pressure to sustain or create social norms." Paul Schiff Berman, "Global Legal Pluralism" (2007) 80 S Cal L Rev 1155 at 1173. Luna & Cassell comment that the adoption of norms is the result of the actions of agents of change, opinion leaders such as norm entrepreneurs. See Erik Luna & Paul G Cassell, "Mandatory Minimalism" (2010) 32 Cardozo L Rev 1 at 16.

52 See Lisbeth Segerlund, *Making Corporate Social Responsibility a Global Concern: Norm Construction in a Globalizing World* (Burlington: Ashgate, 2010) at 1.

53 See *Convention for the Amelioration of the Condition of the Wounded in Armies in the Field*, 22 August 1864, online: ICRC <www.icrc.org/ihl/52d68d14de6160e0c12563da005fdb1b/87a3bb58c1c44f0dc125641a005a06e0>.

54 See Harold Hongju Koh, "Bringing International Law Home" (1998) 35 Hous L Rev 623 at 640-41 [Harold Koh].

55 See Abbott & Snidal, "Values and Interests," *supra* note 8 at 141.

56 See Harold Koh, *supra* note 54 at 648.

57 See Sandeep Gopalan, "Alternative Sanctions and Social Norms in International Law: The Case of Abu Ghraib" (2007) Mich St L Rev 785 at 815 [Gopalan, "Alternative Sanctions"] (in the case of torture awareness). Bodansky contends that in climate change issues, particularly in the United States, corporate non-state actors have been considerably more successful in their lobbying efforts than environmental NGOs. See Bodansky, *The Art and Craft of International Environmental Law*, *supra* note 62 (Chapter 1) at 127.

national and transnational levels.[58] An example of governments engaging transnationally as norm entrepreneurs can be found in the Basel Accords.[59] This refers to a set of non-binding banking regulations that more than 100 central banks have adopted.[60] Arguably, the Basel Accords have standardized banking practices around the world on issues such as capital and liquidity requirements, as well as leverage ratios.[61]

Analyzing the recent conflict in Syria, Çakmak claims that, although with limited success, Turkey has attempted to play the role of norm entrepreneur in establishing the responsibility to protect Syria's own subjects and avoid mass casualties.[62] An agency within a State may also exercise leadership. In the United States, the Federal Trade Commission (FTC) played an instrumental role in addressing privacy issues on the Internet.[63] The active involvement of these agencies becomes desirable, for it shapes the expected types of behavior.[64] Arguably, governments and their agencies can not only trigger norm entrepreneurship but can also block or delay the whole process despite the normative push by industry stakeholders.[65]

---

58 See Steven Hetcher, "The FTC as Internet Privacy Norm Entrepreneur" (2000) 53 Vand L Rev 204; Lee Tien, "Architectural Regulation and the Evolution of Social Norms" (2004) 7 Yale JL & Tech 1 (arguing that the US government acted as a "norm entrepreneur" when adopting the *Communications Assistance for Law Enforcement Act of 1994*, which required telecommunication services providers to provide content of a telephone communication which is subject to wiretapping) at 10; Joseph FC DiMento "Process, Norms, Compliance, and International Environmental Law (2003) 19 J Envtl L & Litig 251 (discussing that governments can promote norm entrepreneurship) at 260.

59 See Charles K Whitehead, "What's Your Sign? International Norms, Signals, and Compliance" (2006) 27 Mich J Int'l L 695 at 720. Similarly, elaborating further on cooperation agencies of different States, Raustiala advances the notion of "transgovernmentalism," which "will supplement, rather than supplant, the traditional tools of international law." Kal Raustiala, "The Architecture of International Cooperation: Transgovernmental Networks and the Future of International Law" (2002) 43 Va J Int'l L 1 at 6.

60 See Whitehead, *supra* note 59 at 723.

61 *Ibid.* Arguably, resorting to soft law instruments presents some advantages. Neuhold points out that soft law is a clear "manifestation of international law" that "may in fact sometimes solve the underlining problems." Hanspeter Neuhold, "Variations on the Theme of Soft International Law" in Isabelle Buffard et al, eds, *International Law between Universalism and Fragmentation* (Leiden: Martinus Nijhoff Publishers, 2008) at 359. For Ellis, instruments of soft law may be "a means of covering up significant divergences and disagreements and creating an appearance of consensus and common purpose." Jaye Ellis, *Soft Law as Topos: The Role of Principles of Soft Law in the Development of International Environmental Law* (DCL Thesis, McGill University Institute of Comparative Law, 2001) [unpublished] at 112. In addition, it may also "govern the interpretation of binding law, and may influence the content of norms." *Ibid.* at 114.

62 See Cenap Çakmak, "Assessing Turkey's Performance as Norm Entrepreneur in Syrian Crisis" (10 September 2013), online: Today's Zaman <www.todayszaman.com/news-325955-assessing-turkeys-performance-as-a-norm-entrepreneur-in-syrian-crisis.html>.

63 See Steven Hetcher, "The FTC as Internet Privacy Norm Entrepreneur" (2000) 53 Vand L Rev 204.

64 See Philip J Weiser, "Beyond Fair Use" (2010) 96 Cornell L Rev 91 1 (advocating more involvement from government to suggest what may be considered fair use in the copyright context) at 21. See also Abbott, *supra* note 34 (commenting that norm entrepreneurs "use techniques of persuasion to enlist States as norm supporters." In this respect, "early adopter States help enlist other until a tipping point is reached, and eventually the norms are widely internalized" at 28).

65 See discussion below.

International organizations may also serve as agents driving for change.[66] Finnemore's seminal work on *National Interest in International Society* describes the norm entrepreneurship role of the World Bank, UNESCO, and the International Committee of the Red Cross in exerting pressure on developed States to address poverty alleviation, impressing the need to develop science bureaucracies at the State level, and addressing the handling of the war wounded, respectively.[67]

7.1.4 Transnational Networks

Another feature of most norm entrepreneurs is the fact that they operate as part of a network. They may also establish relationship with epistemic communities. On international issues, norm entrepreneurs form transnational networks composed of very sophisticated actors.[68] Through these transnational networks, as Thomas points out, norm entrepreneurs may exert significant influence in the adoption of international norms by domestic actors.[69] Powell explains how human rights norm entrepreneurs resorted to their transnational networks (e.g. foreign governments, scholars, foreign human rights activists) to persuade the Bush Administration to factor in human rights considerations into the US policy of detaining suspected terrorists at Guantanamo Bay.[70]

Norm entrepreneurs in international civil aviation exhibit the classical attributes of transnational networks. As discussed throughout this book, one of the key actors in this process has been IATA. Although representing 240 airlines from around the world responsible for 84 percent of the international traffic,[71] and as such a heavy-weight in its own right, IATA has formed the Air Transport Group (ATAG), a broader coalition or a network of industry players to deal with climate change issues. Although led primarily by IATA, ATAG is composed of the trade associations of the world's airports, Airport Council International (ACI), and the world's air navigation services providers, Civil Air Navigation Services Organization (CANSO), and all the major aircraft and engine manufacturers (Boeing, Airbus, Embraer, Bombardier, ATR, Pratt & Whitney, GE Aviation, Honeywell, Royce Rolls), as well as of a number of regional and domestic airlines trade

66 See Martha Finnemore, *National Interests in International Society* (Ithaca: Cornell University Press, 1996) at 22 [Finnemore, *National Interests*].

67 *Ibid.*

68 See Finnemore & Sikkink, *supra* note 17 at 910.

69 See Daniel C Thomas, "Boomerangs and Superpowers: International Norms, Transnational Networks and US Foreign Policy" (2002) 15:1 Cambridge Rev Int'l Affairs 25 (explaining the considerable influence of Soviet dissidents and Eastern European émigrés exerted on US foreign policy with regard to human rights norms of the Helsinki Accords of the 1970s) at 44.

70 See Powell, *supra* note 44 at 47.

71 See IATA, "About Us" online: <www.iata.org/about/Pages/index.aspx>.

associations.[72] The only industry players absent from this network are the low-cost carriers who, for the most part, are not members of IATA.[73] In order to portray an image of a united, inclusive, and wide-ranging industry, all submissions, including working papers and presentations, dealing with climate change issues at ICAO are done through ATAG.[74] By extending its membership to actors other than airlines, ATAG has considerably expanded its reach. It operates as a true transnational network. ATAG's members have played an instrumental role on a number of key issues; for instance, Airbus, a member of ATAG, exerted enormous pressure on European authorities to temporarily suspend the EU ETS.[75] Even members of the European Parliament expressly recognized this pressure.[76] Because ATAG is essentially run by and through IATA, I will concentrate on IATA when addressing the role of industry.[77]

Although it does not have an equally expansive reach as ATAG, the European Union (EU) is another norm entrepreneur that uses its network to advance its quest to regulate GHG emissions from international aviation. Acting through the Commission, the EU typically resorts to academics, international environmental NGOs, and its numerous diplomatic missions around the world in order to advance its objectives.

7.1.5 What Drives Norm Entrepreneurs?

McAdams claims that "beliefs such as respect, prestige and esteem are what guide actor's behavior. Withholding esteem is, under certain conditions, a costless means of inflicting

---

72 See ATAG, "Our Members" online: <www.atag.org/membership/our-members.html>.

73 See Transport & Environment, "Global Deal or No Deal?" *supra* note 72 (Chapter 1).

74 See e.g. ICAO, A38-WP/68 EX/33, *Addressing CO_2 Emissions From Aviation* (presented by the Airports Council International (ACI), the Civil Air Navigation Services Organisation (CANSO), the International Air Transport Association (IATA), the International Business Aviation Council (IBAC), and the International Coordinating Council of Aerospace Industries Associations (ICCAIA)).

75 See Michael Szabo, "Trade War Unavoidable if EU Airline Emissions Plan Blocked – Lawmaker", *Reuters* (20 November 2013) online: <www.reuters.com/article/2013/11/20/eu-airlines-emissions-idUSL5N0J533O 20131120>.

76 Dr. Peter Liese, German Member of the European Parliament and *rapporteur* of the "Stopping the Clock" initiative" has been reported saying the following: "I'm most concerned that parts of the European industry are encouraging other countries to attack EU policy. This is completely unacceptable and I'm sure it will motivate MEPs (to support the commission's proposal). Airbus is the biggest problem, and this view is shared by my colleagues." Ewa Krukowska and Alesandro Vitelli, "EU Lawmaker Liese to Seek Changes to Draft Aviation Carbon Law", *Bloomberg* (20 November 2013) online: <www.bloomberg.com/news/2013-11-20/eu-lawmaker-liese-to-seek-changes-to-draft-aviation-carbon-law.html>. See also Szabo, *supra* note 75.

77 See generally Robert L Thornton, "Governments and Airlines" in Robert O Keohane & Joseph S Nye, Jr., eds, *Transnational Relations and World Politics* (Cambridge: Harvard University Press, 1970) (noting that "[t]he [airline] industry has developed a dynamism of its own, scarcely supranational but possessing sufficient power to occasionally thwart the desires of powerful governments" at 202).

costs on others."[78] Finnemore and Sikkink suggest that values such as "empathy, altruism, and ideological commitment" drive norm entrepreneurs.[79] These authors also contend that "many norm entrepreneurs do not so much act against their interest as they act in accordance with a redefined understanding of their interest."[80] As such, norm entrepreneurs are guided by a logic of appropriateness.[81] Another line of thinking argues that due to societal pressure, firms that manufacture goods, for instance, are more likely to engage in behavioral changes, if they are vulnerable to a reputational loss.[82] Because this may lead directly to financial losses, profit fall, and share prices decline, firms that value their reputation are more likely to engage in norm entrepreneurship.[83] States also care about their reputation.[84]

Sunstein highlights that the fear of social sanctions also drives support for a change of norms.[85] Abbot and Snidal favor a broader conception. In addition to noting that actors are guided by self-interest, they also recognize that other values or "nonmaterial ends," such as "prestige and respect" or doing the right thing, play a significant role in shaping their preferences and decisions.[86] There are also cases were actors are prompted to act due to a combination of self-interests and moral beliefs.[87] As will be explained further below, this is exactly the case of some of the actors in the aviation context.

7.1.6 *The Norm's Life Cycle*

Finnemore and Sikkink were the first scholars to advance an explanation of the life cycle of norms.[88] This classification is important for a better understanding of the different stages that a new norm goes through. It is relevant to examine the different legal norm and extra-legal norm proposals put forward by relevant actors as a means of dealing with GHG emissions from international civil aviation and to establish at which stage of the norm's life they are in order to determine whether they have succeeded or failed. For instance, as

78 See McAdams, *supra* note 7 at 342.
79 See Finnemore & Sikkink, *supra* note 17 at 898.
80 *Ibid.*
81 *Ibid.*
82 See Flohr et al, *supra* note 49 at 33.
83 *Ibid.* at 82.
84 See George W Downs & Michael A Jones, "Reputation, Compliance, and International Law" (2002) XXXI J Leg Stud 95 at 101.
85 See Sunstein, *supra* note 18 at 970.
86 See Abbott & Snidal, "Values and Interests," *supra* note 8 at 145.
87 For instance, Çakmak attributes two factors to Turkey's "normative stance" in the Syrian crisis, namely, a genuine moral belief that the responsibility to protect a State's own subjects should form the basis of an international norm, as well as the threat that such conflict presents for Turkey's national security. See Çakmak, *supra* note 62.
88 See Finnemore & Sikkink, *supra* note 17 at 888.

will be demonstrated, the EU ETS did not succeed in completing the whole norm's life cycle.

Unlike Sunstein who argues that the process of norm entrepreneurship leads "societies [to experience only] norm bandwagons and norm cascades,"[89] Finnemore and Sikkink identify three different stages in the life of norms.[90] These are emergence, cascade, and internalization.[91] Each of these stages is "affected by different social processes and logic of action."[92]

For a norm to appear or emerge, two requirements must be fulfilled. First, there must be a norm entrepreneur promoting the new norm or attempting to redefine an already existing norm that is not necessarily obeyed. During this stage, norm entrepreneurs engage in a process of persuasion and socialization to promote their ideas. Second, norm entrepreneurs also need a platform or an institutional setting to launch their proposal.[93] Normally, "norm entrepreneurs work from standing international organizations that have purposes and agendas other than simply promoting one specific norm."[94] Along these lines, Segerlund notes that "the platforms used by the norm entrepreneurs are often those of already existing international organizations (such as the UN, [ICAO, the International Maritime Organization (IMO)] and the International Labor Organization] where it will be possible to promote the norm. The expertise within and diffusion of information from these organizations are important sources of influence in international relations."[95]

To pass from the simple emergence of a norm to a full process of norm cascade, a tipping point must occur.[96] For Sunstein, the tipping point is a sign that identifies the occurrence of norm bandwagons, a stage characterized by the fact that "the lowered cost of expressing new norms encourages an ever increasing number of people to reject previously popular norms."[97] At this stage, because the new norm has already tipped, "adherence to the old norms [leads to social disapproval]."[98] Given that, to a large extent, Sunstein focuses on an incentive-based structure, his explanation may not capture those situations where actors are guided by factors other than interests, such as values or shared beliefs. Moving away from this conception, Finnemore and Sikkink point out that the tipping point is the moment when a critical mass of actors (e.g. States) opts to accept the norm change (e.g. global MBM

89 Sunstein, *supra* note 18 at 909.
90 See Finnemore & Sikkink, *supra* note 17 at 895.
91 *Ibid.* at 895.
92 Segerlund, *supra* note 52 at 6.
93 See Finnemore & Sikkink, *supra* note 17 at 896. Harold Koh refers to these institutional platforms as "law-declaring *fora*." Harold Koh, *supra* note 54 at 649.
94 Finnemore and Sikkink, *supra* note 17 at 899.
95 Segerlund, *supra* note 52 at 27-28
96 *Ibid.* at 29.
97 See Sunstein, *supra* note 18 at 912.
98 *Ibid.*

scheme).[99] This in turn prompts a new conceptualization or redefinition of the appropriate behavior sought.[100] Similarly, Segerlund claims that the tipping point is characterized by a stage where the norm gathers support from key actors.[101] This in fact legitimates the change in norms.[102]

The question then is how to establish what constitutes the critical mass of actors. Here, Finnemore and Sikkink point out that, in the international context, some model suggests that a critical mass represents a third of the membership of the relevant international organization.[103] Bearing in mind that ICAO has 191 Member States, 64 States will form a critical mass. Likewise, Gilligan and Nesbitt write that "the number of signatories of a convention that codifies a certain norm is the best measure of the number of States that consider that norm legitimate."[104] However, Finnemore and Sikkink also acknowledge that the determination of the critical mass should not be circumscribed to a mathematical formula, but rather by establishing who the critical actors are.[105] Not all States carry equal weight. For instance, it is well known that ICAO is developing a CO_2 emission standard for new aircraft. Regardless of how many States eventually adopt it, a critical mass will only be achieved when those hosting aircraft and engine manufacturers adhere to it (e.g. among others, Brazil, China, Canada, France, Germany, Russian Federation, the United States, and the United Kingdom). Similarly, it is ludicrous to think of a global MBM to address GHG from international aviation without the major emitters on board (e.g. the United States, the United Kingdom, China, Japan, Germany, UAE, and France).[106] Therefore, the determination of the critical mass of authors will inevitably be case-specific.

The second stage is norm cascade. Sunstein characterizes cascading as the period where "society is confronted with rapid shift in norms."[107] At this juncture, law may accelerate or slow down the process of norm cascade.[108] For Finnemore and Sikkink, this does not only refer to the frequency of changes but rather to the norm's "broad acceptance."[109] At this stage, norm entrepreneurs resort to "an active process of international socialization intended to induce norm breakers to become norm followers."[110] For these authors, the

99 Finnemore & Sikkink, *supra* note 17 at 895. See also Michael J Gilligan & Nathaniel H Nesbitt, "Do Norms Reduce Torture" (2009) 38 J Legal Stud 445 at 449.

100 Finnemore & Sikkink, *supra* note 17 at 902.

101 See Segerlund, *supra* note 52 at 28.

102 *Ibid.*

103 See Finnemore & Sikkink, *supra* note 17 at 901.

104 Gilligan & Nesbitt, *supra* note 99 at 450.

105 See Finnemore & Sikkink, *supra* note 17 at 901.

106 See Brown Weiss, *supra* note 89 (Chapter 2) (suggesting that "[i]t is necessary to include in international environmental agreements all those States that are essential for the agreement to be effective" at 691).

107 See Sunstein, *supra* note 18 at 912.

108 *Ibid.* at 968.

109 Finnemore & Sikkink, *supra* note 17 at 895.

110 *Ibid.* at 902.

"combination of pressure for conformity, desire to enhance international legitimation, and the desire of [different actors] to enhance their self-esteem facilitate norm cascade."[111]

The final stage in the norm's life cycle is internalization. This is where the norm achieves consolidation.[112] As Segerlund notes, this implies a "transformative change in substance."[113] At this stage, the norm acquires its "taken-for-granted quality."[114]Finnemore and Sikkink claim that "iterated behavior and habit" may facilitate this process.[115] Because there is a common understanding about its definition, its content, as well as the relevant actors involved,[116] the norm is fully internalized and becomes accepted. Any remaining resistance goes away, and the norm is no longer the subject of debate.[117] Reflecting their constructivist view, Finnemore and Sikkink also argue that through internalization, a norm becomes the "standard of appropriateness against which new norms emerge and compete for support."[118] In other words, the norm sets what actors ought to do. Both norm cascading and internalization are of utmost importance to gain adherence to ICAO's global MBM scheme.

7.1.7 Conditions Inducing Norm Entrepreneurship

Ellickson discusses that any given group comprises actors, enforcers, and members of the audience.[119] Driving for a norm change, and in order to "please their audiences," enforcers "administer informal sanctions to influence the behavior of actors."[120] For Ellickson, a change of norms occurs as a natural response to two specific situations, namely, (i) where there is a change in the economic conditions or incentive structure facing the group or (ii) where the presence of an "external event" or threat makes it necessary to reassess membership of the group.[121] Ellickson assumes that the audience is utilitarian. In other words, the audience would accept a change in behavior to the extent that it provides benefits to the membership as a whole, irrespective of whether there might be individual

111 *Ibid.* at 895.
112 See Segerlund, *supra* note 52 at 29.
113 *Ibid.* at 29.
114 *Ibid.* at 30.
115 Finnemore & Sikkink, *supra* note 17 at 905.
116 See Segerlund, *supra* note 52 at 29.
117 See Finnemore & Sikkink, *supra* note 17 at 895; Segerlund, *supra* note 52 at 29.
118 Finnemore & Sikkink, *supra* note 17 at 895.
119 Ellickson, "Market", *supra* note 49 at 5.
120 *Ibid.* at 6.
121 *Ibid.* at 4. Similarly, Skeel notes that "norm entrepreneurs are most likely to be effective if the costs or benefits of an existing norm, or the membership of the relevant community, have changed – that is, when exogenous factors have undermined an existing norm." David A Skeel, Jr. "Shaming in Corporate Law" (2001) 149 U Pa L Rev 1811 at 1822. Kübler points out that a change in the incentive structure "[changes] the social meaning and reputational value of a norm." Dorothea Kübler "On the Regulation of Social Norms" (2001) 17 JL Econ & Org 449 at 472.

losses.[122] Ellickson notes that social norms are likely to arise in situations where actors stand to obtain a benefit from cooperating and where there is a continuous interaction among them.[123] Ellickson's analysis is extremely relevant when analyzing the role that IATA has played, as well as the rationale behind it. It is also important when examining the role of informal enforcers when attempting to further elaborate on some key design elements of ICAO's global MBM scheme.[124] Other scholars suggest that norm entrepreneurs are more likely to arise where an active transnational public pressure exists and where the home State encourages agents to take social responsibility.[125] The domestic institutional landscape is also relevant. Norm entrepreneurship is likely to occur where the home State, through its different institutional channels, such as consultation processes already established, encourages the interaction between all relevant actors.[126] Given the transnational nature of corporations and the effects of globalization, one would think that the attitude of the home State is immaterial to the formation of norm entrepreneurship. This is hardly the case. The more disengaged the home State remains, the more difficult it is for either firms or individual actors to promote norm cascade within the context of an international organization. For instance, Thailand is a major touristic destination. The country ranks very high in terms of international aviation activity.[127] However, given that the Thai government has not been engaged at ICAO discussions, it will be very difficult for Thai Airways to advocate for normative change on aviation and climate change issues.

7.1.8 Influences Hampering Norm Entrepreneurship

Scholarship has sought to shed light on the conditions under which norm entrepreneurship and change in norms in general are more likely to occur. What the literature has perhaps not paid much attention to are those influences that may foster or hinder the development of norm entrepreneurship or the adoption of new norms. These may be extra-legal and legal influences operating either at the domestic or transnational levels of an endogenous or exogenous nature and affecting both social norms and legal norms. By legal influence, I mean a legal norm exerting significant pressure on the development of another legal norm. In Hart's mind, this would act as a "secondary rule" affecting the "creation or vari-

122 Ellickson, "Market", *supra* note 49 (noting that "in a group in which the fabric of social interactions is thick, a rational member should be able to see the advantage of having a utilitarian bias when appraising a proposed norm change" at 8).
123 *Ibid.*
124 This will be addressed in Chapter 8.
125 See Flohr et al, *supra* note 49 at 52.
126 *Ibid.*
127 "Civil Aviation: 2012 International RTK by State of Operator Certificate", online: ICAO <www.icao.int/Meetings/a38/Documents/International%20Scheduled%20RTK%20(Annual%20Report).PDF>.

ation of duties or obligations."[128] Similarly, non-legal influences are those conditions, other than legal influences, affecting the developing of legal norms.

The right-wing domestic resistance to climate change issues and the country's overall poor economic environment (extra-legal influences) make it extremely difficult for the United States to take a progressive stance in negotiating the post-*Kyoto* regime (legal norm). The EU ETS (legal influence or secondary rule) has been extremely influential in the ICAO Assembly resolution through which States agreed to develop a global MBM for international aviation (legal norm). China's national position (extra-legal influence) has played a significant role in the stance that Chinese state-owned international carriers have taken on aviation and climate change issues. Chinese carriers were always skeptical of the IATA-proposed industry targets (extra-legal social norm). When IATA adopted an industry-wide resolution calling upon ICAO Member States to adopt a global MBM scheme for aviation, Chinese carriers filed a reservation.[129] In addition, it is worth bearing in mind that the coalition of the unwilling that opposed the EU ETS (legal norm) resorted to a number of tactics to derail it. These included legal challenges by the industry and the Council Declaration of 2 November 2011.[130]

Exogenous forces are one specific type of influence that is problematic when addressing GHG emissions from international aviation. These are influences that are completely external to the actors' natural environment.[131] They may be extra-legal or legal influences. Although actors may operate within their own sectors, their actions may have economic, legal, or political repercussions way beyond their domains. Here, the issues are not only mono-sectoral but rather multi-sectoral and may even involve a plethora of different institutions. In these situations, those who are targeted by the norm must not only look at the proposal from the norm entrepreneur's sectoral perspective but must also consider its wider implications. This illustrates the limitations of sectoral norm entrepreneurs. This is particularly true for global industries such as international aviation, and it is even more so for multi-sectoral and multi-institutional subject matter such as climate change. In fact, as will be addressed in the following sections, this is part of the problem in addressing GHG emissions from international civil aviation. As discussed in Chapter 1, this is very much related to the fact that aviation's impact on climate change is a small piece within the global climate change regime as a whole.

---

128 Hart, *supra* note 27 at 79. Hart distinguishes between primary rules, that is to say, those imposing or forbidding a given action upon individuals (e.g. thou shalt not kill), and secondary rules which "are all concerned with the primary rules themselves." *Ibid.* at 92. Secondary rules empower normative change by establishing obligations. *Ibid.* at 79.

129 See IATA Resolution on the Implementation of the Aviation CNG 2020 Strategy [IATA, CNG Resolution].

130 See ICAO Council Declaration of 2 November 2011, *supra* note 153 (Chapter 4).

131 See Oran R Young, "Regime Dynamics: The Rise and Fall of International Regimes" (1982) 36:2 International Organization 277 (discussing that "international regimes quite frequently fall victim to the impact of exogenous forces" at 294).

7.1.9 Norm Internalization

One central aspect of norm entrepreneurship and norms in general is internalization: the process of acceptance that leads to the norm's adoption by the relevant actors. As will be seen in the case of the EU ETS examined below, it is insufficient to promote a norm if the actors that will need to adopt it do not internalize it, either because they do not share or fail to understand the values embedded in such norm or because they do not perceive any incentive structure that will induce them to do so.[132] If there is no internalization, like fashion, the new norm will pass quickly. They may emerge but may not necessarily lead to a change of behavior. This is why celebrities rarely serve as norm entrepreneurs.[133] Without internalization, norm entrepreneurship cannot have a meaningful effect and normative change is unlikely to occur.

In broad terms, there are two distinctive approaches to norm internalization. Rational choice theory suggests that there are situations in which the benefits that actors stand to attain provide sufficient justification to accept or internalize the proposed norm change. For these cases, the norm entrepreneur will seek to work on the incentive structure to trigger a norm cascade that will ultimately lead to full internalization.[134] If we take the ICAO global MBM as an example, States should be better off by joining the scheme. For constructivists, norm internalization is based on the "logic of appropriateness." According to Finnemore and Sikkink, "actors internalize [norms,] roles and rules as scripts to which they conform, not for instrumental reasons – to get what they want – but because they understand the behavior to be good, desirable, and appropriate."[135] Yet norms do not exist in a vacuum.[136] As noted above, there is a strong correlation between values and interests. In real life, they coexist and are often intertwined. As Abbott and Snidal put it, "interest may generate values over time," and "values may [also] shape interests."[137] To this end, it is imperative to consider norm internalization from both perspectives.

To better understand the "internalization" of international norms, I look into Koh's transnational legal process. For Harold Hongju Koh, this refers to "a process whereby an international law rule [or norm] is interpreted through the interaction of transnational actors in a variety of law-declaring fora, then internalized into a nation's domestic legal

---

132 See generally Gopalan, "Changing Norms," *supra* note 49 at 22.

133 *Ibid.* at 32.

134 See Abbott, *supra* note 34 (suggesting that "to influence self-interested actors, norm entrepreneurs must deal with incentives" at 30); Checkel, *supra* note 8 at 477.

135 Finnemore & Sikkink, *supra* note 17 at 912.

136 See Florini, *supra* note 16 at 376.

137 Abbott & Snidal, "Values and Interests," *supra* note 8 at 176.

system."[138] In essence, this is a top-down approach to incorporate international norms into the domestic system.[139]

Koh advances the idea that internalization refers to an "actor's internal acceptance of [a] rule as a guide for behavior."[140] Koh says that a norm may trigger four types of relationships with respect to an actor's conduct: coincidence, conformity, compliance, and obedience.[141] With no causal nexus, an actor may decide to follow a norm out of mere "coincidence."[142] An actor does not necessarily make a conscious assessment of whether it should follow the norm. Likewise, an actor may observe or conform to a norm out of convenience.[143] According to Koh, "compliance" takes place when an actor "consciously accepts" the norm to gain a benefit or to avoid a punishment.[144] This stage would seem to be in line with Sunstein's norm bandwagon stage where the lower transaction costs induce actors to join the norm. It also follows the rational choice model, where actors act because they are induced by changes in the incentive structure. On the other hand, "obedience" occurs not only when the actor internalizes the norm but when it incorporates it as part of its own values.[145] Following the constructivist's view, Koh claims that the internalization of norms should not necessarily emphasize on coercion but rather on obedience.[146] Yet actors would only internalize a norm after two key processes have occurred: interaction and interpretation.[147] Interaction refers to the continuous interface of all the relevant actors. This is extremely important, given that, as Björkdahl says, before adopting new norms, actors such as States tend to engage in internal discussions.[148] Interpretation, on the other hand, is the subjective assessment that actors engage in with respect to the proposed norm

138 Harold Koh, *supra* note 54 at 626.

139 This approach is germane to this book, for we are concerned with international or regional norms or proposals for international norms (e.g. EU ETS, IATA's proposal for a global offsetting scheme, and ICAO's commitment to develop a global MBM for international aviation). At this stage of the process, it is not clear in what form the proposed international norms will impose implementing obligations at the domestic level. What is evident is that they will affect the behavior of States and domestic actors (e.g. airlines).

140 Harold Koh, *supra* note 54 at 628.

141 *Ibid.* (discussing "four kinds of relationship between stated norms and observed conduct" at 627).

142 *Ibid.*

143 *Ibid.* at 628.

144 *Ibid.* A number of commentators describe the role of sanctions, whether legal or non-legal, as one of the characteristics present in norms. For example, McAdams regard social norms as "informal social regularities that individuals feel obliged to follow because of an internalized sense of duty, because of a fear of external non-legal sanctions, or both." See McAdams, *supra* note 7 at 340. Likewise, Ellickson conceives norms as "rules governing an individual's behavior that third parties other than state agents diffusely enforce by means of social sanctions." Ellickson, "Market", *supra* note 49 at 3. However, the relevance of sanctions is perhaps not as important in the international context, for there is no "direct punitive capacity." Finnemore & Sikkink, *supra* note 17 at 893.

145 See Harold Koh, *supra* note 54 at 628.

146 *Ibid.* at 629. Similarly, Florini writes that "[n]orms are obeyed not because they are enforced, but because they are seen as legitimate." Florini, *supra* note 16 at 365.

147 See Harold Koh, *supra* note 54 at 632.

148 See Björkdahl, *supra* note 11 at 13.

change.[149] Moreover, Koh suggests that obedience involves an iterative educational process to teach relevant actors about the right thing to do.[150] For Cooter "internalization specially occurs through socialization."[151] Sharing this line of reasoning, Checkel writes that in many cases, there are no "obvious material incentives."[152] One of the diffusion mechanisms to internalize norms is the learning process of the elites.[153] The teaching of new values seeks to guide actors into a "logic of appropriateness."[154]

As a result of interaction, "[n]orms are internalized and constitute a set of shared intersubjective understandings that make behavioral claims."[155] They are instrumental in leading to cognitive consensus.[156] Although the role of the educational process is fully aligned with the constructivist view, for it seeks to form values and shared understanding of norms, it may also have some application in determining the actors' interests. For a variety of reasons, some States often confront difficulties in defining their own public interest. In practice, it is not rare to find States, particularly small States, not knowing what they want or what their position should be vis-à-vis a given issue. Finnemore writes that "States may not always know what they want and are receptive to teaching about what are appropriate and useful actions to take."[157] For instance, despite all these years since the *Kyoto Protocol* was adopted, a number of ICAO Member States have failed to understand why GHG emissions from international ought to be regulated, they do not perceive the benefits of a global MBM scheme, nor do they have a position on the subject. This leads them to exercise a passive role. Here, the educational process may play a significant role even when dealing with the interests of States.

Where benefits are not necessarily self-evident, as Vandenbergh suggests, influencing the actor's belief with respect to a given subject matter may facilitate the internalization of a norm change.[158] However, in order for this to happen, norm entrepreneurs must raise awareness of the issue in question through the handling of accurate and appropriate information. As we will see in the case of MBMs in international aviation, mishandling of information and a poor communications campaign can lead to opposite results. In fact, I will argue below that this has been one of ICAO's missed opportunities. Individuals, firms,

149 See Harold Koh, *supra* note 54 at 626.

150 *Ibid.* at 629.

151 Robert Cooter, "Normative Failure Theory of Law" (1997) 82 Cornell L Rev 5 947 at 954.

152 Checkel, *supra* note 8 at 487.

153 *Ibid.*

154 *Ibid.* at 477.

155 *Ibid.* Similarly Abbott and Snidal claim that if the targets are actors who based their decisions on values, the strategy should be to work on "incentives on information but by engaging processes of persuasion, education, socialization and legitimation." See Abbott & Snidal, "Values and Interests," *supra* note 8 at 151-52.

156 See Bodansky, *The Art and Craft of International Environmental Law, supra* note 62 (Chapter 1) at 149.

157 Finnemore, *National Interests, supra* note 66 at 11.

158 See Michael P Vandenbergh "Order Without Social Norms: How Personal Norm Activation Can Protect the Environment" (2005) 99 Nw UL Rev 1101 at 1102.

and even States may reject the change in norms. Those with whom the norm is internalized must fully understand why norm entrepreneurs advocate the change. Launching a change crusade is simply not enough. The relevance of Koh's interaction and interpretation processes cannot be underestimated. It is imperative to create a platform where all relevant actors discuss and exchange views.[159] In order to build up trust, the process will only gain legitimacy if it is inclusive. The proposed change of norms should not be perceived as an imposition. If this is the case, although actors may grudgingly accept the norm, it will certainly never form part of their values, making it much more difficult to drive for a genuine change of norms.[160] The EU ETS epitomizes a case where because the normative change is perceived as an imposition, those targeted by the norm simply reject it.

7.1.10 The Relevance of Regime Architectural Design in the Formation of International Norms

The architectural design of the regime developing international norms is extremely important for norm entrepreneurs, since some types may pose significantly more challenges. From the perspective of the direction in which the norm is being proposed, at least two distinctive scenarios may arise: bottom-up or top-down norm formation. However, this may not necessarily be the perspective for those other international actors consenting or opposing to the norm. As we will see further below, the top-down approach (which is most often used in the process of norm formation when dealing with aviation and climate change issues) is much more problematic.

7.1.10.1 Bottom-Up Norm Construction

In a bottom-up norm construction scenario, the legal norm is initiated by domestic actors, either within the private or government sectors or both. A variety of reasons may explain the motives of domestic actors in their pursuit, including, but not limited to, self-interests, societal pressure, and to some extent values and beliefs. After a process of consultation and interaction, the proposed new legal norm is taken from the domestic level to the international area through the intervention of transnational norm entrepreneurs.[161]

159 See also Brunée & Stephen J Toope "International Law and Constructivism: Elements of an Interactional Theory of International Law" (2000) 39 Colum J Transnat'l L 19 (discussing that "law is most persuasive when it is created through processes of mutual construction and a wide range of participants in the legal system" at 19).

160 Critics to norm entrepreneurship point out that the whole process of norm formation may take considerable time to materialize. They also contend that the concept of norm entrepreneurship is applied loosely with no precise understanding of why other actors follow these advocates of change. See Pozen, *supra* note 48 at 310.

161 See Björkdahl, *supra* note 11 (suggesting that "domestic norms may spread to the global level" at 18).

If the norm entrepreneurs are private actors, they will necessarily need the intervention of a State (usually their home State) to sponsor the new norm. Both the private norm entrepreneur and the sponsoring State will resort to an institutional forum where the proposal will be discussed and eventually adopted.[162] At this stage, in most cases, the proposed norm change has already been adopted at the domestic level.[163] This implies a considerable degree of domestic acceptance. In a bottom-up norm construction scenario, domestic actors of the sponsoring States are less likely to interfere with, offer resistance, or lobby against the new norm. On the contrary, they are likely to support it. As such, the chances of domestic, extra-legal influences on legal norms are substantially less.

Two examples may well illustrate this scenario: the adoption of the OECD Anti-bribery Convention and the ICAO standards on reinforced cockpit doors. In the 1970s, the United States experienced a series of notorious scandals, such as Watergate, the illegal contributions to US President Richard Nixon's reelection campaign and the bribing of high-ranking senior foreign officials by Lockheed Martin to secure military aircraft purchases.[164] As a result of this, the US Congress enacted the *US Foreign Corrupt Practices Act* by criminalizing the bribing of foreign officials by US citizens and US corporations.[165] Because the United States was the first country to adopt this type of legislation, US corporations were disadvantaged vis-à-vis, for instance, European firms which did not face any statutory restrictions. In order to level the international playing field, US firms, in particular General Electric, lobbied the US government to push for an international convention that would criminalize this type of behavior.[166] After years of negotiations, and with the strong US influence, in 1997, the OECD finally adopted the Anti-bribery Convention.[167] Similarly, prompted by the 9/11 societal pressure, on 11 January 2002, among other security measures, the US Federal Aviation Administration (FAA) imposed new requirements on reinforced aircraft cockpit doors for both US air carriers and foreign aircraft flying to and from the United States.[168] At the same time, the United States sought to transpose this domestic legal norm to the international arena. Thanks to US norm entrepreneurship, through an expedited procedure, on 21 March 2002, ICAO incorporated this requirement as one of its interna-

162 See Finnemore & Sikkink, *supra* note 17 at 893.

163 *Ibid.* (noting that "international norms must always work their influence through the filter of domestic structures and domestic norms" at 891).

164 See Abbott & Snidal, *supra* note 8 at 161-62.

165 *US Foreign Corrupt Practices Act*, 15 USC § 78 dd-1, 2 (1977). See also Bruce Maloy, "Extraterritorial Application of United States Criminal Statutes to the Gaming Industry" (1997) 1 Gaming L Rev 491; Zerk, *supra* note 10 (Chapter 5) at 33.

166 See Abbott & Snidal, *supra* note 8 at 161-62.

167 *Ibid.*

168 See FAA, Press Release, "FAA Sets New Standards for Cockpit Doors", *FAA* (11 January 2002) online: <www.faa.gov/news/press_releases/news_story.cfm?newsId=5470>.

tional standards.[169] Since then, as a renowned norm entrepreneur, the United States has resorted to the bottom-up approach to norm construction in a large number of aviation security issues. The adoption of domestic legal norms is quickly followed by an attempt to internationalize it at ICAO.[170] Domestic interests serve as influences, either legal or extra-legal, in the adoption of international norms.

7.1.10.2 Top-Down Norm Construction

The adoption of a large number of international treaties follows a top-down norm construction approach. Under this scenario, a norm entrepreneur, either a State or a private actor through a sponsoring State, advances a proposal to adopt a new international, legal norm. The proposal is discussed and negotiated by States at the international forum (e.g. UN specialized agency). In most cases, non-State actors also participate in these discussions. States adopt the international norm, and then each State bears some implementing obligations. The international norm trickles down into the domestic system. The main difference is that in a bottom-up norm formation approach, there is already a certain degree of acceptance at the domestic level. This is not necessarily the case in a top-down scenario.

The State internalizing those implementing obligations that stemmed from the new international norm may face considerable resistance at the domestic level.[171] A classic example of this approach is the adoption of the *Kyoto Protocol* and its subsequent rejection by the United States. To a great extent, the adoption of the *Kyoto Protocol* was the result of the significant US involvement in the text's preparatory stages, as well as the active engagement of the Clinton Administration headed by Vice-President Al Gore, a well-known environmentalist. Yet the US Congress did not necessarily share the Clinton Administration's commitment toward climate change issues. These were never internalized within the US constituency, at least sufficiently enough to ensure US adherence to *Kyoto*. Pressed by strong, domestic lobbying groups, the US Congress was forced to preclude the ratification of the instrument. Consequently, the Clinton Administration did not even attempt to send the instrument for ratification to the US Congress. Years later, US President George W Bush made it explicit that the United States would not ratify the instrument.[172]

---

169 See ICAO Annex 6, *supra* note 99 (Chapter 5), Standard §13.2.2. See also ICAO, News Release, PIO 04/02, "ICAO Council Adopts Stronger in Flight Security Standards" (21 March 2002), online: ICAO <http://legacy.icao.int/icao/en/nr/2002/pio200204_e.pdf>.

170 See generally Checkel, *supra* note 8 (commenting that "societal pressure explains the domestic empowerment of global norms" at 478).

171 Rosendal explains that "implementation" refers to the "deliberate efforts by national authorities to follow up their international commitments in domestic policies within the specific issue area." This may involve "ratification, legislation establishing domestic policies, and programs aimed at following up international commitments at the national level." Rosendal, *supra* note 25 (Chapter 2) 67 at 96.

172 See Kathryn Harrison, "The United States as Outlier: Economic and Institutional Challenges to US Climate Policy" in Kathryn Harrison & Lisa McIntosh Sundstrom, eds, *Global Commons, Domestic Decisions: The Comparative Politics of Climate Change* (Cambridge: MIT Press, 2010) at 93.

7.2 WHO IS THE REAL NORM ENTREPRENEUR?

The previous sections have laid down the foundational grounds of the theory of norms and norm entrepreneurship. In this section, I apply these concepts to the role of the major actors in addressing climate change in international aviation: the EU, IATA, the United States, and ICAO. By doing so, I seek to identify not only their merits and shortcomings but also I purport to identify corrective actions that may contribute to ICAO's global MBM.

7.2.1 *Europe*

7.2.1.1 An Obstacle or a Building Block for an International Agreement?

The EU ETS was described as a "major impediment,"[173] a "roadblock,"[174] "counterproductive,"[175] and "antithetical"[176] toward achieving a global climate regime for international aviation.[177] Milde noted that "the unilateral and hasty action by the EU [constituted] an obstacle to the progress of ICAO that is underway."[178] Sanchez suggested that the scheme undermined the "diplomatic goodwill to build an authentically global aviation emissions framework."[179] Some ICAO Member States also indicated that Europe's unilateral action would "jeopardize global efforts."[180] In a similar vein, the US State Department said that the EU ETS represented a challenge to the ICAO process.[181] More specifically, Todd Stern, the US Special Envoy for Climate Change Negotiations, conjectured that, at the peak of the controversy, the EU ETS would "hold up [the *UNFCCC*] climate change talks."[182]

---

173 Statement of the Honorable Ray LaHood, Secretary of Transportation before the Committee on Commerce, Science, and Transportation of the US Senate (6 June 2012). See also US State Department, "Ninth Meeting of the EU-US Joint Committee Record of Meeting of 22 June 2011" *US Department of State* (22 June 2011) online: <www.state.gov/e/eb/rls/othr/ata/e/eu/192088.htm>; Bart Jansen, "FAA Official Raps EU over Airline Carbon Tax", *USA Today* (13 April 2013), online: <http://travel.usatoday.com/flights/post/2012/04/faa-official-wraps-eu-over-airline-carbon-tax/670800/1>.

174 Nancy N Young, *supra* note 4.

175 "Comment on European Court of Justice Decision", *A4A* (21 December 2011) online: <www.airlines.org/Pages/news_12-21-2011.aspx>.

176 FAA Modernization and Reform Act of 2012 at 97.

177 See Stephanie Koh, *supra* note 45 (Chapter 2) at 145.

178 See Milde, "EU Emissions Trading Scheme," *supra* note 62 (Chapter 4) at 10. See also US, Bill HR 2594, *An Act to Prohibit Operators of Civil Aircraft of the United States from Participating in the European Union's Emissions Trading Scheme, and for other Purposes*, 112th Cong, 2011.

179 Gabriel S Sanchez, "In Defense of Incrementalism for International Aviation Emissions Regulation" (2012) 53:1 VJIL 1 at 4.

180 Nigam Prusty, "EU Airline Charge Hurts Climate Fight-China, India", *Reuters Africa* (14 February 2012) online: <http://af.reuters.com/article/energyOilNews/idAFL5E8DE8AS20120214>.

181 See Karen Walker, "US State Views EU ETS as a Challenge to ICAO Process", *Air Transport World* (23 March 2012) online: <http://atwonline.com/airline-finance-data/news/us-state-views-ets-challenge-icao-process-0323>.

182 *Ibid.*

Similarly, a senior Indian official commented that the EU ETS did "not only [stand] in violation of the principles and provisions of the [*Chicago Convention*], but [it also did not] augur well for the success of future climate change negotiations."[183] To add to the chorus of criticisms, IATA characterized the scheme as a "misguided plan" that undermined ICAO's leadership,[184] and "a polarizing obstacle that was preventing real progress."[185] Mr. Tony Tyler, IATA's Director General, has been reported by the press as saying that Europe's plan to cover only emissions generated within EU's own airspace "has the potential to undermine the goodwill [of other States]."[186] Along with some commentators,[187] the EU attempted to justify the measure by saying that it is "an important contribution to, and a catalyst towards, global [environmental stewardship] rather than an obstacle."[188]

Despite the criticism and the politically charged allegations leveled against the scheme, it is hard to see how pushing the environmental envelope by including aviation into the EU ETS, as Europe did, could become an obstacle toward the achievement of an international agreement on the subject. It is true that, as explained in Chapter 4, Europe's unilateral actions were met with resentment from other States. Yet the thrust of this argument is that the international community spent a considerable amount of time fighting the EU ETS instead of finding an international solution for a policy dilemma that involves a sector that, by its very nature, is global. Judging from the track record of those States that vehemently opposed the EU ETS, it seems quite unlikely that any concrete, binding proposals to control aviation's CO_2 emissions would have been advanced had the EU ETS not existed. In fact, many of those countries not only failed to exercise leadership but acted as "persistent obstructionists" by blocking progress at ICAO.

---

183 Tarun Shukla, "India Seeks Reversal of EU Airline Carbon Levy", *Green Air Online* (9 January 2012) online: <www.livemint.com/2012/01/09214425/India-seeks-reversal-of-EU-air.html>.

184 See Position of the Coalition of the Unwilling, *supra* note 150 (Chapter 4) at 3. See also "Europe Should Abandon 'Misguided' ETS Plans", *Global Travel Industry News* (28 September 2011) online: <www.eturbonews.com/25447/iata-europe-should-abandon-misguided-ets-plans>. See also Position of the Coalition of the Unwilling, *supra* note 150 (Chapter 4) at 4; summary of the discussions of the meeting that took place in Washington, DC, on 31 July through 1 August 2012 (archived on file with the author).

185 IATA, Press Release, 25 (11 June 2012) "IATA Urges ICAO Solution on Economic Measures for Climate Change" online: IATA <www.iata.org/pressroom/pr/Pages/2012-06-11-04.aspx>. See also Bisignani, *Words of Change*, *supra* note 20 (Introduction) (warning that the EU ETS could "undermine global efforts, and distort markets" at 170).

186 Daniel Michaels, "Airlines: EU Plans Put Global Emissions Deal at Risk", *The Wall Street Journal* (17 October 2013), online: <http://blogs.wsj.com/brussels/2013/10/17/airlines-eu-plans-put-global-emissions-deal-at-risk/>.

187 See Mayer, "Defense of the EU ETS", *supra* note 414 (Chapter 5) (underlining that the scheme "pushes for a global regime" at 27); Kulovesi, "Addressing Sectoral Emissions", *supra* note 214 (Chapter 1) (discussing that the scheme "makes useful contribution to the evolution of climate law and toward the implementation of the ultimate objective of the *UNFCCC*" at 202).

188 Letter of Sim Kallas (Vice-President of the European Commission) and Connie Hedegaard (Climate Change Commissioner) on 16 January 2012 addressed to Ms. Hillary Rodham Clinton (US Secretary of State) and Mr. Raymond H LaHood (US Secretary of Transportation). See also Delbeke, *supra* note 7 (Chapter 4).

As the "elephant in the room," the EU ETS pressed both ICAO and the aviation industry to develop concrete measures to regulate CO_2 emissions from international aviation. As examined below, with regard to ICAO, the EU ETS has acted as a legal influence or a secondary rule in Hart's mind on the development of a legal norm (e.g. ICAO Assembly Resolution agreeing to develop a global MBM scheme for international aviation). Yet the EU ETS has also exerted influence on IATA's extra-legal norms (e.g. industry, aspirational targets, and its overall strategy to tackle the sector's climate change impact).

At present, the EU ETS is still the only regulatory measure in place. This leverage is perhaps best exemplified by the fact that, in November 2011, the ICAO Council was "induced" to instruct the Secretary General to carry out an "accelerated work program" to explore the feasibility of a global MBM scheme for international aviation, on the same day that it decided to pass a declaration opposing the EU ETS.[189] Some weeks later, the Secretary General of ICAO announced that the organization would table a draft proposal on globally agreed MBMs by the end of 2012.[190] In June 2012, the ICAO Secretariat along with a group of experts identified 4 potential measures.[191] The Secretariat suggested that "possible options for a global MBM scheme" may be put forward for consideration by the ICAO Council by November 2012.[192] In spite of the Secretariat's proposal, thanks to extensive US lobbying, strongly supported by countries such as China, Brazil, India, Saudi Arabia, and Argentina, any reference to a timeline was deleted.[193] This was a deliberate attempt to slow down the initial rush to develop a global scheme. NGOs were quick to point out that the organization was becoming less ambitious, as compared to its announcement of December 2011.[194] The EU announcement to postpone the enforcement of the EU ETS in November 2012 further weakened the desire to explore MBMs as a means of addressing international aviation's carbon footprint.[195] At this juncture, the development of the global scheme at ICAO was stalled for reasons other than the EU ETS.

189 See ICAO, C-DEC 194/2, *supra* note 154 (Chapter 4). See also ICAO Council Declaration of 2 November 2011, *supra* note 153 (Chapter 4).

190 See "UN Aviation Body Says Emissions Proposal by Year-End", *Reuters* (2 March 2012) online: <www.reuters.com/article/2012/03/02/airlines-emissions-idUSL2E8E2B9720120302>. This announcement was received with cautious optimism. However, some commentators were skeptical about ICAO being able to develop a global scheme through multilateralism. The inability of ICAO Member States to agree on a harmonized aviation trade regime on issues such as market access, competition, and fair and equal opportunities may be illustrative of the fact that fragmentation of a patchwork of different regimes will continue to coexist. See Sanchez, *supra* note 179 at 6.

191 See ICAO, C-WP/ 13798, *supra* note 188 (Chapter 2).

192 *Ibid.*

193 On 9 March 2012, the ICAO Council "requested the Secretary General…to continue further work on the framework for MBMs and the exploration of the feasibility of a global MBM scheme." ICAO, C-DEC 195/9, *Council, 195th Session, Decision of the Ninth Meeting.*

194 See Open Letter from 25 NGOs Addressed to ICAO dated 7 November 2012 [unpublished, archived on file with the author].

195 For instance, the Moscow Declaration urged EU Member States "to work constructively forthwith in ICAO on a multilateral approach to address international civil aviation emissions." However, once the EU ETS

Outside of Europe, there was almost unanimous consensus that only ICAO (and not the EU) should have the competence to address emissions from international aviation. This notion is to be found in almost every single speech condemning the EU ETS.[196] Objectors, however, disagreed with respect to what exactly the organization was supposed to do to address the problem. There were many references to a "global MBM scheme" or a "framework of MBMs" under the auspices of ICAO, yet not much substance was provided as to how these concepts will eventually work. As will be explained below, it took a considerable amount of time for IATA to provide details and design elements for the global scheme. Quite often, these rhetorical statements were not supported with concrete actions.[197] Obviously, this was not ICAO's fault. Rather, it was the result of a serious lack of political commitment on the part of ICAO's Member States.[198] This explains why the European Commission, while acknowledging the international discomfort with the EU ETS, often asked what the international community envisioned under the dogmatism of working at ICAO.[199] Most of the time, responses were evasive.[200] In this sense, the EU ETS was not an obstacle but rather a fierce force (legal influence) that directly challenged the status quo.[201]

was suspended, opponents pushed back. All of a sudden, there was no urgent and immediate need to address the issues. See Joint Declaration of the Moscow Meeting on the Inclusion of International Civil Aviation in the EU ETS of 20-21 February 2012.

196 See generally "Stop the European Union from Unilaterally Applying a Cap-and-Trade Tax to U.S. Airlines and Passengers" *A4A* online: <www.airlines.org/Pages/EU-ETS-Fact-Sheet.aspx>; "ETS Should be About Sustainability, not Politics", *AEA* online: <http://files.aea.be/News/PR/Pr11-020.pdf>; "Emissions Trading Concerns Will Be Resolved Through ICAO", *AEA* online: <http://files.aea.be/News/PR/Pr11-025.pdf>.

197 This is not an exclusive problem of ICAO or its Member States vis-à-vis the organization's work program. For instance, Noah Sacks comments that there is also a significant "gap between States' rhetorical commitment to enhancing the role of tort in international law and what States actually have accomplished." Noah Sachs, "Beyond the Liability Wall: Strengthening Tort Remedies in International Environmental Law" (2008) 55 UCLA L Rev 837 at 843.

198 As Andrew Guzman points out, "international institutions [such as ICAO] are created and controlled by States." Andrew T Guzman, *How International Law Works: A Rational Choice Theory* (Oxford: Oxford University Press, 2008) at 19.

199 For instance, at the 38th Assembly, Vietnam presented a working paper addressing GHG emissions from international aviation. Although Vietnam indicated that the issue should be "addressed through a global approach under the leadership of ICAO," it failed to provide any guidelines or concrete recommendations for such scheme. See ICAO, A38-WP/350 EX/120, *To Reduce CO_2 Emissions from Aviation* (presented by Vietnam) at 2.

200 An interesting discussion between the United States and the EU occurred during the occasion of the ninth meeting of the EU-US Joint Committee that took place on 22 June 2011. At this meeting, the EU asked a number of questions to understand US policy and aviation and climate change and how the United States "envisaged a global agreement through ICAO." The United States responded that the subject matter of the discussion was the EU ETS, but not US policy. See US State Department, "Ninth Meeting of the EU-US Joint Committee Record of Meeting of 22 June 2011" <www.state.gov/e/eb/rls/othr/ata/e/eu/192088.htm>.

201 See Petsonk, "Testimony", *supra* note 98 (Chapter 4) (saying that it is "no coincidence that serious discussions at ICAO would finally occur in tandem with the advent of the EU-ETS" at 15).

A report prepared by the US Congressional Research Service even acknowledges the influence of the scheme on ICAO's decision to "accelerate" its work.[202]

The legislation including international aviation in the EU ETS conferred a permanent mandate upon the Commission and EU Member States to continue engaging in negotiations with third countries.[203] It also acknowledged that bilateral arrangements on linking other market-based regulatory initiatives with the EU ETS, as well as the recognition of equivalent measures, could serve as building blocks toward a global agreement.[204] The Commission repeatedly stated that it was open to discuss with other States, but was not willing to amend its legislation (although it ultimately was forced to), unless a binding global agreement had been reached at ICAO.[205] As the Aviation Directive clearly suggested,[206] the global scheme was certainly the preferred regulatory option, for it would take the discontent and the international pressure away from Europe.

When the *Kyoto Protocol* was adopted, European States were largely responsible for keeping the subject of MBMs on ICAO's agenda. In the words of Finnemore and Sikkink, as a norm entrepreneur, Europe has "[called] attention to the issue."[207] Using the norm life cycle advanced by these authors, we can see that Europe has been responsible for the "emergence" of the norm. Like it or not, this stubbornness on the part of Europe, coupled with the inability of the international community to commit to binding emission reduction targets for aviation as shown to date, placed tremendous pressure on ICAO Member States and the aviation industry as a whole.[208] As will be addressed below, IATA's climate change targets were only announced after the Commission made public its intention to include foreign aircraft operators into the EU ETS.

---

202 See Jane A Leggett, Bart Elias & Daniel T Shedd, "Aviation and the European Union's Emission Trading Scheme", Congressional Research Service (15 May 2012) online: <www.fas.org/sgp/crs/row/R42392.pdf> at 3.

203 AFP, "EU 'Open' to Talks on Airlines Tax, but Won't Change Law", *Business Recorder* (4 March 2012) online: <www.brecorder.com/top-news/1-front-top-news/48229-eu-open-to-talks-on-airlines-tax-but-wont-change-law-.html>.

204 See Aviation Directive, *supra* note 17 (Introduction) at preambular clauses. Even the CJEU noted that the Commission should "seek an agreement on global measures" for international aviation. See *ATA Decision*, *supra* note 63 (Chapter 4) at para 12; Bogojovic, *supra* note 134 (Chapter 4) at 348.

205 "Commission seeks to speed up EU carbon market remedy", *Reuters* (5 October 2012) online: <www.reuters.com/article/2012/10/05/eu-ets-idUSL6E8L5HJE20121005>.

206 See Aviation Directive, *supra* note 17 (Introduction) (stressing that Europe "should continue to seek an agreement on global measures to reduce" GHG emissions from international civil aviation" at preambular clauses, para 17).

207 Finnemore & Sikkink, *supra* note 17 at 897.

208 Commentators also assert that for the European climate change policy to succeed, a global carbon market is a *condition sine qua non*. A global system may contribute to this objective. For this to happen, engagement of key global actors, such as the United States, becomes necessary. See Wolfgang Sterk & Joseph Kruger, "Establishing a Transatlantic Carbon Market" (2009) 9 Climate Pol'y 389 at 397; Gattini, *supra* note 135 (Chapter 4) (noting that the EU ETS has served as a "bargaining tool in the ICAO negotiations towards a global market-based mechanism for aviation" at 990).

Just as the United States was an unquestionable leader in the environmental revolution of the 1970s, which paved the way for the adoption of the *Montreal Protocol* in the late 1980s, the Commission and its Member States have, in recent times, been instrumental normative actors in advancing climate change issues.[209] Despite being one of the main drivers behind the early diplomatic negotiations that later led to the adoption of the *Kyoto Protocol*, by not ratifying it, the United States unconsciously created a magnificent opportunity for others to exercise a more predominant role.[210] In fact, some commentators contend that the US renunciation of the *Kyoto Protocol* was precisely the turning point that has allowed the EU to become a leader in this field.[211] As a "frontrunner" on climate change issues,[212] the Commission has clearly played an active role in the adoption of the scheme.[213] The leadership role that the EU Member States and the Commission have played on climate change issues is beyond question.[214] Through its ETS, Europe has exerted significant exogenous pressure over ICAO.[215]

Drawing from Sunstein's theory, EU Member States and the Commission have been "interested in changing social norms."[216] Acting as norm entrepreneurs, they have sought to achieve positive change. Arguably, the scheme was intended to shape corporate and state behavior (i.e. the behavior of airlines and ICAO Member States). In addition to being a political priority, mitigating international aviation's carbon footprint is part of a set of values that these actors as a group or society adhere to. Some experts have indicated that,

209 See Vogler, *supra* note 33 (Chapter 2) at 836-37; Angela Liberatore, "The European Union: Bridging Domestic and International Environmental Policy-Making" in Miranda A Schreurs & Elizabeth C Economy, eds, *The Internationalization of Environmental Protection* (Cambridge: Cambridge University Press, 1997) (describing the EU as a "visible actor in international environmental negotiations" at 204); Miranda A Schreurs & Yves Tiberghien, "European Union Leadership in Climate Change: Mitigation through Multilevel Reinforcement" in Kathryn Harrison & Lisa McIntosh Sundstrom, eds, *Global Commons, Domestic Decisions: The Comparative Politics of Climate Change* (Cambridge: MIT Press, 2010) 27 (suggesting that the EU has positively influenced climate change discussions both at the domestic and international levels) at 45.

210 See Vogler, *supra* note 33 (Chapter 2) at 840.

211 See Hardy, *supra* note 4 (Chapter 4) at 305.

212 See Yue-Jun Zhang & Yi-Ming Wei, "An Overview of Current Research on EU ETS: Evidence from its Operating Mechanism and Economic Effect" (2010) 87 Applied Energy 1804.

213 See Jon Birger Skjærseth & Jørgen Wettestad, "Fixing the EU Emissions Trading System? Understanding the Post-2012 Changes" (2010) 10:4 Global Enviro Politics 101 at 112.

214 See Liberatore, *supra* note 209 (describing the EU as a "visible actor in international environmental negotiations" at 204); Schreurs & Tiberghien, *supra* note 209 (suggesting that the EU has positively influenced climate change discussions both at the domestic and international levels) at 45; Barnes, *supra* note 8 (Chapter 4) at 55; Gattini, *supra* note 135 (Chapter 4) at 977. Europe's leadership on environment may be attributed to a number of factors, including, but not limited to, the growing presence of green parties into EU politics, the lobbying efforts of environmental NGOs, the active engagement of the EU Parliament, as well as the role played by key countries such as Germany, the United Kingdom, the Netherlands, Demark, Austria, Finland, Sweden, and Luxemburg. See Schreurs & Tiberghien, *supra* note 209 at 42-43.

215 See Gregory Polek, "Climate Experts Challenge Anti-ETS 'Myths'", *AIN Online* (15 October 2012) online: <www.ainonline.com/aviation-news/ain-air-transport-perspective/2012-10-15/climate-experts-challenge-anti-ets-myths>.

216 See Sunstein, *supra* note 18 at 909.

because of Europe's significant market power, it intends to use the EU ETS to persuade the formation of norms in other regions.[217] In reality, however, the EU ETS has not triggered a norm bandwagon nor led to a cascade effect. In fact, no tipping point has occurred. There is no indication that other states have decided or are about to take concrete action in regulating GHG emissions from aviation as a result of the EU ETS. Similarly, no State has indicated to the Commission that it intends to avail itself of the scheme's equivalent measures provision to attain a partial exemption while taking its own measure.[218] Opponents have raised numerous legal concerns, thereby questioning the scheme's legitimacy.[219]

7.2.1.2 Why Has the EU ETS Not Led to Norm Cascading?

There are at least seven factors that may explain Europe's failure to trigger norm cascading. First, the Commission has shown very little flexibility to accommodate the concerns of non-EU actors. As mentioned previously, the scheme suffers from notorious Euro-centrism. International aviation has been dragged into the general ETS with little or almost no consideration for the particularities of the sector. Europe lost almost every opportunity to engage its opponents. Developing countries did not necessarily see any benefit in allowing their aircraft operators to participate in the scheme, neither was there an incentive structure to encourage them to do so. Lifting the ceiling of credits from the clean development mechanism (CDM) could have been attractive. There was no discussion to include emission reduction credits from projects such as Reducing Emissions from Deforestation, Degradation and Forest Enhancement (REDD-plus).[220] These are issues of interest for developing countries. The Commission discarded these options, for they were not part of the general

217 See Scott & Rajamani, *supra* note 239 (Chapter 6); Kulovesi, "Make Your Own Special Song", *supra* note 134 (Chapter 4) at 542; Bogojovic, *supra* note 137 (Chapter 4) (commenting that Europe uses its bargaining power to force environmental standards like a norm entrepreneur) at 355.

218 See *contra* Fahey, *supra* note 134 (Chapter 4) (suggesting that the scheme constitutes a "successful indirect promulgation of global standards through the acceptance of EU law standards by non-EU external actors" at 1248).

219 Koh writes that social "internalization" of norms only takes place "when [it] acquires so much public legitimacy that there is widespread general adherence to it." See Harold Koh, *supra* note 54 at 642. This is clearly not the case of the EU ETS.

220 Viñuales explains that these projects seek "to channel funds to developing countries in order for them to conserve their forests, which are extremely valuable resources not only for biodiversity purposes but also for the capture and storage of carbon dioxide. The underlying rationale is that conserving forests in developing countries would be an effective and much cheaper way to reduce global emissions of carbon dioxide as compared with other approaches, such as technological changes in the way energy is produced or in the transportation sector." Jorge E Viñuales, "Managing Abidance by Standards for the Protection of the Environment" in Antonio Cassese, ed, *Realizing Utopia: The Future of International Law*, (Oxford: Oxford University Press, 2012) at 335. See also Colin AG Hunt, *Carbon Sinks and Climate Change: Forests in the Fight against Global Warming* (Cheltenham: Edward Elgar, 2009) (noting the important contribution of REDD in the post-Kyoto environment) at 214. *UNFCCC's* COP/19 reached consensus on a set of internationally agreed standards for REDD. See Stian Reklev, "UN Agrees Multi-Billion Dollar Framework to Tackle Deforestation", *Reuters* (22 November 2013) online: <http://uk.reuters.com/article/2013/11/22/forest-climate-talks-idUKL4N0J72XJ20131122>.

ETS. It was only in September 2013 that the Commission announced a declaration of intent signed with ICAO to offer € 6.5 million toward capacity building in mitigation efforts to reduce GHG emissions in Africa and the Caribbean.[221] Also, on 16 October 2013, the Commission made public its new proposal to re-start the EU ETS but only covering emissions within European airspace. The Commission also suggested that the new EU ETS will exclude a large number of routes to and from developing countries.[222] On the contrary, there were a number of policy and political risks associated with the scheme.[223] Finnemore and Sikkink claim that "new norms never enter a normative vacuum but instead emerge in a highly contested space where they must compete with other norms and perceptions of interest."[224] The Commission would seem to have underestimated the exogenous extra-legal and legal influences on the EU ETS (e.g. the pressure exerted by the climate change regime and the strong positions of some countries in the *UNFCCC*). Secondly, the Commission's communication campaign was, at best, poor.[225] The opponents never really understood how the EU ETS works in practice. As described in Chapter 4, it was extremely easy to portray the scheme as a tax- or as a revenue-generating mechanism. US airlines in particular exploited this flaw. It was only very late in the process – and thanks to the strong intervention of the US-based environmental NGO Environmental Defense Fund (EDF) – that the Commission explained that aircraft operators could completely bypass the auction process, should they wish to do so.[226] This was surprising, given that norm entrepreneurs tend to hold more information than other actors. The communication flaws also made it difficult for aircraft operators to interact and to interpret the proposed norm change. The teaching of the elites, as a diffusion mechanism to internalize norms that Checkel identifies, was either nonexistent or, at best, not effective.

Thirdly, by not targeting other more carbon-intensive sectors such as road transport and maritime sectors, the Commission simply fed critics for whom the scheme was unjustifiably fixed with international aviation due to its great publicity. Under this line of reasoning, the international aviation sector was targeted to show that Europe was visibly

221 See ICAO, "Europe Proposes EUR 6.5 Million for State CO_2 Reduction Through ICAO" online: ICAO <www.icao.int/Newsroom/News%20Doc%202013/COM.34.13.ICAO.EU.FINAL.EN.pdf>.

222 See EC, "Commission Proposal for European Regional Airspace Approach for the EU Emissions Trading for Aviation: Frequently Asked Questions", MEMO/13/905 (Brussels, 16 October 2013).

223 As mentioned earlier, for some developing countries, the EU ETS represents a dangerous precedent for the *UNFCCC* negotiations, given the fact that the scheme does not differentiate between developed and developing countries. In other words, it does not recognize CBDR, as this principle is currently conceived within the *UNFCCC* context.

224 Finnemore & Sikkink, *supra* note 17 at 897.

225 Ruckelshaus underscores the importance of effective communication in addressing climate change issues. He suggests that a clear set of values must be articulated by leaders in both the public and private sectors." William D Ruckelshaus, "Toward a Global Environmental Policy" in Ruth S DeFries & Thomas F Malone, eds, *Global Change and our Common Future* (Washington: National Academy Press, 1989) 3 at 3.

226 See Petsonk, Testimony before US Congress, *supra* note 98 (Chapter 4).

doing something, albeit arguably not in the most effective manner.[227] Also, the Commission failed to adequately explain why the emphasis was on aviation and not on other modes of transportation. This certainly caused a loss of credibility. [228]

Fourthly, the EU ETS did not replace the carbon taxes that EU Member States, such as Austria, Ireland, Germany, and the United Kingdom, levied on passengers departing from their airports.[229] In addition to contributing to the distrust of non-EU actors, this certainly reinforced the perception that Europe was not so much concerned about protecting the environment, but rather in finding ways to raise revenues. Under this view, Abbott and Snidal's opinions are directly on point when they note that, with the EU ETS, the Commission sought to "[press] its economic interests while dressing them up in normative garb."[230] Although it was not *strictu sensus* its fault, the Commission did not do much to remedy the situation.

Fifthly, the EU ETS has been structured as a top-down regime. Almost all the actors participating in the regime, including their home States, not only raise serious legal and policy concerns but also attempt to sabotage the scheme. In this context, gaining acceptance – let alone obedience in Koh's conception – is much more difficult because in most cases top-down regimes are not necessarily built from solid domestic consensus. In the case of the EU ETS, I am referring to support at the domestic level of the participating actors (e.g. foreign airlines and their home States).[231]

Sixthly, the CJEU's *ATA Decision* contributed significantly to impregnate the sentiment within the international community that the EU ETS is a form of EU environmental indoctrination. While paying very little attention to issues of international law, the CJEU declared that the *Chicago Convention* was not binding upon the EU.[232] This is what Fahey calls "exportation of EU values through law."[233] This is linked to the seventh factor. Although

227 See Gattini, *supra* note 135 (Chapter 4) (noting that international aviation 2 percent share of CO_2 emissions is substantially less than the 6-8 percent contribution of light- and heavy-duty vehicles) at 991.

228 See FitzGerald, Europe's Emissions Trading System, *supra* note 96 (Chapter 4) at 211.

229 See ICAO, HGCC/3-IP/5, *supra* note 130 (Chapter 1) at Appendix A.

230 Abbott & Snidal, *supra* note 8 at 144.

231 It is noteworthy that, when the Commission unveiled its proposal to re-arrange the EU ETS to cover both foreign and EU aircraft operators for GHG emissions generated within EU airspace, France, Germany, and the United Kingdom expressed serious concerns, whereas Nordic countries were significantly more sympathetic to the idea. This is indicative that such initiative did not achieve a widespread support. See Barbara Lewis, "RPT-UK, France, Germany Attack EU Aircraft Carbon Plan", *Reuters* (6 December 2013) online: <www.reuters.com/article/2013/12/06/eu-aviation-idUSL5N0JL1MN20131206>; Mathew Carr, "EU May Extend Aviation Carbon Freeze to 2020, Adviser Says", *Bloomberg* (5 December 2013) online: <www.bloomberg.com/news/2013-12-05/europe-may-extend-aviation-carbon-freeze-to-20-eu-adviser-says.html>.

232 See *ATA Decision*, *supra* note 63 (Chapter 4).

233 Fahey, *supra* note 134 (Chapter 4) at 1266. In a way, this imposition of EU values into the broader international aviation community resembles the Tuna-Dolphin dispute. Bodansky notes that in this case the United States put in place unilateral trade measures to protect dolphins largely as a result of US policy, as opposed to a requirement stemmed from international law. See Bodansky, "Unilateralism", *supra* note 3 (Chapter 6) at

Mayer argues that the EU ETS is not in and of itself "imperialistic,"[234] since it seeks to advance the widely accepted *UNFCCC* objectives, the truth is that the scheme has been overwhelmingly rejected by all non-EU ICAO Member States. It has been perceived as an imposition of European environmental values, which are clearly not shared by the rest of the world.

7.2.1.3 EU ETS and Norm Internalization

Values cannot be forced. If values could be forced, participants would never internalize norms and would ultimately opt to reject it. Applying Koh's approach toward the internalization of international norms, foreign aircraft operators complied "under protest" with the EU ETS rules to avoid sanctions.[235] These actors never really obeyed the norms.

In addition, for the vast majority of non-EU Member States, the EU ETS was not necessarily part of their own set of core values. The incentive structures were never made very clear. The processes of interaction and interpretation of norms were not conducive. As some commentators have noted, in the absence of a process of internalization, norms pass quickly and do not form part of a set of values shared by the actors upon whom they are imposed.[236] This is a central component of norm creation.[237] It is not only sufficient to take steps to bring about change or to propose a new standard of behavior on a given issue. Norm emergence is one of the stages in the process of norm formation. Norm entrepreneurship will never achieve its objectives if the tipping point that prompts the norm cascade does not occur. In the absence of broad acceptance, internalization of the norm becomes part of a utopia. Norm internalization is unlikely to occur if the change in norm is either perceived as an imposition or if it evolves without the participation of the relevant actors. The rejection of the *Kyoto Protocol* by the United States stands out as a classic example.[238]

---

342. On the other hand, the *Shrimp-Turtle* dispute was substantially different. In this case, the international community had already declared sea turtles as endangered species. *Ibid.* at 343.

234 Mayer, Case -366/10, *supra* note 133 (Chapter 1) at 1138.

235 Mendes de Leon, "Enforcement of the EU ETS", *supra* note 220 (Chapter 4) at 287.

236 See Gopalan, "Changing Norms", *supra* note 49 at 32.

237 As Koh notes, if the purpose of the policy measure is to improve participants' compliance, as understood in its general context, far more emphasis should be placed on the "internalization" of norms. See Harold Koh, *supra* note 54 at 642.

238 The US decision to forego participation in the *Kyoto Protocol* was largely attributed to the lack of substantial mitigation commitments of major developing countries such as China, India, and Brazil, but it was also due to the fact that the US constituency did not necessarily share the same values and the perception, rightly or wrongly, that the Clinton Administration negotiated a deal without having involved all relevant actors. See Kathryn Harrison, "The United States as Outlier: Economic and Institutional Challenges to US Climate Policy" in Kathryn Harrison & Lisa McIntosh Sundstrom, eds, *Global Commons, Domestic Decisions: The Comparative Politics of Climate Change* (Cambridge: MIT Press, 2010) 67 at 93.

7.2.1.4 EU ETS: Significant Contribution or Missed Opportunity?

Despite the policy imperative to regulate GHG emissions from international aviation stemming from the *Kyoto Protocol*,[239] Europe must be credited for the re-emergence of the norm. In this sense, the EU ETS is a contribution. Having gone through a trial-and-error period, the Commission and its Member States have gained hands-on experience in running an MBM for international civil aviation. Arguably, all these elements will be extremely valuable when ICAO develops its global scheme for international civil aviation. The numerous studies that the Commission has conducted may serve as valuable input. This does not mean, however, that emissions trading will be the regulatory policy instrument of choice. Although the Aviation Directive asserted that "the Community scheme may serve as a model for the use of emissions trading worldwide,"[240] as will be explained below, from a purely architectural design perspective, States will most likely opt for a measure other than emissions trading, despite the fact that the scheme has served as useful experiment. In this respect, the EU ETS will likely be a constraint upon the future design of the global MBM under ICAO. Notwithstanding its positive aspects, the EU ETS also represents a missed opportunity to trigger meaningful action. Had European authorities been more pragmatic, and more mindful of the concerns of the international community, norm cascading may have taken place. Environmental results would have been considerably better, and the European scheme could have paved the way for the ICAO global MBM.

7.2.2 IATA

7.2.2.1 A Reactive Response: The IATA Four-Pillar Strategy

Although international aviation has been under the spotlight since 1997 when the *Kyoto Protocol* tasked ICAO to address international aviation's GHG emissions,[241] climate change – let alone MBMs – was not necessarily an issue of immediate concern for the world's leading airlines trade association.[242] Prior to 2005 – the year the Commission made public

239 See *Kyoto Protocol*, *supra* note 12 (Introduction) at Art 2.2.

240 Aviation Directive, *supra* note 17 (Introduction) at preambular clauses, para 17.

241 See *Kyoto Protocol*, *supra* note 12 (Introduction) at Art 2.2.

242 In order to have an idea of the aviation industry's priorities over the last 12 years, it is worth perusing the book published by IATA's former Director General Giovanni Bisignani. This contains a collection of his most famous (and sometimes contested and provocative) speeches. They are illustrative of the industry's priorities during his tenure at the helm of IATA. The reader quickly learns that when Bisignani came on board to lead IATA in 2002, the industry's challenges, as he himself described, geared toward "safety, security, long-term profitability, fair competition between carriers in the different regions of the world, increased customer satisfaction, and environmental consciousness." Bisignani, *Words of Change*, *supra* note 20 (Introduction) at 6. By "environmental consciousness," Bisignani did not mean addressing the sector's carbon footprint, but rather attempting to demystify the misconception on aviation as a notorious polluter. This purported to highlight the sector's improvements in its environmental performance. Airlines have always been quick to measure improvements in fuel efficiency in terms of revenue tonne-kilometers. For instance,

its intention to include international aviation into its ETS – IATA focused on fuel-saving campaigns addressing operational measures.[243] Once the Commission made the announcement, IATA made it very clear that the airline industry "[would] be tough in opposing any local or regional solutions."[244] Two years later, in 2007, as the EU ETS became unavoidable, IATA launched its so-called four-pillar strategy to address international aviation's carbon footprint.[245] IATA identified four key areas where emission reductions may be achieved. These were (i) technology,[246] (ii) operations,[247] (iii) infrastructure,[248] and (iv) economic measures.[249] In trying to exercise the role of norm entrepreneurs, actors and

---

Bisignani noted that, on average, airlines consumed 2.5 liters of fuel per 100 passenger-kilometers flown, which is roughly the same consumption as that of a small compact car. *Ibid.* at 52. Similarly, A4A, the US airline trade association, claims that its members have "improved fuel efficiency 120 percent since 1978, reducing CO_2 emissions by 3.3 billion metric tons, savings equivalent to taking 22 million cars off the road annually." Nancy Young, "Airlines, Getting Greener", Letter to the Editor, *The New York Times* (3 February 2013) online: <www.nytimes.com/2013/02/04/opinion/airlines-getting-greener.html?_r=1&>. In spite of this, other sources report that US airlines' "jet fuel use is projected to be 16 percent higher in 2020 [as compared to 2012] and 42 percent higher in 2030." Jake Schmidt, "US Signs into Law Bill Calling for Global Solution to Aviation's Carbon Pollution", *Switchboard NRDC* (27 November 2012) online: <http://switchboard.nrdc.org/blogs/jschmidt/ball_in_the_hands_of_proponent.html>. By centering their performance on RTKs, airlines are a more environmentally friendly mode of transportation. What this statistic ignores is the fact that, despite of the efficiency gains, the sector's net emissions continue to grow. RTKs do not factor the sector's growth into the equation. As Lyle indicates, despite of the fact that "per unit fuel consumption for air transport will continue to fall over the coming years, in the order of about 1.5 [percent] per annum worldwide," because of the sector's consistent yearly growth of 4-5 percent, "there would still be a substantial increase in absolute emissions." Chris Lyle, "Will ICAO States at Last Deliver a Meaningful Global Agreement on Mitigating International Aviation Emissions", *Green Air Online* (2 September 2013) online: <www.greenaironline.com/photos/Will_ICAO_deliver_on_aviation_emissions_mitigation_-_Chris_Lyle_Sept_2013.pdf> at 2 [Lyle, "Meaningful Global Agreement"].

243 See Bisignani, *Words of Change*, *supra* note 20 (Introduction) (describing the "save a minute and the environment" project, as well as criticizing European authorities for their procrastination on the implementation of the Single European Sky) at 47. In spite of the fact that operational initiatives ultimately reduce CO_2 emissions, as compared to a business-as-usual scenario, they do not in any way seek to establish a cap or require that the sector offset its emissions elsewhere. This is a fundamental difference as compared with subjecting the sector's carbon footprint to an MBM scheme.

244 Bisignani, *Words of Change*, *supra* note 20 (Introduction) at 52.

245 *Ibid.* at 97.

246 The technology component encompasses new aircraft designs and the development of alternative fuels. In fact, industry claims that alternative fuels, such as those derived from algae, jatropha, and camelina, have the potential to reduce 80 percent of aviation's carbon footprint. See "The Right Flight Path to Reduce Aviation Emissions", *ATAG* online: <www.enviro.aero/Content/Upload/File/AviationPositionPaper_COP16_normalprinter.pdf>; "Beginners Guide to BioFuels", *ATAG* (2 September 2011) online: <www.enviro.aero/content/upload/file/beginnersguide_biofuels_webres.pdf>.

247 Among others, operational measures include efficient flight procedures, reduction in the usage of auxiliary power units, and aircraft weight reduction.

248 Infrastructure involves better use of air traffic management.

249 Economic or market-based measures may include measures such as emissions trading, offsetting, and environmental charges or levies. See ATAG, "The Right Flight Path to Reduce Aviation Emissions", *supra* note 246.

lobbying groups often act in response to perceived attacks or external threats as a reactive tactic.[250]

7.2.2.2 The Fear of Patchwork Regulations

Fearing a potential patchwork of different regulatory measures, the aviation industry saw the need to exercise leadership.[251] Here, IATA followed a "rational-actor perspective."[252] Ellickson explains that this takes place when an agent or a norm entrepreneur realizes that the immediate cost involved in accepting the change of norms far outweighs the long-term costs of rejecting them.[253] Similarly, other scholars have pointed out that those actors "exposed to heterogeneous regulatory environments are likely to engage in norm setting and norm development processes to reduce the costs of regulatory diversity."[254] Higher transaction costs increases the likelihood of actors engaging in norm entrepreneurship roles.[255] The potential quilt of different regulatory approaches would have been much more expensive for the airline industry.[256] In Ellickson's conceptualization, the EU ETS acted as an exogenous shock (legal influence) affecting the airlines' incentive structure: an appropriate situation to exercise norm entrepreneurship by advocating new norms.[257]

7.2.2.3 The Long Road to the IATA Industry Targets

In addition to its four-pillar strategy, IATA has gone through a process of evolution to define the industry targets. In June 2007, at its 63rd Annual Meeting held in Vancouver, IATA announced that airlines would be 25 percent more fuel efficient by 2020 as compared to a 2005 baseline and that the sector should aim at becoming "an industry that does not

250 See Geoffrey P Miller "Norms and Interests" (2003) 32 Hofstra L Rev 637 at 647.

251 See Steele, *supra* note 206 (Chapter 1). The fear of patchwork regulatory measures is not uncommon in the environmental field. In the 1980s, when the Montréal Protocol on ozone-depleting substance was being negotiated, the US industry was also very concerned for the potential "patchwork of varying state regulations." Richard Elliot Benedick, "The History of the Montréal Protocol" in Donald Kaniaru, ed, *The Montréal Protocol: Celebrating 20 Years of Environmental Progress* (London: Cameron May Ltd, 2007) at 45. Similarly, the international maritime industry has made clear its preference for "unified rules to a patchwork of regional and unilateral measures." Ronald Mitchell et al, "International Vessel-Source Oil Pollution" in Young, *supra* note 6 at 51.

252 Sunstein points out that that "exogenous factors, including social norms," determine an actor's rational choices. See Sunstein, *supra* note 18 at 956.

253 Ellickson, "Market", *supra* note 49 at 11.

254 See Flohr et al, *supra* note 49 at 33.

255 *Ibid.* at 164.

256 In the maritime context, industry also favors a global approach. See generally Nadine Heitmann & Setareh Khalilian, "Accounting for Carbon Dioxide Emissions from International Shipping: Burden Sharing under Different UNFCCC Allocation Options and Regime Scenarios" (2011) 35 Marine Pol'y 682.

257 See Charlie Leocha, "Airlines Embrace the Environment: It is Good for their Bottom Line", *Consumer Traveler* (13 September 2013) online, <www.consumertraveler.com/today/airline-embrace-the-environment-its-good-for-their-bottom-line/> (suggesting that a pro-environment approach improves airlines' profitability).

pollute."[258] A few months later, in September 2007, at the 36th Assembly, IATA unveiled its vision "to put aviation on a gradual path towards carbon neutral growth and, eventually, a zero-carbon future."[259] However, IATA did not provide a timeline for the implementation of those goals. Relying heavily on technological developments, Bisignani (then IATA's Director General) said that the real challenge was to "build a zero-emissions aircraft in the next 50 years."[260] At the time, this was received with skepticism, given that it was not necessarily self-evident how the industry would achieve these goals. There was no indication that aircraft manufacturers could produce such aircraft or that alternative fuels would deliver such results. Years later, Bisignani would admit that, at the time, IATA did not "have a fully developed plan" to achieve these goals but that such a ground-shaker was necessary to lay down the organization's roadmap with respect to climate change.[261] Although Bisignani's adventurous style may be the subject of criticism, his audacity must also be praised. His leadership was instrumental in setting a vision for the industry that included climate change goals, albeit unclear.[262] In a way, Bisignani himself was a norm entrepreneur within the airline industry.

Forced to adopt more realistic targets, IATA lowered its expectations. Three years later, in 2010, at its 66th Annual Meeting held in Berlin, IATA established three aspirational reduction targets for its member airlines: a fuel efficiency improvement of 1.5 percent yearly to 2020,[263] carbon neutral growth (CNG) from 2020 at 2020 levels, and a 50 percent CO_2 reduction by 2050 compared to 2005 levels.[264]

IATA claims that "[n]o other industry [has set such] ambitious targets."[265] Although the statement is accurate, it fails to recognize that, with the notable exception of the international maritime transportation industry, most other industries fall within the *Kyoto Protocol* targets. In some countries, they have been subjected to some form of regulatory oversight, whereas this has not been the case with respect to international aviation. In

---

258 Bisignani, *Words of Change*, *supra* note 20 (Introduction) at 96-98.

259 See ICAO, A36-WP/85 EX/33, *Towards a Carbon Neutral and Eventually Carbon Free Industry* (presented by the International Air Transport Association) [ICAO, A36-WP/85 EX/33].

260 Bisignani, *Words of Change*, *supra* note 20 (Introduction) at 98.

261 *Ibid.* at 2.

262 While delivering his State of the Air Transport Industry speech, at the 63rd IATA Annual General Meeting, Bisignani said: "I don't have all the answers, but our industry started with a vision that we could fly. The Wright brothers turned that dream into reality and look at where we are now. We can see potential building blocks for a carbon-free future." *Ibid.* at 98.

263 *Ibid.*

264 See Bisignani, Words of Change, *supra* note 20 (Introduction) at 122-25; IATA, "Responsibly Addressing Climate Change," online: <www.iata.org/policy/environment/Documents/policy-climate-change.pdf> [IATA, Policy Measures for Climate Change].

265 Bisignani, Words of Change, *supra* note 20 (Introduction) at 122; IATA, Press Release, "Remarks of Tony Tyler", *supra* note 141 (Chapter 1). See also Nancy Young (presentation delivered at ATW's 5th Annual Eco-Aviation Conference, 21 June 2012) online: A4A <www.airlines.org/Pages/A-Look-Back-and-Forward-from-Rio-Ever-Sustainable-Aviation.aspx> (describing the aviation industry's strategy on climate change as an "aggressive set of measures").

other words, other industries need not form part of a global system. The IATA targets have also been strongly criticized. For instance, Lyle notes that the yearly fuel efficiency improvement goal constitutes a mere "business-as-usual" scenario for most airlines.[266] Lyle also suggests that the CNG and the 50 percent reduction by 2050 place overly optimistic expectations on breakthrough developments in technology and unrestricted availability of alternative fuels in sufficient quantities to meet the sector's needs.[267] Similarly, environmental NGOs have pointed out that this goal is based on a high expectation of "massive infusions of biofuels."[268] The fact that IATA had to correct its unrealistic zero-emissions vision to the more modest 50 percent emission reductions by 2050 goal also created doubt as to whether these targets are, in fact, achievable.

In spite of these criticisms, it is fair to acknowledge that the IATA targets have been largely relevant in framing ICAO's aspirational goals.[269] In fact, ICAO has called upon its Member States to work with industry to achieve these goals.[270] Given the predominant role that IATA plays at ICAO, as explained in Chapter 3, it is unlikely that ICAO will ever adopt targets that are more stringent than those proposed by the industry it seeks to regulate.[271]

Arguably, one of the most overlooked characteristics of the IATA targets concerns the fact that they are aspirational, non-binding, and non-attributable to its member airlines. These collective soft goals have not necessarily been internalized by each member airline. They bear no penalty for not taking action. In this instance, IATA opted for a drastic departure from other successful industry practices such as e-ticketing[272] and its operational

---

266 "Aviation and Climate Change: What now for a Global Approach", *Green Air Online* (24 January 2011) online: <www.greenaironline.com/news.php?viewStory=1039> [Green Air, "Global Approach"]. IATA's leniency on its fuel efficiency targets should not come as a surprise. In the past, IATA had already set targets that were not "especially onerous" for its constituents. For instance, in the late 1990s, IATA sought to achieve a 26 percent improvement in liters per 100 revenue tonne-kilometer. By the end of 2000, British Airways already attained a 20 percent fuel reduction. See Peter Morrell, "An Evaluation of Possible EU Air Transport Emissions Trading Scheme Allocation Methods" (2007) 35 Ener Pol'y 5562 at 5563.

267 See Green Air, "Global Approach", *supra* note 266.

268 Transport & Environment, "Global Deal", *supra* note 72 (Chapter 1).

269 See ICAO, A37-19, *supra* note 207 (Chapter 1) at paras 4 & 6; ICAO, A38-18, *supra* note 21 (Introduction) at paras 5, 7. See also ICAO, A36-22, *supra* note 17 (Chapter 2) at Appendix K.

270 See ICAO, A37-19, *supra* note 207 (Chapter 1) at para 6; ICAO, A38-18, *supra* note 21 (Introduction) at paras 5 and 7.

271 Hardeman claims that "[p]roactive positioning from the aviation industry paid off and saw the sector's own proposals largely reflected in the [climate change] Resolution" adopted by the 37th Assembly Session. Hardeman, "Reframing Aviation Climate Politics and Policies", *supra* note 44 (Chapter 2) at 9.

272 Since aviation's early beginnings, each passenger itinerary required the printing of numerous paper-based flight coupons. At its peak, the printing of each ticket cost airlines an average of 9 US dollars. See Bisignani, *Words of Change*, *supra* note 20 (Introduction) at 40. With the advent of the Internet in the 1990s, most major airlines saw the opportunity to obtain significant savings by issuing electronic tickets (e-tickets) on their flights. In 2004, IATA announced that by 2008 e-tickets would be mandatory for its member airlines when interlining – that is to say, when an airline issues a ticket for another airline. At the time, IATA printed

safety audit program (IOSA),[273] where compliance is a mandatory requirement to either retain or accede to IATA membership.[274] The latter has proved to be an invaluable driver and enforcer of norm change. According to Ellickson, "a new norm wins acceptance only when most actors conform to it and enforcers routinely levy sanctions to support it."[275] Aspirational targets do not provide for an enforcement mechanism. Yet in IATA's defense, aspirational, non-binding, voluntary,[276] and sometimes non-attributable commitments (e.g. ICAO aspirational goals) do not seem to be the exception in the wider context of the climate change regime. Rajamani et al correctly point out that "there is a shift from a prescriptive, quantitative, time-bound, compliance-backed approach to one that rests on self-selection of targets and actions, and a robust reporting system."[277] Similarly, Mehling claims that "voluntary international commitments coupled with domestic legislation may be all that many parties to the international climate regime are willing to accede to."[278] In this context, the IATA approach is not a surprise but rather a reflection of an emerging trend in climate change issues.

7.2.2.4 A Global, Sectoral Approach

It would take IATA a number of years to propose a global, sectoral approach to address GHG emissions from international aviation. In fact, in 2007, at the 36th Assembly, IATA, more concerned with limiting the EU ETS, acknowledged "emissions trading may have a

some 300 million interline tickets. With this initiative, IATA hoped that its member airlines would have saved roughly 3 billion US dollars per year. *Ibid.*

273 Another interesting example of IATA's self-imposed industry standards was its Operational Safety Audit (IOSA) program. Two main factors prompted IOSA. First, as a result of mega alliances (i.e. Star Alliance, One-World, and Sky Team), airlines were forced to audit themselves to ensure common safety practices and avoid potential liabilities. Second, it became evident that the safety track record of airlines from some regions of the world fell short of industry standards. As a result, IOSA came into existence as the "first global benchmark for airline operational safety management." *Ibid.* at 61. By the end of 2009, all IATA member airlines were IOSA-certified. *Ibid.* at 2. See also David Hodgkinson, "IOSA: The Revolution in Airline Safety Audits" (2005) 30:4/5 Air & Space L 302.

274 See IATA, *Articles of Association* online: <www.iata.org/about/Documents/articles-of-association.pdf> at Art 5, para 1 (ii) (requiring that applicants to the IATA membership "[m]antain a valid IATA Operational Safety Audit (IOSA) Registration or equivalent as it may be renamed from time to time") [IATA, *Articles of Association*].

275 Ellickson, "Market", *supra* note 49 at 10.

276 With regard to voluntary measures, Harrison writes that "it is also difficult to draw conclusions about effectiveness because so few rigorous program evaluations are available. Indeed, it is quite remarkable how little we know about the environmental effectiveness of voluntary program." Kathryn Harrison, "Talking with the Donkey: Cooperative Approaches to Environmental Protection" in Peter M Haas, ed, *International Environmental Governance* (Burlington: Ashgate, 2008) 339 at 354.

277 Lavanya Rajamani, Jutta Brunnée, & Meinhard Doelle, "Introduction: The Role of Compliance in an Evolving Climate Regime" in Jutta Brunnée, Meinhard Doelle & Lavanya Rajamani, *Promoting Compliance in an Evolving Climate Regime* (Cambridge: Cambridge University Press, 2012) 1 at 7 [Rajamani et al].

278 Michael Mehling, "Enforcing Compliance in an Evolving Climate Regime" in Jutta Brunnée, Meinhard Doelle & Lavanya Rajamani, *Promoting Compliance in an Evolving Climate Regime* (Cambridge: Cambridge University Press, 2012) 194 at 214.

role to play, but only as part of a package of measures involving technology, operations and infrastructure."[279] In 2009, at ICAO's High-Level Meeting on International Aviation and Climate Change (HLM-ENV), in addition to presenting its industry targets, IATA urged States to develop a "framework for measures addressing CO_2 emissions from aviation."[280] Although IATA also called for a global sectoral approach[281] for aviation, little detail was provided on the required design elements.[282] This was perhaps not too surprising given that, at the time, it was evident that there was little appetite for such a global scheme. As a result, for a long time thereafter, the notion of a global sectoral approach for international aviation emissions remained an abstract, symbolic concept without much substance. In 2010, at the 37th Assembly, IATA proposed that States should agree on a global framework for MBMs.[283] Although IATA did elaborate guiding principles in support of this proposal, States could not agree on the actual content of the framework. IATA also failed to properly explain what exactly was envisaged by a global sectoral approach and how this concept differed from the framework for MBMs. IATA's failure to provide clarification on this topic had profound consequences, for both concepts remained confusing for a considerable period of time even in the eyes of the most important actors.

In 2013, IATA urged ICAO Member States to "adopt a global MBM" but only as an interim measure "to fill any gap[s]" in the effort to reach CNG by 2020 that could not be met through technological, operational, and infrastructure measures.[284] IATA is willing to grudgingly swallow the global MBM pill, but only "as transitional and temporary measures until other measures deliver sufficient reductions to reach medium and long term targets."[285] Because technology needs time to develop, CNG is not achievable without MBMs. However, IATA and its member airlines strongly believe that "technology, operations and infrastructure are the long-term solutions for aviation's sustainable growth."[286] As such, they will be able to halve their emissions by 2050 by resorting to these types of

279 See ICAO, A36-WP/85 EX/33, *supra* note 259 at 2.

280 See ICAO, HLM-ENV/09-WP/19, *A Global Sectoral Approach for Aviation* (presented by the Airports Council International (ACI), Civil Air Navigation Services Organisation (CANSO), International Air Transport Association (IATA), and International Coordinating Council of Aerospace Industries Associations (ICCAIA)) at 2.

281 Barrett favors the adoption of sectoral approaches to address the broader climate change issues. He further notes that by dismantling the wide-web complexity of the different array of climate change subjects into smaller components, regulators could apply the "best means for enforcing each piece individually." Barrett, *Why Cooperate*, *supra* note 57 (Chapter 1) at 208.

282 *Ibid.*

283 See ICAO, A37-WP/217 EX/39, *supra* note 112 (Chapter 1).

284 IATA, CNG Resolution, *supra* note 129.

285 Steele, *supra* note 206 (Chapter 1).

286 See IATA, Policy Measures for Climate Change, *supra* note 264.

measures. Some States seem to support industry's views.[287] For Europe and the environmental NGOs, the scenario is quite different. Although they acknowledge that alternative fuels may play a role, they still believe that MBMs will be necessary.

In a way, by emphasizing the idea that MBMs should only be applied temporarily to international civil aviation, and by establishing high expectations in the realm of alternative fuels, IATA and its member airlines have also contributed to de-emphasize the need for these measures. If MBMs are only an interim solution, why not concentrate on the root of the problem and try to develop long-term options? Certainly, industry seems to believe that measures other than MBM will provide the cure. In part, this explains why a number of States still question the need to embrace MBMs. This is, unfortunately, one of the unintended consequences of IATA's qualified support for MBMs.

7.2.2.5 Why Not Binding Industry Self-Regulation?

The idea of a self-imposed, aviation industry global MBM is certainly not new.[288] The literature suggests that firms gain credibility when they set "[collective] binding rules and norms and may even seek compliance by *de facto* coercion, using the structural power they have at their disposal."[289] Perez underscores the "emergence of private environmental governance as an important transnational phenomenon."[290] Moreover, Perez is of the view that "[t]he expansion of private regulation over the last decade represents a robust social process, which is likely to further expand in the [future]."[291] Berman highlights the increasing role of non-state actors in adopting international self-regulation.[292] Falkner says that "[p]rivate governance has become [an undeniable] reality in global environmental politics."[293] Similarly, George writes that these actors constantly engage in rule making.[294] In his seminal work *Order without Law*, Ellickson explains how, in light of high transaction costs and the complexity of the legal system, close-knit groups such as ranchers of a Cali-

287 See ICAO, A38-WP/272, *Position of African States on Climate Change* (describing ICAO's global MBM scheme as a "transitional measure and complementary to" operational and technological initiatives) [ICAO, A38-WP/272].

288 Tsai & Petsonk, *supra* note 213 (Chapter 1) (suggesting that IATA could develop an industry-led ETS where penalty for non-compliance would be losing membership of the organization) at 799.

289 Flohr et al, *supra* note 49 at 203.

290 Oren Perez, "Private Environmental Governance as Ensemble Regulation: A Critical Exploration of Sustainability Indexes and the New Ensemble Politics" (2011) 12 Theoretical Inq L 543 at 545.

291 *Ibid.* at 579.

292 See Paul Schiff Berman, "A Pluralist Approach to International Law" (2007) 32 Yale J Int'l L 301 (mentioning the Internet Corporation for Assigned Names and Numbers, the credit ratings of Moody's and Standard & Poor's, Motion Picture Association of America) at 312-13.

293 Robert Falkner, "Private Environmental Governance and International Relations: Exploring the Links" in Peter M Haas, ed, *International Environmental Governance* (Burlington: Ashgate, 2008) 281 at 293.

294 See Erika R George, "The Place of the Private Transnational Actor in International Law: Human Rights Norms, Development Aims, and Understanding Corporate Self-Regulation as Soft Law" in Roundtable: A Multiplicity of Actors and Transnational Governance (2007) 101 Am Soc'y Int'l L Proc 469 at 476.

fornian county have resorted to social norms to settle their disputes more expeditiously.[295] Arguably, as McAdams suggests, "[extra-legal] norms [may] govern behavior irrespective of the legal rule, making the choice of a formal rule surprisingly unimportant."[296]

Had IATA opted for binding targets or a self-regulated MBM, the global system under the auspices of ICAO would have been much less relevant and perhaps not even necessary. However, this has not been the case. Therefore, it begs the question why IATA opted for non-binding approach.[297] Four possible reasons may underline this approach.

First, the regulatory uncertainty created by the EU ETS and the inability of ICAO to set up a global system made it more difficult for airlines to adopt industry standards.[298] There was certainly no guarantee that States would recognize self-imposed industry standards. Government intervention sometimes is necessary "to manipulate the norm-making process."[299]

Second, the adoption of aspirational goals is reflective of the internal difficulties that IATA faced within its diverse constituents. For fast-growing carriers, such as those from Asia,[300] the Middle East, and Latin America, the IATA industry targets are extremely onerous, whereas for carriers experiencing little growth, such as those in the mature US or Canadian markets, a scheduled fleet renewal under a business-as-usual scenario is sufficient. In other words, these carriers may meet the aspirational targets, at least initially, solely through the implementation of technological and operational measures. Arguably, at the time IATA members were not ready to accept binding emission reductions commitments.

Third, although IATA has actively advocated the need to address the sector carbon's footprint ever since the European Commission announced its intention to include international aviation into its ETS, the fact still remains that this is yet to be of concern to a large number of its members which do not operate flights to and from Europe. Likewise, the passive or unconcerned role of the home States of these carriers, for whom climate change mitigation does not rank high on their policy agendas, makes it harder for norm

---

295 See Ellickson, *Order without Law*, *supra* note 31 at 282.

296 McAdams, *supra* note 7 at 340.

297 See generally Mehling, *supra* note 278 (discussing that States "will be compelled to explore alternative options to secure compliance" at 213).

298 See generally Christian Engau & Vokker H Hoffmann, "Effects of Regulatory Uncertainty on Corporate Strategy: An Analysis of Firms' Responses to Uncertainty about Post-Kyoto Policy" (2009) 12 Enviro Science & Pol'y (discussing that regulatory uncertainty does not necessarily postpone industry investments) at 766-777.

299 Ellickson, "Market", *supra* note 49 at 4.

300 At the 38th Assembly, the Republic of Korea reported that its seven air carriers have experienced an 8 percent yearly increase in GHG emissions. See ICAO, A38-WP/268 EX/88, *Agreement of Voluntary Activity for GHG Reduction in the Republic of Korea* (presented by the Republic of Korea) at 2. This rate is a much higher GHG increase rate than that of other regions such as North America and Europe. *Ibid.*

internalization to occur. As mentioned above, the role of States and their governmental agencies cannot be ignored in this process (extra-legal influences).

Fourth, as a norm change advanced by IATA, environmental protection may not necessarily form part of the values of its constituent airlines. This is not the case with other issues such as safety, which is an imperative value to the world's airlines and the international civil aviation community. Mr. Tony Tyler, IATA's Director General, is frequently reported stating that safety is the "airlines' top priority."[301] The fact that the IOSA audit program is a mandatory IATA requirement serves as a true testament of how deeply rooted this value is within its member airlines. In 2012, following IOSA's trend, IATA established an environmental assessment program (IEnvA), a *voluntary* audit program that seeks to ascertain the environmental performance of airlines.[302] At the time of this writing, the IEnvA registry revealed that only South African Airways had completed the first stage of the program.[303] Although the situation may change in the near future, at least to date, the low level of participation and the fact that IEnvA remains optional are indicative of a lower degree of consideration given to environmental protection as an institutional value at IATA. It should then not be a surprise that IATA did not opt for a binding system for its member airlines. From the perspective of reducing GHG emissions from international aviation, this has been unfortunate. A binding industry-developed system of self-regulation could have facilitated the adoption of a global MBM for international aviation.

7.2.2.6 The "Historic" Resolution: Shaping the Future?

Perhaps the most interesting industry development on climate change took place in June 2013, at IATA's 69th Annual General Meeting held in Cape Town, South Africa. With only four votes against and two abstentions from its member airlines,[304] IATA passed what has been described as a "historic"[305] resolution to implement carbon neutral growth (CNG) by 2020, urging ICAO Member States to adopt a global offsetting mechanism at its 38th Assembly (Sep/Oct 2013).[306] For the first time ever, through the resolution, IATA and its member airlines advance some key design principles to operationalize the global scheme, such as an industry baseline, special adjustments for new entrants, and provisions addressing fast-growing and early-mover carriers.[307] The IATA CNG Resolution is historic

301 See IATA, "Strategic Partnerships: Building Relationships" *IATA* online: <www.iata.org/about/sp/Pages/index.aspx>.

302 See IATA, "IATA Environmental Assessment" (IEnvA), *IATA* online: <www.iata.org/whatwedo/environment/Pages/environmental-assessment.aspx>.

303 See IATA, "IATA Environmental Assessment (IEnvA) Registry" online: <www.iata.org/whatwedo/environment/Pages/ienva.aspx?Query=all>.

304 See Steele, *supra* note 206 (Chapter 1).

305 See IATA, Press Release, no 34, "Historic Agreement on Carbon-Neutral Growth", *IATA* (3 June 2013) online: <www.iata.org/pressroom/pr/Pages/2013-06-03-05.aspx> [IATA, "Historic Agreement"]).

306 See IATA, CNG Resolution, *supra* note 129.

307 See *ibid.* at Appendix 1.

for at least three reasons. First, this is the first time that IATA provided some specificity to the broad and sometimes vague notion of a global MBM scheme. As Ellickson suggests, norm entrepreneurs possess superior "technical knowledge relevant to the norms within [their] specialty" that are of utmost importance in driving for norm change.[308] In the past, IATA resorted to statements lacking in content, such as "global sectoral approach" or "working through ICAO." In those limited circumstances where details were put forward, they were usually advanced by regional trade associations and not IATA.[309] This became counterproductive, because it made the transit from mere symbolism to concrete action much more difficult.

A number of the design elements proposed in the IATA CNG Resolution were adopted by the 38th ICAO Assembly that agreed to develop a global MBM scheme.[310] In fact, to a certain extent, the IATA CNG Resolution will likely shape the future design characteristics of the eventual global MBM scheme. For instance, the industry's preference for a global mandatory offsetting scheme[311] rules out the possibility of other options such as a global ETS or any mechanism that would involve a revenue-generating mechanism. Arguably, despite the fact that ETS as an instrument choice provides more flexibility (e.g. possibility of windfall profits[312] stemming from allowances), a global mandatory offsetting scheme is much simpler to administer and it is certainly less politically sensitive, as it is not perceived to be one of the policy options advanced by the EU. The preference for a global mandatory offsetting scheme marks a drastic departure from previous industry statements where IATA favored a "global emissions trading scheme."[313]

Secondly, with the IATA CNG Resolution, the airlines declared unequivocally that they favor a global MBM scheme developed by ICAO. Although in the past IATA always strove to portray a sense of unity within the aviation sector, doubts lingered as to whether all of its member airlines in fact shared the IATA environmental values and favored a

---

308 Ellickson, "Market", *supra* note 49 at 15; Dorothea Kübler "On the Regulation of Social Norms" (2001) 17 JL Econ & Org 449 (claiming that norm entrepreneurs handle "superior information" at 453).

309 One notable example was the proposal of the Association of European Airlines (AEA) to accommodate the principle of CBDR in designing a global MBM mechanism. See AEA, Position Paper, "AEA Contribution to a Global Approach for International Aviation Emissions", *AEA* (15 December 2008) online: <http://files.aea.be/Downloads/gap_paper_final.pdf> [AEA, "Global Approach"].

310 See ICAO, A38-18, *supra* note 21 (Introduction) ("[deciding] to develop a global MBM scheme for international aviation" at para 18). The references to adjustments for fast growth carriers, provisions addressing new entrants, and early movers that appear in A38-18 were taken to a large extent from IATA's CNG Resolution. See IATA, CNG Resolution, *supra* note 129 at Appendix 1.

311 Analysts suggest that if international aviation is subject to a mandatory offsetting scheme, airlines could be required to buy CDM credits. This in turn could boost the ailing carbon market. See "UN Pins Hopes on Govts, Airlines to Bail out CDM", *Reuters* (4 December 2012) online: <www.pointcarbon.com/nes/1.2086199>.

312 See Transport & Environment, "Global Deal", *supra* note 72 (Chapter 1).

313 Bisignani, *Words of Change*, *supra* note 20 (Introduction) at 98. See also ICAO, A36-WP/85 EX/33, *supra* note 260.

global MBM for international aviation.[314] For instance, early in 2012, while fighting the EU ETS, Airlines for America (A4A) – the US airline trade association – told the US press that initiatives should "[primarily] focus on getting further fuel efficiency and emissions savings through new aircraft technology, sustainable alternative aviation fuels and air traffic management and infrastructure improvements."[315] A4A also said that it would only accept an MBM if its members failed to meet the industry's aspirational goals.[316] Clearly, at the time, A4A's main focus was to stop the EU ETS. The US airline trade association was not particularly sympathetic to the idea of a global scheme, partly due to their assumption that CNG was attainable without MBMs.[317] A4A's views had a tremendous impact on the national position adopted by the United States with respect to the global scheme.[318] A44's conduct may be regarded as an endogenous, extra-legal influence affecting the development of an extra-legal norm (e.g. IATA targets) and even on legal norms (ICAO's global MBM scheme). To this end, it can be said that the IATA CNG Resolution consolidated a unified industry position in support of a global MBM.

Finally, by using its transnational network, IATA aligned its position with A4A to uncompromisingly support the development of a global MBM scheme under the auspices

---

314 Aakre and Hovi argue that economic, political, and cultural factors determine whether a participant acts by *motus proprio* or pushed by norm changers. See Stine Aakre & Jon Hovi, "Emission Trading: Participation Enforcement Determines the Need for Compliance Enforcement" (2010) 11 European Union Politics 427 at 430.

315 "EU ETS Remains Bad News for Airlines", *A4A* (11 February 2013) online: <www.airlines.org/Pages/EU-ETS-Remains-Bad-News-For-U.S.-Airlines.aspx>.

316 Elizabeth Rosenthal, "Your Biggest Carbon Sin May Be Air Travel", *The New York Times* (27 January 2013) online: <www.nytimes.com/2013/01/27/sunday-review/the-biggest-carbon-sin-air-travel.html>. According to environmental NGO Transport & Environment, "A4A continues to argue that MBMs are only a gap filler and that sometime around 2030 technological and operational innovations, along with biofuels, will render them unnecessary." Transport & Environment, "Global Deal", *supra* note 72 (Chapter 1). See also ICAO, C-MIN 199/13, *Council, 199th Session, Minutes of Thirteenth Meeting* (statement of Colombia proposing that "MBMs should be transitory and complementary [gap filler] of the other measures to address aviation's CO_2" at 5).

317 A4A's preference for operational and technical measure is understandable. The yearly growth rate of US airlines is almost insignificant. The sector is still going through significant consolidation (e.g. American Airlines/US Airways merger). As a result, a reduction of capacity is likely to occur. In addition, it is worth bearing in mind that the US fleet is one of the oldest of the industry. Therefore, US carriers may achieve the industry fuel efficiency targets by just renewing aircraft.

318 Even when describing the results of the 38th Assembly on climate change, A4A suggested that the organization's major achievement was agreeing on "technology, operations and infrastructure measures as primary means for addressing" the sector's GHG emissions. See "A4A Applauds ICAO Climate Change Resolution", *A4A* (4 October 2013) online: <www.airlines.org/Pages/A4A-Applauds-ICAO-Climate-Change-Resolution.aspx> [A4A, "A4A Applauds ICAO Climate Change Resolution"]. A4A reported that ICAO agreed to "works toward a global [MBM] measure to fill the gap, should industry not be able to achieve carbon neutral growth from 2020 through concerted industry and government efforts." *Ibid.* ICAO, however, "agreed to develop a global MBM," not to work toward developing one. Although the difference is subtle, A4A's characterization weakens the political agreement. Also, there is simply no reference in the ICAO resolution that would describe the global MBM scheme as gap filler and interim measure. This has rather been A4A's position on the matter.

of ICAO. IATA was also able to exert a significant influence on State behavior (e.g. the position of the US government). This stands out as one of IATA's greatest accomplishments in recent times. IATA's role exemplifies, as Thomas suggests, "the power of transnational networks" in influencing domestic positions when their "campaigns are linked to international norms." [319] As explained in Chapter 4, in the past, the United States like other States was more concerned with restricting the geographical scope of the EU ETS. This represents a fundamental step forward, since the active engagement of the United States is a *conditio sine qua non* for a meaningful global scheme for international aviation.[320] In this context, the IATA CNG Resolution was instrumental in ICAO's agreement to develop a global scheme by 2016.[321]

7.2.2.7 Exogenous Influences: Some Problems for Sectoral Norm Entrepreneurs

Despite IATA's drive for norm change, it is clear that a number of States are still reluctant to accept the idea of a global MBM for international aviation. This would seem counterintuitive, for a norm entrepreneur (airlines) is asking for regulations (global MBM scheme) and regulators (States) shy away from this request.[322] In fact, with its CNG Resolution, IATA urged States to adopt a global mandatory offsetting scheme for international aviation.[323] In response, at its 38th Assembly Session, ICAO Member States only "agree[d] to develop" it. Arguably, ICAO and its Member States are a few steps behind the industry.

The regulators' reluctance to adopt a global MBM scheme for international civil aviation cannot be attributed to those advocating the norm change. There are a number of exogenous factors exerting direct and indirect pressure on States to resist regulations. These factors act as both extra-legal and legal influences. Since the sector is only a small piece within the bigger climate change puzzle, international aviation cannot be disassociated from the broader climate change context. As explained in Chapter 1, a number of States approach the issue not from the sectoral aviation perspectives, but rather from a *UNFCCC* perspective. This is what guides their interests. States consider that whatever is decided at ICAO may set precedents for future *UNFCCC* negotiations, irrespective of the actual contents of ICAO Assembly resolutions. As Teitel and Howse write, "politics [also] spills over across [special-

---

319 Thomas, *supra* note 69 at 44.

320 See generally Olav Schram Stokke, Lee G Anderson, & Natalia Mirovitskaya, "The Barents Sea Fisheries" in Oran R Young, ed, *The Effectiveness of International Environmental Regimes: Causal Connections and Behavioral Mechanism* (Cambridge: MIT Press, 1999) 91 (suggesting that a meaningful climate change regime ought to alter "the behavior of actors who are relevant to the problem" at 276).

321 See ICAO, A38-18, *supra* note 21 (Introduction).

322 Yet this is not at all unprecedented in international relations. The US chemical manufacturers were early champions of the negotiations that led to the adoption of the international ozone regime. See Benedick, *supra* note 251.

323 See IATA, CNG Resolution, *supra* note 129.

ized regimes such as ICAO]."[324] Accepting the industry request may suggest willingness to consent to a wider global scheme in the climate change context. In waiving the CBDR principle under the pretext that it is incompatible with ICAO's non-discrimination principle, there is a risk of losing the bipolar differentiated structure that has guided the architecture of climate change discussions over the past 20 years. In addition, supporting a global sectoral approach as advanced by the airline industry also places immense pressure on domestic climate change action. This is particularly sensitive for a country such as the United States where the political will to implement domestic carbon legislation has thus far been severely lacking.[325]

These exogenous factors are not present when one has to deal with intra-sectoral issues. For instance, when IATA puts forward a proposal for a norm change relating to an operational issue, such as the requirements for aircraft certification in extended-range twin-engine operations (ETOPS), all the relevant actors are within the sector. Normally, these actors will gather at ICAO to discuss these issues. IATA may confront conflicting views, but the forces are predominately endogenous. The process of learning or empowerment of new norms is certainly not as difficult as it may be when dealing with multi-sectoral issues such as the case of aviation and climate change, where significant exogenous influences emerge. In these situations, the battleground becomes considerably more challenging for norm entrepreneurs. As Finnemore and Sikkink underscores, NGOs such as IATA cannot coerce but only persuade.[326] The role of norm entrepreneurs is pivotal, but as Florini notes, sometimes it may just not be sufficient to trigger norm cascade, let alone norm internalization.[327]

In addition, perhaps one of IATA's main flaws in addressing aviation's climate change impact has been its failure to directly tackling some of these exogenous forces. After all these years, IATA has not provided a solution to deal with the apparent tension between *UNFCCC's* CBDR principle and ICAO's non-discrimination principle. Here, the world airlines trade association has left the issue for States to resolve. In its CNG Resolution, IATA stated that only States may decide whether to take into account the special circumstances and respective capabilities of States when designing a global MBM scheme for

324 Rutti Teitel & Robert Howse, "Patterns, Possibilities and Problems: Cross Judging: Tribunalization in a Fragmented but Interconnected Global Order" (2009) 41 NYU J Int'l L & Pol 959 at 990.

325 See Stephanie Monjon & Philippe Quirion "Addressing Leakage in the EU ETS: Border Adjustment or Output-based Allocation? (2011) 70 Ecological Economics 1957 (discussing that the United States is not expected to pass cap-and-trade regulation anytime soon) at 70; Robert N Stavins, "Addressing Climate Change with a Comprehensive US Cap-And-Trade System" (2008) 34:2 Oxford Rev Econo Pol'y 298 (suggesting that by passing domestic carbon legislation, the United States will gain international credibility) at 318.

326 See Finnemore & Sikkink, *supra* note 17 at 900.

327 See Florini, *supra* note 16 at 375. See also Sunstein, *supra* note 18 (discussing that "sometimes private groups are unable to produce desirable change on their own" at 947).

international aviation.[328] This is most unfortunate, given that, as a norm entrepreneur with superior knowledge and technical expertise, IATA is in a much better position to elaborate a proposal to reconcile these principles in an implementable manner. After all, it is industry that best understands the operational repercussions of any given initiative that may attempt to establish a differentiated scheme without leading to discriminatory treatment and ultimately to market distortions.

7.2.2.8 An Assessment of IATA's Norm Entrepreneurship Role

Although it has taken IATA a number of years to respond to the climate change crisis, pressure from the EU ETS has led IATA to play an active role in addressing GHG emissions from international aviation. IATA has exercised the role of norm entrepreneur in its own right. It has clearly pushed for the adoption of industry norms such as the IATA aspirational targets and the call upon States to adopt a global MBM. The IATA targets have been influential in the adoption of ICAO's aspirational goals. Its push for a global MBM scheme has shaped the position of influential States and, to some extent, is responsible for ICAO having agreed to develop a global MBM scheme for international aviation. This would not have been possible without the active involvement of the international aviation industry. In this regard, IATA's aspirational industry and its CNG Resolution have exerted a strong influence on ICAO.

One cannot help but wonder what would have happened if IATA had embarked upon this progressive approach right from the moment when the *Kyoto Protocol* assigned the mandate to ICAO to regulate aviation emissions. It took IATA more than 10 years to come up with industry targets and a plan to address its carbon footprint. And this was only primarily (if not exclusively) the result of the significant pressure imposed by the EU ETS. Environmentalists would argue that IATA should have used the EU ETS as a building block for a global scheme. Unfortunately, as mentioned above, Europe's own imperialistic attitude was not of much help.

It is also clear that, in advancing its norms, IATA was subject to numerous endogenous and exogenous influences. Reaching consensus among its own constituents was difficult. In a way IATA's internal struggles constitute a microcosm of the complexities of different State interests represented at ICAO. As a norm entrepreneur, IATA could have exercised a much more active role in educating States as to why these norms are in fact necessary. Finnemore is of the view that norms "may provide [S]tates, individuals, and other actors with understandings of what is important or valuable and what are effective and/or legitimate means of obtaining those valued goods."[329] This could have substantially facilitated the process of norm internalization. A binding self-regulation regime conditional to

328 See IATA, CNG Resolution, *supra* note 129 at Appendix 1.
329 Finnemore, *National Interests*, *supra* note 66 at 15.

maintaining or acceding to the IATA membership could have also accelerated the development and implementation of a global MBM scheme for international aviation.

7.2.3 ICAO

It is evident that ICAO faces formidable exogenous pressure.[330] Those outside aviation remain unimpressed with the organization's result. Although ICAO has done a fantastic job in technical and operational measures, as well as disseminating climate change action plans for its Member States, it has thus far failed to put in place a system to regulate GHG emissions.[331] These continue to be essentially unregulated.[332] This section analyzes whether ICAO can be regarded as a norm entrepreneur. I also consider whether ICAO's institutional setting has contributed to or been an obstacle to achieving progress on aviation and climate change issues.

7.2.3.1 ICAO: A Norm Entrepreneur or an Institutional Platform?

Despite the ICAO Secretariat's tireless efforts to advance the organization's work on climate change issues, the failure to achieve meaningful results can primarily be attributed to the lack of commitment and political will on the part of its Member States. In many occasions, the Secretariat has put forward aggressive proposals only for these to be turned down or diluted by the interests and political preferences of States.[333] This has prevented ICAO from serving as a norm entrepreneur. Driving for change and pushing for norm formation has become extremely problematic. Strictly speaking and as an international organization, although ICAO has international legal personality, it does not have a personality of its

---

330 Exogenous forces have always been influential in inducing ICAO's action on environmental issue. For instance, in 1971, the 18th Assembly adopted the first-ever resolution dealing with environmental issues. See ICAO, A18-11, *ICAO Position at the International Conference on the Problems of the Human Environment (Stockholm, June 1972)* [ICAO, A18-11]. As its name indicates, the resolution was adopted to present ICAO's position to the Stockholm environmental conference. It was an exogenous factor that prompted action. See also Assad Kotaite, *My Memoirs: 50 Years of International Diplomacy and Conciliation in Aviation* (Montréal: ICAO, 2013) at 215.

331 See Lyle, "Meaningful Global Agreement", *supra* note 242 at 2.

332 See Kulovesi, "Make Your Own Special Song", *supra* note 132 (Chapter 4).

333 The ambitious work program and timeline that the Secretariat proposed to the Council when the Expert Group on MBMs was attempting to carve out the main design elements for a framework and global scheme is illustrative. In 2012, the Secretariat presented three reports to the Council with concrete future actions. By an overwhelming majority, on all three occasions, the Council declined to agree on any real timeline. It is interesting to compare the Working Papers presented by the Secretariat and their proposed course of action vis-à-vis the final decisions that the Council adopted. See ICAO, C-WP/ 13798, *Study on Market-Based Measures (MBMs)* (presented by the Secretary General); C-WP/13858 *Environmental Protection: Recent Developments at ICAO*) (presented by the Secretary General); ICAO, C-WP/13894, *supra* note 193 (Chapter 3); ICAO C-DEC 195/9, *supra* note 193; C-DEC 196/7, *supra* note 216 (Chapter 3); C-DEC 197/6, *supra* note 218 (Chapter 3).

own in the psychological sense.[334] It cannot act in a manner that is disassociated from its Member States' interests.[335] Lionel Alain Dupuis, former Representative of Canada to the Council, put it brilliantly:

> ICAO does not have a mind of its own: it reflects the will of its membership. Accusing ICAO of not moving forward on the Environment is an unfair statement: ICAO will reflect always what its membership is able to produce on the basis of consensus; and consensus is indeed a Chicago requirement for decision-making.[336]

Although Finnemore provides various examples where international organizations have served as "agents of change," in the case of ICAO, instead of acting as a norm entrepreneur, it rather resembles a battleground or an institutional platform where changes in norms are discussed. The fact that the last three assemblies have been almost entirely dominated by either the EU ETS or aviation and climate change in general, to the point where all other issues were put on hold at one time, supports this line of thinking.[337]

7.2.3.2 Barriers to Norm Entrepreneurship

There are a number of factors that have made it more difficult for ICAO to exercise the role of a norm entrepreneur, as well as that are required for norms to cascade and to be internalized by the organization's constituents.

The Tendency to Overemphasize Achievements

ICAO seizes every opportunity to make headlines in the news and highlight its achievements, even where these may just be purely symbolic.[338] For instance, despite the fact that the Resolution adopted by the 37th Assembly received the largest number of reservations

334 See generally Jan Klabbers, *An Introduction to International Institutional Law* (Cambridge: Cambridge University Press, 2002) at 42.

335 See Guzman, *supra* note 198 (discussing that "international institutions are normally created and controlled by States" at 19); Koremenos et al, *supra* note 4 (Chapter 2) (suggesting that "States use international institutions to further their own goals, and they design institutions accordingly" at 762).

336 Lionel Alain Dupuis, Discours d'adieu du Représentant permanent du Canada au Conseil de l'OACI, Doyen des Membres du Conseil (29 June 2011) (archived on file with the author). See also See ICAO, C-MIN 181/22, *supra* note 37 (Chapter 2) at 265 (Statement of Canada).

337 Michel Wachenheim, President of the 38th Assembly, has recognized that climate change issues "were again at the centre of the Assembly's debates." Michel Wachenheim, "Interview with Michel Wachenheim – President of the 38th Session of the ICAO Assembly" in The European Civil Aviation Conference Magazine, *ECAC News* 50 Winter 2013/2014 at 10-11.

338 Symbolic achievements are not the sole purview of ICAO. For instance, Noah Sachs underlines that States, as part of their rhetorical strategies, often pledge symbolic commitments on climate change issues. See Sachs, *supra* note 197 at 843.

in the organization's history,[339] ICAO labeled it a historic agreement.[340] Major stakeholders such as the EU,[341] the United States, and the international aviation industry praised the organization.[342] Arguably, both the EU and the United States thought that a vote of confidence to ICAO was necessary.[343] In October 2013, following the adoption of a new climate change resolution in the face of severe opposition, the same actors would repeat the same story.[344]

Another interesting example is the CO_2 standard. In February 2013 ICAO widely reported CAEP's progress on a new CO_2 standard for aircraft engines.[345] This is highly illustrative given the fact that the standard has not yet been adopted by the ICAO Council. This is only expected to occur sometime in 2016. CAEP, the Council's technical advisory body on environmental matters, had merely reached agreement on the metric for a CO_2 standard and yet this was publicized as a major achievement. ICAO does not necessarily report to the media routine progress in other areas of its technical activities.

---

339 See "Stiff Challenge Facing ICAO After Unprecedented Number of Reservations on Assembly Climate Change Resolution", *Green Air Online* (29 January 2011) online: <www.greenaironline.com/news.php?viewStory=1043>.

340 ICAO, Press Release, 14/10, "ICAO Member States Agree to Historic Agreement on Aviation and Climate Change" (8 October 2010) online: ICAO <www.icao.int/Newsroom/Pages/icao-member-states-agree-to-historic-agreement-on-aviation-and-climate-change.aspx>.

341 See European Commission, Press Release, MEMO/10/482, "Breakthrough in Climate Change talks at UN Aviation Body" (9 October 2012) online: <http://europa.eu/rapid/press-release_MEMO-10-482_en.htm>.

342 See IATA, Press Release, no 46, "IATA Applauds ICAO Agreement on Aviation and Climate Change" (8 October 2010) online: IATA <www.iata.org/pressroom/pr/pages/2010-10-08-01.aspx>; Airports Council International, Media Release, "ACI Welcomes Historic ICAO Assembly Resolution" (9 September 2010) online: Airports Council International <www.google.ca/url?sa=t&rct=j&q=&esrc=s&source=web&cd=1&ved=0CCsQFjAA&url=http%3A%2F%2Fwww.aci.aero%2FMedia%2Faci%2Ffile%2FPress%2520Releases%2F2010%2FPR_091010%2520_ICAO_Assembly_Resolution.pdf&ei=dAh-U6WSDvPZ8AGTnY-DgBQ&usg=AFQjCNGv3hQRK11J8bbBosWXTmP0aoVtsQ&sig2=3fn27_c1VvTvDU9V_OSVYQ&bvm=bv.67229260,d.b2U>.

343 While discussing the merits and shortcomings of ICAO's climate change resolution that the 37th Assembly Session adopted, Hardeman says that "[w]hile superlative praise was initially heaped onto governments for clinching a deal on the new ICAO Resolution, subsequent closer inspection quickly revealed the many shortcomings of the agreement." Hardeman, "Reframing Aviation Climate Politics and Policies", *supra* note 44 (Chapter 2) at 9.

344 See ICAO, News Release "Dramatic MBM Agreement and Solid Global Plan Endorsements Help Deliver Landmark ICAO 38th Assembly" (4 October 2013) online: ICAO <www.icao.int/Newsroom/Pages/mbm-agreement-solid-global-plan-endoresements.aspx> (describing the new ICAO resolution on climate change as a "historic milestone for air transport"); IATA, Press Release, no 55, "Landmark Agreement on Climate Change at ICAO: Major Progress on Environment, Safety, Security, Operations and Regulation" (4 October 2013) online: IATA <www.iata.org/pressroom/pr/Pages/2013-10-04-01.aspx>; "A4A Applauds ICAO Climate Change Resolution", *supra* note 318.

345 See ICAO, Press Release, COM 4/13, "ICAO Environmental Protection Committee Delivers Progress on Aircraft CO_2 and Noise Standards" (4 February 2013) online: ICAO <www.icao.int/Newsroom/Pages/ICAO-environmental-protection-committee-delivers-progress-on-new-aircraft-CO2-and-noise-standards.aspx>.

The action plans is another telling story.[346] The 37th Assembly suggested that states develop action plans to mitigate aviation's climate impact.[347] In addition to providing technical assistance, the Secretariat organized numerous seminars and workshops in order to encourage Member States to submit these action plans.[348] The Secretariat has pointed out that, as of 30 June 2013, "61 Member States, representing over 78.89 percent of global international air traffic prepared and submitted action plans."[349] The way in which the Secretariat presents this statistic constitutes an overemphasis of the fact that aviation's major players have presented action plans. It ignores that this only represents about 32 percent of the total ICAO membership. Currently, only 24 State action plans had been made public.[350] That is less than 50 percent of those submitted to ICAO. From those that were made public, it is not easy to infer that States have regulations in place to control or reduce GHGs from aviation. Most plans describe operational and technological measures where fuel savings are reported in comparison with a business-as-usual scenario. The sector's growth is often not accounted for, and therefore, it is not possible to establish that net reductions in GHG emissions have in fact been achieved. In spite of all this criticism, one must also recognize that, as discussed in Chapter 3, as part of a learning process, however, the action plans are an invaluable tool to engage Member States to start addressing the issue for the first time.

ICAO's tendency to exaggerate its achievements leads to unintended consequences. Member States are led to believe that the organization has attained substantial progress where in fact most of its achievements are just symbolic. It is considerably more challenging for ICAO Member States to determine the organization's shortcomings. The tendency also negatively affects the sense of urgency to act. Although one could understand that this tendency is the natural response to the strong exogenous pressure, the environment is not conducive for norm cascading, let alone internalization. A more realistic presentation of events and facts is required.

Reaching Out to a Broader Audience

The borderless nature of climate change requires the engagement of a broader audience than just those State members to the Council. The low number of action plans submitted is once again an indicator that for a large number of States, climate change does not rank

346 See ICAO, A38-WP/30, EX/25, *States' Action Plans for CO_2 Emissions Reduction Activities* (presented by the Council of ICAO) [ICAO, A38-WP/30, EX/25].

347 See ICAO, A37-19, *supra* note 207 (Chapter 1) at para 9.

348 See ICAO, "ICAO Symposium on State Action Plans" online: ICAO <www.icao.int/Meetings/Green/Pages/default.aspx>.

349 ICAO, A38-WP/30, *State's Action Plans for CO_2 Emissions Reduction Activities* (presented by the Council of ICAO).

350 See ICAO, "Climate Change: Action Plan" online: ICAO <www.icao.int/environmental-protection/Pages/action-plan.aspx>.

high enough on their policy agendas. ICAO's failure to trigger wider participation is not a new issue. I have discussed this structural deficiency when addressing the organization's institutional setting, which is strongly dominated by States that are members of the Council. For instance, Qatar and Thailand rank in the top 20 countries in terms of international air traffic. However, because they are not members of the Council, they have very little interaction with the organization on these issues.[351] Yet this structural deficiency is not limited to States. I have also noted the symbiotic relationship between ICAO and industry.[352] Already, there have been attempts to restrict the role of industry observers within the organization.[353] Yet one additional issue which has not been subjected to much analysis is the absence of low-cost carriers. Most notably, missing from the IATA membership are the low-cost airlines. Lately, many of these carriers have experienced some of the highest growth rates in the world. In Europe, low-cost carriers account for almost 40 percent of the intra-European traffic.[354] These carriers are also making significant strides in Asia, the Middle East, and Latin America. It is uncontested that they are becoming major players in international civil aviation. An environmental NGO, Transport & Environment, points out that these carriers' views on aviation and climate change are not always aligned with that of IATA.[355] At ICAO, low-cost carriers are not represented. This raises a key question: should ICAO reach out to low-cost carriers as well?

Another group of actors that would like to have more access to ICAO is the environmental NGOs. Although, in the context of climate change discussions at ICAO, some environmental NGOs have gradually gained more access (e.g. by participating in ICAO symposia, workshops, conferences, ICAO's Committee on Aviation Environmental Protection, delivering informal briefings to the Council), these actors still face considerable hurdles when it comes to accessing relevant information. This has to do with the fact that Council documentation, as well as that relating to high-level meetings and special working groups, is primarily posted on secured websites (e.g. ICAO Portal). Environmental NGOs usually do not have permanent representatives accredited to the organization. Therefore, gaining access to such information is much more difficult for these actors. It is no surprise that they often complain about the lack of transparency in the ICAO process.[356] More

351 See ICAO, "Civil Aviation: 2012 International RTK by State of Air Operator Certificate" online: ICAO <www.icao.int/Meetings/a38/Documents/International%20Scheduled%20RTK%20(Annual%20Report).PDF>.

352 Implicitly acknowledging this relationship, Assad Kotaite's, former President of the ICAO Council for almost 30 years, writes that "ICAO "relies on the industry to reach a carbon-neutral growth from 2020 and a 50 percent emission-reduction by 2050, compared to 2005 levels." Kotaite, *supra* note 330 at 215. It is worth pointing out that halving emissions from 2050 at 2005 levels is an industry aspirational goal that has yet to be embraced by ICAO Member States. *Ibid.*

353 See Chapter 3.

354 See Transport & Environment, "Global Deal", *supra* note 72 (Chapter 1).

355 *Ibid.*

356 *Ibid.*

transparency may reduce incentives to defect, free-ride or cheat and promote compliance.[357] Environmental NGOs bring a completely different perspective to the issue of aviation and climate change and a much needed balance. To this end, it is in the best interest of ICAO and its Member States to reach out and provide more access to these actors.

In highlighting the need to reach out to a broader audience, Lyle notes that ICAO has done a poor job in engaging those who will be most affected by climate change, namely, small island developing States (SIDS) and least developed land-locked States (LLDCS).[358] This significantly differs from the *UNFCCC* context where these actors have gained considerable participation.[359] Here, the words of Bodansky et al. resonate strongly: "[f]or a global regime building, the important lesson is that shared normative understandings must be gradually cultivated and deepened and that regimes must be designed so as to maximize the opportunities for normative interaction."[360] Interaction, as Koh notes, plays a much lesser role in the process of norm internalization if the proposed changes in norms are only discussed by a selected handful of actors. In such an environment, norm entrepreneurship is extremely unlikely to occur. According to Koh, interaction is what leads transnational actors to obey international rules or norms. As such, the institutional design should "empower more actors to participate in the process."[361] Otherwise, as Powell suggests, lack of interaction diminishes not only the "normative power of law"[362] but also of those extra-legal norms.

The Exculpation Argument

ICAO and the aviation industry have placed formidable emphasis on the fact that international aviation's carbon share is only 2 percent of worldwide CO_2 emissions. As discussed in Chapter 1, this exculpating argument misses the point. The problem does not lie in the relative small contribution that aviation makes to global GHG emissions, but rather the exponential growth associated with the industry. Also, from a regulatory point of view, it is inconceivable to demand that all sectors should reduce their emissions, but exclude aviation on the basis that, for the time being, it is not a major emitter. By placing enormous emphasis on its 2 percent contribution, ICAO and the aviation industry have created what I call the "leave me alone syndrome." Given that the only 2 percent contribution exculpation argument has been repeatedly made, a number of States have been led to believe that the

---

357 See Oran R Young & Marc A Levy, "The Effectiveness of International Environmental Regimes" in Young, *supra* note 6 at 23; Oran R Young, "Regime Effectiveness: Taking Stock" in Young, *supra* note 6 at 269.

358 See Green Air "Global Approach", *supra* note 266.

359 See Carlarne, *supra* note 153 (Chapter 2) (noting that "the interest and voices of specific groups – for example, indigenous peoples in the Arctic region and citizens of the Association of Small Island States – that represent minority or marginalized populations increasingly infiltrate international debate" at 472).

360 Bodansky et al, *Oxford Handbook*, *supra* note 66 (Chapter 2) at 12.

361 Harold Koh, *supra* note 54 at 677.

362 Powell, *supra* note 44 at 54.

sector is such a small emitter and therefore should be left alone. Its contribution will not make any difference to the overall picture.

The All-Mighty Technological Defenses

ICAO and industry have also said that part of the problem is due to a reputational crisis. The sector has poorly communicated its continuous environmental improvements. After all, its track record speaks for itself. Therefore, both ICAO and industry present a notorious tendency to overemphasize the sector's technological and operational achievements. For instance, the climate change resolution adopted by the 38th Assembly states that "aircraft produced today [are] about 80 percent more fuel efficient per passenger kilometer than in the 1960s."[363] As one State has noted, although technically accurate, by dissociating the statement from the net growth of CO_2 emissions during such period, it gives the false impression that the sector may attain its aspirational goals by just resorting to technological measures.[364] If this is the case, and if aviation's share is only 2 percent of the world's CO_2 problem, why does the sector need MBMs? It is worth bearing in mind that a number of States continue to argue that MBMs are not needed and that ICAO should rather focus on operational and technological measures.[365] The way in which the problem has been framed does not necessarily contribute to raising a sense of awareness and urgency for Member States to act.[366] This sets the stage for our next subject: the presentation of information and data and its relevance to norm internalization.

---

363 ICAO, A38-18, *supra* note 21 (Introduction) at preambular clauses.

364 See ICAO, A38, WP/258 EX 85 EX/85, *UAE's Views on Aviation and Climate Change* (presented by the United Arab Emirates). Similarly, environmental NGO Transport and Environment points out that "[i]ndustry and ICAO have taken the view for many years that new and better aircraft, technology and operational improvements together with the arrival of drop-in alternative fuels, primarily biofuels, will be enough to arrest [the sector's] growth in emissions thus ensuring that aviation itself can continue to grow unconstrained." Transport & Environment, "Global Deal", *supra* note 72 (Chapter 1).

365 See ICAO, A38-WP/183 EX/72, *Achieve Emissions Reduction Through Technical and Operational Measures: What China Has Done* (presented by the People's Republic of China) (calling ICAO Member States to "[recommend] that resolutions of ICAO Assembly, as a policy tool, should specify technical and operational measures as the priority means for achieving international aviation emission reduction, and that such measures be treated as ICAO's focus of work in the future"); ICAO, A38-WP/272, *Position of African States on Climate Change* (presented by 54 African States) (describing ICAO's global MBM scheme as a "transitional measure and complementary to" operational and technological initiatives); ICAO, A38-WP/176 EX/67, *Views of Saudi Arabia* (suggesting that ICAO should not focus on MBMs, but rather on operational and technological measures at 4); ICAO, A38-WP/250 EX/83, *Market-Based Measures as the Factor of an Increase of Greenhouse Gas Emissions in the Sector of International Civil Aviation* (presented by the Russian Federation) (underscoring the negative effects of MBMs on international aviation).

366 In his fascinating 392-page memoir recalling more than 50 years at ICAO, Assad Kotaite dedicates less than 2 pages to the subject of the environment. It is very telling that, in addition to emphasizing that international aviation's carbon footprint is roughly 2 percent of global emissions, as the man who ran ICAO for an unprecedented 30 years, Kotaite stresses that "[o]ver the last 40 years, fuel-efficiency has improved by 70 percent." There is no reference however to the growth of emissions during the same period. Kotaite, *supra*

Information, Data, and Norm Internalization

I am not suggesting that ICAO has engaged in sharing inaccurate data with its constituents. What I argue is that the framing of the information and data has not been appropriate. In fact, in many situations, such as those described above, ICAO has unintentionally misled its Member States. The way in which the information is presented to actors is extremely important. For instance, Vandenbergh comments that, because eco-labeling campaigns have struggled to communicate the intended message in a way that it is easy for the audience to understand, they have had almost insignificant influence in persuading actors to change behavior.[367] Sunstein points out that "choices, meaning, roles, and norms are commonly based on beliefs about relevant facts."[368] Similarly, "[w]hen the belief shifts, the norm [also] shifts."[369] However, if the facts are either wrong or misrepresented, actors will not engage in the process of norm change. Neither will they see the need for nor the benefits in doing so. Sunstein notes that "[n]orms about behavior are interpenetrated with beliefs about harm and risk."[370] In other words, it is very unlikely to cure a disease if the patient does not know that he or she is ill and is not cognizant about the risks associated with the illness. Information must be properly presented.[371] Otherwise, it becomes extremely difficult to influence the actors' beliefs over a proposed norm change.

Educational Process

Expanding and enhancing education to its constituents and its audience at large constitutes another area of opportunity for ICAO. International organizations and governments have different tools at their disposal to promote a change of norms. Education is certainly one of them.[372] Although ICAO has organized numerous symposia, conferences, workshops, and high-level meetings to discuss aviation and climate change issues, this has not necessarily been the most effective way to spread the message that climate change requires the sector's collective attention. South Korea has stressed the need to develop education programs on aviation environmental issues.[373] Similarly, Argentina has pointed out that ICAO's regional office should be involved in disseminating the outcomes of the works of ICAO's Committee on Aviation and Environmental Protection (CAEP) through the States that

note 330 at 214-15. These types of statements serve to illustrate the mind-set of a large number of policy makers involved in international aviation.

367 See Vandenbergh, *supra* note 158 at 1136.

368 Sunstein, *supra* note 18 at 930.

369 *Ibid.*

370 *Ibid.*

371 See generally Koremenos et al, *supra* note 4 (Chapter 2) (discussing that "States are risk-adverse and worry about possible adverse effects in its dealings with international institutions" at 782).

372 See Sunstein, *supra* note 18 at 949.

373 See ICAO, A38-WP/270 EX 90, *Development of Aviation Environmental Education Program* (presented by the Republic of Korea) at 2.

are not committee members.[374] Norm formation is not an abstract process. It is the result of an iterative process in which States need to engage and participate. However, engagement and participation are useless processes unless States, their elite decision makers, as well as relevant actors understand what is at stake. It cannot be assumed that all States have a clear and profound understanding of the issues, neither are they able to define by themselves their own State interests. Education is of paramount importance in the process of norm formation, both for the definition of interests as well as values. Its absence certainly makes things much more complicated.

Regime Architectural Design: The Problems of Top-Down Norm Formation

At the international level, norms are discussed and negotiated, for the most part, at ICAO. Although one cannot but recognize the exogenous influence that IATA, its member airlines, as well as other organizations exert, the fact still remains that norm formation takes place from top to bottom. It is not necessarily built on domestic acceptance.[375] This is how the global MBM is being developed. As such, this poses significantly more challenges for norm formation in the international civil aviation context.

To a large extent, the vast majority of international norms relating to aviation and climate change issues are adopted through a top-down approach. In fact, the climate change regime has been built on this basis. However, some commentators have already spotted that there is a tendency to focus more and more on bottom-up approaches.[376] A number of measures take place at the domestic side. Mehling suggests that "the locus of enforcement efforts may shift to the domestic level."[377] Torres and Pinho are of the view that "local and individual actions are essential to mitigate [climate change's impact]."[378] In light of the difficulties to agree on the post-Kyoto climate change regime at the global level, a number of countries and local governments within those countries have taken measures. Time will only tell whether bottom-up will replace the top-down or whether both approaches may complement each other.

---

374 See ICAO, A38-WP/318 EX 110, *Environmental Protection, CAEP and the ICAO Regional Offices* (presented by Argentina) at 3.

375 This is why Björkdahl suggests that the internalization at the domestic level of international norms becomes much easier when they form part of a "heightened domestic interest." Björkdahl, *supra* note 11 at 18. For the most part, this is not necessarily the case when addressing international civil aviation and climate change issues.

376 See Jacob Werksman, "Compliance and the Use of Trade Measures" in Jutta Brunnée, Meinhard Doelle & Lavanya Rajamani, *Promoting Compliance in an Evolving Climate Regime* (Cambridge: Cambridge University Press, 2012) 262.

377 Mehling, *supra* note 278 at 213.

378 Miguel Torres & Paulo Pinho, "Encouraging Low Carbon Policies through a Local Emissions Trading Scheme" (LETS) (2011) 28 Cites 576.

7.2.3.3 Assessing ICAO's Role

Ever since the *Kyoto Protocol* was adopted, ICAO has attempted to demonstrate results. This however has not been possible due, primarily, to the lack of political will on the part of its Member States. It is clear that rather than acting as a norm entrepreneur, ICAO has become the institutional platform where relevant actors gather to discuss aviation-related climate change issues. As discussed in Chapter 3, a number of factors relating to the organization's institutional setting and its governance structure make it much more difficult for norm entrepreneurship to arise. The organization should place significantly more emphasis in reaching out to a broader audience, promoting interaction and participation, as well as enhancing its educational campaign. In addition, ICAO should make a better effort in the diffusion of appropriate information. The framing of sensitive information and data has not been useful in conveying the message that the international aviation community must act to tackle climate change. On the contrary, it has served as ammunition for those who oppose the change in norms. These factors will be even more relevant when the organization enters into the design and implementation phases of its intended global MBM scheme for international civil aviation.

7.2.4 The United States

7.2.4.1 Not Just Another Player

Segerlund writes that in international relations, "not all actors have equal weight."[379] With the largest domestic aviation market and the largest share of international traffic,[380] this is certainly the case of the United States. Although traffic patterns indicate that the sector's growth is moving to other regions,[381] and the balance of power will change in the next 30-40 years, the United States remains the most critical actor in international civil aviation. Whatever the United States decides to endorse or oppose bears significant consequences. In today's world, no scheme could work without the backing of the United States. A number of treaties in international air law have either failed to enter into force or bear no practical application because the United States has not ratified or acceded to them.[382] This section examines the role of the United States in the domain of aviation and climate change.

379 Segerlund, *supra* note 52 at 28.

380 ICAO, "Civil Aviation: 2012 International RTK by State of Operator Certificate" online: ICAO <www.icao.int/Meetings/a38/Documents/International%20Scheduled%20RTK%20(Annual%20Report).PDF>.

381 By 2050, Asia is expected to have the largest share of CO_2 emissions. See Transport & Environment, "Global Deal", *supra* note 72 (Chapter 1).

382 Among others, these include the *Convention on Damage Caused by Foreign Aircraft to Third Parties on the Surface*, Oct 7, 1952; *Convention on Compensation for Damage Caused by Aircraft to Third Parties*, May 2, 2009; *Convention on Compensation for Damage to Third Parties, Resulting from Acts of Unlawful Interference Involving* Aircraft, May 2, 2009 [*Unlawful Interference Convention*].

7.2.4.2 The United States and Climate Change

The United States has been instrumental in the development of international environmental law. Many environmental principles of general application have originated from US statutes.[383] It is uncontested that the United States has played the role of a norm entrepreneur in the adoption of numerous international environmental instruments, including the *Montreal Protocol*.[384] Yet its influence on the climate change regime has been far from commendable. The withdrawal of the United States from the *Kyoto Protocol* has created a formidable distrust in the climate regime.[385] In fact, it strongly influences the position of a number of countries.[386] Winkler claims that without serious US involvement in the climate change regime, other countries may not be willing to undertake any mitigation obligations.[387] The US climate change policy may be best characterized by its "defensive approach." In essence, the United States seeks to avoid taking up any binding international commitments, and it has fought to ensure that the differentiated regime under which major developing countries bear no mitigation obligations ceases to exist. All agencies within the US government dealing with climate change issues are fully aware that US Congress has no appetite for an international climate regime imposing mitigation obligations on developed States while leaving emerging developing economies off the hook.[388] Congress poses considerable challenges to reaching an international agreement on climate change.[389] Despite modest attempts by the Obama Administration, the United States has been unable to pass carbon legislation at the federal level. This bears significant consequences for the climate regime as a whole.[390] In other words, the domestic landscape shapes the US position

383 See Philippe Sands, "The Greening of International Law: Emerging Principles and Rules" (1994) 1:2 Ind J Global Leg Stud 293.

384 Benedick, *supra* note 251 at 53.

385 Joyeeta Gupta, "Developing Countries and the Post-Kyoto Regime: Breaking the Tragic Lock-in of Waiting for Each Other's Strategy" in Douma, Massai & Montini, *supra* note 192 (Chapter 1) at 161. Unless they receive side payments, as well as financial and technical assistance, major developing countries are unwilling to accept binding commitments. See Barrett, "Why Cooperate", *supra* note 57 (Chapter 1) at 205; Michal Meidan, "China's Emissions Reduction Policy: Problems and Prospects" in Antonio Marquina, ed, *Global Warming and Climate Change: Prospects and Policies in Asia and Europe* (London: Palgrave Macmillan, 2010) at 317. Similarly, the United States will not participate in any post-*Kyoto* regime unless China and other major developing economies join. This wait-and-see game of conditioning membership to a given regime until others adhere is not exclusive of the climate change discussions. For example, it is well recorded that the United States did not sign into the land mines regime because other important players such as the Russian Federation, China, and India did not join. See Harold Koh, *supra* note 54 at 659.

386 See Tuerk et al, *supra* note 242 (Chapter 2) (noting that Canada "has repeatedly tied its climate policy to the US position" at 6).

387 See Harald Winkler, "An Architecture for Long-term Climate Change: North-South Cooperation Based on Equity and Common but Differentiated Responsibilities" in Frank Biermann, Philipp Pattberg & Fariborz Zelli, eds, *Global Climate Governance Beyond 2012: Architecture, Agency and Adaptation* (Cambridge: Cambridge University Press, 2010) at 105.

388 See Werksman, *supra* note 376 at 262.

389 See Boyd, *supra* note 32 at 469.

390 See Perdan & Azapagic, *supra* note 22 (Chapter 4) at 6051.

internationally. These endogenous forces, which have taken the form of both extra-legal and legal influences, affect the norm formation (both extra-legal and legal norms) process at the international level. As Barrett puts it, "international cooperation needs to be joined with improvements in national governance. The challenges are local and not only global."[391]

7.2.4.3 The United States: Aviation and Climate Change

With minor variations, the United States has transposed its overall position on climate change matters into the field of international aviation. At ICAO, the United States has strongly supported operational and technological measures, but has, at best, been skeptical with respect to MBMs.[392] Lyle claims that for the United States, MBMs are "unnecessary."[393] In fact, the United States' Action Plan contains a full description of operational and technological initiatives aimed at reducing GHG emissions from aviation.[394] Yet there are timid references to potential MBMs to be applied at the domestic level.[395]

To a large extent, and aside from the recent inclusion of some reference to CBDR in the guiding principles of ICAO's climate change resolution,[396] the United States has successfully resisted any attempt to accommodate such principle in the domains of ICAO.[397] As discussed in Chapter 4, the United States spent a considerable amount of time and resources fighting the EU ETS.[398] This battle has been the dominant feature of the aviation environmental agenda since 2005 when EU initially announced the inclusion of international aviation.

The 2012 US Presidential Elections also significantly influenced the US position. In spite of President Obama's well-known environmental credentials, the US government

391 Barrett, *Why Cooperate*, *supra* note 57 (Chapter 1) at 189.

392 The United States believes that it can achieve ICAO's aspirational goals without introducing MBMs. Both the US government and the US aviation industry place great expectations on the role that alternative fuels will play.

393 See Lyle, "Mitigating International Air Transport Emissions through a Global Measure", *supra* note 122 (Chapter 1).

394 See generally Detlef Sprinz & Tapani Vaahtoranta, "The Interest-Based Explanation of International Environmental Policy" in Peter M Haas, ed, *International Environmental Governance* (Burlington: Ashgate, 2008) 131 (discussing that technological factors may lessen actual or anticipated abatement costs and thereby increase the propensity of a country to support international environmental regulation" at 159).

395 ICAO, "US Aviation Greenhouse Gas Emissions Reduction Plan" online: ICAO <www.icao.int/environmental-protection/Documents/ActionPlan/CAEP-U%20S-ClimateActionPlan.pdf>.

396 See ICAO, A38-18, *supra* note 21 (Introduction) (indicating that in designing MBMs for international aviation, States, among others, should take into account the CBDR principle at Appendix A (p)).

397 For the United States, CBDR is a foreign concept to international civil aviation and as such not applicable at all. See ICAO, C-MIN 194/2, *supra* note 86 (Chapter 1) at para 101 (statement of the United States); ICAO, "A38 Resolutions – Reservations" online: ICAO <www.icao.int/Meetings/a38/Documents/Resolutions/United_States_en.pdf> (Reservation of the United States).

398 Likewise, in 2007, the United States led the opposition of ICAO Member States against the EU ETS. See Transport & Environment, "Global Deal", *supra* note 72 (Chapter 1). At the time, the 36th Assembly decided that States should not implement an ETS, unless they obtain the mutual consent of the States involved. See ICAO A36-22, *supra* note 17 (Chapter 2) at Appendix L, para 1 (b).

was reluctant to assume any commitment on aviation and climate change issues throughout 2012. The United States strongly argued that the ICAO's focus should be on developing a framework instead of designing a global MBM scheme.[399] The likelihood is that the United States perceived that the framework would provide a tool to further restrict the geographical scope of the EU ETS. The United States seized every opportunity to dilute any proposal from the ICAO Secretariat on the global scheme.

7.2.4.4 Changing the Approach: From Resistance to Endorsement

In 2013, the attitude of the United States toward aviation and climate change shifted dramatically. Slowly but gradually, the US government started showing signals that it may be supportive of the idea of developing a global scheme under the auspices of ICAO. One can think of a number of factors to explain this change. First, reelection provided the Obama Administration increased freedom to engage in climate change issues. Even at his inauguration, President Obama emphatically referred to the "threat of climate change."[400] Months later, when unveiling his climate change plan, President Obama expressly referred to ICAO by noting that the United States is working "towards agreement to develop a comprehensive global approach."[401] Second, the appointment of Senator John Kerry as Secretary of State may have exerted pressure on the US airline industry and the various agencies within the US government to play a more progressive role on aviation and climate change issues. In the past, while criticizing the position of the US government and US air carriers against the EU ETS, then US Democratic Senator John Kerry had said that the United States has been "one of the most consistent foot-draggers on reducing airline emissions."[402] Third, as explained above, the IATA Resolution and the extensive lobbying activities of both IATA and A4A were largely responsible for the change of position of the US government with respect to the global MBM scheme. Without them, it would have been less likely that that the United States would ever support a proposal at ICAO to develop a global MBM scheme for international aviation.

7.2.4.5 Assessing the United States' Role

For a long time, the United States did not act as the most constructive partner on aviation and climate change issues. Its containment strategy was only geared to stop the application

399 ICAO, HGCC/2-WP/9, *supra* 160 (Chapter 3).

400 The White House, Office of the Press Secretary, "Inaugural Address by President Barack Obama", *The White House* (21 January 2013) online: <www.whitehouse.gov/the-press-office/2013/01/21/inaugural-address-president-barack-obama>.

401 See The White House, *The President's Climate Action Plan, The White House* (June 2013) online: <www.whitehouse.gov/sites/default/files/image/president27sclimateactionplan.pdf>.

402 Laing, *supra* note 337 (Chapter 5); Patrick Goodenough, "Sen. Kerry Blames U.S. for Dispute Over E.U. Airline Carbon Tax: 'We Dragged Our Feet'", *CNS News* (7 June 2012) online: <http://cnsnews.com/news/article/sen-kerry-blames-us-dispute-over-eu-airline-carbon-tax-we-dragged-our-feet>.

or to restrict the scope of the EU ETS. As explained in Chapter 4, the United States successfully led the Coalition of the Unwilling against the EU ETS directly or indirectly. Things would have been completely different had the United States genuinely supported the idea of a global scheme from the moment ICAO was tasked with the mandate of addressing aviation's carbon footprint. Having said this, one must also recognize that the swift change in the US position that occurred in 2013 was vital in order for the achievement of consensus at 38th Assembly with regard to agreeing to develop a global MBM scheme for international aviation.[403]

It is well known that "traditionally, the United States uses its foreign policy to shape the policy of other countries in ways favorable to or in line with US goals and interests."[404] However, as Barrett writes, "when the opportunities to supply a global public good are seized by [a] great [power such as the US, even if it is] motivated only by self-interest, but acting within the framework of the general good, the entire world benefits."[405] Similarly, Florini notes that powerful States have much more opportunities to advance a change in norms.[406] Without the active involvement of the United States, a global MBM scheme for international aviation would remain wishful thinking.[407]

It is fair to say that throughout this process, the United States has not acted as a norm entrepreneur. In fact, for many years, it has resorted to a delay tactic and placed so much emphasis on efforts to derail the EU ETS. Certainly, this cannot be categorized as constructive behavior on the part of a major player in global aviation. Having said this, the recent shift in the position of the United States to endorse the development of a global MBM scheme brings cautious optimism for the international community. Although this is only a commitment to develop a scheme that is expected to be operational in 2020, and even though considerable work is required to define its design elements, the highly constructive and progressive approach that the United States has brought to the table is definitely a game changer.[408]

---

403 See ICAO, A38-18, *supra* note 21 (Introduction) at para 20. See also Ewa Krukowska, "First Global Emissions Market for Airlines Wins Support," *Bloomberg* (3 October 2012) online: <www.bloomberg.com/news/2013-10-04/first-global-emissions-market-for-airlines-wins-support.html>.

404 Guri Bang & Miranda A Schreurs, "A Green New Deal: Framing US Climate Leadership" in Wurzel & Connelly, *supra* note 194 (Chapter 1) 231 at 247.

405 Barrett, *Why Cooperate*, *supra* note 57 (Chapter 1) at 11.

406 See Florini, *supra* note 16 at 374.

407 EDF, Press Release "ICAO Committee Agrees on Path Forward to Limit Aviation Emissions, but U.S. Leadership will be Critical to Delivering Results" (3 October 2012), online: <www.edf.org/media/icao-committee-agrees-path-forward-limit-aviation-emissions-us-leadership-will-be-critical?utm_source=feedburner&utm_medium=feed&utm_campaign=Feed%3A+EnvironmentalDefense%2FPressReleases+%28EDF.org+-+Press+Releases%29> (highlighting the fact that US leadership will be of paramount importance in developing global MBM scheme).

408 See Aysha Al Hamili, "Some Reflections on ICAO's Assembly Historic Achievements" (2013) 6 ICAO J at 13-14 online: <www.icao.int/publications/journalsreports/2013/6806_en.pdf>.

7.3 Conclusion

Norm entrepreneurship theory provides an excellent platform to examine when actors advance a proposed change in norms. It also helps to dissect the conditions and circumstances under which norm entrepreneurs will most likely emerge and the requirements for norm internalization by recipients. The theory is particularly relevant in analyzing aviation and climate change: an area where all key players claim to be ahead of the curve. I have thoroughly examined the role played by Europe and the contribution as well as the flaws of the EU ETS. Europe's quest to advance mitigation efforts in the overall climate context cannot be denied. The inclusion of international aviation into the EU ETS is also an attempt to drive normative change. In all likelihood, Europe understands that international aviation's CO_2 output cannot continue unregulated *per secula seculorum*. One must recognize that Europe's unyielding quest to address climate change issues for international aviation has raised its profile and kept discussions alive. Had it not been for Europe, climate change would not have been a top policy issue for ICAO. Unfortunately, Europe's major sin with respect to its ETS has been its notorious Euro-centrism, stubbornness, and inability to make compromises during the early stages of the game. Europe has shown a remarkable ability to disregard the concerns of the whole international community. Europe has alienated almost the whole membership of ICAO in an unprecedented manner. In this context, the EU ETS has proved sterile grounds for norm cascade and norm internalization. In fact, no State has taken action as a result of the scheme. The European experience also demonstrates the importance of effectively and accurately communicating the whole process of norm change. Failure to do so leads to undesired results.

Before the EU ETS, IATA mainly focused on fuel-saving initiatives. The advent of the regional scheme prompted IATA to adopt aspirational industry targets. Likewise, to avoid a patchwork of multiple and conflicting regulations, IATA advanced the idea of a global sectoral approach. The "proactive" IATA response falls exactly within the conditions described by Ellickson whose conception is that norm entrepreneurs normally arise where exogenous threats appear.[409] By changing the incentive structure, IATA could have required its member airlines to comply with these targets in order for the organization to develop its own industry scheme by making it a condition of membership. For those advancing a change in norms, the sense of belonging to a club such as IATA is an invaluable self-enforcing mechanism and a salient characteristic of norm entrepreneurship. Perhaps, these targets were not ripe for internalization within the broad and diverse IATA constituency. The uncertainty stemming from the regulators did not help much either. Although IATA's

409 One should not underestimate the influence of "exogenous forces," for "international regimes quite frequently fall victim to [their] impact." Oran R Young, "The Rise and Fall of International Regimes" (1982) 36:2 International Organization 277 at 294.

strong opposition to the EU ETS may be criticized, its most significant contribution occurred during IATA's 69th AGM where airlines unanimously called upon ICAO to adopt a global MBM scheme. In addition to exerting tremendous pressure on key players such as the United States, the IATA CNG Resolution screamed loud and clear that air carriers were ready to engage and participate in a global MBM scheme. As Tony Tyler, Director General of IATA, has articulated, "sustainability is aviation's license to grow."[410]

By examining ICAO's work on climate change through the lens of norm entrepreneurship theory, I have identified a number of obstacles. ICAO is an excellent case study for two reasons. First, it reveals the difficulties that an organization faces when it attempts to prompt changes in norms within its membership while its institutional setting is not conducive. Second, it reinforces the importance of accurately conveying the information. However, if the message contains wrong data or if it may be misleading, actors will not necessarily embrace change. As a result of the exogenous pressure, one can understand ICAO's and the industry's tendency to overemphasize usually minor technological and operational improvements. Yet this has also made it more difficult for Member States to fully grasp the reasons why MBMs are needed. Without MBMs, States and the industry will simply not achieve the aspirational goals.[411] A price on carbon may serve as an incentive "to speed up deployment" of non-MBM measures.[412] ICAO's educational campaign could be significantly enhanced. If interaction and interpretation, as explained by Koh, cannot be fully developed, it will be much more difficult to internalize norms.

I have also examined the role of the United States in this process. Its influence should not be underestimated. Although in the past the United States devoted much of its time to stalling progress on climate change issues or fighting the EU ETS, it must be recognized that the more progressive stance taken by the US government since 2013 brings the much needed optimism.

Applying norm entrepreneurship to aviation and climate change also reveals that the theory faces much difficulty in addressing transnational issues where significant exogenous multi-sectoral/multi-institutional forces are present. States do not only care about the interests of their respective aviation sectors, but they are also concerned with potential wider implications in the context of the *UNFCCC* climate change negotiations. The top-down norm formation architectural design that is currently used in international aviation and climate change issues certainly makes it more complicated for norm entrepreneurs to promote new norms, because that approach does not benefit from a solid domestic approval.

---

410 IATA, "Historic Agreement", *supra* note 305.

411 See ICAO, HGCC/3-AIP/6, *supra* note 6 (Introduction) at 2.

412 *Ibid.*

If one follows Finnemore's conceptualization that full norm internalization occurs when the norm in question acquires a taken-for-granted status,[413] then international civil aviation has a long way to go. To a large extent, numerous States still question the need for the norm of regulating GHG emissions from international aviation through an MBM scheme.[414] For example, Cuba has noted that "MBMs [do not] reduce aviation emissions [but rather they do have] the potential to inhibit [its] development."[415] Moreover, "MBMs [are] environmentally irresponsible and [are] aimed at generating revenue."[416] In fact, these States acknowledge that "emissions [will] increase due to the expected growth in international air traffic," but this will be the case "until lower emitting technologies and fuels and other mitigating measures are developed and deployed."[417] These statements point to the apathy among States to wait for technological change to resolve the climate crisis.

These views, however, should not be perceived as an insurmountable obstacle, but rather as a reality check. Relevant actors ought to develop a twofold approach that would contemplate the very basic ideas of both rational choice theorists and constructivists. This should include working on making more attractive the incentive structures for relevant actors. Any scheme should induce participation, encourage compliance, and reduce GHG emissions from international aviation.[418] At the same time, the approach should also place

413 See Finnemore, *National Interests*, *supra* note 66 at 23.

414 See Reservation of the Russian Federation to ICAO A38-18, Consolidated Statement of Continuing ICAO Policies and Practices related to Environmental Protection: Climate Change, online: ICAO <www.icao.int/Meetings/a38/Documents/Resolutions/Russia_en.pdf> (noting that MBMs "reduce the potential of the [aviation]sector to actually lower GHG emissions and will also have a negative impact on overall flight safety indicators, due to slower rates of technological development in the sector"). See also "Russia Urges ICAO to Reconsider Unrealistic Carbon Neutrality Goal and Market Measures for International Aviation", *Green Air Online* (18 November 2013) online: <www.greenaironline.com/news.php?viewStory=1786>. Similarly, during ensuing Council discussions prior to the 38th Assembly Session, Colombia noted that ICAO should concentrate on "measures designed to protect the environment, in particular, those that lead to *an actual reduction of aviation emissions*, such as operational and technological measures and the use of aviation biofuels." ICAO, C-MIN 199/13, *supra* note 316 at 5 (emphasis added). Colombia further commented that MBMs "were complementary and transitory measures." *Ibid.* Peru also indicated that "not everyone was convinced of the appropriateness of [MBMs]." *Ibid.* at 5.

415 ICAO, C-MIN 199/13, *supra* note 316 at 12 (statement of Cuba).

416 *Ibid.*

417 ICAO, A38-WP/427 EX/142 *Proposed Amendments for the Draft Consolidated Statement of Continuing ICAO Policies and Practices Related to Environmental Protection: Climate Change* (presented by Argentina, China, Cuba, India, Islamic Republic of Iran, Pakistan, Peru, the Russian Federation, Saudi Arabia, and South Africa) at 3.

418 See Scott Barrett, "Climate Treaties and the Imperative of Enforcement" (2008) 24 Oxford Rev Econo Pol'y 239 (noting that achieving compliance and participation are two key elements in the design of any scheme to reduce GHG emissions at 244); Bodansky, *The Art and Craft of International Environmental Law*, *supra* note 62 (Chapter 1) (suggesting that a scheme should either lower the cost of participation or raise the cost of not participating) at 181; Felix R FitzRoy & Elissaios Papyrakis, *An Introduction to Climate Change Economics and Policy* (London: Earthscan, 2010) (suggesting that a treaty may not be necessary to reduce GHG

much more emphasis on the processes of interaction, interpretation, education, participation, and engagement in international norm formation. At this stage, it is not clear that these players have considered these elements.[419] Failing to do so will not only delay norm internalization but will also make it more difficult for the sector to respond to climate change.

emissions to the extent that the system provides for strong incentives and penalties for non-compliance) at 121.

419 In November 2013, a group of developing countries proposed the formation of an Environmental Advisory Group (EAG) at ICAO. The EAG would report to the Council. What is interesting about this initiative is that it seeks to ensure broader participation in climate change discussions at the ICAO Council. This is very illustrative of the participation problems that ICAO faces. See "Proposal by BRICS States to Set Up a New Group on a Global MBM for Aviation Approved by ICAO Council", *Green Air Online* (20 December 2013) online: <www.greenaironline.com/news.php?viewStory=1804>.

8 The Way Ahead: Key Considerations in Addressing GHG Emissions from International Aviation in the Future

GHG emissions from international aviation will continue to remain one of the sector's biggest and most complex challenges. ICAO Member States have already decided in 2013 to develop a global MBM scheme to tackle GHG emissions from international aviation.[1] Although the organization has identified potential options for this scheme and in fact has established a road map to achieve this objective,[2] a number of issues remain unresolved. This final chapter examines in more detail some of the elements that must be taken into account in designing this scheme for international aviation. In particular, by attempting to reconcile the principles of CBDR and non-discrimination in a manner compatible with the aviation context, this chapter seeks to capture the concerns of developing countries and establish an incentive structure that would induce broader participation. I explore the valuc of potcntial voluntary commitments and how this could affect the mobilization of a larger mass of major critical emitters.

Two of the most forgotten and certainly less studied elements necessary to adopt and implement the global MBM scheme must be examined in this analysis: the legal vehicle and the enforcement provisions. Although considerable technical work has already been carried out, it is evident that not much consideration has been given to these issues. In addition to analyzing options for adopting the scheme (e.g. standards, assembly resolutions, an international convention), the chapter proposes mechanisms that may be used to enforce the scheme.

I end by suggesting additional actions that may enhance the implementation of the global MBM scheme. I draw attention to certain areas of action where further involvement of ICAO and IATA may be desirable. I also identify some issues that the EU may consider worthy of further examination.

1 See ICAO, A38-18, *supra* note 21 (Introduction).

2 See ICAO, C-DEC 201/3, *Council, 201st Session, Decision of the Third Meeting.*

8.1 Addressing Some of the Required Design Elements of the Global MBM Scheme

8.1.1 Reconciling CBDR and Non-Discrimination

Chapter 2 addressed the conflict between CBDR and non-discrimination in detail. I suggested that the reconciliation of these principles should constitute a fundamental objective in designing the global MBM for international civil aviation.[3] One of the reasons why a concrete measure to reduce or limit GHG emissions has not yet been established is the fact that the sector has only timidly advanced proposals to address this issue.[4] Given the highly political nature of the issue, IATA has left it for States to decide.[5] Similarly, ICAO has hitherto only scratched the surface of the matter.[6] Although the 38th Assembly decided that "special circumstances and respective capabilities of developing countries could be accommodated through...exemptions from, or phased implementation,"[7] the concept has yet to be articulated. By mapping out an incremental, route-based global MBM, the sections below provide certain criteria to select routes to be covered under the scheme and at what stage.[8] It also explores how financial flows may be redirected to developing countries as well as options to address fast-growing participants and those early movers who have already made significant investments in new technology.

8.1.1.1 Route-Based/Phase-In Approach

In order to address equity issues, the global MBM scheme for international aviation could be designed in such a way that it gradually covers specific international routes. These routes may be phased into the scheme in accordance with a specified set of criteria. This is what the literature refers to as "incrementalism."[9] In light of the fact that one of ICAO's aspira-

---

3 In the broader climate change context, commentators have suggested that equity considerations may be addressed in a number of ways. See Jennifer Morgan, "The Emerging Post-Cancun Climate Regime" in Jutta Brunnée, Meinhard Doelle, & Lavanya Rajamani, eds, *Promoting Compliance in an Evolving Climate Regime* (Cambridge: Cambridge University Press, 2012) 17 at 35; Stavins, "Post-Kyoto Era," *supra* note 68 (Chapter 2) (discussing that "cost-effectiveness and distributional equity could both be addressed" at 148).

4 See AEA, "Global Approach," *supra* note 309 (Chapter 7).

5 See IATA, CNG Resolution, *supra* note 129 (Chapter 7).

6 See ICAO, C-WP/13894, *supra* note 193 (Chapter 3) at A-3.

7 See ICAO, A38-18, *supra* note 21 (Introduction) at para 19 (d).

8 While supporting the notion of a route-based approach to accommodate CBDR into the design of an MBM, Lyle suggests other criteria. These may include (1) excluding least developed, landlocked, and small island States; (2) excluding least developed States; and (3) grouping States in accordance with their financial contributions to ICAO. See Lyle, "Mitigating International Air Transport Emissions through a Global Measure," *supra* note 122 (Chapter 1).

9 Penelope Canan & Nancy Reichman, "Lessons Learned" in Donald Kaniaru, ed, *The Montréal Protocol: Celebrating 20 Years of Environmental Progress* (London: Cameron May Ltd, 2007) 107 at 107; Havel & Sanchez, Toward a Global Aviation Emissions Agreement, *supra* note 214 (Chapter 1) at 372; Tuerk et al,

tional goals is the attainment of CNG from 2020 onward, the global MBM scheme could have three phases: 2020 to 2023, 2024 to 2026 and 2027 to 2030. Although in theory each phase could be longer (e.g. five years), this may jeopardize the likelihood of achieving the aspirational goal, as aircraft operators will join the system at a later stage. If this is the case, it may be more difficult to obtain a critical mass of participants in order to capture a significant percentage of GHG emissions from international aviation. This may necessitate a reassessment of ICAO's aspirational goals.

Efforts to introduce differentiation risk market distortion and, as a basic principle, aircraft operators flying on the same route (whether they are from developed or developing States) will be subjected to the same rules. The scheme will only apply where the points of origin and destination of both routes are subject to the same phase. If not, the scheme would not apply to the route. By way of illustration, let us assume that during the first phase of the scheme, only routes to and from Canada and to and from Europe are covered. This will mean that Air Canada flying the route Montreal-Frankfurt-Montreal and Lufthansa flying the same route in reverse will both be subject to the global MBM scheme. Yet, this will also mean that the scheme would be applicable to Air India serving the route Montreal-Frankfurt-Delhi but only with respect to the Montreal-Frankfurt segment. The fact that routes to and from Canada are part of the first phase of the scheme would not mean that all of Air Canada's routes will be covered. Again, both points of origin and destination must be part of the same phase. Therefore, Air Canada's flight to Buenos Aires will not be subject to the scheme as routes to and from Argentina would not be covered under the first phase of the regime.

The key question with this approach is what criteria determine which routes will be covered and when. I advance some ideas on potential options to phase routes into the different phases of the scheme.

Phasing in with a UNFCCC Flavor

Under this option, from 2020 to 2023, the scheme will first cover routes to and from Australia, Canada, New Zealand, Japan, the Russian Federation, the United States, and Member States of the European Civil Aviation Conference (ECAC).[10] This will capture about 50

---

supra note 242 (Chapter 2) at 1. See generally Stavins, "Post-Kyoto Era," *supra* note 68 (Chapter 2) (analyzing, in the broader climate change regime, the possibility of "targets that become more stringent for individual developing countries as those countries become more wealthy" at 147).

10 At the last ICAO assemblies, under the close look of the European Commission, a large number of European States have coordinated and presented positions on aviation and climate change issues under the aegis of the European Civil Aviation Conference (ECAC). At present, ECAC is composed of 44 States, whereas 28 States form part of the EU. For the purpose of designing different phases in alternatives for ICAO's global MBM, I group European States in accordance with ECAC's membership. The following States are members to ECAC: Albania, Armenia, Austria, Azerbaijan, Belgium, Bosnia and Herzegovina, Bulgaria, Croatia, Cyprus, Czech Republic, Denmark, Estonia, Finland, France, Georgia, Germany, Greece, Hungary, Iceland, Ireland, Italy, Latvia, Lithuania, Luxembourg, Malta, Moldova, Monaco, Montenegro, the Netherlands,

States who today are responsible for roughly 57 percent of GHG emissions from international aviation.[11] The justification for including these States in the first stage is twofold.[12] First, it follows the *UNFCCC* principle that developed States should take the lead in addressing GHG emissions. These States bear a historical responsibility for past emissions. Most of these States are Annex 1 parties for the purpose of the *UNFCCC*. Second, the bulk of ECAC Sates are included in this phase because they have been the early champions of aviation and climate change issues. It is also expected that the European constituency will demand that all routes to and from Europe should be covered in the first phase of the scheme.

Norway, Poland, Portugal, Romania, San Marino, Serbia, Slovakia, Slovenia, Sweden, Switzerland, Macedonia, Turkey, Ukraine, and the United Kingdom. See ECAC, Member States, online: European Civil Aviation Conference <www.ecac-ceac.org//about_ecac/ecac_member_states>. It is worth noting that most of the ECAC members are considered as Annex 1 parties in the *UNFCCC*. See UNFCCC, "List of Annex 1 Parties to the Convention," *UNFCCC* online: <https://unfccc.int/parties_and_observers/parties/annex_i/items/2774.php>.

11 For the purpose of this exercise, I assimilate international aviation activity on the basis of RTKs as being directly proportional to the production of GHG emissions from international aviation emissions. In other words, if the US market share of international aviation activity is 12.94 percent of the world's RTK total, such figure also reflects the country's percentage in terms of GHG emissions. See ICAO, "Civil Aviation: 2012 International RTK by State of Air Operator Certificate (AOC)" online: ICAO <www.icao.int/Meetings/a38/Documents/International%20Scheduled%20RTK%20(Annual%20Report).PDF> [ICAO, "2012 RTK Ranking"]. Ideally, a State's carbon footprint should be based on emissions produced by its aircraft operators in international flights and not on the basis of RTKs. While some States, such as the United States, have very accurate data measuring their aircraft operators' carbon footprint, others do not. As of to date, ICAO does not have data measuring aircraft emissions from its Member States. The RTK rankings are probably what it could be most assimilated to listing States on the basis of the emissions generated by their aircraft operators.

12 Havel and Sanchez suggest that an incremental system to address GHG emissions from international aviation could include first the US and EU member States. See Havel & Sanchez, "Toward a Global Aviation Emissions Agreement," *supra* note 214 (Chapter 1). See also "AEA Contribution to a Global Approach for International Aviation Emissions," *AEA* (15 December 2008) online: AEA <http://files.aea.be/Downloads/gap_paper_final.pdf>.

Figure 8.1

Route-Based / Phase-in Approach **PHASE 1**

Phasing in with an UNFCCC Flavor

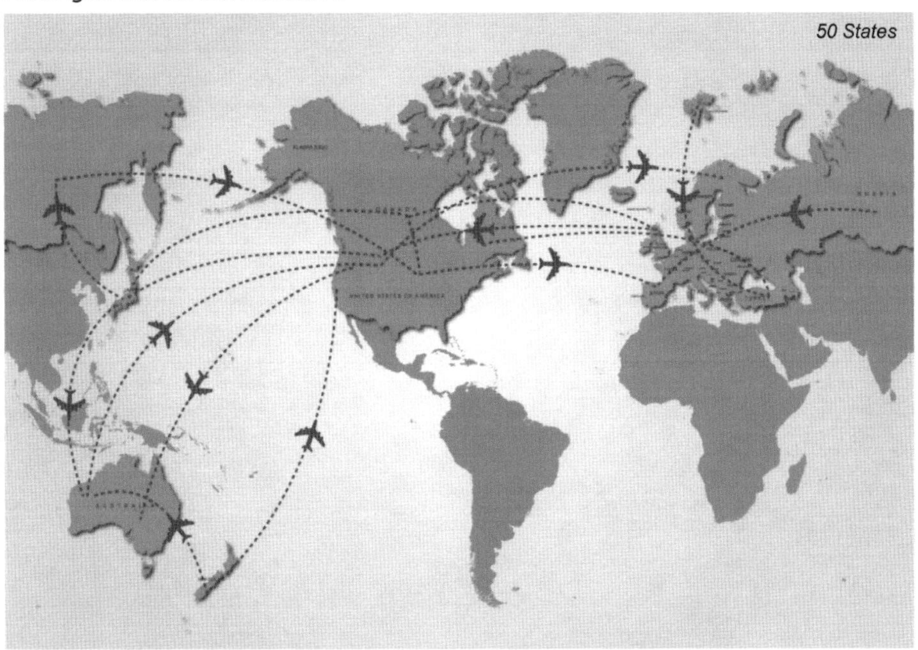

2020-2023 57% GHG emissions from international aviation

Australia, Canada, New Zeland, Japan, Russian Federation, US, and Member States of the European Civil Aviation Conference (ECAC)

I use regional market shares in order to determine which routes would be phased into the scheme during the second and third phases.[13] With 27 and 13 percent of international traffic, routes to and from Asia Pacific and the Middle East, respectively, will be phased into the scheme during the second stage (i.e. between 2024 and 2026).[14] This will probably capture an additional 37 percent of total emissions from international aviation.[15] At this stage, the scheme's overall coverage will be approximately 94 percent of the sector's inter-

13 In essence, this is a form of market block differentiation. See ICAO, C-WP/13894, *supra* note 193 (Chapter 3).

14 ICAO, News Release, "Regional Passenger Traffic and Capacity Growth, Market Shares and Load Factors in 2013" (16 December 2013) online: ICAO <www.icao.int/Newsroom/Pages/2013-ICAO-AIR-TRANSPORT-RESULTS-CONFIRM-ROBUST-PASSENGER-DEMAND,-SLUGGISH-CARGO-MARKET.aspx> [ICAO, "International Traffic Market Share"].

15 See ICAO, 2012 RTK Ranking, *supra* note 11.

national emissions.[16] Routes to and from major aviation powerhouses such as China, UAE, Singapore, Qatar, Korea, Malaysia, Indonesia, India, and Saudi Arabia will be added to the scheme.

Figure 8.2

Route-Based / Phase-in Approach **PHASE 2**

Phasing in with an UNFCCC Flavor

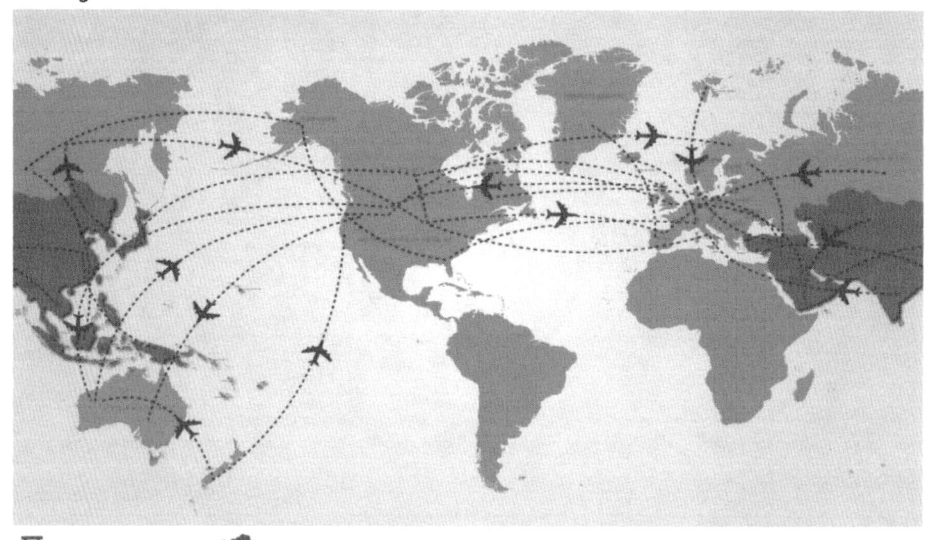

2024-2026 94% GHG emissions from international aviation

States in phase 1 + Asia Pacific & Middle East (e.g. China, USA, Singapore, Qater, Korea, Malaysia, Indonesia, India & Saudi Arabia)

Finally, in the third phase of the scheme (i.e. from 2027 to 2030), routes to and from Latin America and the Caribbean and the African regions, whose international traffic market share is only 4 and 3 percent, respectively,[17] will be phased into the scheme.[18] This will ensure that routes to and from Africa will only join the scheme at a late stage. It may be recalled from Chapter 2 that during the last two ICAO Assemblies, by strongly pushing the inclusion of a de minimis threshold, African States made it abundantly clear that they should not be included in any global or regional MBM.

16 *Ibid.*

17 See ICAO, "International Traffic Market Share," *supra* note 14.

18 *Ibid.*

Figure 8.3

Route-Based / Phase-in Approach **PHASE 3**

Phasing in with an UNFCCC Flavor

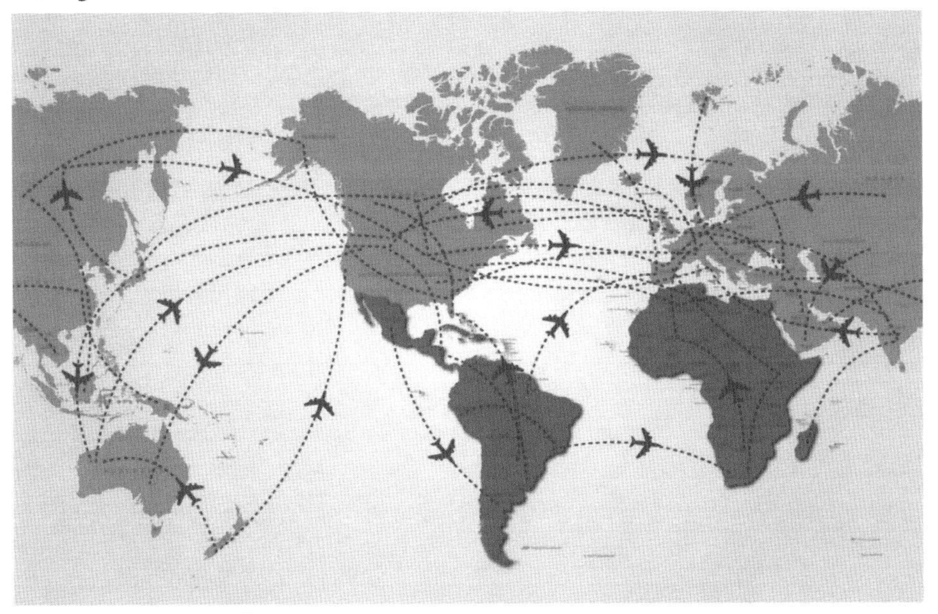

2027-2030 100% GHG emissions from international aviation

States in phases 1 & 2 + Latin America & Caribbean and Africa

Arguably, phasing in routes to and from States on the basis of their market share of international aviation traffic follows a *UNFCCC*-type differentiation between developed and developing States. It does, however, present some obvious problems. While Latin America and Africa have almost insignificant traffic market shares, they are home to some of the world's most prosperous and fastest-growing airlines. It does not seem plausible to shield Brazilian/Chilean airline conglomerate LATAM,[19] Ethiopian Airlines, and South African Airways from carbon responsibility. While Qatar Airways will join the scheme from 2024 onward, LATAM, Ethiopian Airlines, and South African Airways would only form part of it in 2027. Clearly, these carriers are in a much better position to address climate change issues than Air Namibia or Uruguay's BQB Airlines. Some form of market distortions may likely arise.

19 LATAM is the result of the merger between LAN Group of Chile with Brazilian's airline TAM Linhas Aereas. See LATAM Airlines Group, online: LATAM Airlines Group <www.latamairlinesgroup.net/phoenix.zhtml?c=81136&p=irol-home>.

The obvious criticism of this option is that by first phasing in routes to and from developed States, the system incorporates, albeit temporarily, a camouflaged version of the CBDR principle. In essence, the option distinguishes routes to and from developed and developing countries. This could make it difficult for some developed States such as the United States to accept the proposal. The counter argument is that, unlike in the *UNFCCC* context, the differentiation would not be perpetual. It would only be intended to provide temporary headroom. Six years after its adoption, the scheme will cover almost all the relevant routes to and from developed and developing countries alike.

Phasing in on the Basis of International Aviation Activity

Another option is to move completely away from differentiating routes on the basis of whether they belong to a developed or developing country and to exclusively take into account international aviation activity.[20] As an alternative proposal, the scheme could phase in routes on the basis of levels of international aviation activity as measured by ICAO in RTKs. For simplicity, all ECAC Member States will be grouped together. From 2020 to 2023, the first phase of the scheme will cover all routes to and from ECAC Member States and the top 10 States ranked by RTKs. In addition to the European States, this will capture major aviation players such as the United States, China, UAE, Korea, Singapore, Japan, Qatar, Russian Federation, Australia, and Canada.[21] Using this option, 54 States will be held accountable for roughly 80.5 percent of international aviation emissions.[22]

20 See ICAO, 2012 RTK Ranking, *supra* note 11.

21 *Ibid.*

22 *Ibid.*

Figure 8.4

Route-Based / Phase-in Approach **PHASE 1**

Phasing in on the Basis of International Aviation Activity

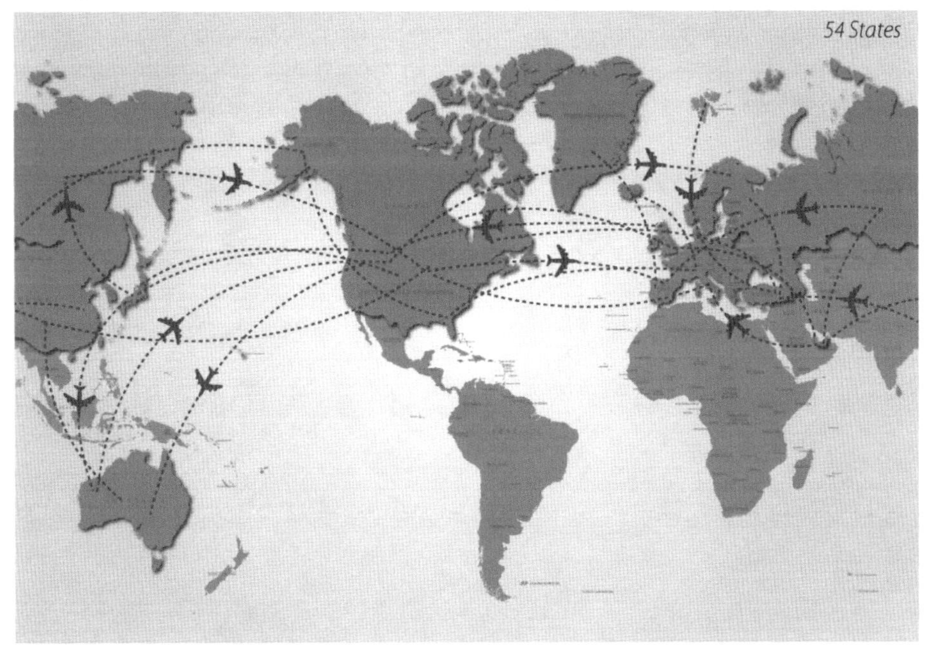

2020-2023 80.5% GHG emissions from international aviation

ECAC Member States + the top 10 States ranked by international RTKs (Europe + US, China, USA, Korea, Singapore, Japan, Qater, Russian Federation, Australia and Canada)

From 2024 to 2026, the second phase will include routes to and from the following 10 non-EU States ranked by RTKs: Thailand, Malaysia, India, Saudi Arabia, Brazil, South Africa, New Zealand, Chile, Ethiopia, and the Philippines.[23] With the inclusion of these States, the scheme's coverage would increase to roughly 90 percent of international aviation emissions.

23 *Ibid.*

Figure 8.5

Route-Based / Phase-in Approach **PHASE 2**

Phasing in on the Basis of International Aviation Activity

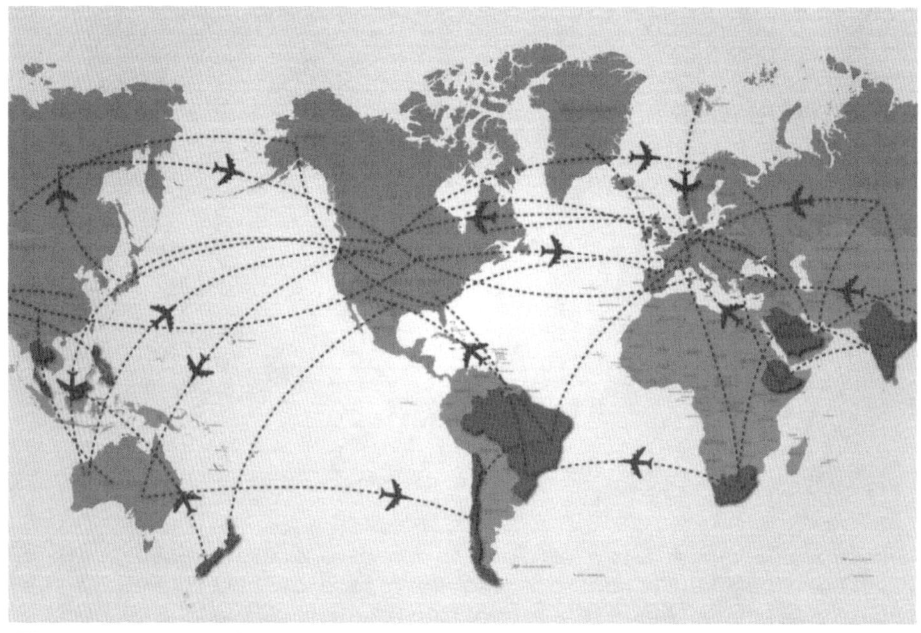

2024-2026 90% GHG emissions from international aviation

States in phase 1 + the following 10 States ranked by international RTKs (Thailand, Malaysia, India, Saudi Arabia, Brazil, South Africa, New Zealand, Chile, Ethiopia & the Philippines)

Finally, from 2027 to 2030, the remaining States will be phased in. This approach ensures that routes to and from major aviation States are phased into the scheme at an early stage in the process. In comparison with phase 1 of the first option, there is a significant increase in the coverage of GHG emissions under this option (57 vis-à-vis 80 percent). Moreover, market distortions could be reduced even further if phases 1 and 2 are merged into one phase. Since a State's level of international aviation activity would be the only factor that matters, none of the *UNFCCC*-related considerations (e.g. historical responsibilities) would be relevant to the determination of which States should be phased into the scheme under this option. In practice, this will mean that routes to and from major emitters will be covered first under the scheme.

Figure 8.6

Route-Based / Phase-in Approach **PHASE 3**

Phasing in on the Basis of International Aviation Activity

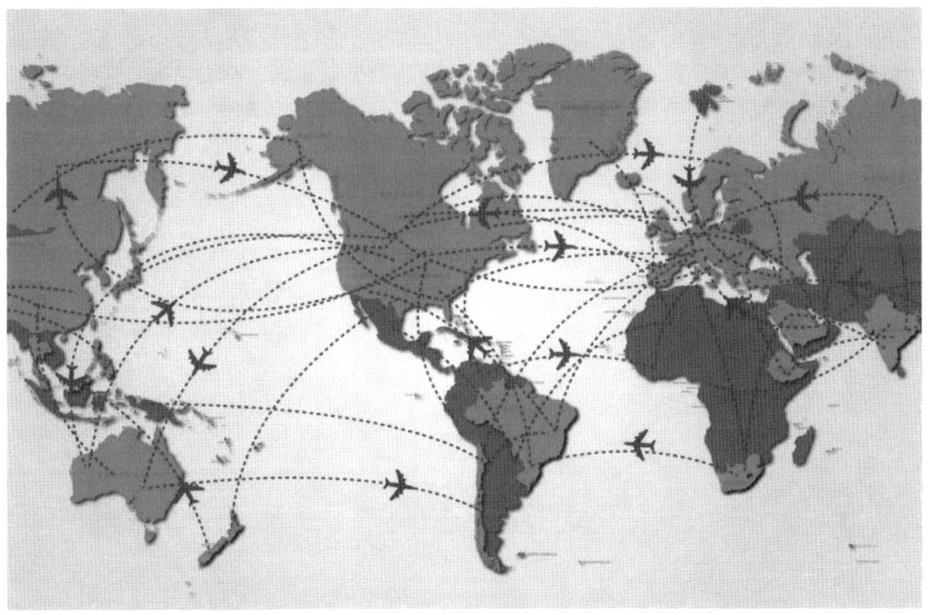

2027-2030 100% GHG emissions from international aviation

States in phases 1 & 2 + all other States

A Mixed Approach to Phasing In

My alternative blends elements present from the first two options. From 2020 to 2023, the scheme will capture routes to and from Australia, Canada, New Zealand, Japan, the Russian Federation, the United States, and ECAC Member States and thereby cover 57 percent of emissions from international civil aviation. The inclusion of these States in the first phase of the scheme would amount to recognition of their responsibility for historical emissions forming part of the overall GHG contribution to climate change.

Figure 8.7

Route-Based / Phase-in Approach **PHASE 1**

Mixed Approach: UNFCCC Flavor with International Aviation Activity

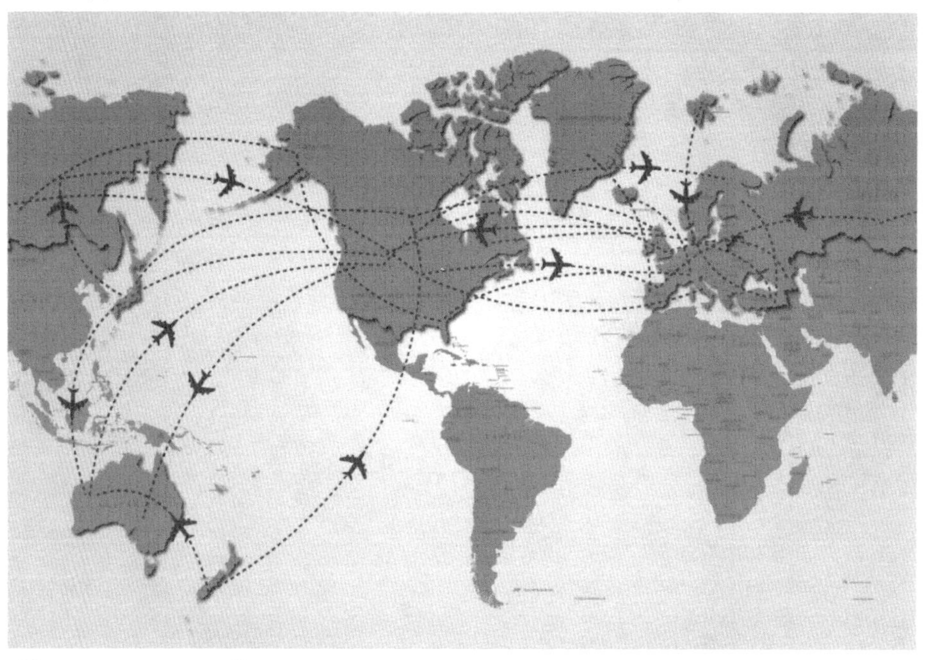

2020-2023 57% GHG emissions from international aviation

Australia, Canada, New Zealand, Japan, Russian Federation, US, and ECAC Member States (UNFCCC Flavor)

For the subsequent phases, the approach would be different. Instead of grouping States according to the regions where they are geographically located or classifying them as developed or developing countries, this option will phase in States in line with their respective levels of aviation activity. Thus, from 2024 to 2026, the second phase will include routes to and from States ranking in the top 20 RTK list which did not join the system in the earlier period.[24] Capturing countries such as China, UAE, Korea, Singapore, Qatar, Thailand, Malaysia, India, Saudi Arabia, Brazil, South Africa, Chile, Ethiopia, Israel, the Philippines, Indonesia, Colombia, Egypt, Mexico, and Panama, this will cover an additional 37.8 percent of GHG emissions from international aviation, totaling 94.8 percent.[25]

24 *Ibid.*
25 *Ibid.*

Figure 8.8

Route-Based / Phase-in Approach **PHASE 2**

Mixed Approach: UNFCCC Flavor with International Aviation Activity

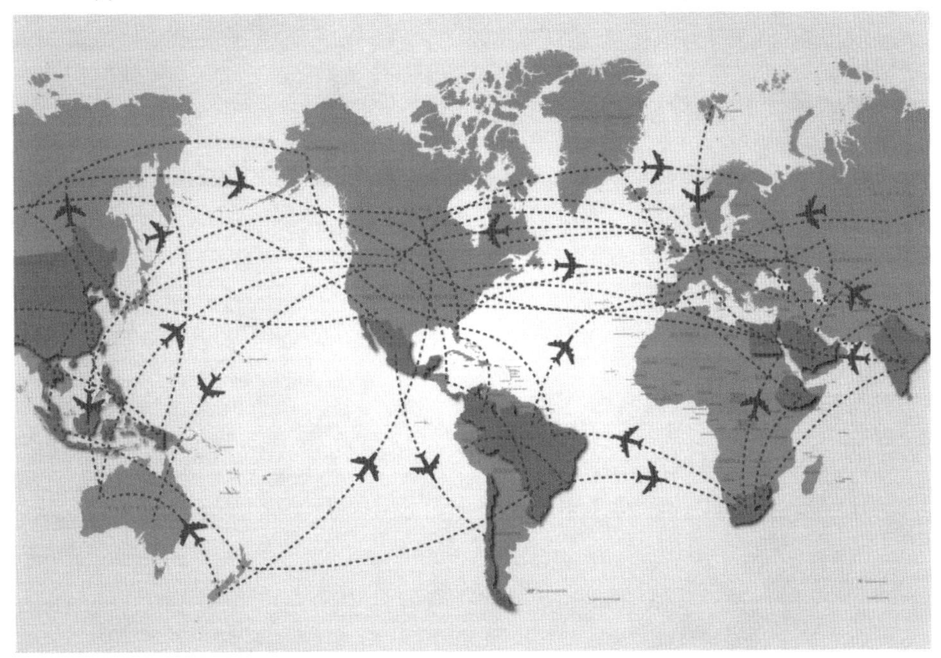

2024-2026 94.8% GHG emissions from international aviation

States in phase 1 + top 20 States ranked by International RTKs (China, USA, Korea, Singapore, Qater, Thailand, Malaysia, India, Saudi Arabia, Brazil, South Africa, Chile, Ethiopia, Israel, Philippines, Indonesia, Colombia, Egypt, Mexico and Panama)

Finally, from 2027 to 2030, the remaining States will join the system. Again, market distortions may be reduced if phases 1 and 2 are merged.

Figure 8.9

Route-Based / Phase-in Approach **PHASE 3**

Mixed Approach: UNFCCC Flavor with International Aviation Activity

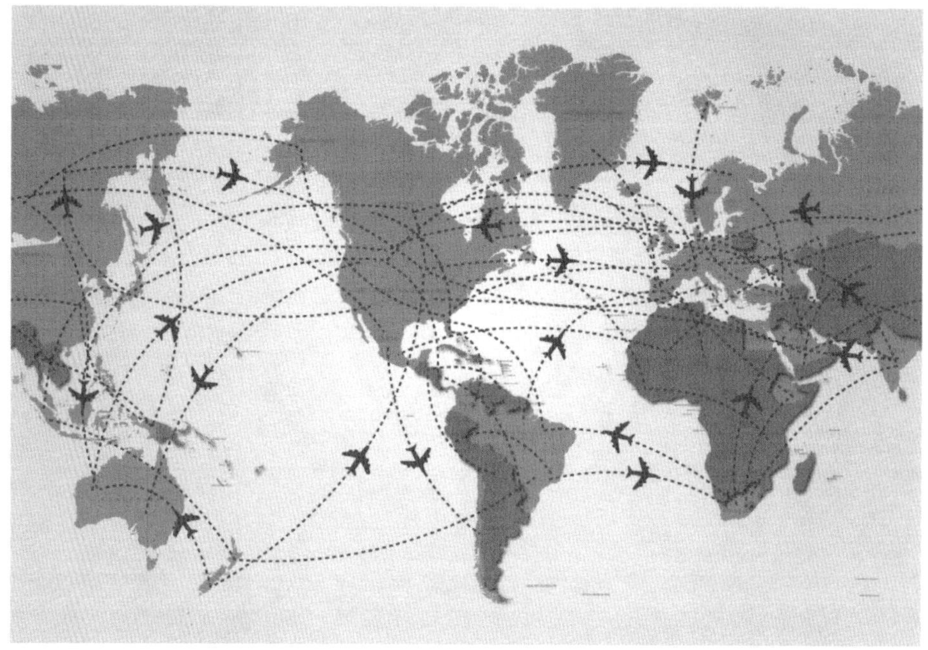

2027-2030 100% GHG emissions from international aviation

States in phases 1 & 2 + all other States

8.1.1.2 The Potential Value of Voluntary Commitments

While binding emission reduction commitments are preferable, the reality is that, for a variety of reasons, not all States are in a position to accept them. One of the salient characteristics of the architectural design of the international climate change regime is that States are increasingly being encouraged to take voluntary measures to address climate change. For instance, the *UNFCCC* has established the concept of nationally appropriate mitigation actions (NAMAs).[26] Various States and groups of States have voluntarily pledged to contribute funds for climate change mitigation and adaptation.[27] The design of ICAO's global MBM should also allow States to make voluntary commitments.

26 NAMAs are actions that developing countries may voluntarily undertake to reduce GHG emissions. The concept recognizes that different States may take different measures. See UNFCCC, *Nama Registry*, *UNFCCC* online: <https://unfccc.int/cooperation_support/nama/items/7476.php>.

27 See, e.g., UNFCCC, Report of the Conference of the Parties on its Fifteenth Session, held in Copenhagen from 7 to 19 December 2009, FCCC/CP/2009/11/Add 1, online: *UNFCCC* <http://unfccc.int/resource/docs/

In the context of a route-based system, a State included in one of the later stages of the scheme may voluntarily decide to phase in routes to and from its territory during an earlier phase (e.g. a State deciding to join the scheme in phase 1 when the system only requires it to do so in phase 2). Voluntary commitments should be seriously explored. Six developing countries (UAE, Republic of Korea, Singapore, Qatar, Thailand, and Malaysia) who have demonstrated a rather progressive attitude toward aviation and climate change issues are cumulatively responsible for 21.6 percent of global GHG emissions from international aviation.[28] Including routes to and from these States in the global MBM scheme at an earlier stage could significantly boost the scheme's scope of coverage.

The obvious argument against this proposition is that there is no apparent incentive for these States to do so. Why would a State expose its aircraft operators to a carbon price when any rational actor would delay it as much as possible? Although this is a valid point, it ignores the fact that these States will also face a serious reputational cost. All of these States host sophisticated, fast-growing, and, in most cases, profitable airlines. It is no secret that these airlines provide far superior services and that their overall performance is significantly higher than that of North American and European air carriers. In this context, delaying the inclusion of routes to and from these countries may be perceived as providing an unfair advantage to these carriers, who have been attacked on similar grounds in connection with other than non-climate change-related issues.[29] In the EU ETS saga, these carriers have already been accused of profiting from carbon leakage.[30] These States may

2009/cop15/eng/11a01.pdf> (where a group of developed states voluntarily "[committed] to a goal of mobilizing jointly USD 100 billion dollars a year by 2020 to address the needs of developing countries").

28 ICAO, 2012 RTK Ranking, *supra* note 11.

29 See, e.g., Fred Lazar, "Multilateral Trade Agreement for Civil Aviation" (2011) 36:6 Air & Space L 379 (suggesting that Emirates has enjoined unfair advantages); Andrew Parker, "Emirates: A Perspective on Issues in Canadian Aviation" (2012) 37:6 Air & Space L 419 (arguing that Emirates has not received subsidies and does not engage in price dumping practices, at 419); Sean McGonigle, "Past its Use-by Date: Regulation 868 Concerning Subsidy and State Aid in International Air Services" (2013) 38:1 Air & Space L 1; AEA, "U.S. Customs Facility in Abu Dhabi Harms the Competiveness of the Transatlantic Aviation Market" online: AEA <http://files.aea.be/News/PR/Pr13-011.pdf>; Susan Carey & Daniel Michaels, "Rise of Middle East Airlines Does not Fly with U.S. Rivals," *The Wall Street Journal* (29 October 2013) online: The Wall Street Journal <http://online.wsj.com/news/articles/SB10001424052702304384104579141732219208604>.

30 Some commentators advanced the idea that the original design of the EU ETS could have produced carbon leakage that disproportionately benefitted airlines from the Gulf region (e.g. Emirates, Etihad Airways, and Qatar Airways). See Ulrich Steppler & Angela Klingmüller, "EU Emissions Trading Scheme and Aviation: Quo Vadis?" (2009) 34:4/5 Air & Space L 253 at 258; *Written Observations of the International Air Transport Association and the National Airlines Council of Canada in the Court of Justice of the European Union*, Reference from the High Court of Justice, London, Case C-366/10, at 81, para 198. Gulf airlines transport a significant amount of traffic from Europe through their home bases in Dubai, Doha, or Abu Dhabi to final destinations in North East Asia (e.g. China, Korea, and Japan), South East Asia (e.g. Thailand, Malaysia, and Singapore) and the South Pacific (e.g. Australia and New Zealand). Gulf carriers compete on routes from Europe to these destinations with European carriers. However, the original design of the EU ETS would have subjected Gulf carriers for their first segment of their respective itineraries, that is to say from a point in Europe to a point in the Gulf region. European air carriers would have to surrender more allowances for

consider showing leadership by voluntarily phasing in routes to and from their territories during the earlier phases of the global MBM scheme, putting this criticism to rest.

8.1.1.3 Potential Criticism

Implementing a global MBM scheme through a route-based approach in different phases may face two main criticisms. First, although the system is designed on the premise that aircraft operators serving the same route are subject to the same carbon rules, it does not completely eliminate the possibility of market distortions. This is particularly so if one considers origin-to-destination passenger itineraries. To illustrate, let us consider the following example. Again, let us assume that routes to and from Canada and Europe are covered in the first phase of the scheme. A passenger is travelling from Toronto to Delhi. He is offered two alternatives: (i) Toronto Delhi via Frankfurt on Air Canada/Lufthansa and (ii) Toronto Delhi via Dubai on Emirates. The segment Toronto-Frankfurt is subject to the global MBM, whereas the second travel option is completely exempted, at least for the time being. The passenger may find that Air Canada and Lufthansa would attempt to pass on the cost of participating in the global MBM to him, whereas such cost is simply absent on the Emirates ticket price. Depending on the actual price of carbon, such cost may provide an incentive to pick one airline over another. The fact that routes to and from the UAE are not covered in the first phase of the scheme creates a perception that an unfair advantage is given to Emirates.

Although there is some truth in this argument, for a number of reasons, market distortions are already present in international aviation. For instance, Asian carriers enjoy significantly lower labor costs than many European airlines. Similarly, an airline from a landlocked/non-oil-producing country probably bears much higher fuel costs as compared to those from the Gulf region. These are just some examples. Since the infancy of international aviation, aircraft operators have experienced different costs in different parts of the world. This is a natural consequence of operating in a global environment. A global MBM implemented on the basis of a route-based approach in different phases will not change this trend, but nor will it aggravate it.

Although market distortions exist in practice, a different question is whether the design of the global MBM scheme should induce them. Clearly, the objective should be to reduce

---

a non-stop flight, for instance, London to Beijing, than the requirements imposed on Gulf carriers for the same itinerary through a point in the Gulf. Although it is undeniable that in this case European airlines would have borne a higher carbon cost, this does not necessarily mean that the original design of the scheme induced carbon leakage scenarios to arise. Flights to the Gulf region would have occurred regardless of whether the EU ETS was in place. These flights are the direct consequence of market demands and the unique advantages of hubs in those regions. They are not triggered by the imposition of a carbon price. In any case, the cost increase of the EU ETS was not significant enough to induce any change in the behavior of aircraft operators. See Annela Anger & Jonathan Köhler, "Including Aviation Emissions in the EU ETS: Much ado About Nothing?" (2010) 17 Transp Pol'y 38 at 44.

market distortions. It is however simply naïve to expect that they could be completely eliminated. The fact that the last two ICAO Assemblies have recognized that minimizing market distortions should be a principle when designing MBMs implicitly acknowledges that they do exist.[31] The probability that some market distortion may occur should not prevent the adoption of a fundamental element of the global MBM scheme for international aviation. As discussed in Chapter 3, the cost of participation in the scheme is expected to have only a marginal effect on aircraft operators.

The second criticism concerns the political dimension. Regardless of the technical justification, differentiating and phasing routes into the scheme involve complex political compromises that pose challenges to adopting and implementing the scheme. Although one cannot help but accept that these elements add to the complexity of the scheme, it is also unrealistic to expect that from day one the scheme will cover all aircraft operators on all routes. It is precisely because of the inherently political nature of aviation and climate change issues that some form of temporary differentiation is required. CBDR must be reconciled with non-discrimination in a manner compatible with the international aviation environment. This is necessary to enhance political acceptance and ensure that the system is widely implemented. Despite the difficulties, the route-based/phase-in approach is one way to achieve this objective.

8.1.2 *Redirecting Financial Flows*

This project has maintained that one of the ways to effectively implement CBDR could be to divert revenues to developing countries.[32] As explained in Chapter 3, one of the options for ICAO's global MBM envisages an offsetting scheme with a built-in revenue generation mechanism that will collect funds and later redistribute them to developing countries. The literature also describes this alternative as a rebate mechanism. In fact, some NGOs have put forward concrete proposals in this respect.[33] Although this seems like a reasonable way to address equity concerns, this proposal is unlikely to succeed for three main reasons. First, the airline industry has rejected this option as it is perceived as another tax on the industry. Second, the inclusion of a revenue generation mechanism increases the complexity of the scheme. The handling of revenues not only presents operational but also legal challenges. As discussed below, if this is the case, a new international convention will have to be adopted to implement the scheme. This will in turn delay progress considerably. And finally, it is very unlikely that major developed countries still enjoying the largest share of

31 See ICAO, A37-19, *supra* note 207 (Chapter 1) at Annex (g); ICAO, A38-18, *supra* note 21 (Introduction) at Annex (g).

32 See Karim & Alam, *supra* note 24 (Chapter 2) at 135.

33 See WWF, "Aviation Report," *supra* note 91 (Chapter 1) at 10.

international traffic will ever support the idea of levying a fee on their passengers to support mitigation and adaptation to climate change in developing countries. In addition, one must bear in mind that ICAO's previous experiences in setting international funds have not been successful.[34]

There are more subtle ways to transfer financial flows to developing countries. It is clear that both emissions trading and offsetting will require that aircraft operators purchase either allowances or ECUs to fulfill their compliance obligations. For the near future, intra-sector emission reductions are still expected to be much more expensive than purchasing credits elsewhere. Therefore, the design of the global scheme could require that aircraft operators acquire a large percentage of these credits from climate change-related projects carried out in developing countries such as those implemented under the Clean Development Mechanism (CDM) and Reducing Emissions from Deforestation, Degradation and Forest Enhancement (REDD-plus).

One could also envision linking ECUs to actual participation in the scheme by a State. For instance, an air carrier would only be allowed to buy ECUs or CDM-related credits from projects in countries that have already joined the ICAO global MBM scheme. This could serve as a huge incentive to avoid the problem of free riders. As suggested in Chapter 2, the scheme may also include concrete provisions on financial assistance, technical cooperation, and technology transfer.[35] These options may affect the incentive structures of both participants and their respective States.

---

34 For instance, in the process of modernizing the Warsaw Convention of 1929 on air carrier's liability, in 1971, ICAO Member States adopted the Guatemala City Protocol. See *Protocol to Amend the Convention for the Unification of Certain Rules Relating to International Carriage by Air,* 28 September 1955, 478 UNTS 371, ICAO Doc 7632. The instrument allowed States to implement a supplementary fund to pay compensation to victims for damages sustained as a result of the contract of carriage. *Ibid.* at Art XIV. The instrument never entered into force. See Paul S Dempsey & Michael Milde, *International Air Carrier Liability: The Montreal Convention of 1999* (Montreal: Centre for Research in Air & Space Law McGill University, 2005). Similarly, in 2009, ICAO Member States adopted the *Unlawful Interference Convention.* This instrument, which has yet to enter into force, establishes a fund compensating victims suffering damages on the ground as a result of an act of unlawful interference. The fund is formed by the contributions of passengers and shippers. The fund foresees levying a fee from passengers and cargo departing from the territory of a State party. See *Unlawful Interference Convention, supra* note 382 (Chapter 7) at Arts 8-17. One of the reasons why this instrument may never enter into force is precisely that States with significant international traffic are not willing to subsidize the fund to compensate damages that may take place elsewhere. In other words, policy makers are not necessarily convinced that, for instance, a US passenger should contribute to the fund to provide compensation for damages sustained as a result of an unlawful interference incident happening in Europe. The sense of international collective responsibility is not present. In a way, this is very similar to climate change.

35 See Hardeman, "Reframing Aviation Climate Politics and Policies," *supra* note 44 (Chapter 2) at 27. See also Gilbert Bankobeza, "Compliance Regime of the Montreal Protocol" in Donald Kaniaru, ed, *The Montréal Protocol: Celebrating 20 Years of Environmental Progress* (London: Cameron May Ltd, 2007) 75 (stressing the importance of financial assistance and compliance mechanism in the Montreal Protocol, at 83). It is self-evident that any scheme designed to address GHG emissions from international civil aviation will require a level of cooperation among States. See generally Kerr, *supra* note 54 (Chapter 1) (noting that any global

8.1.3 Addressing Fast Growers and Early Movers

Another way to address equity considerations and take into account the special needs of developing countries is through the notions of fast growers and early movers. Chapter 1 highlights the fact that in the next 20 years a large proportion of the projected growth in international air transport will occur in developing countries. As ICAO has already noted, one of the objectives to be achieved in designing the global MBM scheme is to avoid "unduly [burdening] fast-growing participants and markets."[36]

Another relevant factor concerns those participants who have already taken early action by investing heavily in new technology or greener operational procedures in an effort to reduce their carbon footprint: the so-called early movers.[37] As explained in Chapter 4, in a way, the allowance allocation methodology of the EU ETS did factor in early movers. Aircraft operators flying new technology did substantially better in terms of allowances received than their competitors operating aging fleet. This is an important issue for a large number of developing countries whose fleets are quite new as compared to those of, for instance, US aircraft operators. The interests of fast-growing and early mover participants may be addressed through the allocation of allowances (in the case of an emissions trading scheme) or through the baseline establishment process (in the case of offsetting).[38]

8.2 The Challenge of Finding a Legal Vehicle to Enforce Compliance with the Global MBM Scheme

By agreeing to develop a global MBM scheme for international civil aviation, the 38th Assembly has set the expectation that ICAO will adopt such a scheme in 2016 and that it will be operational by 2020.[39] ICAO has carried out a considerable amount of technical work to identify potential options, yet significantly less attention has been paid to the legal vehicle through which the scheme will be implemented.[40] Although experts have identified standards, Assembly resolutions, and international conventions as possible legal vehicles,[41] very little analysis has been carried out. The issue was absent from ICAO's latest resolution

ETS will necessarily require "cooperative agreement and trading amongst sovereign States" at 3); Barrett, *Why Cooperate, supra* note 57 (Chapter 1) (noting that "international cooperation can benefit every country – and every person living in every country" at xi).

36 ICAO, C-WP/13894, *supra* note 193 (Chapter 3) at A-3.

37 Scott D Deatherage, *Carbon Trading: Law and Practice* (Oxford: Oxford University Press, 2011) at 36.

38 See also ICAO, A38-18, *supra* note 21 (Introduction) at para 19.

39 *Ibid.*

40 See ICAO, C-WP/13894, *supra* note 193 (Chapter 3); ICAO, C-WP/13861, *supra* note 164 (Chapter 3); ICAO, Global MBM Assessment, *supra* note 89 (Chapter 3).

41 See ICAO, C-WP/13894, *supra* note 193 (Chapter 3) at A4.

on climate change.[42] In this section, I discuss some of the pros and cons of each of these options. It is clear that the final selection of the legal vehicle will depend on a number of technical and political considerations.

8.2.1 ICAO Standards

As indicated above, the *Chicago Convention* entrusts ICAO with the mandate to develop standards addressing "matters concerned with the safety, regularity and efficiency of air navigation as may from time to time appear appropriate."[43] It is the Council that adopts standards as part of its mandatory functions.[44] In theory at least, one could foresee that a global MBM for international aviation could be implemented through ICAO's standard setting process. In this regard, provisions for this scheme could be developed through standards that would form part of a new Annex to the *Chicago Convention* (i.e. *Annex 20 to the Chicago Convention – Global MBM for International Aviation*). This option presents two obvious advantages. First, the development and implementation of a set of standards may be significantly less time-consuming than an international treaty. A set of standards for a global MBM scheme or even an entirely new Annex may be developed in less than two years. In most cases, unless a majority of ICAO Member States object (something which has yet to happen in the annals of ICAO), a standard enters into effect 90 days after its adoption by the Council.[45] The second advantage of this option concerns the fact that it would permit States to bypass their respective legislatures in incorporating the global MBM within their domestic legal systems. Although standards may require that States adopt implementing domestic regulation, this does not automatically mean that they must obtain legislative approval. ICAO's standard setting process is part of its law-making function as set out in the *Chicago Convention* which States are parties to.[46]

The use of standards as a legal vehicle for a global MBM scheme, however, comes with three clear disadvantages. First, hitherto, standards have dealt primarily with technical issues. Although the subject matter of standards should not be construed in a restrictive

42 See ICAO, A38-18, *supra* note 21 (Introduction).

43 *Chicago Convention*, *supra* note 36 (Chapter 2) at Art 37.

44 *Ibid.* at Art 54 (l). In order to adopt standards, the *Chicago Convention* requires a vote of two-thirds of the Council (24 votes). *Ibid.* at Art 90.

45 See *Chicago Convention*, *supra* note 36 (Chapter 2) at Art 90. In order to facilitate implementation by Member States, ICAO may nonetheless provide some headroom and defer the applicability date of the standard to a later date. The following example may better illustrate the different dates involved. The Council adopts a standard on 10 July 2014. No State expressed disapproval. The standard enters into effect on 10 October 2014. However, given the technical and operational requirements involved to implement it, such standard will only be applicable as of 1 January 2016.

46 See *Chicago Convention*, *supra* note 36 (Chapter 2) at Art 37.

manner,[47] they have traditionally not dealt with economic measures such as those envisaged in a global MBM scheme. Therefore, the incentives to free ride may be enormous especially if the price of carbon rises.

Second, the *Chicago Convention* allows States to file differences with the organization in situations where they do not find it practicable to bring their domestic practices in conformity with the standards.[48] In practice, this functions as an opt-out mechanism for those States who find it impracticable to comply with a standard.[49] Nowadays, this may even be done electronically.[50] Although one could argue that a State may also withdraw from a treaty, doing so certainly entails a much more difficult process, and it implies other considerations (e.g. the country's reputational loss, international relations considerations, etc.). It will be much easier to opt out of a climate change-related standard than it is to do so with those relating to safety and aviation security. As explained below, there are external enforcing mechanisms that (e.g. bilateral air services agreement and the *Chicago Convention* itself) make it much harder for States to deviate from safety and security standards. But this is not at all the case with climate change.

Third, ICAO has no powers to enforce standards. It is true, however, that through its safety and aviation security audits, ICAO oversees State compliance with standards. This has only been partially extended to cover environment-related issues. Should the global MBM be adopted through standards, they should form an integral part of the ICAO audit program.

Standards may be a useful legal vehicle to develop certain technical elements of ICAO's global MBM scheme, including but not limited to a common monitoring, reporting, and verification (MRV) system, rules on "fuel consumption reporting requirements, recording, surrendering, cancelling and acquiring of [ECUs]."[51]

---

47 For instance, in 1974, although not expressly mentioned in the *Chicago Convention* as part of ICAO's core aims and objectives, the organization developed standards relating to aviation security issues. See *Ibid.* at Annex 17.

48 *Ibid.* at Art 38.

49 *Ibid.*

50 See ICAO, *Electronic Filing of Differences (EFOD) System and Filing of Differences Task Force (FDTF)* [unpublished, archived on file with the author].

51 ICAO, C-WP/13894, *supra* note 193 (Chapter 3). See generally Paolo Contini & Peter H Sand, "Methods to Expedite Environment Protection: International Ecostandards" in Haas, *supra* note 40 (Chapter 7) 3 at 56; Scott Barrett, *Environment & Statecraft: The Strategy of Environmental Treaty-Making* (Oxford: Oxford University Press, 2003) (recognizing the role of technical standards in international climate change issues, at 398).

8.2.2 ICAO Assembly Resolution

Just like UN General Assembly resolutions, ICAO Assembly resolutions are per se not binding upon its Member States.[52] They are nonetheless indicative of the organization's policies.[53] In theory, a global MBM scheme could be adopted through an Assembly resolution.[54] In all likelihood, this option would be considerably less time-consuming than an international treaty, which requires ratification at the domestic level. Another advantage is that non-binding agreements are flexible and can be easily changed when the circumstances so demand.[55]

The main challenges associated with this option are the difficulties in enforcing the global scheme and in imposing penalties upon non-compliant participants (e.g. aircraft operators). Admittedly, Assembly resolutions are significantly weaker vehicles for carrying out these functions. But one could argue that the vehicle used to implement the global MBM scheme has nothing to do with these challenges. After all, this is an intrinsic problem involving all international organizations and not only ICAO. Regardless of the vehicle chosen, it is simply unrealistic to expect that ICAO will ever impose penalties on non-compliant participants.[56] While this point has merit, an international convention provides more options when it comes to enforcement and imposition of penalties.

In theory, an Assembly resolution could delegate enforcement functions and the imposition of penalties to States. Each State will be in charge of enforcing the scheme on its aircraft operators.[57] This, however, may pose serious challenges. States may exercise different levels of enforcement (e.g. different levels of monetary sanctions). In fact, Mendes de Leon warns us about the risks of different levels of enforcement.[58] Given that it could lead to market distortions, States may also be reluctant to impose sanctions where other States fail to take measures over their own aircraft operators.

52 See Lyle, "Mitigating International Air Transport Emissions through a Global Measure," *supra* note 122 (Chapter 1) (commenting that "ICAO resolutions have no definitive binding nature").

53 Guzman suggests that violation of soft law instruments also represents a reputational cost for States. See Andrew T Guzman, "A Compliance-Based Theory of International Law" (2002) 90 CLR 1823 at 1880 [Guzman, "Compliance-Based Theory"].

54 See ICAO, C-WP/13894, *supra* note 193 (Chapter 3) at A-4.

55 See Bodansky, *The Art and Craft of International Environmental Law*, *supra* note 62 (Chapter 1) at 156.

56 For instance, through the auspices of ICAO, States have adopted 7 international conventions in the field of aviation security. These instruments criminalize certain types of behavior as criminal offenses. None of these instruments, however, contained specific penalties. Following standard practice of counterterrorism conventions, the adoption of "severe penalties" is left to States.

57 See generally Brownlie, *supra* note 31 (Chapter 5) (discussing that "[i]n the context of environmental problems, the way forward lies in the deployment of effective enforcement systems fully integrated into legal and administrative systems of individual states" at 287).

58 See Mendes de Leon, Enforcement of the EU ETS, *supra* note 220 (Chapter 4) at 7.

8.2.3 International Convention

A new international convention may also be used as a legal vehicle to adopt a global MBM for international aviation.[59] Through a treaty, it may be easier to address issues such as enforcement, sanctions, and dispute resolution mechanism.[60] A treaty may also establish a new international entity with its own legal personality to handle specific issues such as the scheme's registry or the management of revenues and the imposition of sanctions.[61] In essence, the treaty may empower either ICAO or the newly created entity to carry out functions which are not necessarily provided for in the *Chicago Convention* and air services agreements or for which ICAO's Assembly resolutions are not suitable. For instance, a scheme with a revenue generation mechanism will most likely require the adoption of a new treaty given that decisions will have to be made on who pays what and for what purposes. The *Chicago Convention* does not provide ICAO with specific powers to manage funds other than those that form part of its regular budget.[62] The new instrument could set up sanctions for non-compliance such as operational bans to be implemented by States. At present, this is not contemplated in bilateral or multilateral air services agreements.

A new international convention also presents some disadvantages. Its binding nature may pose some challenges for some States. This is particularly so with regard to climate change where international aviation is not necessarily disassociated from the *UNFCCC* context.[63] The most notorious disadvantage of this option is that the mere negotiation and adoption of the instrument may take a number of years.[64] In fact, this may be the perfect

59 See generally Scott Barrett, *Why Cooperate*, *supra* note 57 (Chapter 1) (discussing that the main objective of a treaty on climate change should be to tackle incentive structures to avoid free riders, at 192).

60 See Andrew T Guzman, "Compliance-Based Theory," *supra* note 53 at 1873.

61 These may include an international registry, conference of parties, or an international fund. Several international instruments developed under the auspices of ICAO have established new "international entities." In this regard, the *Cape Town Convention* has set up an International Registry and a Supervisory Authority. Similarly, the *Marking of Plastic Explosives* has created the International Explosives Technical Commission. *Convention on International Interests in Mobile Equipment*, 16 November 2001, 2307 UNTS 41143 [*Cape Town Convention*]; *Protocol to the Convention on International Interest in Mobile Equipment on Matters Specific to Aircraft Equipment*, 16 November 2011 [*Cape Town Protocol*]; *Marking of Plastic Explosives Convention*, *supra* note 11 (Chapter 5).

62 See *Chicago Convention*, *supra* note 36 (Chapter 2) at Art 44.

63 For instance, it is well known that the US Congress is unlikely to pass any legislation imposing binding emission reduction commitments on US enterprises, unless a number of factors are met. The withdrawal of the United States from the *Kyoto Protocol* and the reluctance of the US Congress to ratify it was a clear example. If the ICAO-led global aviation MBM is adopted through an international instrument, the United States will have to go through the US Congress – a daunting challenge.

64 Developing a new instrument normally requires placing the subject matter as part of the working program of ICAO's Legal Committee by either the Council or the Assembly. Council then establishes a Secretariat working group. Formed by a limited number of States, depending on the complexity of the matter, the group meets several times and conducts the preliminary work. It may also propose draft text. A rapporteur may also be appointed. The Secretariat working group then recommends that Council convene a meeting of the Legal Subcommittee. The Council will then extend invitations to 20-25 States to take part in the subcommittee

recipe for States to engage in dilatory tactics. It is true, however, as Milde points out, that some treaties such as the *Marking of Explosives Convention* were developed over a period of less than two years.[65] This is the exception rather than the rule. The most recent international instruments adopted under the auspices of ICAO seem to confirm this view.[66] As a way of an example, the negotiation process that led to the adoption of the Beijing Convention and its Protocol in September 2010 lasted roughly 9 years.[67] Obtaining the required ratifications or accessions for the instrument to enter into force may also be a lengthy process. For instance, the last 5 international conventions adopted under the auspices of ICAO since 2008 are yet to enter into force. Moreover, it does not seem likely that any of these instruments will enter into force in the near future.[68] Although the Montreal Convention of 1999 on air carrier liability entered into force roughly 4 years after its adoption,[69] given its complexity and the multiplicity of exogenous elements, it is unlikely that an aviation climate change treaty will enter into force in such a short period of time if ever adopted. By way of comparison, it is instructive to note that the *Kyoto Protocol* entered into force almost 8 years after its date of adoption.[70] Moreover, the fact that the *Kyoto Protocol* did enter into force may, to a large extent, be attributed to Europe's very strong norm entrepreneurship. It is submitted that if the global MBM scheme is adopted through a treaty, a norm entrepreneur will have to champion the cause and lobby around the world

---

meetings. In most cases, the subcommittee meets at least twice. Once the draft text of the new instrument is sufficiently mature, the Council will convene a meeting of the Legal Committee. Only at this stage is the full membership of ICAO invited to participate in the process. The last three international instruments adopted under the auspices of ICAO only require one meeting of the Legal Committee. If there is sufficient consensus, the Legal Committee recommends and the Council convenes a Diplomatic Conference. Just the process of developing the instrument may take several years. See also ICAO Doc 7669-LC/139/5, *Legal Committee – Constitution – Procedure for Approval of Draft Conventions – Rules of Procedure*, 5th ed (1998).

65 See Milde, *International Air Law and ICAO, supra* note 59 (Chapter 2) at 253-55.

66 For instance, in 2009, IATA presented a proposal to examine the convenience of modernizing the *Tokyo Convention 1963*. The Council formed a Secretariat Study Group. After two meetings which took place in 2011, the group recommended that the instrument should be updated. Upon the convening of the Council, the Legal Subcommittee met twice in 2012. The Legal Committee met once in 2013. Finally, on 4 April 2014, the Diplomatic Conference adopted a protocol introducing substantial amendments to the *Tokyo Convention 1963*. The whole process of developing the amendments lasted more than 4 years. See *Protocol to Amend the Convention on Offences and Certain Other Acts Committed on Board Aircraft*, signed in Tokyo on 14 September 1963, 4 April 2014 (not yet in force). See also Alejandro Piera & Michael Gill, "Unruly & Disruptive Passengers: Do We Need to Revisit the International Legal Regime?" (2010) XXXV Ann Air & Sp L 355 at 363.

67 See Alejandro Piera & Michael Gill, "Will the New ICAO-Beijing Instruments Build a Chinese Wall for International Aviation Security?" (2014) 47:1 Vand J Transnat'l L 145 at 152 [Piera & Gill, "Beijing Instruments"].

68 *Ibid.* at 235.

69 At present, 105 States are parties to the Montreal Convention 1999. It was adopted on 28 May 1999, and it entered into force on 4 November 2003. See *Hague Protocol, supra* note 254 (Chapter 5).

70 Adopted on 11 December 1997, the Kyoto Protocol entered into force on 16 February 2005. See UNFCCC, "Status of Ratification of the Kyoto Protocol," *UNFCCC* online: <https://unfccc.int/kyoto_protocol/status_of_ratification/items/2613.php>.

to expedite the ratification or accession processes.[71] Without such a norm entrepreneur, it is unlikely that an aviation and climate change treaty will enter into force any time soon.[72]

8.2.4 *Enforcement*

As Hart notes, international law lacks a "centrally organized effective system of sanctions."[73] In fact, just like most international organizations, ICAO has no enforcement authority in a strict legal sense.[74] This, however, does not mean that the design of the global MBM scheme should not envisage mechanisms to facilitate enforcement. Without such mechanisms, it is ludicrous to expect that the scheme will bear any meaningful effect in reducing or limiting GHG emissions from international aviation. As mentioned before, the incentives to free ride may be significant. This section provides suggestions to facilitate enforcement.

8.2.4.1 The Value of Transparency

An argument could be made that, although important, enforcement and penalties *strictu sensus* do not constitute insurmountable challenges for the adoption of a global scheme, not even where it is adopted through a set of standards or an Assembly resolution. ICAO has other means to induce compliance.[75] One of this is transparency. For instance, ICAO

71 The success story of both the Cape Town Convention and the Cape Town Protocol may be attributed to the norm entrepreneurship of the Aviation Working Group (AWG) – a trade association representing aircraft manufacturers, engine manufacturers, major leasing enterprises (lessors), as well as major financial institutions involved in the financing of aircraft equipment. See AWG, "AWG Purpose," *AWG* online: <www.awg.aero/inside/purpose>. Not only was AWG extremely active in the negotiation process that led to the adoption of the instrument, but it has also been instrumental in promoting their ratifications by States. To date, 54 States are parties to the *Cape Town Convention* and its Protocol. See *Cape Town Convention, supra* note 61; *Cape Town Protocol, supra* note 61. Without AWG's involvement, these instruments would have never had any practical effect.

72 Along with some States, IATA has taken a much more active role in promoting the ratification of the Montreal Convention of 1999. See ICAO, A38-WP/170 LE/6, *Promotion of the Convention for the Unification of Certain Rules for International Carriage by Air* (*Montreal Convention of 1999*) (presented by the United Arab Emirates, the Air Crash Victims Families Group (ACVFG), and the International Air Transport Association (IATA)); ICAO, ATConf/6-WP/70, *Promotion of the Convention for the Unification of Certain Rules for International Carriage by Air (Montreal Convention of 1999)* (presented by Canada, Germany, United Arab Emirates, the United States, and the International Air Transport Association (IATA)). See also ICAO, A38-20, *Promotion of the Montreal Convention 1999.*

73 Hart, *supra* note 27 (Chapter 7) at 3.

74 See Lyle, "Mitigating International Air Transport Emissions through a Global Measure," *supra* note 122 (Chapter 1).

75 See generally Bodansky, *The Art and Craft of International Environmental Law*, *supra* note 62 (Chapter 1) (discussing that non-compliance undermines trust in the system at 226); Mostafa K Tolba & Iwona Rummel-Bulska, "The Story of the Ozone Layer" in Donald Kaniaru, ed, *The Montréal Protocol: Celebrating 20 Years of Environmental Progress* (London: Cameron May Ltd, 2007) 27 (describing the non-compliance provisions of the Montreal Protocol as "unprecedented" at 41). Redgwell writes that "the primary objective in establishing compliance procedures is to provide, within a multilateral context, encouragement to states to comply with their treaty obligations, and, in the event of non-compliance, to provide a softer system to address non-

has implemented audit programs on safety and aviation security.[76] By overseeing adherence to ICAO standards, these audits have been instrumental in raising awareness of safety and security issues as well as contributing to higher state compliance levels. One of the reasons of this success story is that audit results have progressively been made public.[77] As Blumenkron has noted, transparency has served as a "quasi-enforcement" mechanism.[78] Likewise, the President Emeritus of the ICAO Council, Assad Kotaite, notes that "[through transparency ICAO may be] able to achieve results not through enforcement, but through moral persuasion."[79]

Transparency must therefore be a cornerstone design element of the global MBM scheme regardless of the legal vehicle chosen for its adoption. Transparency should allow the public to know aircraft operators' verified fuel consumption levels, CO_2 emissions produced over a compliance period, whether the emissions produced by aircraft operators have exceeded their limits, and the number of offsets, ECUs, or allowances required to comply with their obligations. ICAO should establish a public, central registry (e.g. International Aviation Climate Change Registry) where this information will be processed. This registry will record all transactions involving participants. These may include exchanges of allowances, ECUs, and offset purchases. The ICAO registry should also be linked to the

---

compliance than that afforded by traditional dispute settlement procedures under general international law." Catherine Redgwell, "Facilitation of Compliance" in Jutta Brunnée, Meinhard Doelle, & Lavanya Rajamani, *Promoting Compliance in an Evolving Climate Regime* (Cambridge: Cambridge University Press, 2012) 177 at 177.

76 See ICAO, A38-15, *Consolidated Statement of Continuing ICAO Policies related to Aviation Security*, Appendix E; ICAO, A32-11, *Establishment of an ICAO Universal Safety Oversight Audit Program* (establishing "regular, mandatory, systemic and harmonized safety audits"); ICAO, A37-5, *The Universal Safety Oversight Audit Program (USOAP) Continuous Monitoring Approach* [ICAO, A37-5].

77 ICAO, A37-5, *supra* note 76 ("[r]ecognizing that transparency and the sharing of safety information is one of the fundamental tenets of a safe air transport system"). This resolution instructed ICAO' Secretary-General "to make all safety oversight-related information... available to all Contracting States through the ICAO restricted website." *Ibid.* at para 6. It also instituted the notion of "significant security concerns" (SSCs). *Ibid.* SSCs constitute a set of serious findings of non-compliance with ICAO standards arising from the safety audits. SSCs imply that a State has serious deficiencies in its safety oversight that do not permit to properly ensure compliance with ICAO standards. See "Safety Audit Information," online: ICAO <www.icao.int/safety/Pages/USOAP-Results.aspx> [ICAO, SSCs Public]. States with SSCs are placed on a special list (MARB) and monitored on a regular basis. ICAO shares information on States with SSCs with stakeholders. In 2012, the Council decided to make available to the general public, as of 1 January 2014, SSCs of its Member States. See ICAO, C-DEC 197/4, *Council, 184th Session, Decision of the Fourth Meeting.*

78 See Jimena Blumenkron, "Implications of Transparency in the International Civil Aviation Organization's Universal Safety Oversight Audit Program" (2009) XXXIV Ann Air & Sp Law 31. Blumenkron writes: "Transparency in the [Universal Safety Oversight Audit Program (USOAP)] has been used as a "quasi-enforcement" power by ICAO. Today, States, international organizations, financial institutions, and passengers have access to the safety oversight audit results of most States. Transparency used as a "quasi-enforcement" power has improved aviation safety, especially among States that previously did not have the political will to comply. International organizations lack legal mechanisms to ensure compliance with international law developed by them. Transparency has proven to be an efficient "quasi-enforcement" tool to ensure compliance with the [standards and recommended practices (SARPS)] at 69."

79 Kotaite, *supra* note 330 (Chapter 7) at 212.

UNFCCC's International Transaction Log.[80] This link would ensure that carbon credits generated outside aviation (e.g. CERs) are properly recorded, cancelled if necessary, and not accounted for twice.

Transparency will not only allow the public to identify those non-compliant participants, but it will also place significant "blame and shame" on them. This may expose them to reputational losses. However, making the system fully transparent will be a challenging task due to concerns over disclosure of commercially sensitive information.

The obvious criticism to placing such high expectation on the scheme's transparency as a quasi-enforcement mechanism is that the subject matter is completely different from those involving safety and aviation security issues. It is one thing to expose a State and an aircraft operator on the basis of their poor safety or security records than to blame them for their lack of environmental commitment. In cases involving safety and security, one can certainly expect that the travelling public will not only be concerned but may also opt for another destination or another airline. Clearly, the public values much more safety and security than it cares for the aviation sector's contribution to climate change. This is the case in most countries. Although this criticism is right, it is expected that through education environmental issues will gradually attain a degree of importance that, to date, it is certainly not present in the agendas of most aviation policy makers. When this happens, transparency will play a major role.

8.2.4.2 External Enforcers

Although transparency has been an invaluable tool to induce compliance with ICAO's standards, this should not be examined in isolation. Transparency by itself may not be sufficient. In practice, the ICAO audits have also had "external enforcers." Actors outside of ICAO – either industry stakeholders or States – have indirectly or directly induced compliance. Close consideration of this factor is therefore extremely relevant in the design of the global MBM scheme.

Almost all air services agreements through which States have exchanged traffic rights among themselves contemplate the prerogative of revoking the authorization granted to a foreign aircraft operator when the granting State "has reasonable ground to believe" that the other State is not observing ICAO safety or aviation security-related standards.[81] In addition, the *Chicago Convention* has a built-in mechanism whereby a State may not rec-

80 See UNFCCC, "International Transaction Log," *UNFCCC* online: <http://unfccc.int/kyoto_protocol/registry_systems/itl/items/4065.php>.

81 See *Multilateral Agreement on the Liberalization of Air Transportation (MALIAT)*, 1 May 2001, online: MALIAT <www.maliat.govt.nz/agreement/article9.shtml> at Arts 6 and 7 [MALIAT Agreement]; US Department of State, "Current Model Open Skies Agreement Text," *US Department of State* (12 January 2012) online: US Department of State <www.state.gov/e/eb/rls/othr/ata/114866.htm> at Arts 6 and 7 [US Model Open Skies Agreement]; *Air Transport Agreement between the United States of America and 27 Member States of the European Union* [2007] OJ, L 134/11 at Arts 8 and 9 [EU-US Open Skies Agreement].

ognize certificates of airworthiness, competency, and licenses issued by another State which were rendered not in conformity with ICAO standards on these issues.[82] On this basis, some States have banned foreign aircraft operators in cases where their home States allegedly were not in compliance with ICAO standards (e.g. issuance of licenses and certificates of airworthiness not in compliance with the *Chicago Convention*).[83] In view of the language used in the *Chicago Convention*, however, this would only be applicable to situations in which non-compliance involves standards dealing with certificates and licenses. It would not cover environment-related standards or, more specifically, technical standards establishing a global MBM scheme. Both the air services agreements and the *Chicago Convention* have served as additional enforcers for ICAO standards. In their present forms, neither of these could be interpreted to address the issues presented by the global MBM scheme.[84]

As mentioned above, it seems much easier to address enforcement issues through the adoption of a new treaty. If, on the other hand, the chosen vehicle is an Assembly resolution, things get much more complicated as the assistance derived from "external enforcers" would become much more relevant. The design of the global scheme should therefore carefully consider this element. For instance, IATA could make compliance with ICAO's global MBM mandatory for its member airlines. These carriers should be ICAO compliant in order to be eligible for or to maintain the IATA membership. It is true though that this will not capture the full universe of airlines. However, it will ensure that a great number of them are already on board. IATA could impose on its member airlines mandatory participation in its environmental audits. States could also gradually start introducing amendments to their air services agreements to allow for the imposition of operational bans upon aircraft operators of other States to the extent that they are not in compliance with ICAO's global MBM. ICAO could also develop guidance material to assist States in this respect.

8.2.4.3 Reporting Non-Compliance to the Assembly: Article 54 (j)

The built-in reporting mechanism of the *Chicago Convention* may also serve as an appropriate tool to enforce ICAO's global MBM scheme. Article 54 (j) mandates the Council to report non-compliant States to the Assembly.[85] The scope of this provision is twofold.[86]

82 See *Chicago Convention, supra* note 36 (Chapter 2) at Art 33.

83 See FAA, *International Aviation Safety Assessments (IASA) Program, FAA* online: <www.faa.gov/about/initiatives/iasa>.

84 The fact that IATA has also instituted a mandatory safety audit for its member airlines has also indirectly facilitated compliance by States with ICAO' safety-related standards.

85 See *Chicago Convention, supra* note 36 (Chapter 2) at Art 54 (j). See also Jon Bae, "Review of the Dispute Settlement Mechanism under the International Civil Aviation Organization: Contradiction of Political Body Adjudication" (2012) Journal of International Dispute Settlement 1 at 8.

86 In theory, Art 54 (j) and Art 84 (dispute settlement mechanism) of the *Chicago Convention* serve different purposes. The former is a reporting mechanism which the Council may ex officio exercise against any Member State. It is not required that a State report the alleged infringing State to Council. On the other

First, it deals with an infraction of the *Chicago Convention* by a Member State. Second, it also involves situations in which a State fails to implement a Council "recommendation" or "determination."[87] The *Chicago Convention* does not provide a definition of what would constitute an "infraction." However, in a seminal legal opinion rendered in 1999, ICAO's Legal Bureau clarified that the term should be given its ordinary meaning,[88] that is to say any breach, violation, or infringement of any of the articles of the *Chicago Convention.*[89] Non-compliance with a global MBM scheme is unlikely to constitute a violation of the *Chicago Convention.* The constating instrument of the global MBM scheme - such as an Assembly resolution - could however mandate the Council to report non-compliant participants to the Assembly. The Council will inform affected States that some aircraft operators are not in compliance with the scheme (amounting to a *determination* in the words of the *Chicago Convention*). Such States will be given reasonable time to take corrective action. If such corrective actions are not implemented, the Council will then report the matter to the Assembly. In theory, this procedure could act as a strong deterrence to induce compliance. However, the reality is that the Council has never formally resorted to this process.[90]

---

hand, the latter allows a Member State to settle a dispute with another Member State over the application and interpretation of the *Chicago Convention* and its Annexes. Here a State brings a complaint to the Council against another Member State. See Weber, *supra* note 59 (Chapter 2) at 40. In 1957, however, Czechoslovakia brought a complaint to the Council that serves to demonstrate that the distinction between the objectives and scope of the two mechanisms is rather blurring. More specifically, Czechoslovakia complained that the US Air Force had flown "espionage balloons" over its airspace. The United States denied the allegations. Czechoslovakia expressly requested the Council to invoke Art 54 (j). Without deciding whether to accept the applicant's petition and declare that an infraction against the *Chicago Convention* had been committed, the Council opted to instruct the Secretary-General to undertake a study. Without attributing responsibility, the study revealed that an infraction against the *Chicago Convention* was in fact committed. The Council was not able to trigger Art 54 (j) and report the infraction by a Member State. The controversy was later settled through diplomatic channels. Later, in 1960 and given that the controversy between the parties was already moot, the Council issued a general declaration on the use of uncontrolled balloons as being hazards to air navigation. This, however, did not include the particular details of the case that Czechoslovakia had brought to the attention of the Council, and it did not constitute an exercise of its mandatory functions under Art 54 (j). See Buergenthal, *supra* note 5 (Chapter 3) at 133. Czechoslovakia's complaint resembles a dispute that should have been brought under Art 84.

87 See *Chicago Convention*, *supra* note 36 (Chapter 2) at Art 54 (j).

88 See ICAO, C-WP/11186, *Infractions of the Convention on International Civil Aviation* (presented by the Secretary-General) [ICAO, C-WP/11186].

89 *Ibid.* The reporting of an infraction is in fact a complex process. Although the *Chicago Convention* is silent on this issue, basic principles of due process would dictate that, in order to establish that a State has committed an infraction, the body exercising the adjudicatory function ought to hear both the claimant and the respondent. As in any other legal proceeding, the alleged infringing party must be given an opportunity to defend itself. *Ibid.* A basic investigation ought to take place. Otherwise, any finding of the Council would constitute an arbitrary act. In addition, an inherent problem with this mechanism is that it implicitly requires a legal assessment of the facts. The Council, the body in charge, is not a judicial organ.

90 *Ibid.* Although the Council has never resorted expressly to Art 54 (j), in a number of cases involving the unlawful interference of aircraft or the destruction of aviation facilities, the Council has indirectly found a violation of the principles of the *Chicago Convention.* This is somewhat similar to establishing the occurrence

8.3 Additional Actions

This section briefly discusses some additional actions that ICAO, IATA, and the EU may take to address GHG emissions from international civil aviation. Actions by these actors bear significant implications for the global MBM scheme.

8.3.1 ICAO

Although a number of structural amendments to address GHG emissions could be recommended to the *Chicago Convention*, in practice, these are unlikely to occur in the near future due to a variety of reasons. Therefore, instead of focusing on how to introduce theoretical amendments to the *Chicago Convention*, it is submitted that ICAO should take a more realistic approach by tackling three specific challenges: (i) strengthening membership participation, (ii) moving beyond certain industry stakeholders, and (iii) enhancing communications with States about the relevance of climate change issues and promoting transparency.[91]

ICAO faces formidable problems in attracting States other than those who are part of the Council to participate in climate change deliberations. In addition, the membership of CAEP – ICAO's technical advisory group that deals with aviation environmental protection – is heavily imbalanced. Notably, developing countries are underrepresented at CAEP.

These deficiencies may be addressed in a number of ways. One option would be to empower ICAO's 7 regional offices to further engage their constituent States in their respective regions. In this regard, the regional offices could facilitate the dissemination of CAEP's work as suggested by Argentina.[92] These regional offices could also enhance communication between States and ICAO by keeping them abreast with developments

of an "infraction." See ICAO, C-WP/11186, *supra* note 88. See, for instance, ICAO, *Council Resolution on Libyan Civil Aircraft of 4 June 1973* (following the shooting down by an Israeli fighter aircraft of a Libyan civil aircraft which lost its course over the occupied Egyptian territory of Sinai on 21 February 1973, resulting in 108 fatalities, the Council found no justification for Israel's actions and urged Israel to comply with the purposes of the *Chicago Convention*); ICAO, *Council Resolution of 28 February 1986* (1986) (condemning Israel for the interception and forced diversion of a Libyan Arab Airlines over the high seas by an Israeli military aircraft as a violation of the *Chicago Convention*); ICAO, *Council Resolution adopted at the Twentieth Meeting of its 148th Session on 27 June 1996* (on the shooting down of two US-registered private civil aircraft off the cost of Cuba); ICAO, *Council Resolution adopted on 13 March 2002 relating to the Destruction of Gaza International Airport* (2002) (condemning the Israel's bombing of Gaza International Airport as contrary to the principles of the *Chicago Convention*). Weber argues that these statements of Council constitute an implicit invocation of the prerogatives granted by the *Chicago Convention* under Art 54 (j)). See Weber, *supra* note 59 (Chapter 2) at 41.

91 See Lyle, "Mitigating International Air Transport Emissions through a Global Measure," *supra* note 122 (Chapter 1).

92 See ICAO, A38-WP/318 EX/110, *supra* note 374 (Chapter 7).

on climate change issues. Most of the non-Council Member States do not necessarily follow discussions on these issues. However, in 2016, at the 39th Assembly, these States will be asked to consider a proposal for a global MBM scheme. In all likelihood, this proposal will bear serious implications for both States and aircraft operators. Yet, these States would not have been involved in the elaboration of the proposal.

Another way to prepare States would be to follow the model that ICAO itself implemented for the High-Level Aviation Security Conference held in September 2012.[93] Before this event, ICAO organized a number of regional meetings to discuss the agenda of the conference as well as its intended outcome with groups of States. These regional meetings proved extremely successful in building trust which ultimately led to the attainment of consensus. Aviation and climate change issues urgently require a similar process. Although the Secretariat has proposed to establish "Global Aviation Dialogues" (GLADs) to engage a broader audience in the design of a global MBM scheme,[94] the Council has not approved it as of the time of this writing.[95] Another obvious alternative to forge participation is by promoting capacity building through regional workshops, seminars, and training in States that are not normally privy to aviation and climate change discussions. The biggest challenge here is the availability of financial resources. Although its budget for environmental protection has increased in recent years, it is still pretty insignificant for the herculean task assigned to ICAO.[96] This is where partnering with industry stakeholders will be necessary.

ICAO's relationship with industry stakeholders also needs to be re-examined. Although industry's input is fundamental in ensuring that proposed regulations take into account their concerns, these interests cannot be the only criteria against which the extent of such regulations may be assessed. Industry must participate in, but certainly not guide, the process. Representatives of the international civil society should also be given a fair opportunity to participate in the process. Arguably, these organizations are not on the same footing as industry stakeholders.[97] The following example may be illustrative of this point. In 2014, ICAO unveiled a road map and an action plan to develop a global MBM scheme for international aviation.[98] It also instituted an Environment Advisory Group

93 See ICAO, C-MIN 201/3 *Council, 201st Session, Minutes of the Third Meeting.*

94 In essence, the GLADs are intended to be a number of regional seminars where States will discuss a proposal for a global MBM for international aviation. The GLADs seek to obtain feedback from States other than those who usually participated in aviation and climate change discussions at ICAO. See ICAO, C-WP/14101, *Review of Assembly Resolutions and Decisions: Environmental Protection* (presented by the Secretary-General) at Appendix B; ICAO, C-WP/14102, *Establishment of the Environment Advisory Group (EAG)* (presented by the President of the Council). See also ICAO, Action Plan for the Development of a Global MBM Scheme (Council Informal Briefing, 31 January 2014) [unpublished, archived on file with the author].

95 See ICAO, C-DEC 201/3, *supra* note 2.

96 See ICAO, Budget 2014-2016, *supra* note 28 (Chapter 3).

97 See Lyle, "Mitigating International Air Transport Emissions through a Global Measure," *supra* note 122 (Chapter 1).

98 See ICAO, C-WP/14101, *supra* note 94 at Appendix B.

(EAG),[99] a group of Council Representatives tasked with overseeing the work on developing the scheme. During the process of setting out EAG's terms of reference, some States requested that IATA be given observer status.[100] Although the United States supported the inclusion of "appropriate" environmental NGOs as observers to EAG, the Council ultimately decided to only invite IATA.[101]

Considerably more emphasis has been placed on the rather insignificant percentage of the sector's contribution to global GHG emissions. In addition, by magnifying the attainment of technological efficiencies, ICAO has inadvertently removed from its Member States the sense of urgency to act. Much more emphasis should be placed on current growth trends and the simple fact that without regulation international civil aviation cannot develop in a sustainable manner.

8.3.2 IATA

IATA will continue to play a key role in the design and implementation of the global MBM scheme. In fact, it is doubtful that ICAO would have ever agreed to develop such a scheme but for IATA's active involvement. There are at least three additional areas where IATA could further make a difference in the ultimate global MBM scheme. Firstly, IATA ought to continue playing a leading role with regard to raising awareness about the importance of addressing aviation's impact on climate change. It is true that the most sophisticated airlines have actively engaged in this process. However, this is not necessarily the case for smaller aircraft operators from developing countries. This will become particularly challenging during the implementation of the global MBM. IATA is more suitable to engage in this because it possesses substantially more resources than ICAO.

Secondly, IATA could lobby its member airlines to persuade their home States to undertake voluntary commitments to phase their routes into the scheme at an earlier stage. In expanding the scope of coverage of the scheme, this could have a significantly positive impact. Thirdly, perhaps IATA's major challenge going forward is to ensure that its membership remains fully on board in designing and implementing a global MBM scheme. One could expect that 2016 will be a challenging year not only because the 39th Assembly will consider the global MBM proposal that Council will table but also because it will be a presidential election year in the United States. A change of administrations in the White House may imply a completely different approach to climate change issues. This could either reconfirm support for a global MBM or frustrate its implementation. After all, we should not forget that a previous US administration denied that climate change was an

99 See ICAO, C-DEC 200/4, *supra* note 76 (Chapter 3).
100 See ICAO, C-MIN 201/3, *supra* note 93.
101 ICAO, C-DEC 201/3, *supra* note 2.

issue.[102] In 2013, IATA and A4A were instrumental in lobbying the US government to support the global MBM scheme that the 38th Assembly agreed to develop. They should therefore remain very vigilant.

8.3.3 Europe

This project has highlighted the leading role that Europe has played in the wider climate change context.[103] To a certain extent, the inclusion of foreign aircraft operators into the EU ETS exerted significant pressure on ICAO eventually inducing Member States to agree to develop a global MBM scheme for international aviation. This leads to an obvious question: what should be the EU's role in spearheading the adoption and implementation of ICAO's global MBM scheme? Havel and Sanchez suggest that the EU could provide some incentives to some States to join a global agreement in exchange for further market access to the EU.[104] It is well known that carriers such as Emirates, Etihad Airways, Qatar Airways, and Singapore Airlines are desirous of obtaining more traffic rights to and from Europe. Routes to and from these countries account for 13.77 percent of international aviation global GHG emissions.[105] One could speculate that these carriers may be tempted to join the scheme at an earlier stage should the EU offer traffic rights in return. It is to be expected that European airlines will oppose any such deals.[106] In light of this, Havel and Sanchez rightly ask: "if the [EU] is truly wedded to its publicly expressed commitments to carbon reduction, should it not be willing to forego market protectionism to expand coverage for an aviation emissions reduction agreement?"[107]

8.4 Conclusion

Although the 38th Assembly has agreed to develop a global MBM scheme for international aviation, numerous unresolved issues remain. Perhaps one of the most challenging issues has been how to articulate the CBDR principle in a manner that is compatible with international civil aviation. This is one area in which ICAO and industry have achieved very

102 See Faiz Shakir, "Bush Ignores Science, Claims 'There Is A Debate' Over The Cause of Global Warming," *Think Progress* (26 June 2006) online: Think Progress <http://thinkprogress.org/politics/2006/06/26/5997/bush-debate-climate/#>.

103 Vogler & Bretherton point that Europe was instrumental in the entering into force of the *Kyoto Protocol* and that it was willing to make concessions that may have affected the competitiveness of the European private sector. See Vogler & Bretherton, *supra* note 194 (Chapter 1) at 17.

104 See Havel & Sanchez, Toward an Emissions Agreement, *supra* note 214 (Chapter 1) at 381.

105 ICAO, 2012 RTK Ranking, *supra* note 11.

106 See Havel & Sanchez, "Toward an Emissions Agreement," *supra* note 214 (Chapter 1) at 381.

107 *Ibid.*

little progress. By introducing a route-based/phase-in approach, the global MBM scheme could accommodate the CBDR principle to some extent. In this regard, several options to phase in routes have been advanced, all of which demonstrate that special consideration may be given to routes to and from developing countries without infringing the non-discrimination principle.

Yet the route-based/phase-in approach is not the only means by which the interests of developing countries could be accommodated in the design of the scheme. Establishing which ECU participants will be allowed to purchase could also act as an indirect means of redirecting financial flows. Addressing the concerns of fast growers and early movers must also be given serious consideration. Similarly, the possibility of voluntary commitments should not be discarded.

It is true that adopting any of these options will require a political decision. Given that potential implications go well beyond the aviation domain, compromises may be expected. Yet these are necessary if the global MBM is expected to be operational and, more importantly, to have a meaningful effect in limiting GHG emissions from international aviation.

Perhaps one of the elements that has received the least attention thus far is the legal vehicle that will be used to adopt the global scheme. Although a new international convention would seem to provide more options, its lengthy adoption and ratification processes may well prove to be insurmountable obstacles. The international aviation community is under immense pressure to adopt a scheme in 2016 to be implemented in 2020. In this context, the next best option would be to adopt a scheme through an Assembly resolution. But enforcement will be key. Transparency should serve as a quasi-enforcement mechanism. Additionally, a number of elements could serve as the scheme's external enforcers. This may include developing special clauses in air services agreement whereby a State may impose operational bans to those aircraft operators that are not compliant with ICAO's global MBM. IATA could also subject the membership of its airlines to participating in the scheme. ICAO could also resort to its never-used Article 54 (j) reporting mechanism. ICAO's standards seem to be the appropriate vehicle to incorporate some of the technical elements of the scheme, such as monitoring, reporting, and verification systems (MRVs).

Finally, I have suggested some additional roles that may be played by ICAO, IATA, and the EU in addressing GHG emissions in the future. It is of paramount importance that ICAO tackles the problem with State participation. Similarly, IATA could substantially facilitate the enforcement of the scheme by making it mandatory for its member airlines to join the global MBM scheme. Furthermore, just as it brought about the entry into force of the *Kyoto Protocol*, the EU may provide some additional incentives for major aviation emitters to join the global MBM scheme during its initial stage.

9 Concluding Remarks

This book has examined aspects of the legal framework underlying the aviation and climate change discourse with a view to providing some recommendations that may facilitate the adoption, implementation, and, ultimately, compliance with ICAO's global MBM to limit GHG emissions from international aviation. In this regard, it has focused on the following key areas: (i) the setting in which the aviation and climate change discourse has evolved and the research problem; (ii) the interplay between the *Chicago Convention* and the climate change regime; (iii) ICAO's institutional setting and its specific engagement in climate change concerns; (iv) the EU ETS; (v) the roles played by the main actors involved – their merits, shortcomings, and missed opportunities; and (vi) the global MBM scheme itself.

9.1 Setting the Aviation and Climate Change Discourse and the Research Problem

It has been established that international aviation grows at a rate of roughly 5 percent per year. Its GHG emissions are forecasted to increase by 3-4 percent per year. At present, international aviation's contribution to global GHG emissions is approximately 2 percent. However, international aviation's GHG emissions are expected to increase considerably if steps are not taken. Although the sector has introduced technological and operational measures, by themselves, these efforts will not be sufficient to offset the emissions expected from its projected growth. The sector has great expectations from the prospects of using alternative fuels as a means of significantly reducing its carbon footprint. At this stage, however, there is no indication that alternative fuels will be available in sufficient quantities and at reasonable prices to have a meaningful impact in the short to medium term. Therefore, regulation is warranted. As Lyle notes, "[t]he need for some form of MBM as a fundamental part of an emissions mitigation package has by now been well demonstrated."[1]

The sector also faces tremendous exogenous pressure, as those outside aviation circles remain skeptical about the progress thus far achieved. Increasingly, aviation is identified as a potential means of financial contribution to address the wider climate change problem. In addition, on environmental grounds, various States have levied taxes on passengers departing from their territories on international flights. Failure to limit or reduce its carbon footprint may be significantly more costly for the aviation sector.

1 Lyle, "Mitigating International Air Transport Emissions through a Global Measure", *supra* note 122 (Chapter 1).

9.2 The Interaction between International Aviation and the Climate Change Regime

In 1997, the *Kyoto Protocol* enjoined developed States to take steps to limit or reduce GHG emissions from aviation by working through ICAO. States have interpreted this implicit yet vague mandate in furtherance of their parochial national interests. For some influential developing countries, ICAO's involvement in climate change issues must be on the basis of the *UNFCCC's* principle of CBDR, which these States interpret as meaning that their aircraft operators should not be subject to mitigation obligations, at least in the initial phase of any scheme that is adopted. It is up to developed States to take the lead. On the other hand, developed States argue that CBDR directly contradicts the *Chicago Convention*'s non-discrimination principle. As such, CBDR has no place at ICAO since the *Chicago Convention* governs the climate change discourse.

In fact, a large number of States and industry stakeholders consider CBDR to be the major stumbling block that has halted progress on aviation and climate change issues, a principle that is simply inapplicable in the aviation context and the perfect excuse for the "do nothing" approach. Caught in the middle, ICAO has resorted to the notion of special circumstances and respective capabilities (SCRC), a principle that takes into account the special requirements of some States without weighing in the notion of historical responsibilities to the climate change problem and without differentiating, at least in its latest iteration, between developed and developing countries. Unfortunately, SCRC has never been made operational; it remains an abstract concept that has not achieved its intended results.

As discussed in Chapter 2, it is evident that the interplay between the *Chicago Convention* and the climate change regime involves principles that are in tension. Reconciling the CBDR and non-discrimination relationship is central to the design of the global MBM for international aviation. Failure to do so risks either delaying the adoption of the scheme or simply not attracting a critical mass of participants therein even if it is adopted. In light of the foregoing, I attempted to resolve this tension by using the theory of fragmentation of international law, and in particular, ILC's seminal work in this field where significant consideration is given to the conflict rules codified in the *VCLT*. This notwithstanding, the *VCLT* forces the interpreter to choose one principle over the other. Although these rules provide a solution, it is not something that could be implemented. The purpose is not to say that CBDR trumps non-discrimination or vice versa, but rather how these apparently conflicting principles could work in harmony with each other. By examining the principles of systemic integration and Judge Weeramantry's enlightening conceptualization in the *Gabcíkovo-Nagymaros Project* case advocating the reconciliation of principles that appear to be in conflict, the relationship between CBDR and non-discrimination takes on a new dimension. The reconciliation of these principles facilitates their co-existence.

To this end, CBDR should not be conceived as a static, immutable principle, but rather as a dynamic and evolving one. CBDR's conceptualization in the wider climate change context cannot be the only manner of interpretation of the principle. The application of CBDR cannot serve to defy the foundational grounds under which international civil aviation has been structured.

The reconciliation of CBDR and non-discrimination requires a degree of flexibility. To reconcile these principles, we must first reformulate them in light of current circumstances and the objective sought to be achieved. The objective is very clear: to develop a scheme to limit or reduce GHG emissions from international aviation. This also sets out the bounds within which these principles will co-exist. None of the principles should be articulated as antithetical to the rules of the other regime. The basic elements in both principles are: (i) equity, historical responsibility, and the ability to respond to one's own capabilities for CBDR and (ii) prohibiting the treatment of States and aircraft operators arbitrarily for non-discrimination. As a basic notion, aircraft operators flying on the same route should be subject to the same rules irrespective of their nationality.

9.3 ICAO and Climate Change

In Chapter 3, I examined ICAO's institutional setting and its governing structure, as well as their suitability to address climate change issues. It is evident that environmental protection does not form part of ICAO's objectives as set out in the *Chicago Convention*. To carry out its mandate, ICAO has adopted a number of strategic objectives through various proposals presented by its Secretary General, approved by the Council, and endorsed by the Assembly. Gradually, environmental protection has been incorporated into these strategic objectives. Although this approach is a quick fix to correct the existing gap in ICAO's constitutional framework, it is not immune to problems. Such problems typically arise when States contest the application of a given environmental measure on the ground that such measure is contrary to the organization's objectives as set out in the *Chicago Convention*.

In the effort to address climate change issues, ICAO's governing structure presents some serious challenges when the need arises to engage a broader audience beyond those States that form the Council. This is particularly the case for the global MBM scheme. As the Council performs most of the organization's important functions, it is extremely difficult for States that are not members of the Council to participate in some of the organization's most important policy deliberations. Therefore, it is no surprise that various States have repeatedly proposed an increase in the membership of the Council.

While in theory the *Chicago Convention* could be amended, the reality is that there is no political will to introduce substantial amendments. In 2009, the Council considered this option and the view of the majority was that the septuagenarian instrument provides adequate responses to today's international civil challenges. One could only speculate as to the real motives behind such a decision. Therefore, a more pragmatic approach is needed. ICAO could consider more realistic measures to further engage its membership and address the challenges arising from its institutional setting. Amongst others, this could include empowering its seven regional offices and enhancing its communication campaign.

Clearly, communications is an area where there is significant room for improvement. By stressing the efficiencies introduced as a result of technological advancement and over-emphasizing the sector's hitherto relatively small contribution to the global share of GHG emissions (e.g. the only 2 percent argument), ICAO has unintentionally discarded the sense of urgency to act. It is not uncommon to hear those within aviation circles complaining that it is unfair to single out the sector when the contribution of the road transportation sector, for instance, is much larger. Arguably, this sentiment is symptomatic of a defensive mechanism that seeks to shift the responsibility elsewhere. They also do not take into account the sector's continuously rising growth rate and how this affects the overall increase in GHG emissions.

ICAO should also permit organizations representing civil society at large such as NGOs to play a more significant role in its climate change discussions. Although one could understand ICAO's natural tendency to favor industry participation, other stakeholders should also be placed on an equal footing. Apart from being non-conducive to advancing the process, this imbalance also raises questions of transparency and legitimacy on the part of ICAO.[2]

It is unquestionable that CAEP has made a significant contribution in advancing ICAO's technical work on climate change issues. In fact, the Council has recently entrusted CAEP with two key tasks in the design phase of the global MBM, namely, the elaboration of MRV standards and the determination of a sustainability criteria for offset credits. The only concern with regard to CAEP lies in its imbalanced membership. The committee continues to be dominated by developed States with an overwhelming industry influence. There is little participation from developing countries. This will inevitably pose challenges when the technical work reaches the political arena and decisions have to be made. This will be extremely relevant to the success or failure of the global MBM scheme. It is highly advisable that ICAO attempts to correct this imbalance as soon as possible.

Ever since the *Kyoto Protocol* enjoined States to limit or reduce GHG emissions from aviation working through ICAO, the organization has faced enormous pressure. It is, however, incorrect to suggest that ICAO has done nothing in this connection. An impressive amount of technical work has been carried out and the ICAO Secretariat has advanced numerous proposals. It is true that a long time has passed since the adoption of

2 *Ibid.*

the *Kyoto Protocol*, and GHG emissions from international aviation remain essentially unregulated. But this is not ICAO's fault. Rather, it reflects the lack of political commitment from a large number of States, including some of the most influential players.

More recently, ICAO has made concrete progress toward establishing a scheme to regulate GHG emissions. The CO_2 standard is on track for adoption. Although this will not be a major breakthrough, it constitutes a significant achievement, if one takes into account the numerous hurdles and the strong resistance that needed to be overcome. ICAO has engaged its membership in developing voluntary State action plans. This is perhaps one of the least noticed but most significant achievements of the organization in the climate change arena. For the first time, States are strongly encouraged to (i) consider climate change issues, (ii) measure their impact, (iii) report GHG emissions, and (iv) develop mitigation proposals. A number of regional, technical workshops have been organized by ICAO to assist States in this respect. Although it is not as yet possible to quantify GHG emission reductions as a result of these plans, the sole fact that States are discussing these issues is an accomplishment in and of itself. In the near future, these plans should be mandatory and must be included in the ICAO audit process in the same manner as aviation security and safety. This will facilitate compliance with the global MBM scheme.

9.4 The EU ETS

The EU ETS is central to the purpose of this academic endeavor. Having dominated the aviation and climate change landscape for almost ten years now, it is beyond question that the EU ETS has served as an *agent provocateur* and exerted enormous pressure on States, industry stakeholders, and ICAO. As a policy issue, climate change would have not attracted much consideration had it not been for the advent of the EU ETS. Yet the scheme has also been its own worst foe.

The legality of the EU ETS has been challenged on numerous grounds. The extraterritorial reach inherent in the original geographical scope of the EU ETS is in fact impermissible. Certainly, the principles of international law on jurisdiction do not justify it nor does the doctrine of State responsibility. Similarly, the effects doctrine as developed in both the United States and Europe does not provide much relief either. International law does not authorize the unilateral exercise of extraterritorial jurisdiction to protect scattered common interests. These legal allegations have called into question the scheme's legitimacy. Although the scheme has also been labeled as contravening WTO rules and the CBDR principle, these allegations remain unconvincing.

Those opposing the EU ETS have also devoted considerable effort to portray it as an impermissible tax that runs contrary to the provisions of the *Chicago Convention* and ASAs. These actors have taken full advantage of the poor communication campaign launched by the European Commissions to explain the main features of the scheme. These allegations, mostly based on political rather than legal grounds, ignore the very basic premise upon which emissions trading as a regulatory option was conceived: a mechanism other than taxes to allow the internalization of environmental externalities. Aside from the legal arguments against the EU ETS, the rather impressive international coalition that was assembled to defy it serves as an uncontested indication that the international aviation community would never accept the inclusion of foreign aircraft operators into the scheme.

Despite its demise, the EU ETS provides at least four invaluable lessons that should be taken into account in developing ICAO's global MBM scheme. Firstly, a complex scheme such as an MBM to tackle aviation's GHG emissions cannot be imposed, but it should rather arise from a shared understanding of why it is needed and what it seeks to achieve. The world outside Europe has always perceived the EU ETS as a Brussels-made imposition or a form of eco-imperialism that the international aviation community was not ready to tolerate. Drawing from the theory of norms in general and norm entrepreneurship, it is clear that a norm (e.g. EU ETS) emerged. Yet it has not led to norm cascading, let alone norm internalization. Opponents and even some key actors within Europe did not really accept the scheme, which never achieved its taken-for-granted status. For the success of the ICAO global MBM scheme it is critical to ensure that those participating in the scheme (e.g. aircraft operators) as well as Member States fully support it. If this is not the case, the scheme will face serious enforcement problems, given the enormous incentives to free ride and defect.

Secondly, the EU ETS was designed from an environmental perspective. It is instructive to remember that international aviation was included into Europe's general ETS. In practice, this meant that the scheme did not necessarily take into account the peculiarities of the aviation sector. The Commission underestimated the degree and strength of the international opposition against the scheme. The Commission did not attempt to make the scheme more attractive to those opposing it. In other words, there was no incentive structure provided to promote engagement with the scheme. Lifting the ceiling of CDM's CERs or allowing for REDD-plus credits could have eased the apprehension of developing countries. Such an incentive structure must be present in ICAO's global MBM. In the absence of such a structure, most of the States will simply not join the scheme.

Thirdly, communicating how the scheme works, what it seeks to achieve, what it allows, and what it forbids, as well as why it is needed, is of paramount importance. Only very late in the process did the Commission clarify a number of misconceptions regarding the EU ETS, particularly those revolving around issues such as auctions and allowances. Opponents, such as A4A exploited this communication gap. The ICAO global MBM cannot afford such a communication malfunction.

Finally, the EU ETS illustrates the difficulties faced by a top-down norm construction architectural design. By championing the EU ETS, the Commission acted as a genuine norm entrepreneur. The scheme however never really gained significant acceptance at the domestic level even within Europe, let alone the industry complaints. This strengthened the position of those opposing the scheme and weakened the role of the Commission. In designing and implementing the global MBM scheme, ICAO should be mindful of the importance of gaining domestic acceptance among the constituencies of Member States.

9.5 THE GLOBAL MBM SCHEME

Although some scholars remain skeptical,[3] the unanimous agreement at the 38th ICAO Assembly to develop a global MBM scheme for international aviation was a major achievement for ICAO and the international aviation community as a whole. It is true that this only represents a commitment to developing the scheme, but it is also a major step forward if one considers that 11 months prior to the Assembly, the only concern for the vast majority of States was the development of a framework of MBMs to restrict the geographical scope of the EU ETS. At that time, a global scheme seemed rather farfetched.

Considerable technical work has already been carried out in furtherance of the global MBM scheme, but it is evident that much more needs to be done. Although ICAO has explored other options, it is ultimately expected that the Global MBM scheme will involve an offsetting scheme given the preferences expressed by the industry and the animosity that emissions trading *per se* has attracted as a result of the EU ETS saga. One could view this as an unfortunate development given that, as a policy option, emissions trading offers much more flexibility for participants. The reality, however, is that managing and administering such a scheme could be more complex than offsetting. The presence of allowances exponentially raises the level of complexity of the scheme. In theory at least, it might have been desirable to further explore an MBM with a revenue-generating mechanism. These funds could have been re-channeled to address the concerns of some developing countries. In practice however this feature would create an additional level of administrative, legal, and political complexity.

To a great extent the success or failure of ICAO's global MBM scheme will depend on the reconciliation of CBDR and non-discrimination, the incentive structure provided for States to join the scheme, enforcement and compliance, and the simplicity and cost-effectiveness of the scheme. CBDR and non-discrimination could be reconciled through the adoption of a route based/phase in approach where routes from different countries to a specific destination could be integrated into the scheme at different stages in accordance with a specified set of criteria. In fact, I advanced some elements that could be included in such criteria. In addition, the determination of which ECU participants will be allowed

3 *Ibid.*

to purchase, grace periods, technical assistance, and technology transfers have also been identified as potential mechanisms that could effectively be used to recognize CBDR in the design of the global MBM scheme.

As discussed in Chapter 8, another issue that should not be forgotten is the legal vehicle that will be used to adopt the global scheme. It would seem that all actors presuppose that, in 2016, the 39th ICAO Assembly will adopt the scheme, which will subsequently be implemented as of 1 January 2020, and that everything will work smoothly. Surprisingly, States, ICAO, and the industry have paid very little attention to this issue thus far. Such an omission could be extremely dangerous because it may be exploited by those who deny the fact of climate change to either delay the adoption of the global MBM scheme or to derail the whole negotiation process. Three options have been identified as potential legal vehicles to adopt and implement the scheme. These are standards, an Assembly resolution, and a new international treaty. In deciding which option to use, States should be mindful of their respective advantages and limitations. Also, adopting a global MBM scheme is substantially different from adopting an aviation safety or security audit program. The *Chicago Convention* and most if not all pre-existing ASAs contain built-in provisions that facilitate the enforcement of safety or security related provisions. In part, this explains why audits are easily implemented through an Assembly resolution. Here, States have the legal tools to enforce the safety and security requirements against non-compliant participants. In the case of climate change, these built-in enforcement mechanisms are nowhere present.

A major element of the global MBM scheme should be the educational process. As discussed in Chapter 7, a number of actors within aviation circles still question the need of MBMs. For these actors, MBMs do not seek to limit or reduce GHG emission from international aviation; they are rather an excuse to raise money by levying taxes on airlines. They claim that emission reductions may be achieved through the implementation of operational and technological measures. The industry qualification of MBMs as gap-fillers or interim measures has not been helpful either. It is unlikely that the ICAO global MBM will be successful, unless all concerned actors understand why such a measure is needed. ICAO, Member States, and industry stakeholders ought to place considerably more emphasis on the educational process.

Lastly, one should also be very mindful of the limitations of the ICAO global MBM scheme. It will certainly be a major achievement if the scheme is eventually adopted in 2016 and implemented in 2020. However, it is unlikely that the scheme will ever be more stringent that what airlines are willing to commit to. At present at least, the political will is just not there. In other words, the European goal that by 2020 GHG emissions from international aviation should be reduced by 10 percent compared to 2005 levels is no more than wishful thinking. Again, in light of the divergent positions of States, it will be quite remarkable if international civil aviation could put in place a scheme seeking to achieve CNG from 2020 at 2020 levels. In this context, given the sector's growth trends and the political realities, one can be modestly optimistic and expect emissions to be contained but certainly not reduced.

9.6 THE ROLE PLAYED BY SOME OF THE MAIN ACTORS

In addition to the influential roles of ICAO and the EU for the global MBM scheme, I also identified the predominance position of IATA and the United States. IATA has been extremely influential in the agreement reached at the 38th Assembly to develop a global MBM scheme. The US endorsement to such scheme may be largely attributed to the extensive lobbying efforts of IATA and its sister trade association A4A. This is in fact a major accomplishment on the part of IATA. In addition, the aspirational goals that ICAO has established to tackle GHG emissions from international aviation have been largely framed by the IATA's initial proposals. In putting together the design elements of such global scheme, one may expect that IATA will continue to play a significant role. Notwithstanding the foregoing, it is also desirable that IATA strengthens the environmental performance of its member airlines. I identified two quick ways in which IATA could make even further contribution to the global MBM scheme. These involve making it mandatory for its member airlines to go through IATA's environmental audits (IEnvA) and subject them to ICAO's global MBM scheme. Imposing these requirements as a condition to maintain or accede to the IATA membership will serve as a huge incentive for airlines.

The US role in the global MBM scheme cannot be underestimated. There cannot be a meaningful scheme without effective and constructive US participation. If, for whatever reasons, the political landscape changes and the United States is forced to withdraw its endorsement, the global MBM scheme will never see the light of day. If, on the other hand, the US support for such a scheme remains, the prospects of success will be considerably higher.

9.7 THE ROAD AHEAD

Climate change will continue to be one of the most contentious policy issues that international aviation will face in the coming years. While the aviation sector has relied on a number of defensive tactics, the exogenous and endogenous pressures on the sector will not go away. Failure to address the issue will most likely create onerous burdens for international aviation. If this is the case, it is quite likely that an extraneous solution will be imposed on the sector. Not only is it very likely that taxation will be proposed as an alternative solution, but it may well be the case that "IATA's worst nightmare of a patchwork of inconsistent, duplicative and anticompetitive national and regional schemes" may become a reality.[4] It is certainly in aviation's best interest to tackle its climate change impact in a meaningful way as soon as possible.

The agreement by ICAO Member States to develop a global MBM scheme offers a great opportunity. ICAO, its Member States, and industry stakeholders bear an enormous responsibility. The international aviation community as a whole must rise to this challenge. The unsuccessful EU ETS has provided invaluable lessons for ICAO's global MBM scheme. The organization and its Member States must capitalize on the European experience. It would be quite unfortunate if the international aviation community repeats the same mistakes.

I have strongly advocated for the articulation of CBDR with non-discrimination in a manner compatible with international civil aviation, paying close attention to legal vehicles and enforcement mechanisms necessary to adopt such a scheme and examining the roles of some of the main actors with a view to further contribute to addressing the sector's carbon footprint. It is hoped that policy and decision-makers take these considerations into account in the design and implementation of ICAO's global MBM scheme.

4 *Ibid.*

Bibliography

Treaties and Other International Agreements

a) *International Air Law Instruments*

Convention Relating to the Regulation of Aerial Navigation, 13 October 1919, 11 UNTS 173.

Convention for the Unification of Certain Rules Relating to International Carriage by Air, 10 October 1929, 3145 LON 137 13 LoN-3145.

Convention on International Civil Aviation, 7 December 1944, 15 UNTS 295, ICAO Doc 7300.

International Air Services Transit Agreement, 7 December 1944, 84 UNTS 389, ICAO Doc 7500.

Convention on Damage Caused by Foreign Aircraft to Third Parties on the Surface, 7 October 1952, 310 UNTS 181, ICAO Doc 7364.

Protocol Relating to Certain Amendments to the Convention on International Civil Aviation - Articles 48 (a), 49 (e) and 61, 14 June 1954.

Protocol to Amend the Convention for the Unification of Certain Rules Relating to International Carriage by Air, 12 October 1929, 28 September 1955, 478 UNTS 371, ICAO Doc 7632.

Protocol Relating to the Amendments of Articles 48 (a), 49 (e) and 61 of the Convention on International Civil Aviation, 12 December 1956, ICAO DOC 7300.

Convention on Offences and Certain Other Acts Committed on Board Aircraft, 14 September 1963, 704 UNTS 219, ICAO Doc 8364.

Convention for the Suppression of Unlawful Seizure of Aircraft, 16 December 1970, 860 UNTS 105, ICAO Doc 8920.

Convention for the Suppression of Unlawful Acts against the Safety of Civil Aviation, 23 September 1971, 974 UNTS 177, ICAO Doc 8966.

Protocol Relating to an Amendment to the Convention on International Civil Aviation Article 50 (a), 26 October 1990, 2216 UNTS 483, ICAO Doc 9561.

Convention on the Marking of Plastic Explosives for the Purpose of Detection, 1 March 1991, 2122 UNTS 359, ICAO Doc 9571.

Agreement on the Application of Sanitary and Phytosanitary Measures, 15 April 1994, 1867 UNTS 493.

Agreement on Technical Barriers to Trade, 15 April 1994, 1868 UNTS 120.

Convention for the Unification of Certain Rules for International Carriage by Air, 28 May 1999, 2242 UNTS 309, ICAO Doc 9740.

Convention on International Interest in Mobile Equipment, 16 November 2001, 2307 UNTS 285, ICAO Doc 9793.

Protocol to the Convention on International Interest in Mobile Equipment on Matters Specific to Aircraft Equipment, 16 November 2001, 2367 UNTS 517, ICAO Doc 9794.

Convention on Compensation for Damage to Third Parties, Resulting from Acts of Unlawful Interference Involving Aircraft, 2 May 2009, ICAO Doc 9919.

Convention on the Suppression of Unlawful Acts Relating to International Civil Aviation, 10 September 2010, ICAO Doc 9960.

Protocol Supplementary to the Convention for the Suppression of Unlawful Seizure of Aircraft, 10 September 2010, ICAO Doc 9959.

Protocol to Amend the Convention on Offenses and Certain and Order Acts Committed on Board Aircraft, signed in Tokyo on 4 April 2014.

b) *International Air Law Agreements*

Agreement between the Government of the United Kingdom and the Government of the United States of America Relating to Air Services between Their Respective Territories, 11 February 1946, 253 UNTS 3.

Air Transport Agreement between the United States of America and 27 Member States of the European Union, [2007] OJ, L 134/11.

Multilateral Agreement on the Liberalization of Air Transportation (MALIAT), 1 May 2001, online: MALIAT <www.maliat.govt.nz/agreement/article9.shtml>.

Memorandum of Cooperation (MOC) between the International Civil Aviation Organization (ICAO) and the European Aviation Safety Agency (EASA) Regarding Safety Oversight Audit and Related Matters, signed on 21 March 2006.

Memorandum of Cooperation between the International Civil Aviation Organization and the European Community Regarding Security Audits/Inspections and Related Matters, signed on 17 September 2008.

Memorandum of Understanding between the World Customs Organization and the International Civil Aviation Organization, signed on 24 June 2011.

Memorandum of Cooperation between the International Civil Aviation Organization and the World Tourism Organization, signed on 10 October 2010.

c) *Other International Treaties and Instruments*

Convention for the Amelioration of the Condition of the Wounded in Armies in the Field, 22 August 1864, online: ICRC <www.icrc.org/ihl/52d68d14de6160e0c12563da005fdb1b/87a3bb58c1c44f0dc125641a005a06e0>.

Charter of the United Nations, 26 June 1945, Can TS 1945 No 7.

Statute of the International Court of Justice, 26 June 1945, online: ICJ <www.icj-cij.org/documents/index.php?p1=4&p2=2&p3=0&>.

CE, *Treaty Establishing the European Economic Community*, 25 March 1957, online: Eur-Lex <http://eur-lex.europa.eu/LexUriServ/LexUriServ.do?uri=CELEX:11957E/TXT:EN:NOT>.

Treaty on Principles Governing the Activities of States in the Exploration and Use of Outer Space, Including the Moon and Other Celestial Bodies, 27 January 1967, 610 UNTS 205, 6 ILM 386.

Vienna Convention on the Law of Treaties, 23 May 1969, 1155 UNTS 331, 8 ILM 679, 63 AJIL 875.

International Convention Relating to Intervention on the High Seas in Cases of Oil Pollution Casualties, 29 November 1969, 970 UNTS 211.

International Convention for the Prevention of Pollution from Ships, adopted on 2 November 1973, online: IMO <http://goo.gl/Akl2NV>.

Convention on the Prevention and Punishment of Crimes against Internationally Protected Persons, Including Diplomatic Agents, 14 December 1973, 1037 UNTS 167.

Protocol of 1978 relating to the International Convention for the Prevention of Pollution from Ships, 1973, 17 February 1978, 10 February 1983, 1340 UNTS 61.

International Convention against the Taking of Hostages, 17 December 1979, 1316 UNTS 205.

Convention on the Physical Protection of Nuclear Material, 26 October 1979, 1456 UNTS 101.

Convention for the Conservation of Salmon in the North Atlantic, 2 March 1982, 1338 UNTS 33.

United Nations Convention on the Law of the Sea, 10 December 1982, 1833 UNTS 3.

Convention on the Succession of States in Respect of State Property, Archives and Debts, 8 April 1983, online: United Nations <http://legal.un.org/ilc/texts/instruments/english/conventions/3_3_1983.pdf>.

Vienna Convention for the Protection of the Ozone Layer, 22 March 1985, 1513 UNTS 323.

Montreal Protocol on Substances that Deplete the Ozone Layer, 16 September 1987, 522 UNTS 3, 26 ILM 1550 (1987).

Convention for the Suppression of Unlawful Acts against the Safety of Maritime Navigation, 10 March 1988, 1678 UNTS 201.

Protocol to the Convention for the Suppression of Unlawful Acts against the Safety of Fixed Platforms Located on the Continental Shelf, 10 March, 1988, 1678 UNTS 201.

Rome Statute of the International Criminal Court, done at Rome on 17 July 1988, 2187 UNTS 3.

The Hague Declaration on Environment, 11 March 1989, 28 ILM 308.

United Nations Framework Convention on Climate Change, 9 May 1992, 1771 UNTS 107.

Convention on Biological Diversity, 5 June 1992, 1760 UNTS 79.

Agreement on Trade-Related Aspects of Intellectual Property Rights, 14 April 1994, 1869 UNTS 299.

General Agreement on Trade in Services, 15 April 1994, 1869 UNTS 183.

General Agreement on Tariffs and Trade, 1947, 55 UNTS 194.

Kyoto Protocol to the United Nations Framework Convention on Climate Change, 11 December 1997, 2303 UNTS 162.

International Convention for the Suppression of Terrorist Bombings, 15 December 1997, 2149 UNTS 256.

International Convention for the Suppression of the Financing of Terrorism, 9 December 1999, 2178 UNTS 197.

International Convention for the Suppression of Acts of Nuclear Terrorism, 13 April 2005, 2445 UNTS 89.

CE, *Treaty on the Functioning of the European Union*, [2012] OJ, C 326/47.

d) United Nations/United Nations Bodies Resolutions and Decisions

Marking of Plastic or Sheet Explosives For the Purpose of Detection, SC Res 635, UN SC, 2869 Mtg, UN Doc S/RES/635 (14 June 1989).

The Crime of Genocide, GA Res 96(I), UNGA, 1st Sess, UN Doc A/RES/1/96(I) (11 December 1946).

World Charter for Nature, GA Res 7, UNGA, 37th Sess, UN Doc A/RES/37/7 (28 October 1982).

United Nations Conference on Environment and Development, GA Res 228, UNGA, 44th Sess, UN Doc A/RES/44/228 (22 December 1989).

Articles on Responsibility of States for Internationally Wrongful Acts, GA Res 83, UNGA 83d Sess, UN Doc A/RES/56/83 (28 January 2002).

In-depth Review of Ongoing Work on Alien Species That Threaten Ecosystems, Habitats or Species, COP Dec 9, UNEP CBD, UNEP/CBD/COP/DEC/IX/4 (9 October 2008).

Invasive Alien Species, COP Dec 38, UNEP CBD, UNEP/CBD/COP/DEC/X/38 (29 October 2010).

Legislation

a) Australia

Passenger Movement Charge Act of 1978, Act No 118, online: Australian Government <www.comlaw.gov.au/Details/C2012C00605>.

Passenger Movement Charge, online: Australian Custom and Border Protection Service <www.customs.gov.au/site/page6068.asp>.

b) Canada

Criminal Code, RSC, 1985, c C-46.

c) *Paraguay*

Law No 1331/88, Which Modifies Law No 85, 16 December 1991.

Presidential Decree No 793/2008, Which Establishes the International Embarkation Tax by Air for Each Passenger, 10 November 2008.

d) *United States*

i) **Codes**

Commerce and Trade, 15 USC.

Crime and Criminal Procedure, 18 USC.

Logan Act, 18 USC § 953 (1799).

Internal Revenue Code, 26 USC.

Judiciary and Judicial Procedure, 28 USC.

Labour, 29 USC.

Age Discrimination in Employment Act, 29 USC § 623 (1967).

Mineral Lands and Mining, 30 USC 40.

Oil Pollution Act 33 USC ch 40 § 2701 (1990).

Securities Exchange Act of 1934, § 10 (b), 15 USCA § 78j (b); 28 US CA § 1331.

US Foreign Corrupt Practices Act, 15 USC § 78 dd-1 (1977).

The Public Health and Welfare, 42 USC.

Transportation, 49 USC.

ii) Session Law

Antiterrorism and Effective Death Penalty Act of 1996, Pub L No 104-132, 110 Stat 1214 (1997).

Sarbanes-Oxley Act of 2002, Pub L 107-204, 116 Stat 745 (2002).

FAA Modernization and Reform Act of 2012, Pub L 112-95 126 Stat 11 (2012).

iii) Bills & Resolutions

US, H Res 86, 106th Cong, 1999.

US, Bill HR 2594, *An Act to prohibit operators of civil aircraft of the United States from participating in the European Union's emissions trading scheme, and for other purposes*, 112th Cong, 2011.

US, S 1956, *An Act to prohibit operators of civil aircraft of the United States from participating in the European Union's emissions trading scheme, and for other purposes*, 112th Cong, 2012.

iv) Other

Restatement (Third) of Foreign Relations Law of the United States (1987).

US DOT Order 2011-12-10, Docket OST-2011-0230.

Smoking Aboard Aircraft, 14 CFR Part 252.

46 FR 48109, Exec Order No 12324, 1981 WL 404170 (Pres).

57 FR 23133, Exec Order No 12807, 1992 WL 12135821 (Pres).

Jurisprudence

a) Permanent Court of International Justice

The Case of SS "Lotus" (France v Turkey) (1927), PCIJ (Ser A) No 70.

b) *International Court of Justice*

i) **Judgments**

Fisheries Case (United Kingdom v Norway), [1951] ICJ Rep 116.

Nottebohm Case (Liechtenstein v Guatemala), [1955] ICJ Rep 4.

South West Africa Cases (Ethiopia v South Africa; Liberia v South Africa) Second Phase, [1966] ICJ Rep 6.

Case Concerning The Barcelona Traction, Light and Power Company, Limited (Belgium v Spain) Second Phase, [1970] ICJ Rep 3.

Fisheries Jurisdiction Cases (United Kingdomv Iceland), [1973] ICJ Rep 3.

Fisheries Jurisdiction Cases (United Kingdomv Iceland), [1974] ICJ Rep 3.

Case Concerning Military and Paramilitary Activities in and Against Nicaragua (Nicaragua v United States of America), [1986] ICJ Rep 14.

Case Concerning East Timor (Portugal v Australia), [1995] ICJ Rep 90.

Case Concerning Application of the Convention on the Prevention and Punishment of the Crime of Genocide (Bosnia and Herzegovina v Serbia and Montenegro), [1996] ICJ Rep 595.

Case Concerning the Gabčikovo-Nagymaros Project (Hungary v Slovakia), [1997] ICJ Rep 7.

Application of the Convention on the Prevention and Punishment of the Crime of Genocide (Bosnia and Herzegovina v Serbia and Montenegro), [2007] ICJ Rep 47.

ii) **Advisory Opinions**

Reparations for Injuries Suffered in the Service of the United Nations, Advisory Opinion, [1949] ICJ Rep 174.

Legal Consequences for States of the Continued Presence of South Africa in Namibia (South West Africa) Notwithstanding Security Council Resolution 276 (1970), Advisory Opinion, [1971] ICJ Rep 16.

The Legality of the Threat or Use of Nuclear Weapons, Advisory Opinion, [1996] ICJ Rep 226.

Legal Consequences of the Construction of a Wall in the Occupied Palestinian Territory, Advisory Opinion, [2004] ICJ Rep 136.

Case Concerning the Arrest Warrant of 11 April 2000 (Democratic Republic of the Congo v. Belgium), [2002] ICJ Rep 3.

c) *International Criminal Tribunal*

Prosecutor v Anto Furundžija, IT-95-17/1-T, Judgment (10 December 1998), (International Criminal Tribunal Yugoslavia, Trial Chamber), online: ICTY <www.icty.org/x/cases/furundzija/tjug/en/fur-tj981210e.pdf>.

d) *World Trade Organization (WTO)*

United States – Restrictions on Imports of Tuna (3 September 1991), GATT DS21/R (Panel Report), unadopted, online: World Trade Law <www.worldtradelaw.net/reports/gattpanels/tunadolphinI.pdf>.

United States – Restrictions on Imports of Tuna (16 June 1994), GATT DS29/R (Panel Report), unadopted, online: Stanford University <http://sul-derivatives.stanford.edu/derivative?CSNID=91790155&mediaType=application/pdf>

United States – Standards for Reformulated and Conventional Gasoline (20 May 1996), WTO Doc WT/DS2/9 (Appellate Body Report), online: WTO <www.wto.org/english/tratop_e/dispu_e/2-9.pdf>.

EC Measures Concerning Meat and Meat Products (Hormones) (13 February 1998), WT/DS26/AB/R (Appellate Body Report), online: WTO <www.wto.org/english/tratop_e/dispu_e/hormab.pdf>.

United States – Import Prohibition of Certain Shrimp and Shrimp Products (12 October 1998), WT/DS58/AB/R (AppellateBodyReport), online: WTO <https://docs.wto.org/dol2fe/Pages/FE_Search/FE_S_S006.aspx?Query=(@Symbol=%20wt/ds58/ab/r*%20not%20rw*)&Language=ENGLISH&Context=FomerScriptedSearch&languageUIChanged=true#>.

United States – Import Prohibitions of Certain Shrimp and Shrimp Products, Recourse to Article 21.5 of the DSU by Malaysia (22 October 2001), WT/DS58/AB/RW (Appellate Body Report) online: WTO <www.wto.org/english/tratop_e/dispu_e/58abrw_e.doc>.

e) *European Court of Justice*

International Fruit Company NV and Others v Produktschap voor Groenten en Fruit, C-21 to 24/72, [1972] ECR I-01219.

Braathens Sverige AB v Riksskatteverket, C-346/97, [1999] ECR I-3433.

The Queen, on the Application of International Association of Independent Tanker Owners (Intertanko) and Others v Secretary of State for Transport, C-308/06, [2008] ECR I-04057.

Commune de Mesquer v Total France SA and Total International Ltd, C-188/07, [2008], ECR I-04501.

Irène Bogiatzi, Married Name Ventouras v Deutscher Luftpool and Others, C-301/08, [2009] ECR I-10185.

TNT Express Nederland BV v AXA Versicherung AG, C-533/08, [2010] ECR I-04107.

Air Transport Association of America and Others v Secretary of State for Energy and Climate Change, C-366/10, [2011] ECR I-13755.

Opinion of Advocate General Kokott, delivered on 6 October 2011, Case C-366/10, online: Europa <http://ec.europa.eu/clima/news/docs/2011100601_case_c366_10_en.pdf>.

Written Observations of the International Air Transport Association and the National Airlines Council of Canada in the Court of Justice of the European Union, Reference from the High Court of Justice, London, Case C-366/10.

f) *Germany*

K v Schleswig-Hoslstein, (1951) Bundesgerichtshof (Supreme Court of the Federal Republic of Germany), Entscheidungen des Bundesgerichtshofes in Zivilsanchen, Vol 4, No 30, p 266 in "Succession of States and Governments", *Yearbook of the International Law Commission 1963*, vol 2 (New York: UN, 1964) (A/CN.4/SER.A/1963/Add 1) 147.

ST v The Land N, (1952) Bundesgerischtshof (Supreme Court of the Federal Republic of Germany), Entscheidungen des Bundesgerichtshofes in Zivilsachen, Vol 8, No 22, p 169 in "Succession of States and Governments", *Yearbook of the International Law Commission 1963*, vol 2 (New York: UN, 1964) (A/CN.4/SER.A/1963/Add 1) 147.

g) United States

American Banana Co v United Fruit Co, 213 US 347 (1909).

United States v Pacific & Arctic Railway and Navigation Company, 228 US 87 (1913).

United States v Bowman, 260 US 94 (1922).

United States v Sisal Sales Corporation, 274 US 266 (1927).

Blackmer v United States, 284 US 421 (1932).

United States v Aluminum Co of America, 148 F2d 416 (2d Cir 1945).

Foley Bros, Inc v Filardo, 226 US 281 (1949).

Steelev Bulova Wath Co, 344 US 280, 73 S Ct 252, 97 L Ed 319 (1952).

Tomoya Kawakita v United States, 343 US 72 S Ct 950 (1952).

Continental Ore Co v Union Carbide & Carbon Corp, 370 US 690 (1962).

Pacific Seafarers, Inc v Pacific Far East Line, Inc, 404 F 2d 804 (Ct App DC1968).

Leasco Data Processing Equipment Corporation, 468 F 2d 1326 (2d Cir 1972).

Timberlane Lumber Co v Bank of America, 549 F 2d 597 (9th Cir 1976).

Mannington Mills, Inc v Congoleum Corporation, 595 F 2d 1287 (3d Cir 1979).

Montréal Trading Ltd v Amax Inc, 661 F 2d 864 (10th Cir 1981).

Laker Airways Limited v Sabena, 731 F 2d 909 (DC Cir 1983).

Beattie v United States, 756 F 2d 91 (DC Cir 1984).

Pfeiffer v Wm Wrigley Jr Co, 755 F 2d 554, Court of Appeals, (7th Cir 1985).

Matsushita Electric Industrial Co v Zenith Radio Corp, 475 US 574 (1985).

EEOC v Arabian Am Oil Co, 499 US 244 (1991).

Amlon Metals, Inc v Fmc Corporation, 775 F Supp 668 (1991).

United States v Yunis, 924 F 2d 1086 (DC Cir 1991).

Lujan v Defenders of Wildlife, 112 S Ct 2130 (1992).

Sale v Haitian Centers Council, Inc, 509 US 155, 113 S Ct 2549 (1993).

Hartford Fire Insurance Co v California, 509 US 764 (1993).

Robinson v TCI/US West Communications Inc, 117 F 3d 900 (5th Cir 1997).

Massachusetts v Environmental Protection Agency, 549 US 497 (2007).

Microsoft Co v AT&T Co, 550 US 437 (2007).

Chafin v Chafin, 133 S Ct 1017 (2013).

h) Arbitral Awards

Award between the United States and the United Kingdom Relating to the Rights of Jurisdiction of United States in the Bering's Sea and the Preservation of Fur Seal (15 August 1893) vol XXVIII, p 263, online: UN <http://legal.un.org/riaa/cases/vol_XXVIII/263-276.pdf>.

Air Services Agreement Case (France v United States) (1978) 18 RIAA 416, online: IILJ <http://iilj.org/courses/documents/AirServicesCase.pdf>.

International Civil Aviation Organization (ICAO) Documents

a) Annexes to the Chicago Convention

Annex 1, *Personnel Licensing*, 11th edn (2011).

Annex 2, *Rules of the Air*, 10th edn (2005).

Annex 3, *Meteorological Service for International Air Navigation*, 18th edn (2013).

Annex 5, *Units of Measurement to be Used in Air and Ground Operations*, 5th edn (2010).

Annex 6, *Operations of Aircraft – International Commercial Air Transport – Aeroplanes*, 9th edn (2010).

Annex 7, *Aircraft Nationality and Registration Marks*, 5th edn (2003).

Annex 8, *Airworthiness of Aircraft*, 11th edn (2010).

Annex 9, *Facilitation*, 13th edn (2011).

Annex 10, *Aeronautical Telecommunications, Volume III – Communication Systems*, 2nd edn (2007).

Annex 11, *Air Traffic Services*, 13th edn (2001).

Annex 13, *Aircraft Accident and Incident Investigation*, 10th edn (2010).

Annex 14, *Aerodromes, Volume I – Aerodrome Design and Operations*, 6th edn (2013).

Annex 16, *Environmental Protection, Volume II – Aircraft Engine Emissions*, 3rd edn (2008).

Annex 16, *Environmental Protection, Volumes I & II – Aircraft Engine Emissions*, 6th edn (2011).

Annex 17, *Security: Safeguarding International Civil Aviation Against Acts of Unlawful Interference*, 9th edn (2011).

b) Assembly Working Papers

A32-WP/147 P/25, *Preventing the Introduction of Invasive Alien Species* (Presented by the United States of America).

A33-WP/11 EC/6, *Report by the Council on Progress in Implementation of Resolution A32-9: Preventing the Introduction of Invasive Alien Species.*

A35-WP/12 EC/4, *Report by the Council on Progress in Implementation of Resolution A33-18: Preventing the Introduction of Invasive Alien Species.*

A36-WP/19 EC/4, *Implementation of Resolution A35-19: Preventing the Introduction of Invasive Alien Species* (Presented by the President of the Council of ICAO).

A36-WP/85 EX/33, *Towards a Carbon Neutral and Eventually Carbon Free Industry* (Presented by the International Air Transport Association).

A36-WP/88 EX/36, *Viewpoint of the Arab Republic of Egypt as a Developing Country on Emissions Trading for Civil Aviation* (Presented by Egypt).

A36-WP/130 EX/50, *Viewpoint of the Latin American Civil Aviation Commission on the Aviation Emissions Trading Scheme* (Presented by the 22 Member States of the Latin American Civil Aviation Commission).

A36-WP/235 EX/76, *Addressing Aviation Emissions Based on the Principle "Common But Differentiated Responsibilities"* (Presented by China).

A36-WP/251 EX/82, *Environment and Emission Trading Charges* (Presented by Nigeria on behalf of African States).

A36-WP/258 EX/86, *Increasing ICAO Council Membership to a Minimum of 39 Seats* (Presented by the Arab Civil Aviation Commission (ACAC)).

A36-WP/284 EX/91 Rev 1, *Proposal for a Study of Policy and Programme with Respect to Examining the International Governance of Civil Aviation* (Presented by Antigua and Barbuda, the Bahamas, Barbados, Belize, Canada, Dominica, Grenada, Guyana, Haiti, Hungary, India, Jamaica, Pakistan, the Republic of Korea, Saint Kitts and Nevis, Saint Lucia, Saint Vincent and the Grenadines, South Africa, Suriname, Trinidad and Tobago, the United Arab Emirates and the United Kingdom).

A36-WP/285 EX/92, *Chile's Position on the Inclusion of Civil Aviation in Emissions Trading* (Presented by Chile).

A37-WP/23 EX/6, *Aviation and Alternative Fuels* (Presented by the Council of ICAO).

A37-WP/108 EX/26, *Addressing Aviation's Environmental Impacts Through: A Comprehensive Approach* (Presented by Belgium on behalf of the European Union and its Member States and by the other States Members of the European Civil Aviation Conference, and by Eurocontrol).

A37-WP/174 EX/31, *Statement Regarding Mitigation of Greenhouse Gas Emissions from Air Passenger Transport* (Presented by the World Tourism Organization, UNWTO).

A37-WP/181 EX/32, *Addressing Global Climate Change within the Framework of Sustainable Development of International Aviation* (Presented by the People's Republic of China).

A37-WP/185 EX/34, *Sustainable Alternative Aviation Fuels* (Presented by the United States of America).

A37-WP/186 EX/35, *A More Ambitious, Collective Approach to International Aviation Greenhouse Gas Emissions* (Presented by Canada, Mexico and the United States).

A37-WP/217 EX/39, *Development of a Global Framework for Addressing Civil Aviation Co_2 Emissions* (Presented by the International Air Transport Association (IATA), on behalf of ACI, CANSO, IATA, IBAC and ICCAIA).

A37-WP/402 P/66, *Report of the Executive Committee on Agenda Item 17 (Section on Climate Change)* (Presented by the Chairman of the Executive Committee).

A38-WP/12 EX/7, *Outcomes of the High-Level Conference on Aviation Security (HLCAS)* (Presented by the Council of ICAO).

A38-WP/17 EX 12, *Proposal to Amend Article 50 a) of the Convention on International Civil Aviation so as to Increase the Membership of the Council to 39* (Presented by the Council of ICAO).

A38-WP/18 EX/13, *Declaration on Aviation Security and the ICAO Comprehensive Aviation Security Strategy (ICASS)* (Presented by the Council of ICAO).

A38-WP/30 EX/25, *States' Action Plans for CO_2 Emissions Reduction Activities* (Presented by the Council of ICAO).

A38-WP/34 EX/29, *Consolidated Statement of Continuing ICAO Policies and Practices Related to Environmental Protection – Climate Change* (Presented by the Council of ICAO).

A38-WP/68 EX/33, *Addressing Co_2 Emissions from Aviation* (Presented by the Airports Council International (ACI), the Civil Air Navigation Services Organisation (CANSO), the International Air Transport Association (IATA), the International Business Aviation Council (IBAC) and the International Coordinating Council of Aerospace Industries Associations (ICCAIA)).

A38-WP/83 EX/38, *A Comprehensive Approach to Reducing the Climate Impacts of International Aviation* (Presented by Lithuania on behalf of the European Union and Its Member States and the other Member States of the European Civil Aviation Conference).

A38-WP/93 AD/13, *Draft Scales of Assessment for 2014, 2015 and2016* (Presented by the Secretary General).

A38-WP/108 EX/44, *Sustainable Alternative Jet Fuels* (Presented by the United States).

A38-WP/170 LE/6, *Promotion of the Convention for the Unification of Certain Rules for International Carriage by Air (Montréal Convention of 1999)* (Presented by the United Arab Emirates, the Air Crash Victims Families Group (ACVFG) and the International Air Transport Association (IATA)).

A38-WP/176 EX/67, *Expectations and Desirable Objectives of the 38th Session of the Assembly Relating to International Aviation and Climate Change: Perspective of the Kingdom of Saudi Arabia* (Presented by the Kingdom of Saudi Arabia).

A38-WP/183 EX/72, *Achieve Emissions Reduction through Technical and Operational Measures: What China Has Done* (Presented by the People's Republic of China).

A38-WP/234 EX/79, *Addressing the Climate Impacts of Aviation* (Presented by the United States).

A38-WP/250 EX/83, *Market-Based Measures as the Factor of an Increase of Greenhouse Gas Emissions in the Sector of International Civil Aviation* (Presented by the Russian Federation).

A38-WP/258 EX/85, *UAE's Views on Aviation and Climate Change* (Presented by the United Arab Emirates).

A38-WP/268 EX/88, *Agreement of Voluntary Activity for GHG Reduction in the Republic of Korea* (Presented by the Republic of Korea).

A38-WP/270 EX/90, *Development of Aviation Environmental Education Program* (Presented by the Republic of Korea).

A38-WP/272 EX/92, *Position of African States on Climate Change* (Presented by 54 African States).

A38-WP/318 EX/110, *Environmental Protection, CAEP and the ICAO Regional Offices* (Presented by Argentina).

A38-WP/350 EX/120, *To Reduce CO_2 Emissions from Aviation* (Presented by Vietnam).

A38-WP/424 EX/139, *Consolidated Statement of Continuing ICAO Policies and Practices Related to Environmental Protection: Climate Change* (Presented by Argentina, Brazil, China, Cuba, Guatemala, India, Islamic Republic of Iran, Pakistan, Peru, Russian Federation, Saudi Arabia, and South Africa).

A38-WP/425 EX/140, *Proposed Amendments for the Draft Consolidated Statement of Continuing ICAO Policies and Practices Related to Environmental Protection: Climate Change* (Presented by Argentina, Brazil, China, Cuba, Guatemala, India, Islamic Republic of Iran, Pakistan, Peru, Russian Federation, Saudi Arabia, and South Africa).

A38-WP/427 EX/142, *Proposed Amendments for the Draft Consolidated Statement of Continuing ICAO Policies and Practices Related to Environmental Protection: Climate Change* (Presented by Argentina, China, Cuba, India, Islamic Republic of Iran, Pakistan, Peru, the Russian Federation, Saudi Arabia, and South Africa).

c) Assembly Resolutions

A18-11, *ICAO Position at the International Conference on the Problems of the Human Environment (Stockholm, June 1972).*

A28-7, *Aeronautical Consequences of the Iraqi Invasion of Kuwait.*

A29-15, *Smoking Restrictions on International Passenger Flights.*

A32-11, *Establishment of an ICAO Universal Safety Oversight Audit Program.*

A32-8, *Consolidated Statement of Continuing ICAO Policies and Practices Related to Environmental Protection.*

A32-9, *Preventing the Introduction of Invasive Alien Species.*

A33-18, *Preventing the Introduction of Invasive Alien Species.*

A33-7, *Consolidated Statement of Continuing ICAO Policies and Practices Related to Environmental Protection.*

A35-5, *Consolidated Statement of Continuing ICAO Policies and Practices Related to Environmental Protection.*

A35-19, *Preventing the Introduction of Invasive Alien Species.*

A36-6, *State Recognition of the Air Operator Certificate of Foreign Operators and Surveillance of Their Operations.*

A36-15, *Consolidated Statement of Continuing ICAO Policies in the Air Transport Field.*

A36-17, *Consolidated Statement of ICAO Policies on Technical Cooperation.*

A36-21, *Preventing the Introduction of Invasive Alien Species.*

A36-22, *Consolidated Statement of Continuing ICAO Policies and Practices Related to Environmental Protection.*

A37-5, *The Universal Safety Oversight Audit Program (USOAP) Continuous Monitoring Approach.*

A37-15, *Consolidated Statement of Continuing ICAO Policies and Associated Practices Related Specifically to Air Navigation, Appendix M, Delimitation of Air Traffic Services Airspaces.*

A37-19, *Consolidated Statement of Continuing ICAO Policies and Practices Related to Environmental Protection – Climate Change.*

A37-20, *Consolidated Statement of Continuing ICAO Policies in the Air Transport Field.*

A37-22, *Consolidated Statement of Continuing ICAO Policies in the Legal Field.*

A38-14, *Consolidated Statement of Continuing ICAO Policies in the Air Transport Field.*

A38-15, *Consolidated Statement of Continuing ICAO Policies Related to Aviation Security.*

A38-17, *Consolidated Statement of Continuing ICAO Policies and Practices Related to Environmental Protection: General Provisions, Noise and Local Air Quality.*

A38-18, *Consolidated Statement of Continuing ICAO Policies and Practices Related to Environmental Protection: Climate Change.*

A38-20, *Promotion of the Montréal Convention 1999.*

A38-26, *Assessments to the General Fund for 2014, 2015, and 2016.*

d) Council Working Papers

C-WP/1169, *Request of the Government of India to the Council of ICAO.*

C-WP/1222, *India/Pakistan Case Telegram Received from the Government of Afghanistan 16 June, 1952.*

C-WP/7696, *Draft Resolution* (Presented by Australia, Canada, Denmark, France, Federal Republic of Germany, Italy, Japan, Kingdom of the Netherlands, Spain, United Kingdom, and United States).

C-WP/7697, *Draft Resolution* (Presented by the Union of Soviet Socialist Republics).

C-WP/7698, *Draft Resolution Interception of Civil Aircraft* (Presented by France).

C-WP/11186, *Infractions of the Convention on International Civil Aviation* (Presented by the Secretary General).

C-WP/12498, *Taxes on Airline Tickets to Finance Development* (Presented by France).

C-WP/12986, *Draft Assembly Working Paper: Civil Aviation and the Environment* (Presented by the Secretary General).

C-WP/13385, *Progress Report on the Group on International Aviation and Climate Change (GIACC)* (Presented by the Secretary General).

C-WP/13399, *Legal Committee: Participation of Observers and Election of Officers* (Presented by the Chair of the Working Group on Governance (Policy) – WGOG).

C-WP/13416, *Review of International Governance (Chicago Convention)* (Presented by the Chair of the Working Group on Governance (Policy) (WGOG)).

C-WP/13520, *Membership in the Committee on Aviation Environmental Protection (CAEP)* (Presented by the Secretary General).

C-WP/13761, *European Emissions Trading Scheme* (*ETS*) (Presented by the Secretary General).

C-WP/13790, *Inclusion of International Civil Aviation in the European Union Emissions Trading Scheme* (*EU ETS*) *and Its Impact* (Presented by Argentina, Brazil, Burkina Faso, Cameroon, China, Colombia, Cuba, Egypt, Guatemala, India, Japan, Malaysia, Mexico, Morocco, Nigeria, Paraguay, Peru, Republic of Korea, Russian Federation, Saudi Arabia, Singapore, South Africa, Swaziland, Uganda, the United Arab Emirates and the United States).

C-WP/13798, *Study on Market-Based Measures* (*MBMs*) (Presented by the Secretary General).

C-WP/13799, *Study on Market-Based Measures* (*MBMs*) (Presented by the Chairman of the Air Transport Committee).

C-WP/13858, *Environmental Protection: Recent Developments in ICAO* (Presented by the Secretary General).

C-WP/13861, *Market-Based Measures* (*MBMs*) (Presented by the Secretary General).

C-WP/13894, *Market-Based Measures (MBMs): Evaluation of Options for a Global MBM Scheme* (Presented by the Secretary General).

C-WP/14101, *Review of Assembly Resolutions and Decisions: Environmental Protection* (Presented by the Secretary General).

C-WP/14102, *Establishment of the Environment Advisory Group (EAG)* (Presented by the President of the Council).

e) Council Decisions

ICAO, *Council Resolution on Libyan Civil Aircraft of 4 June 1973.*

ICAO, *Council Resolution of 16 September 1983.*

ICAO, *Council Resolution of 28 February 1986.*

ICAO, *Council Resolution Adopted at the Twentieth Meeting of Its 148th Session on 27 June 1996.*

ICAO, *Council Resolution of 5 February 1999.*

ICAO, *Consolidated Council Resolution on Taxation of International Air Transport of 24 February 1999.*

ICAO, *Council Declaration on Influenza A (H1N1) of 19 May 2009.*

C-DEC 156/3, *Council, 156th Session, Decision of the Third Meeting.*

C-DEC 176/14, *Council, 176th Session, Decision of the Fourteenth Meeting.*

C-DEC 187/14, *Council, 187th Session, Decision of the Fourteenth Meeting.*

C-DEC 188/6, *Council, 188th Session, Decision of the Sixth Meeting.*

C-DEC 192/6, *Council, 192nd Session, Decision of the Sixth Meeting.*

C-DEC 194/2, *Council, 194th Session, Decision of the Second Meeting.*

C-DEC 195/9, *Council, 195th Session, Decision of the Ninth Meeting.*

C-DEC 196/7, *Council, 196th Session, Decision of the Seventh Meeting.*

C-DEC 197/2, *Council, 197th Session, Decision of the Second Meeting.*

C-DEC 197/4, *Council, 184th Session, Decision of the Fourth Meeting.*

C-DEC 197/6, *Council, 197th Session, Decision of the Sixth Meeting.*

C-DEC 199/13, *Council, 199th Session, Decision of the Thirteenth Meeting.*

C-DEC 200/4, *Council, 200th Session, Decision of the Fourth Meeting.*

C-DEC 201/3, *Council, 201st Session, Decision of the Third Meeting.*

f) Council Minutes

C-MIN 156/3, *Council, 156th Session, Minutes of the Third Meeting.*

C-MIN 181/21, *Council, 181st Session, Minutes of the Twenty-First Meeting.*

C-MIN 181/22, *Council, 181st Session, Minutes of the Twenty-Second Meeting.*

C-MIN 175/15, *Council, 175th Session, Minutes of the Fifteenth Meeting.*

C-MIN 176/13, *Council, 176th Session, Minutes of Thirteenth Meeting.*

C-MIN 176/14, *Council, 176th Session, Minutes of the Fourteenth Meeting.*

C-MIN 187/5, *Council, 187th Session, Minutes of the Fifth Meeting.*

C-MIN 188/6, *Council, 188th Session, Minutes of the Sixth Meeting.*

C-MIN 194/2, *Council, 194th Session, Minutes of the Second Meeting.*

C-MIN 196/7, *Council, 196th Session, Minutes of the Seventh Meeting.*

C-MIN 196/8, *Council, 196th Session, Minutes of the Eighth Meeting.*

C-MIN 197/6, *Council, 197th Session, Minutes of the Sixth Meeting.*

C-MIN 199/13, *Council, 199th Session, Minutes of the Thirteenth Meeting.*

C-MIN-200/4, *Council, 200th Session, Minutes of the Fourth Meeting.*

C-MIN 201/3, *Council, 201st Session, Minutes of the Third Meeting.*

g) ICAO Documents

ICAO Doc 7475/2, *Working Arrangements between the International Civil Aviation Organization and the World Meteorological Organization*, 2nd edn (1963).

ICAO Doc 7669-LC/139/5, *Legal Committee – Constitution – Procedure for Approval of Draft Conventions – Rules of Procedure*, 5th edn (1998).

ICAO Doc 8335, *Manual of Procedures for Operations Inspection, Certification and Continued Surveillance*, 5th edn (2010).

ICAO Doc 8632, *ICAO's Policies on Taxation in the Field of Air Transport*, 3rd edn (2000).

ICAO Doc 9082, *ICAO's Policies on Charges for Airports and Air Navigation Services*, 9th edn (2012).

ICAO Doc 9501 AN/929, *Environmental Technical Manual*, 1st edn (2010).

ICAO Doc 9562, *Airport Economics Manual*, 2nd edn (2006).

ICAO Doc 9574 AN/934, *Manual on a 300 m (1000 ft) Vertical Separation Minimum between FL 290 and FL 410 Inclusive*, 3rd edn (2012).

ICAO Doc 9585-JS/681, *Agreement on the Joint Financing of Certain Air Navigation Services in Greenland (1956) as Amended in 1982 and 2008* (2010).

ICAO Doc 9586-JS/682, *Agreement on the Joint Financing of Certain Air Navigation Services in Iceland (1956) as Amended in 1982 and 2008* (2010).

ICAO Doc 9626, *Manual on the Regulation of Air Transport*, 2nd edn (2004).

ICAO Doc 9718, *Handbook on Radio Frequency Spectrum Requirements for Civil Aviation*, 5th edn (2010).

ICAO Doc 9731, *International Aeronautical and Maritime Search and Rescue (IAMSAR) Manual. Volume I Organization and Management*, 8th edn (2010).

ICAO Doc 9734 AN/959, *Safety Oversight Manual*, 2nd edn (2006).

ICAO Doc 9735 AN/960, *Universal Safety Oversight Audit Program Continuous Monitoring Manual*, 3rd edn (2011).

ICAO Doc 9829 AN/451, *Guidance on the Balance Approach to Aircraft Noise Management*, 2nd edn (2008).

ICAO Doc 9859 AN/474, *Safety Management Manual (SMM)*, 3rd edn (2013).

ICAO Doc 9885, *Guidance on the Use of Emissions Trading for Aviation*, 1st edn (2008).

ICAO Doc 9886, *Committee on Aviation Environmental Protection, Report of the Seventh Meeting, Montréal 5-16 February 2007* (2007).

ICAO Doc 9895, *Budget for the Organization 2008-2009-2010.*

ICAO Doc 9926 LC/194, *Legal Committee - Report of the 34th Session, Montréal 9-17 September 2009* (2009).

ICAO Doc 9929, *Report of the High-Level Meeting on International Aviation and Climate Change* (2010).

ICAO Doc 9931 AN/476, *Continuous Descent Operations (CDO) Manual*, 1st edn (2010).

ICAO Doc 9935, *Report of the High-Level Safety Conference 2010* (2010).

ICAO Doc 9938, *Committee on Aviation Environmental Protection, Report of the Eighth Meeting, Montréal 1-12 February 2010* (2010).

ICAO Doc 9949, *Scoping Study of Issues Related to "Linking" Open Emissions Trading Systems Involving International Aviation*, 1st edn (2011).

ICAO Doc 9952, *Annual Report of the Council – 2010* (2010).

ICAO Doc 9955, *Budget for the Organization 2011-2013-2013* (2010).

ICAO Doc 9976, *Flight Planning and Fuel Management Manual*, 1st edn (2012).

ICAO Doc 9977 AN/489, *Manual on Civil Aviation Jet Fuel Supply*, 1st edn (2012).

ICAO Doc 9988, *Guidance on the Development of States' Action Plans on CO_2 Emissions Reductions Activities*, 1st edn. (2013).

ICAO Doc 9990, *High-Level Conference on Aviation Security* (2012).

ICAO Doc 9992 AN/494, *Manual on the Use of Performance-Based Navigation (PBN) in Airspace Design*, 1st edn (2013).

ICAO Doc 9993 AN/495, *Continuous Climb Operations (CCO) Manual*, 1st edn (2013).

ICAO Doc 9997 AN/498, *Performance-Based Navigation (PBN) Operational Approval Manual*, 1st edn (2013).

ICAO Doc 10001, *Annual Report of the Council 2012.*

ICAO Doc 10004, *2014-2016 Global Aviation Safety Plan*, 1st edn (2013).

ICAO Doc 10007, *Twelfth Air Navigation Conference* (2012).

ICAO Doc 10009, *Sixth Worldwide Air Transport Conference: Sustainability of Air Transport* (2013).

ICAO Doc 10018, *Report of the Assessment of Market-Based Measures*, 1st edn (2013).

ICAO, Doc 10012, *Committee on Aviation Environmental Protection, Report of the Ninth Meeting, Montréal 4-15 February 2013* (2013).

ICAO Doc 10030, *Budget for the Organization 2014-2015-2016* (2013).

h) Other Working and Information Papers

AT-WP/2064, *Adoption of Amendment 22 to Annex 9 – Facilitation* (Presented by the Secretary of the Air Transport Committee).

ATConf/6-WP/33, *Views on Advancing ICAO's Work on Air Transport Liberalization* (Presented by the United Arab Emirates).

ATConf/6-WP/70, *Promotion of the Convention for the Unification of Certain Rules for International Carriage by Air (Montréal Convention of 1999)* (Presented by Canada, Germany, United Arab Emirates, United States and the International Air Transport Association (IATA)).

CAEP/1-WP/97, *Committee on Aviation Environmental Protection (CAEP) First Meeting.*

CAEP/2-WP/1, *Committee on Aviation Environmental Protection (CAEP) Second Meeting.*

CAEP/2-WP/73, *Committee on Aviation Environmental Protection (CAEP) Second Meeting.*

CAEP/3-WP/86, *Committee on Aviation Environmental Protection (CAEP) Third Meeting.*

CAEP/3-WP/101, *Committee on Aviation Environmental Protection (CAEP) Third Meeting.*

CAEP/3-WP/102, *Committee on Aviation Environmental Protection (CAEP) Third Meeting.*

CAEP/4-WP/8, *Committee on Aviation Environmental Protection (CAEP) Fourth Meeting.*

CAEP/5-WP/86, *Committee on Aviation Environmental Protection (CAEP) Fifth Meeting.*

CAEP/8-WP/80, *Committee on Aviation Environmental Protection (CAEP) Eighth Meeting, Montréal, 1-12 February 2010.*

CSG-LAEC/1, *Council Special Group on Legal Aspects of Emissions Charges, Montréal, 6-9 September 2005.*

FAL/12-WP/14, *A Standard to Require the Issuance of Machine Readable Passports* (Presented by the Secretariat).

FAL/12-WP/50, *A Standard to Require the Issuance of Machine Readable Passports* (Presented by the African Civil Aviation Commission (AFCAC)).

FALP/6-WP/7, *Proposal to Amend the Health-Related SARPS of Annex 9 – Facilitation* (Presented by Secretary).

HGCC/1-WP/2, *Role of the Framework for Market-Based Measures* (MBMs) (Presented by the Secretary of the HGCC).

HGCC/1-WP/3, *Designation of Coverage of a Framework for Market-Based Measures (MBMs)* (Presented by the Secretary of the HGCC).

HGCC/1-WP/5, *Means to Accommodate the Special Circumstances and Respective Capabilities (SCRC) of States* (Presented by the Secretary of the HGCC).

HGCC/2-WP/4, *Outline of a Framework for MBMs* (Presented by the Secretary of the HGCC).

HGCC/2-WP/6, *States' Action Plans* (Presented by the Secretary of the HGCC).

HGCC/2-WP/9, *Suggested Elements of the Framework for MBMs* (Presented by the United States).

HGCC/3-WP/4, *Views of the UAE on a Framework for MBMs* (Presented by the United Arab Emirates).

HGCC/3-WP/5, *Proposed Elements of Resolution on the Basket of Measures, Including With Respect To, Inter Alia, the Development of a Market-Based Measures Framework and Other Non Market-Based Measures* (Presented by India).

HGCC/3-WP/7, *CO_2 Emissions Coverage of the Geographic Scope Options for the Framework for* MBMs (Presented by Belgium, France and the United Kingdom).

HGCC/3-IP/5, *Addressing Carbon Emissions from Aviation: Industry Views* (Presented by the Air Transport Action Group (ATAG) on behalf Airports Council International (ACI), Civil Air Navigation Services Organisation (CANSO), International Air Transport Association (IATA), International Business Aviation Council (IBAC), International Coordinating Council for Aerospace Industries Associations (ICCAIA)).

HGCC/3-IP/6, *Views of the Environmental NGO Community* (Presented by the International Coalition for Sustainable Aviation, ICSA).

HGCC/3-AIP/6, *Presentation of the International Coalition for Sustainable Aviation at the Third Meeting of the High Level Group on International Aviation and Climate Change*, 25-27 March 2013.

HLM-ENV/09-WP/19, *A Global Sectoral Approach for Aviation* (Presented by the Airports Council International (ACI), Civil Air Navigation Services Organisation (CANSO), International Air Transport Association (IATA) and International Coordinating Council of Aerospace Industries Associations (ICCAIA)).

HLM-ENV/09-WP/28, *African Position on the GIACC Programme of Action* (Presented by Nigeria on behalf of African States).

Document Presented by Belgium, Denmark, France, Germany, Italy, Slovenia, Spain, and the United Kingdom for consideration of the Council when discussing Item No 25 of the Work Programme of the Council for the 194th Session ("European Emissions Trading Scheme (ETS)") [Unpublished, archived on file with the author].

i) *Other Materials*

ICAO, PRES RK/2166.

ICAO, State Letter AN 5/17.4-09/75.

ICAO, State Letter EC 6/21-02/78.

ICAO, Cir 134-AN/94, *Control of Aircraft Engine Emission*, 1st edn (1977).

ICAO, Cir 292-AT/124, *Economic Contribution of Civil Aviation* (2006).

ICAO, Cir 303 AN/176, *Operational Opportunities to Minimize Fuel Use and Reduce Emissions* (2004).

ICAO, Cir 313, *Air Transport Outlook to the Year 2025* (2007).

ICAO, Cir 334-AN/184, *Guidelines for the Implementation of Lateral Separation Minima* (2010).

ICAO, *Proceedings of the International Civil Aviation Conference, 1 November to 7 December 1944* (Washington: The Department of State, US Government Printing Office, 1948).

ICAO, "Shared Vision on International Aviation and Climate Change Poznan, Poland, 1-10 December 2008" online: ICAO <www.icao.int/environmental-protection/Documents/STATEMENTS/Awglca4_2008_Submission_SharedVision.pdf>.

Al Hamili, Aysha. "Aviation & Climate Change: The UAE Perspective" (2010) 65:2 ICAO Journal 40.

Al Hamili, Aysha. "Can ICAO Act As a Catalyst in the Development of Alternative Fuels Among Member States? (2012) 1 ICAO Journal at 30.

Al Hamili, Aysha. "Some Reflections on ICAO's Assembly Historic Achievements" (2013) 6 ICAO Journal 14.

Dupuis, Lionel Alain. "Discours d'adieu du Représentant permanent du Canada au Conseil de l'OACI, Doyen des Membres du Conseil" (29 June 2011) [unpublished, archived on file with the author].

Eggleston, Simon. "IPCC Guidelines for Estimating National Greenhouse Gas Inventories" in ICAO, *ICAO Environmental Report 2007* (Montréal: ICAO, 2007) 116.

Hardeman, Andreas & Kalle Keldusild. "Overview of ICAO Guidance on Emissions Trading"in ICAO, *ICAO Environmental Report 2007* (Montréal: ICAO, 2007) 149.

Hupe, Jane. "Report Overview" in ICAO, *ICAO Environmental Report 2007* (Montréal: ICAO, 2007) 2.

ICAO Secretariat, "Aircraft Noise Overview" in ICAO, *ICAO Environmental Report 2007* (Montréal: ICAO, 2007) 20.

ICAO, "Global Emissions Overview" in ICAO, *ICAO Environmental Report 2007* (Montréal: ICAO, 2007) 104.

ICAO, *Report of the Twelfth Session of the Facilitation Division, Which Was Held in Cairo, 22 March to 1 April 2004* [unpublished, archived on file with the author].

ICAO, *Group on International Aviation and Climate Change (GIACC) Report* (1 June 2009) [unpublished, archived on file with the author].

ICAO, *Assessment of the Impact of Market-Based Measures (MBMs), Council Informal Briefing* (11 June 2013) [unpublished, archived on file with the author].

ICAO, *Committee on Aviation Environmental Protection – CAEP – Informal Briefing to the Council* (31 January 2014) [unpublished, archived on file with the author].

ICAO, *Council Informal Briefing, Action Plan for the Development of a Global MBM Scheme* (31 January 2014) [unpublished, archived on file with the author].

ICAO, *MBM Activities Chronology* [unpublished, archived on file with the author].

ICAO, *Electronic Filing of Differences (EFOD) System and Filing of Differences Task Force (FDTF)* [unpublished, archived on file with the author].

Runge-Metzger, Artur. "Aviation and Emissions Trading", ICAO Council Briefing (29 September 2011) [unpublished, archived on file with the author].

Steele, Paul. "IATA AGM Resolution Climate Change", Informal Briefing Delivered to the Council of ICAO" (11 June 2013) [unpublished, archived on file with the author].

j) Electronic Sources

"34th Session (Extraordinary) of the Assembly – List of Working Papers by Number", online: ICAO <http://legacy.icao.int/icao/en/assembl/a34/wpno.htm>.

"2013 ICAO Air Transport Results Confirm Robust Passenger Demand, Sluggish Cargo Market", online: ICAO <www.icao.int/Newsroom/Pages/2013-ICAO-AIR-TRANSPORT-RESULTS-CONFIRM-ROBUST-PASSENGER-DEMAND,-SLUGGISH-CARGO-MARKET.aspx>.

"Assembly 37th Session – Working Papers by Numbers", online: ICAO <www.icao.int/cgi/a37.pl?wp;LE>.

"CAEP Terms of Reference", online: ICAO <www.icao.int/environmental-protection/Documents/CAEP/Images/CAEPToR.jpg>.

"CAEP Members and Observers", online: ICAO <www.icao.int/ENVIRONMENTAL-PROTECTION/Pages/CAEP.aspx>.

"Civil Aviation: 2012 International RTK by State of Operator Certificate (AOC)", online: ICAO <www.icao.int/Meetings/a38/Documents/International%20Scheduled%20RTK%20(Annual%20Report).PDF>.

"Climate Change: Action Plans", online: ICAO <www.icao.int/environmental-protection/Pages/action-plan.aspx>.

"Composition of the High-Level Group on International Aviation and Climate Change: HGCC", online: ICAO <https://portal.icao.int/HGCC/Membership/HGCC.Composition%20-%20Members%20and%20Advisors.%20%2025%20March%202013.Revised.pdf>.

"Dramatic MBM Agreement and Solid Global Plan Endorsements Help Deliver Landmark ICAO 38th Assembly", online: ICAO <www.icao.int/Newsroom/Pages/mbm-agreement-solid-global-plan-endoresements.aspx>.

"Former Presidents of the Council", online: ICAO <www.icao.int/icao/en/biog/pres.htm>.

"Guidance Material for the Development of States' Action Plans" online: ICAO <www.icao.int/environmental-protection/Documents/GuidanceMaterial_DevelopmentActionPlans.pdf>.

"Guiding International Civil Aviation into the 21st Century" online: ICAO <www.icao.int/Documents/strategic-objectives/sap1997_en.pdf>.

"ICAO, Assembly 36th Session" online: ICAO <www.icao.int/Meetings/AMC/36th/Pages/default.aspx>.

"ICAO Strategic Objectives 2011-2012-2013", online: ICAO <http://goo.gl/uWMbvZ>.

"ICAO Strategic Objectives 2013-2016" online: ICAO <www.icao.int/about-icao/Pages/Strategic-Objectives.aspx>.

"ICAO Symposium on State Action Plans", online: ICAO <www.icao.int/Meetings/

Green/Pages/default.aspx>.

"Reservations to A37-19", online: ICAO <http://goo.gl/YcXlz>.

"Reservations to Resolution A38-18", online: ICAO <www.icao.int/Meetings/a38/Pages/resolutions.aspx>.

"Reservation of the Russian Federation to ICAO A38-18, *Consolidated Statement of Continuing ICAO Policies and Practices Related to Environmental Protection: Climate Change*", online: ICAO <www.icao.int/Meetings/a38/Documents/Resolutions/Russia_en.pdf>.

"Safety Audit Information", online: ICAO <www.icao.int/safety/Pages/USOAP-Results.aspx>.

"Strategic Objectives of ICAO for 2005-2010", online: ICAO <www.icao.int/Documents/strategic-objectives/strategic_objectives_2005_2010_en.pdf>.

"Summary Listing of Reservations to Resolution A38-18", online: ICAO <www.icao.int/Meetings/a38/Documents/Resolutions/summary_en.pdf>.

"United States Aviation Greenhouse Gas Emissions Reduction Plan", online: ICAO <www.icao.int/environmental-protection/Documents/ActionPlan/CAEP-U%20S-Climate-ActionPlan.pdf>.

Burleson, Carl. "International Aviation and Emissions Trading: A US Perspective" (Presentation delivered at the ICAO Colloquium on Aviation and Climate Change, 13 May 2010) online: ICAO <www.icao.int/Meetings/EnvironmentalColloquium/Documents/2010-Colloquium/5_Burleson_Faa.pdf>.

Kershaw, Andy & Mark Watson. "Global Sectoral Approach for International Aviation Emissions" (Presentation delivered at the ICAO Colloquium on Aviation and Climate Change, 13 May 2010) online: ICAO <www.icao.int/CLQ10/Docs/5_Kershaw-Watson.pdf>.

Locke, Maryalice. "Aviation & the Environment: Issues for Considering Global Aviation Emissions" (Presentation delivered at the ICAO Workshop on Aviation Carbon Markets, 19 June 2008) online: ICAO <www.icao.int/Meetings/EnvironmentalWorkshops/Documents/WACM-2008/5_Locke.pdf>.

Rutherford, Daniel. "The Role of Aviation Alternative Fuels in Climate Change Mitigation" (Presentation delivered at the ICAO Alternative Fuels Workshop, 10-12 February 2009, Montréal, Quebec – Canada) online: ICAO <http://legacy.icao.int/WAAF2009/Presentations/4_Rutherford.pdf>.

Scott, Joan & Lavanya Rajamani, "EU Climate Change Unilateralism: International Aviation in the European Emissions Trading Scheme" online: <www.indiaenvironmentportal.org.in/files/file/EU%20Climate%20Change%20Unilateralism.pdf>.

Third World Network, "Common but Differentiated Responsibilities under Threat", *TWN* (13 June 2012) online: TWN <www.twnside.org.sg/title2/rio+20/news_updates/TWN_update2.pdf>.

k) Press Releases

ICAO, News Release, PIO 15/83, New Release, "ICAO Council Takes Action on Korean Air Lines Incident" (19 September 1983) online: ICAO <http://legacy.icao.int/icao/en/nr/1983/pio198315_e.pdf>.

ICAO, News Release, PIO 04/02, "ICAO Council Adopts Stronger in Flight Security Standards" (21 March 2002) online: ICAO <http://legacy.icao.int/icao/en/nr/2002/pio 200204_e.pdf>.

ICAO, 14/10, "ICAO Member States Agree to Historic Agreement on Aviation and Climate Change" (8 October 2010) online: ICAO <www.icao.int/Newsroom/Pages/icao-member-states-agree-to-historic-agreement-on-aviation-and-climate-change.aspx>.

ICAO, COM 4/13, "ICAO Environmental Protection Committee Delivers Progress on Aircraft CO_2 and Noise Standards" (14 February 2013) online: ICAO <www.icao.int/News room/Pages/ICAO-environmental-protection-committee-delivers-progress-on-new-aircraft-CO2-and-noise-standards.aspx>.

ICAO, News Release, "Europe Proposes EUR 6,5 Million for State CO_2 Reduction Through ICAO" (24 September 2013) online: ICAO <www.icao.int/Newsroom/News%20Doc%20 2013/COM.34.13.ICAO.EU.FINAL.EN.pdf>.

ICAO, "Regional Passenger Traffic and Capacity Growth, Market Shares and Load Factors in 2013" (16 December 2013) online: ICAO <www.icao.int/Newsroom/Pages/2013-ICAO-

AIR-TRANSPORT-RESULTS-CONFIRM-ROBUST-PASSENGER-DEMAND,-SLUG-GISH-CARGO-MARKET.aspx>.

International Law Commission Documents

Report of the International Law Commission on the Work of Its Twenty-eight Session 3 May-23 July 1976, (UN Doc.A/31/10) in *Yearbook of the International Law Commission 1976*, vol 2, part 2 (New York: UN, 1977) A/CN.4/SER.4/1976/Add 1, part 2).

Report of the International Law Commission on the Work of Its Forty-Eighth Session, 6 May-26 July 1996, (UN Doc A/51/10) in *Yearbook of the International Law Commission 1996*, vol 2, part 2 (New York/Geneva: UN, 1998) (A/CN.4/SER.A/1996/Add l, part 2).

Hafner, Gerhard, "Risks Ensuing from Fragmentation of International Law", in *Report of the International Law Commission to the General Assembly on the Work of Its Fifty-Second Session*, (UN Doc A/55/10) in *Yearbook of the International Law Commission 2000*, vol 2, part 2 (New York/Geneva: UN, 2006) (A/CN.4/SER.A/2000/Add 1 (Part 2)/Rev 1) 144.

"Draft Articles on Responsibility of States for Internationally Wrongful Acts", in *Report of the International Law Commission*, Fifty-Third Session, 23 April-1 May and 2 July-10 August 2001, General Assembly Official Records, Fifty-Sixth Session, Supplement No 10, UN Doc A/56/10.

"Draft Report of the Study Group of the International Law Commission, Fragmentation of International Law: Difficulties Arising from the Diversification and Expansion of International Law, Fifty-Eighth Session, 1 May-9 June and 3 July-11 August 2006", UN Doc A/CN.4/L.682 (13 Apr 2006, as corrected UN Doc A/CN.4/L.682/Corr.1 (11 Aug. 2006) (Finalized by Martti Koskenniemi).

"Report of the Study Group of the International Law Commission, Fragmentation of International Law: Difficulties Arising from the Diversification and Expansion of International Law, Fifty-Eighth Session, 1 May-9 June and 3 July-11 August 2006", UN Doc A/CN.4/L.702 (18 July 2006).

INTERGOVERNMENTAL STATEMENTS AND DECLARATIONS

Rio Declaration on Environment and Development, 3-14 June 1992, UN Doc A/CONF.151/26 vol 1.

Joint Statement between the Civil Aviation Administration of the People's Republic of China and the Ministry of Transport of the Russian Federation on the European Union's Inclusion of Aviation into the European Union Emission Trading Scheme, 27 July 2011.

LACAC, Cartagena Declaration, 28 July 2011, online: LACAC <http://clacsec.lima.icao.int/Decisiones2012/Cap05/DEC/DecCartagena2011.pdf>.

New Delhi Joint Declaration of 30 September 2011, online: zoek <https://zoek.officielebekendmakingen.nl/blg-133280.pdf>

Joint Declaration of the Moscow Meeting on the Inclusion of International Civil Aviation in the EU ETS, 20-21 February 2012.

OTHER INTERGOVERNMENTAL MATERIALS

Report of the Secretary-General's High-Level Advisory Group on Climate Change Financing (5 November 2010) online: UN <www.un.org/wcm/webdav/site/climatechange/shared/Documents/AGF_reports/AGF%20Report.pdf>.

International Monetary Fund and the World Bank, *Market-Based Instruments for International Aviation and Shipping as a Source of Climate Finance, Background Paper to the G20 on the Mobilizing Sources of Climate Finance* (November 2011) online: IMF <www.imf.org/external/np/g20/pdf/110411a.pdf>.

ECAC, "Member States", online: ECAC <www.ecac-ceac.org//about_ecac/ecac_member_states>.

UNFCCC, *Report of the Conference of the Parties on Its Fifteenth Session, Held in Copenhagen from 7 to 19 December 2009*, FCCC/CP/2009/11/Add 1, online: *UNFCCC* <http://unfccc.int/resource/docs/2009/cop15/eng/11a01.pdf>.

UNFCCC, *Proposals by India for Inclusion of Additional Agenda Items in the Provisional Agenda of the Seventeenth Session of the Conference of the Parties*, FCCC/CP/2011/INF.

2/Add 1, online: *UNFCCC* <http://unfccc.int/resource/docs/2011/cop17/eng/inf02a01.pdf>.

UNFCCC, "List of Annex 1 Parties to the Convention", online: *UNFCCC* <https://unfccc.int/parties_and_observers/parties/annex_i/items/2774.php>.

UNFCCC, "Nama Registry", online: *UNFCCC* <https://unfccc.int/cooperation_support/nama/items/7476.php>.

UNFCCC, "Status of Ratification of the Kyoto Protocol", online: *UNFCCC* <https://unfccc.int/kyoto_protocol/status_of_ratification/items/2613.php>.

UNFCCC, "International Transaction Log", online: *UNFCCC* <http://unfccc.int/kyoto_protocol/registry_systems/itl/items/4065.php>.

UNFCCC, "Status of Ratifications of the Convention", online: *UNFCCC* <http://unfccc.int/essential_background/convention/status_of_ratification/items/2631.php>.

UNFCCC, "Parties to the Convention and Observer States", online: *UNFCCC* <http://unfccc.int/parties_and_observers/parties/items/2352.php>.

World Bank, "Income per Capita", *World Bank* online: <http://data.worldbank.org/indicator/NY.GDP.PCAP.CD>.

WHO, "Global Alert and Response (GAR) No Rationale for Travel Restrictions", *WHO* online: <www.who.int/csr/disease/swineflu/guidance/public_health/travel_advice/en>.

European Union Materials

a) Directives

EC, *Directive 92/81 of 19 October 1992 on the Harmonization of the Structures of Excise Duties on Mineral Oils*, [1992] OJ, L 316/12.

EC, *Directive 2002/30/EC of the European Parliament and of the Council of 26 March 2002 on the Establishment of Rules and Procedures With Regard to the Introduction of Noise-Related Operating Restrictions at Community Airports*, [2002] OJ, L 85/40.

EC, *Parliament and Council Directive 2003/87 of 13 October 2003 on Establishing a Scheme for Greenhouse Gas Emission Allowance Trading Within the Community and Amending Council Directive 96/61/EC*, [2003] OJ, L 275/32.

EC, *Council Directive 2003/96/EC of 27 October 2003 Restructuring the Community Framework for the Taxation of Energy Products and Electricity*, [2003] OJ, L 283/51.

EC, *Parliament and Council Directive 2004/101 of 27 October 2004 Amending Directive 2003/87/EC Establishing a Scheme for Greenhouse Gas Emission Allowance Trading Within the Community, in Respect of the Kyoto Protocol's Project Mechanisms*, [2004] OJ, L 338/18.

EC, *Directive 2005/35/EC of the European Parliament and of the Council of 7 September 2005 on Ship-Source Pollution and on the Introduction of Penalties for Infringements*, [2005] OJ, L 255/11.

EC, *Parliament and Council Directive 2008/101 of 19 November 2008 Amending Directive 2003/87/EC so as to Include Aviation Activities in the Scheme for Greenhouse Gas Emission Allowance Trading Within the Community*, [2008] OJ, L 8/3.

EC, *Parliament and Council Directive 2009/29 of 23 April 2009 Amending Directive 2003/87/EC so as to Improve and Extend the Greenhouse Gas Emission Allowance Trading Scheme of the Community*, [2009] OJ, L 140/63.

b) Regulations

EC, *Council Regulation (EC) 3094/86 of 7 October 1986 Laying Down Certain Technical Measures for the Conservation of Fishery Resources*, [1986] OJ, L 288/1.

EC, *Council Regulation (EC) 2271/96 of 22 November 1996 Protecting Against the Effects of the Extra-Territorial Application of Legislation Adopted by a Third Country, and Actions Based Thereon or Resulting Therefrom*, [1996] OJ, L 309/1.

EC, *Council Regulation (EC) 925/1999 of 29 April 1999 on the Registration and Operation within the Community of Certain Types of Civil Subsonic Jet Aeroplanes Which Have Been Modified and Recertificated as Meeting the Standards of Volume I, Part II, Chapter 3 of Annex 16 to the Convention on International Civil Aviation, Third Edition* (July 1993), [1999] OJ, L 115/1.

EC, *Regulation 417/2002 of the European Parliament and of the Council of 18 February 2002 on the Accelerated Phasing-in of Double Hull or Equivalent Design Requirements for Single Hull Oil Tankers and Repealing Council Regulation (EC) 2978/94*, [2002] OJ, L 64/1.

EC, *Decision 1600/2002/EC of the European Parliament and of the Council of 22 July 2002 laying down the Sixth Community Environment Action Programme*, [2002] OJ, L 242/1.

EC, *Regulation (EC) 1726/2003 of the European Parliament and of the Council of 22 July 2002 Amending Regulation (EC) 417/2002 on the Accelerated Phasing-in of Double-Hull or Equivalent Design Requirements for Single-Hull Oil Tankers*, [2003] OJ, L 249/1.

EC, *Commission Decision of 18 July 2007 Establishing Guidelines for the Monitoring and Reporting of Greenhouse Gas Emissions Pursuant to Directive 2003/87/EC of the European Parliament and of the Council*, [2007] OJ, L 229/1 at 229/30.

EC, *Commission Decision of 16 April 2009 Amending Decision 2007/589/EC as Regards to Monitoring and Reporting Guidelines for Emission and Tonne-Kilometre Data from Aviation Activities*, [2009] OJ, L 103/10.

c) Other Materials

EC, *Communication from the Commission to the Council, the European Parliament, the European Economic and Social Committee and the Committee of the Regions Reducing the Climate Change Impact of Aviation*, COM (2005) 459 (Brussels, 27 September 2005).

EC, *Proposal for a Directive of the European Parliament and the Council Amending Directive 2003/87/EC so as to Include Aviation Activities in the Scheme for Greenhouse Gas Emission Allowance Trading Within the Community*, COM (2006) 818 (Brussels, 20 December 2012).

EC, *Proposal for a Directive of the European Parliament and of the Council Amending Directive 2003/87/EC Establishing a Scheme for Greenhouse Gas Emission Allowance Trading Within the Community, in View of the Implementation by 2020 of an International Agreement Applying a Single Global Market-Based Measure to International Aviation Emissions*, COM (2013) 722 final (Brussels, 16 October 2013).

EC, *Commission Staff Working Document, Scaling Up International Climate Finance After 2012*, SEC (2011) 487 final (Brussels, 8 April 2011).

EC, European Commission, "Our Director General" online: European Commission <http://ec.europa.eu/clima/about-us/director/index_en.htm>.

EC, European Commission, "Reducing Emissions from Aviation" online: European Commission <http://ec.europa.eu/clima/policies/transport/aviation>.

d) Press Releases

EC, "Breakthrough in Climate Change Talks at UN Aviation Body", MEMO/10/482 (Brussels, 9 October 2010).

EC, "Stopping the Clock of ETS and Aviation Emissions Following Last Week's International Civil Aviation Organization (ICAO) Council", MEMO/12/854 (Brussels, 12 November 2012).

EC, "Commission proposal for European Regional Airspace Approach for the EU Emissions Trading for Aviation: Frequently Asked Questions", MEMO/13/905 (Brussels, 16 October 2013).

EC, European Commission Press Release, IP/12/477, "Second year of emissions reporting from aircraft operators with very high level of compliance", *Europa* (15 May 2012) online: <http://europa.eu/rapid/pressReleasesAction.do?reference=IP/12/477&format=HTML&aged=0&language=EN&guiLanguage=en>.

EC, Commission Proposes Applying EU ETS to European Regional Airspace from 1 January 2014 (16 November 2013) online: European Commission <http://ec.europa.eu/clima/news/articles/news_2013101601_en.htm>.

Government Documents

a) Australia

Australian Custom and Border Protection Service, "Passenger Movement Charge" online: Australian Government <www.customs.gov.au/site/page6068.asp>.

b) *United Kingdom*

UK House of Lords, "Including the Aviation Sector in the European Union Emissions Trading Scheme", European Union Committee, 21st Report of Session 2005-2006, online: UK Parliament <www.publications.parliament.uk/pa/ld200506/ldselect/ldeucom/107/107.pdf>.

Elana Ares, "EU ETS and Aviation SN/SC/5533", *House of Commons Library* (23 May 2012) online: UK Parliament <www.parliament.uk/briefing-papers/SN05533>.

c) *United States*

i) **Committee Hearings**

Jos Delbeke, Director-General, DG Climate Action, European Commission, "Written Testimony for Senate Committee on Commerce, Science, and Transportation Hearing on the European Union's Emissions Trading System" (6 June 2012) online: Europa <http://ec.europa.eu/clima/news/docs/testimony_senate_hearing.pdf>.

Ray Lahood, "Statement of the Honorable Ray Lahood, Secretary of Transportation before the Senate Committee on Commerce, Science and Transportation of the United States Senate" (6 June 2012).

Annie Petsonk, International Counsel Environmental Defense Fund (EDF), "Summary of the testimony of Annie Petsonk before the Senate Committee on Commerce, Science, and Transportation United States Senate" (6 June 2012) online: EDF <www.edf.org/sites/default/files/EDF-Petsonk-Senate-Testimony-EU-ETS-Aviation-060612.pdf>.

Nancy Young, "Statement of Nancy Young Vice President of Environmental Affairs Airlines for America (A4A) before the Senate Committee on Commerce, Science and Transportation" (6 June 2012) online: A4A <www.airlines.org/Pages/A4A-Oral-Testimony-of-Nancy-Young,-VP-for-Environmental-Affairs.aspx>.

Ed Bolen, President and CEO, National Business Aviation Association, "Statement before the Senate Committee on Commerce, Science and Transportation U.S. Senate" (6 June 2012) online: NBAA <www.nbaa.org/advocacy/testimony/20120606-bolen-testimony-senate-eu-ets.pdf>.

Reports, Statements & Documents George Bush, "Statement on Signing the Bill Amending the Clean Air Act", *The American Presidency Project* (15 November 1990) online: UC Santa Barbara <www.presidency.ucsb.edu/ws/index.php?pid=19039>.

Bureau of Transportation Statistics, "Airline Fuel Cost and Consumption", *US Department of Transportation* online: Bureau of Transportation Statistics <www.transtats.bts.gov/fuel.asp?pn=1>.

Environmental Protection Agency, "The Clean Air Act Amendments of 1990", *Environmental Protection Agency* online: <www.epa.gov/air/caa/caaa_overview.html>.

Memorial of the United States of America, *Disagreement Arising under the Convention on International civil Aviation Done at Chicago on December 7, 1944* (14 March 2000).

Kathleen S Swendiman & Nancy Lee Jones, "The 2009 Influenza Pandemic: Selected Legal Issues", *Congressional Research Service* (15 June 2009) online: University of North Texas <http://digital.library.unt.edu/ark:/67531/metadc26152/m1/1/high_res_d/R40560_2009Jun15.pdf>.

Jane A Leggett, Bart Elias & Daniel T Shedd, "Aviation and the European Union's Emission Trading Scheme", *Congressional Research Service* (15 May 2012) online: FAS <www.fas.org/sgp/crs/row/R42392.pdf>.

US State Department, "Ninth Meeting of the EU-US Joint Committee Record of Meeting of 22 June 2011", *US Department of State* (22 June 2011) online: <www.state.gov/e/eb/rls/othr/ata/e/eu/192088.htm>.

US Department of State, "Current Model Open Skies Agreement Text", *US Department of State* (12 January 2012) online: <www.state.gov/e/eb/rls/othr/ata/114866.htm>.

Meeting on International Aviation Emissions, Chair's Summary, 31 July-1 August 2012, Washington D.C. [unpublished, archived on file with the author].

Summary of the Discussions of the Meeting That Took Place in Washington, D.C. on 31 July through 1 August 2012 [unpublished, archived on file with the author].

The White House, "Inaugural Address by President Barack Obama", *The White House* (21 January 2013) online: <www.whitehouse.gov/the-press-office/2013/01/21/inaugural-address-president-barack-obama>.

The White House, *The President's Climate Action Plan*, *The White House* (June 2013) online: <www.whitehouse.gov/sites/default/files/image/president27sclimateactionplan.pdf>.

TSA, "Secure Flight Program", *TSA* online: <www.tsa.gov/stakeholders/secure-flight-program>.

Federal Aviation Administration (FAA)

Security Program of Foreign Air Carriers, 63 Fed Reg 64764 (1998).

FAA-98-4758-4 (Comments by IATA), 24 May 1999.

FAA-98-4758-5 (Comments by Vancouver International Airport).

FAA-98-4758-14 (Comments by the French Embassy in Washington).

FAA-98-4758-13 (Comments by the British Embassy in Washington).

FAA, *International Aviation Safety Assessments (IASA) Program*, *FAA* online: <www.faa.gov/about/initiatives/iasa>.

FAA, Press Release, "FAA Sets New Standards for Cockpit Doors", *FAA* (11 January 2002) online: <www.faa.gov/news/press_releases/news_story.cfm?newsId=5470>.

Secondary Sources: Monographs

Ali, Paul AU & Kanako Yano. *Eco-Finance: The Legal Desgin and Regulation of Market-Based Environmental Instruments* (The Hague: Kluwer Law International, 2004).

Arnell, Paul. *Law across Borders: The Extraterritorial Application of United Kingdom Law* (London: Routledge, 2012).

Barlow, Patricia M. *Aviation Antitrust: The Extraterritorial Application of the United States Antitrust Laws and International Air Transportation* (Boston: Kluwer Law & Taxation Publishers, 1988).

Barrett, Scott. *Environment & Statecraft: The Strategy of Environmental Treaty-Making* (Oxford: Oxford University Press, 2003).

Barrett, Scott. *Why Cooperate? The Incentive to Supply Global Public Goods* (Oxford: Oxford University Press, 2007).

Bartels, Lorand. *The Inclusion of Aviation in the EU ETS: WTO Law Considerations*, Trade and Sustainable Energy Series, Issue Paper No 6 (Geneva: International Centre for Trade and Sustainable Development Switzerland, 2012).

Becker, Howard. *Outsiders: Studies in the Sociology of Deviance* (New York: The Free Press of Glencoe, 1963).

Bell, Stuart & Donald McGillivray. *Environmental Law*, 6th edn (London: Oxford, 2006).

Bodansky, Daniel. *The Art and Craft of International Environmental Law* (Cambridge: Harvard University Press, 2010).

Brownlie, Ian. *Principles of Public International Law*, 5th edn (Oxford: Oxford University Press, 1998).

Brunnée, Jutta & Stephen J Toope. *Legitimacy and Legality in International Law* (Cambridge: Cambridge University Press, 2010).

Buergenthal, Thomas. *The Law-Making in the International Civil Aviation Organization* (Syracuse: Syracuse University Press, 1969).

Buffard, Isabelle & James Crawford, Alain Pellet & Stephan Wittich, eds, *International Law between Universalism and Fragmentation* (Leiden: Martinus Nijhoff Publishers, 2008).

Chen, Kuan-Wei. *The Legality of the Use of Space Weapons: Perspectives from Environmental Law* (LLM Thesis, McGill University Institute of Air and Space Law, 2012) [unpublished].

Chen, Kuan-Wei. *In Search of the International Community (of States)* (LLM Thesis, Leiden University, 2008) [unpublished].

Cheng, Bin. *The Law of International Air Transport* (London: Stevens & Sons Limited, 1962).

Chichilnisky, Graciela & Kristen A Sheeran. *Saving Kyoto: An Insider's Guide to the Kyoto Protocol, How It Works, Why It Matters and What It Means for the Future* (London: New Holland, 2009).

Cryer, Robert, Håkan Friman, Darryl Robinson & Elizabeth Wilmshurst. *An Introduction to International Criminal Law and Procedure*, 2nd edn (Cambridge: Cambridge University Press, 2010).

Dabbah, Maher M. *The Internationalisation of Antitrust Policy* (Cambridge: Cambridge University Press, 2003).

Dales, JH. *Pollution, Property & Prices* (Toronto: University of Toronto Press, 1968).

Deatherage, Scott D. *Carbon Trading: Law and Practice* (Oxford: Oxford University Press, 2011).

Dempsey, Paul Stephen. *Public International Air Law* (Montréal: Institute and Centre for Research in Air & Space Law, 2008).

Dempsey, Paul S & Michael Milde. *International Air Carrier Liability: The Montréal Convention of 1999* (Montréal: Centre for Research in Air & Space Law McGill University, 2005).

Dessler, Andrew E & Edward A Parson. *The Science and Politics of Global Climate Change: A Guide to the Debate*, 2nd edn (Cambridge: Cambridge University Press, 2009).

Diederiks-Verschoor, IH. Ph. *An Introduction to Air Law*, 8th edn (The Netherlands: Kluwer Law International, 2006).

Dupuy, Pierre-Marie & Luisa Vierucci, eds. *NGOs in International Law: Efficiency in Flexibility?* (Cheltenham: Edward Elgar, 2008).

Ellickson, Robert C. *Order without Law: How Neighbors Settle Disputes* (Cambridge: Harvard University Press, 1991).

Ellis, Jaye. *Soft Law as Topos: The Role of Principles of Soft Law in the Development of International Environmental Law* (DCL Thesis, McGill University Institute of Comparative Law, 2001) [unpublished].

Faure, Michael, John Vervaele, & Albert Weale, eds, *Environmental Standards in the European Union in an Interdisciplinary Framework* (Antwerpen: Maklu, 1994).

Finnemore, Martha. *National Interests in International Society* (Ithaca: Cornell University Press, 1996).

Fitz Roy, Felix R & Elissaios Papyrakis. *An Introduction to Climate Change Economics and Policy* (London: Earthscan, 2010).

Flohr, Annegret, Lothar Rieth, Sandra Schwindenhammer & Klaus Dieter. Wolf, *The Role of Business in Global Governance: Corporations as Norm-Entrepreneurs* (New York: Palgrave Macmillan, 2010).

Fossungu, Peter Ateh-Afac. *A Critique of the Powers and Duties of the Assembly of the International Civil Aviation Organization* (LLM Thesis, McGill University Institute of Air & Space Law, 1996) [unpublished].

Friedrich, Jürgen. *Compliance with International Law: The Kyoto Protocol's Compliance Mechanism as an Effective Tool to Promote Compliance* (LLM Thesis, McGill University Institute of Comparative Law, 2003) [unpublished].

Fuller, Lon L. *The Morality of Law* (New Haven: Yale University Press, 1964).

Goldsmith, Jack L & Eric Posner. *Limits of International Law* (Oxford: Oxford University Press, 2005).

Gondek, Micahel. *The Reach of Human Rights in a Globalizing World: Extraterritorial Application of Human Rights Treaties* (Oxford: Intersentia, 2009).

Gratjios, George A. *Airport Noise Pollution – Legal Aspects* (LLM Thesis, McGill University Institute of Air & Space Law, 1990) [unpublished].

Guzman, Andrew T. *How International Law Works: A Rational Choice Theory* (Oxford: Oxford University Press, 2008).

Hans, Jan H. *European Environmental Law* (Oxford: Europa Law Publishing, 2000).

Hart, HLA. *The Concept of Law* (Oxford: The Clarendon Press, 1967).

Huang, Jiefang. *Aviation Safety and ICAO* (The Hague: Kluwer Law International, 2009).

Hunt, Colin AG. *Carbon Sinks and Climate Change: Forests in the Fight against Global Warming* (Cheltenham: Edward Elgar, 2009).

IEA. *World Energy Outlook 2006*, (Paris: OECD/IEA, 2006).

Jackson, Tim. *Efficiency without Tears: No-Regrets Energy Policy to Combat Climate Change* (London: Friends of the Earth, 1992).

Kågeson, Par. *Getting the Prices Right: A European Scheme for Making Transport Pay Its True Costs* (Stockholm: European Federation for Transport and Environment, 1993).

Kerr, Suzi, ed. *Global Emissions Trading: Key Issues for Industrialized Countries* (Northampton: Edward Elgar, 2000).

Klabbers, Jan. *An Introduction to International Institutional Law* (Cambridge: Cambridge University Press, 2002).

Kotaite, Assad. *My Memoirs: 50 Years of International Diplomacy and Conciliation in Aviation* (Montréal: ICAO, 2013).

Krämer, Ludwig. *EC Environmental Law* (London: Sweet & Maxwell, 2007).

Kratochwil, Friedrich V. *Rules, Norms, and Decisions: On the Conditions of Practical and Legal Reasoning in International Relations and Domestic Affairs* (Cambridge: Cambridge University Press, 1989).

Latif, Mojib. *Climate Change: The Point of No Return* (London: Haus Publishing, 2007).

Lavranos, Nikolaos. *Jurisdictional Competition: Selected Cases in International and European Law* (Groningen: Europa Law Publishing, 2009).

Lyall, Francis & Paul B Larsen. *Space Law: A Treatise* (Burlington: Ashgate Publishing Company, 2009).

Lowe, Vaughan A, ed. *Extraterritorial Jurisdiction: An Annotated Collection of Legal Materials* (Cambridge, Grotius Publications, 1983).

Macedo, Stephen, Project Chair. *The Princeton Principles on Universal Jurisdiction Princeton Project on Universal Jurisdiction* (Princeton: Program in Law and Public Affairs, Princeton University, 2001).

Macrory, Richard, Ian Havercroft & Ray Purdy, eds. *Principles of European Environmental Law: Proceedings of the Avosetta Group of European Environmental Lawyers* (Groningen: Europa Law Publishing, 2004).

Mann, FA. *Further Studies in International Law* (Oxford: Claredon Press, 1990).

Mendes de Leon, Pablo. *Cabotage in Air Transport Regulation* (Leiden: M Nijhoff, 1992).

Metz, Bert et al. *Climate Change 2007: Mitigation. Contribution of Working Group III to the Fourth Assessment Report* (Cambridge: Cambridge University Press, 2007).

Milanovic, Marko. *Extraterritorial Application of Human Rights Treaties: Law, Principles, and Policy* (Oxford: Oxford University Press, 2011).

Milde, Michael. *International Air Law and ICAO*, 2nd edn (The Hague: Eleven International Publishing, 2008).

Nelissen, Dagmar & Jasper Faber. *Cost and Benefits of Stopping the Clock: How Airlines Could Profit from Changes in the EU ETS* (Delft: CE Delft, 2012).

Nyampong, Yaw Otu Mankata. *Insuring the Air Transport Industry Against Aviation War and Terrorism Risks and Allied Perils: Issues and Options in a Post-September 11, 2001 Environment* (London: Springer, 2013).

OECD. *The Polluter Pays Principle: Definition Analysis Implementation* (Paris: OECD, 1975).

OECD. *Transport Outlook 2008* (Paris: OECD, 2008).

Paskal, Cleo. *Global Warring: How Environmental Economic and Political Crises Will Redraw the World Map* (Toronto: Key Porter Books Limited, 2010).

Pauwelyn, Joost. *Conflict of Norms in Public International Law. How WTO Law Relates to Other Rules of International Law* (Cambridge: Cambridge University Press, 2003).

Penner, Joyce E et al, eds. *Aviation and the Global Atmosphere.* Published for the Intergovernmental Panel on Climate Change (Cambridge: Cambridge University Press, 1999)

Pigou, Arthur Cecil. *The Economics of Welfare*, 4th edn (London: Macmillan and Co, 1932).

Posner, Eric A. *Law and Social Norms* (Cambridge: Harvard University Press, 2000).

Posner, Eric A & David Weisbach. *Climate Change Justice* (Princeton: Princeton University Press, 2010).

Potts, Jason. *The Legality of PPMs under the GATT: Challenges and Opportunities for Sustainable Trade* (Winnipeg: International Institute for Sustainable Development, 2008).

Price, Gareth. *The EU ETS and Unilateralism within International Air Transport* (LLM Thesis, McGill University Institute of Air & Space Law, 2009) [unpublished].

Ragazzi, Maurizio. *The Concept of International Obligations Erga Omnes* (Oxford: Claredon Press, 1997).

Ralf, Antes, Bernd Hansjürgens & Peter Letmathe, eds. *Introduction in Ralf Antes Bernd Hansjürgens & Peter Letmathe, Emissions Trading and Business* (Heidelberg: Physica-Verlag, 2006).

Ripinski, Sergey & Peter van den Bossche. *NGO Involvement in International Organizations* (London: British Institute of International and Comparative Law, 2007).

Rosenthal, Douglas E & William M Knighton. *National Law and International Commerce: The Problem of Extraterritoriality* (London: Routledge & Kegan Paul, 1982).

Ryngaert, Cedric. *Jurisdiction in International Law* (Oxford: Oxford University Press, 2009).

Segerlund, Lisbeth. *Making Corporate Social Responsibility a Global Concern: Norm Construction in a Globalizing World* (Burlington: Ashgate, 2010).

Shaw, Malcom. N. *International Law*, 3rd edn (Cambridge: Cambridge University Press, 1991).

Skogly, Sigrun I. *Beyond National Borders: States' Human Rights Obligations in International Cooperation* (Oxford: Intersentia, 2006).

Slomanson, William R. *Fundamental Perspectives on International Law*, 4th edn (United States, Thomson West: 2003).

Soltau, Friedrich. *Fairness in International Climate Change Law and Policy* (Cambridge: Cambridge University Press, 2009).

Speth, James Gustave. *The Bridget at the Edge of the World: Capitalism, the Environment, and Crossing from Crisis to Sustainability* (New Heaven: Yale University Press, 2008).

Tanaka, Yumiko. *The World Trade Organization and Dispute Over Extraterritorial Application:The Effectiveness and Function of the World Trade Organization Dispute Settlement Body in International Law* (LLM Thesis, McGill University Institute of Comparative Law, 2001) [unpublished].

Waitz, Ian et al. *Aviation & Environment: A National Vision Statement, Framework for Goals and Recommended Actions* (Cambridge: Massachusetts Institute of Technology, 2004).

Weber, Ludwig. *International Civil Aviation Organization: An Introduction* (The Hague: Kluwer Law International, 2007).

Woolley, David, John Pugh-Smith, Richard Langham & William Upton, eds. *Environmental Law* (Oxford: Oxford University Press, 2000).

Xue, Dong. *A General Study of the Extraterritoriality of Criminal Forfeiture Law: Canada and China* (MA Thesis, Simon Fraser University School of Criminology, 1999) [unpublished].

Zerk, Jennifer. A. *Extraterritorial Jurisdiction: Lessons for the Business and Human Rights Sphere from Six Regulatory Areas* (Cambridge: John F Kennedy School of Government, Harvard University, 2010).

Secondary Sources: Articles

Aakre, Stine & Jon Hovi. "Emission Trading: Participation Enforcement Determines the Need for Compliance Enforcement" (2010) 11 European Union Politics 427.

Abate, Randall S. "Dawn of a New Era in the Extraterritorial Application of U.S. Environmental Statutes: A Proposal for an Integrated Judicial Standard Based on the Continuum of Context" (2006) 31 Colum J Envtl L 87.

Abbott, Kenneth W. "Toward a Richer Institutionalism for International Law and Policy" (2005) 1 J Int'l L & Int'l Rel 9.

Abbott, Kenneth W & Duncan Snidal."Values and Interests: International Legalization in the Fight against Corruption" (2002) 31 J Legal Stud 141.

Abi-Saab, Georges. "Fragmentation or Unification: Some Concluding Remarks" (1999) 31 NYU J Int'l L & Pol'y 919.

Abramovsky, Abraham. "Extraterritorial Abductions: America's 'Catch and Snatch' Policy Run Amok" (1991) 31:2 Va J Int'l L 151.

Addy, Naa Adoley. "Aviation Safety & Security in West Africa: Legal and Regulatory Issues Inherent in Aerospace Activities – Focus on Nigeria" (2008) 33 Ann Air & Sp L 1.

Adler, Emanuel. "Constructivism in International Relations: Sources, Contributions, and Debates" in Walter Emmanuel Carlsnaes & Beth A Simmons, eds, *Handbook of International Relations* (London: Sage Publications, 2013) 112.

Adut, Ari. "Scandal as Norm Entrepreneurship Strategy: Corruption and the French Investigating Magistrates" (2004) 33:5 Theory and Society 529.

Akehurst, Michael. "Jurisdiction in International Law" in Joseph Weiler & Alan T Nissel, eds, *International Law* (London: Routledge, 2011) 175.

Alberts, Ascha, Jan-André Bühne & Heiko Peters. "Will the EU ETS Instigate Airline Network Reconfiguration?" (2009) 15 J Air Transport Management 1.

Ambos, Kai. "Accountability for the Torture Memo: Prosecuting Guantanamo in Europe: Can and Shall the Masterminds of the 'Torture Memos' Be Held Criminally Responsible on the Basis of Universal Jurisdiction" (2009) 42 Case W Res J Int'l L 405.

Anderson, Belina. "Unilateral Trade Measures and Environmental Protection Policy" (1993) 66 Temp L Rev 751.

Andrew, Jane, Mary A Kaidonis & Brian Andrew. "Carbon Tax: Challenging Neoliberal Solutions to Climate Change" (2010) 21 Critical Perspectives on Accounting 611.

Andrews, Mark J et al. "International Transportation Law" (2010) 44 Int'l L 379.

Andrus, Katherine B. "Beyond Aircraft Emissions: The European Court of Justice's Decision May Have Far-Reaching Implications" (2012) 24:4 Air & Space L 1.

Anger, Annela."Including Aviation in the European Emissions Trading Scheme: Impacts on the Industry, CO_2 Emissions and Macroeconomic Activity in the EU" (2010) 16 Journal of Air Transport Management 100.

Anger, Annela & Jonathan Köhler. "Including Aviation Emissions in the EU ETS: Much ado About Nothing?" (2010) 17 Transport Policy 38.

Bae, Jon. "Review of the Dispute Settlement Mechanism under the International Civil Aviation Organization: Contradiction of Political Body Adjudication" (2012) Journal of International Dispute Settlement 1.

Bales, Jennifer S. "Transnational Responsibility and Recourse for Ozone Depletion" (1996) 19 BC Int'l & Comp L Rev 259.

Bang, Guri & Miranda A Schreurs. "A Green New Deal: Framing US Climate Leadership" in Rüdiger KW Wurzel & James Connelly, eds, *The European Union as a Leader in International Climate Change Politics* (London: Routledge, 2011) at 235.

Bankobeza, Gilbert. "Compliance Regime of the Montréal Protocol" in Donald Kaniaru, ed, *The Montréal Protocol: Celebrating 20 Years of Environmental Progress* (London: Cameron May Ltd, 2007) 75.

Barlik, Martin. "The Extension of the European Union's Emission Trading Scheme to Aviation Activities" (2009) 34 Ann Air & Sp L 151.

Barnes, Pamela M. "The Role of the Commission of the European Union" in Rüdiger K W Wurzel & James Connelly, eds, *The European Union as a Leader in International Climate Change Politics* (London: Routledge, 2011) 41.

Barrett, Scott. "Climate Treaties and the Imperative of Enforcement" (2008) 24 Oxford Rev Econ Pol'y 239.

Barton, Jane. "Including Aviation in the EU Emissions Trading Scheme: Prepare for Take-Off" (2008) 5:2 JEEPL 183.

Bassiouni, M Cherif. "Universal Jurisdiction for International Crimes: Historical Perspectives and Contemporary Practice" (2001) 42 Va J Int'l L 81.

Bell, Barbara A. "The Extraterritorial Application of United States Antitrust Law and International Aviation: A Comity of Errors" (1988) 54 J Air L & Com 533.

Bender, Philip. "A State of Necessity: IUU Fishing in the CCAMLR Zone" (2008) 13 Ocean & Costal LJ 233.

Benedick, Richard Elliot. "The History of the Montréal Protocol" in Donald Kaniaru, ed, *The Montréal Protocol: Celebrating 20 Years of Environmental Progress* (London: Cameron May, 2007) 43.

Bennett, Allyson. "That Sinking Feeling: Stateless Ships, Universal Jurisdiction, and the Drug Trafficking Vessel Interdiction Act" (2012) 37 Yale J Int'l L 433.

Benvenisti, Eyal & George W Downs. "The Empire's New Clothes: Political Economy and the Fragmentation of International Law" (2007) 60 Stan L Rev 595.

Benwell, Richard. "Linking as Leverage: Emissions Trading and the Politics of Climate Change" (2008) 21:4 Cambridge Rev Int'l Affairs 545.

Berman, Paul Schiff. "A Pluralist Approach to International Law" (2007) 32 Yale J Int'l L 301.

Berman, Paul Schiff. "Global Legal Pluralism" (2007) 80 S Cal L Rev 1155.

Betsill, Michele M & Elisabeth Corell. "NGO Influence in International Environmental Negotiations: A Framework for Analysis" in Peter M Haas, ed, *International Environmental Governance* (Burlington: Ashgate, 2008) 453.

Beyerlin, Ulrich. "Different Types of Norms in International Environmental Law: Policies, Principle, and Rules" in Daniel Bodansky, Jutta Brunnée & Ellen Hey, eds, *The Oxford Handbook of International Environmental Law* (Oxford: Oxford University Press, 2006) 425.

Biermann, Frank, Fariborz Zelli, Philipp Pattberg & Harro van Asselt. "The Architecture of Global Climate Governance: Setting the Stage" in Biermann, Philipp Pattberg & Fariborz Zelli, eds, *Global Climate Governance Beyond 2012: Architecture, Agency and Adaptation* (Cambridge: Cambridge University Press, 2010) 15.

Bilder, Richard B. "The Role of Unilateral Action in Preventing International Environmental Injury" (1981) 14:1 Vand J Transnat'l L 51.

Birger, Jon Skjærseth & Jørgen Wettestad. "Fixing the EU Emissions Trading System? Understanding the Post-2012 Changes" (2010) 10:4 Global Environmental Politics 101.

Bisset, Mark & Georgina Crowhurst. "Is the EU's Application of Its Emissions Trading Scheme to Aviation Illegal"(2011) 23:3 Air & Space L 1.

Björkdahl, Annika. "Norms in International Relations: Some Conceptual and Methodological Reflections" (2002) 15:1 Cambridge Review of International Affairs 9.

Blakesley, Christopher L. "United States Jurisdiction over Extraterritorial Crimes" (1982) 73:3 J Crim L & Criminology 1109.

Blumenkron, Jimena. "Implications of Transparency in the International Civil Aviation Organization's Universal Safety Oversight Audit Program" (2009) 34 Ann Air Sp L 31.

Bodansky, Daniel. "What Is So Bad About Unilateral Action to Protect the Environment" (2000) 11:2 EJIL 339.

Bodansky, Daniel, Jutta Brunnée & Ellen Hey. "International Environmental Law: Mapping the Field" in Daniel Bodansky, Jutta Brunnée & Ellen Hey, eds, *The Oxford Handbook of International Environmental Law* (Oxford: Oxford University Press, 2006) 1.

Boemare, Catherine & Philippe Quirion. "Implementing Greenhouse Gas Trading in Europe: Lessons from Economic Literature and International Experiences" (2002) 43 Ecological Economics 213.

Bogojovic, Sanja. "Legalizing Environmental Leadership: A Comment on the CJEU's Ruling in C-366/10 on the Inclusion of Aviation in the EU Emissions Trading Scheme" (2012) 24:2 J Envtl L 345.

Bohm, Peter. "Making Carbon Emission Quota Agreements More Efficient: Joint Implementation Versus Quota Tradability" in Ger Klaassen & Finn R Førsund, eds, *Economic Instruments for Air Pollution Control* (Dordrecht: Kluwer Academic Publishers, 1994) 187.

Boisson de Chazournes, Laurence. "Unilateralism and Environmental Protection: Issues of Perception and Reality of Issues" (2000) 11:2 EJIL 315.

Boon, Kristen E. "Regime Conflicts and the U.N. Security Council: Applying the Law of Responsibility" (2010) 42 Geo Wash Int'l L Rev 787.

Borger, Gudo. "All Things Not Being Equal: Aviation in the EU ETS" (2012) 3 Climate Law 265.

Bortscheller, Mary J. "Equitable but Ineffective: How the Principle of Common but Differentiated Responsibilities Hobbles the Global Fight against Climate Change" (2010) Sustainable Development L & Pol'y 49.

Botchway, Francis N. "Threads in Fragments" (2012) Cardozo J Int'l & Comp L 639.

Bowen, Alex & James Rydge. "The Economics of Climate Change" in David Held, Angus Hervey & Marika Theros, eds. *The Governance of Climate Change: Science, Economics, Politics & Ethics* (Cambridge: Polity Press, 2011) 68.

Bowett, D W. "Jurisdiction: Changing Patterns of Authority over Activities and Resources" (1982) 53 Brit YB Int'lL 1.

Bows, Alice, Kevin Anderson & Anthony Footitt. "Aviation in a Low-Carbon EU" in Stefan Gossling & Paul Upham, eds, *Climate Change and Aviation: Issues, Challenges and Solutions* (London: Earthscan, 2009) 89.

Boyd, William. "Climate Change, Fragmentation, and the Challenges of Global Environmental Law: Elements of Post-Copenhagen Assemblage" (2010) 32 U Pa J Int'l L 457.

Brand, Christian & Brenda Boardman."Taming of the Few: The Unequal Distribution of Greenhouse Gas Emissions from Personal Travel in the UK" (2008) 36 Energy Policy 224.

Brand, Christina & John M Preston."60-20 Emission: The Unequal Distribution of Greenhouse Gas Emissions from Person, Non-Business Travel in the UK" (2010) 17 Transport Policy 9.

Braun, Marcel. "The Evolution of Emissions Trading in the European Union: The Role of Policy Networks, Knowledge and Policy Entrepreneurs" (2009) 34 Accounting, Organizations & Society 469.

Brilmayer, Lea. "Extraterritorial Application of American Law: A Methodological and Constitutional Appraisal" (1987) 50 Law & Contemp Probs 11.

Broderick, John. "Voluntary Carbon Offsetting for Air Travel" in Stefan Gossling & Paul Upham, eds, *Climate Change and Aviation: Issues, Challenges and Solution* (London: Earthscan, 2009) 329.

Broomhall, Bruce. "Towards the Development of an Effective System of Universal Jurisdiction for Crimes under International Law" (2001) 35 New Eng L Rev 399.

Broude, Tomer. "Principles of Normative Integration and the Allocation of International Authority: The WTO, the Vienna Convention and the Law of Treaties, and the Rio Declaration" (2008) 6 Loy U Chi Int'l L Rev 173.

Brown, Bartram S. "The Evolving Concept of Universal Jurisdiction" (2001) 35 New Eng L Rev 383.

Brown, Chester. "The Cross-Fertilization of Principles Relating to Procedure and Remedies in the Jurisprudence of International Courts and Tribunals" (2008) 30 Loy LA.Int'l & Comp L Rev 219.

Brownlie, Ian. "Problems Concerning the Unity of International Law" in *Le droit international à l'heure de sa codification : études en l'honneur de Roberto Ago* (Milan: A Giuffre, 1987) 153.

Brunee, Jutta & Stephen J Toope."International Law and Constructivism: Elements of an Interactional Theory of International Law" (2000) 39:1 Colum J Transnat L 19.

Burke-White, William W. "International Legal Pluralism" (2004) 25 Mich J Int'l L 963.

Burleson, Carl. "The EU Emissions Trading System Proposal" (2007) 21:3 Air & Space L 1, 23.

Burns, Charlotte & Neil Carter. "The European Parliament and Climate Change: From Symbolism to Heroism and Back Again" in Rüdiger K W Wurzel & James Connelly, *The European Union as a Leader in International Climate Change Politics* (London: Routledge, 2011) 58.

Bushey, Douglas & Sikima Jinnah. "Evolving Responsibility? The Principle of Common but Differentiated Responsibility in the UNFCCC" (2010) 6 Berkeley J Int'l L Publicists 1.

Cabranes, Jose A. "Our Imperial Criminal Procedure: Problems in the Extraterritorial Application of U.S. Constitutional Law" (2009) 118 Yale LJ 1660.

Calamita, N Jansen. "Countermeasures and Jurisdiction: Between Effectiveness and Fragmentation" (2011) 42 Geo J Int'l L 233.

Canan, Penelope & Nancy Reichman. "Lessons Learned" in Donald Kaniaru, ed, *The Montréal Protocol: Celebrating 20 Years of Environmental Progress* (London: Cameron May Ltd, 2007) 107.

Carbone, Jared C, Carsten Helm & Thomas F Rutherford. "The Case for International Emission Trade in the Absence of Cooperative Climate Policy" (2009) 58 J Enviro Econo & Management 266.

Carlarne, Cinnamon. "The Kyoto Protocol and the WTO: Reconciling Tensions between Free Trade and Environmental Objectives" (2006) 17 Colo J Int'l Envtl L & Pol'y 45.

Carlarne, Cinnamon Piñon. "Good Climate Governance: Only a Fragmented System of International Law Away?" (2008) 30:4 Law & Pol'y 450.

Carminati, M Vittoria Giugi. "Clean Air & Stormy Skies: The EU-ETS Imposing Carbon Credit Purchases on United States Airlines" (2010) 37 Syracuse J Int'l L & Com 127.

Carraro, Carlo & Alice Favero."The Economic and Financial Determinants of Carbon Prices" (2009) 59:5 Czech Journal of Economics and Finance 396.

Carroll, Anthony J. "The Extraterritorial Enforcement of U.S. Antitrust Laws and Retaliatory Legislation in the United Kingdom and Australia" (1983) 13 Denv J Int'l L & Pol'y 377.

Cassel, Douglass. "Empowering United States Courts to Hear Crimes within the Jurisdiction of the International Criminal Court" (2001) 35 New Eng L Rev 421.

Chambers, W Bradnee. "International Trade Law and the Kyoto Protocol: Potential Incompatibilities" in W Bradnee Chambers, *Inter-Linkages: The Kyoto Protocol & the International Trade & Investment Regimes* (New York: United Nations University Press, 2001) 88.

Chambers, W Bradnee, Joy A Kim & Claudia ten Have. "Institutional Interplay and the Governance of Biosafety" in Oran R Young et al, eds, *Institutional Interplay* (New York: United Nations University Press, 2008) 3.

Charney, Jonathan I. "The Impact on the International Legal System of the Growth of International Courts and Tribunals" (1999) 31 NYU J Int'l L & Pol'y 697.

Charnovitz, Steve. "The Law of Environmental PPMs in the WTO: Debunking the Myth of Illegality" (2002) 27 Yale J Int'l L 59.

Charnovitz, Steve. "Nongovernmental Organizations and International Law" (2006) 100:2 Am J Int'l L 348.

Checkel, Jeffrey T. "International Norms and Domestic Politics: Bridging the Rationalist: Constructivist Divide" (1997) 3 European Journal of International Relations 473.

Cheyne, Ilona. "Environmental Unilateralism and the WTO/GATT System" (1995) 24 Ga J Int'l & Comp L 433.

Chinn, Lily N. "Can the Market Be Fair and Efficient? An Environmental Justice Critique of Emissions Trading" (1999) 26 Ecology LQ 80.

Choi, Inho. "Global Climate Change and the Use of Economic Approaches: The Ideal Design Features of Domestic Greenhouse Gas Emissions Trading with an Analysis of the

European Union's CO_2 Emissions Trading Directive and the Climate Stewardship Act" (2005) 45 Nat Resources J 865.

Clark, Timothy A. "Environmental Plaintiff Standing and Extraterritoriality in the Endangered Species Act: Lujan v. Defenders of Wildlife" (1991) 7 J Min L & Pol'y 273.

Clò, Stepano. "Grandfathering, Auctioning and Carbon Leakage: Assessing the Inconsistencies of the New ETS Directive" (2010) 38 Energy Policy 2420.

Coase, Ronald. "The Problem of Social Cost" (1960) 3 J L & Econ 1.

Cohen, Maurie J. "Destination Unknown: Pursuing Sustainable Mobility in the Face of Rival Societal Aspirations" (2010) 39.

Cohen, Miriam. "The Analogy between Piracy and Human Trafficking: A Theoretical Framework for the Application of Universal Jurisdiction" (2010) 16 Buff Hum Rts L Rev 201.

Colangelo, Anthony J. "The New Universal Jurisdiction: In Absentia Signaling Over Clearly Defined Crimes" (2005) 36 Geo J Int'l L 537.

Colangelo, Anthony J. "The Legal Limits of Universal Jurisdiction" (2006) 47 Va J Int'l L 149.

Colangelo, Anthony J."Universal Jurisdiction as an International False Conflict of Laws (2009) 30 Mich J Int'l L 881.

Cole, Daniel. "What's Property Got to do With It? A Review Essay by David M. Driesen of Pollution & Property: Comparing Ownership Institutions for Environmental Protection" (2003) 30 Ecology LQ 1003.

Collier, Paul, Gordon Conway & Tony Venables."Climate Change and Africa" (2008) 24:2 Oxford Rev Econ Pol'y 337.

Condon, Bradly J. "GATT Article XX and Proximity-Of-Interest: Determining the Subject Matter of Paragraphs B and G" (2004) 9 UCAL J Intl'l L & For Aff 137.

Connor, John M & Darren Bush. "How to Block Cartel Formation and Price Fixing: Using Extraterritorial Application of the Antitrust Laws as a Deterrence Mechanism" (2008) 112 Penn State L Rev 813.

Contini, Paolo & Peter H Sand. "Methods to Expedite Environment Protection: International Ecostandards" in Peter M Haas, ed, *International Environmental Governance* (Burlington: Ashgate, 2008) 3.

Cook, Meredith Poznanski. "The Extraterritorial Application of Title VII: Does the Foreign Compulsion Defense Work?" (1996) 20 Suffolk Transnat'l L Rev 133.

Coombes, Karinne. "Universal Jurisdiction: A Means to End Impunity or a Threat to Friendly International Relations?" (2011) 43 Geo Wash Int'l L Rev 419.

Cooter, Robert. "Normative Failure Theory of Law" (1997) 82:5 Cornell L Rev 947.

Coria, Jessica & Thomas Sterner. "Tradable Permits in Developing Countries: Evidence from Air Pollution in Chile" (2010) 19:2 J Enviro & Develop 145.

Coyle, John F. "The Treaty of Friendship, Commerce and Navigation in the Modern Era" (2013) 51 Colum J Transnat'l L 302.

Crawford, James. "Responsibility to the International Community as a Whole" in James Crawford, ed, *International Law as an Open System* (London: Cameron May Ltd, 2002) 341.

Crawford, James. "International Crimes of States" in James Crawford, Alain Pellet & Simon Olleson, eds, *The Law of International Responsibility* (Oxford: Oxford University Press, 2010) 405.

Crespo, Daniel Calleja & Mike Crompton. "The European Approach to Aviation and Emissions Trading" (2007) 21:3 Air & Space L 1.

Czaplinsky, Wladyslaw. "State Succession and State Responsibility" (1990) 28 Can YB Int'l L 339.

Dai, Xiudian & Zhiping Diao. "Towards a New World Order for Climate Change: China and the European Union's Leadership Ambition" in Rüdiger K.W. Wurzel & James Connelly, eds, *The European Union as a Leader in International Climate Change Politics* (London: Routledge, 2011) 252.

Daley, Ben & Holly Preston. "Aviation and Climate Change: Assessment of Policy Options" in Stefan Gossling & Paul Upham, eds, *Climate Change and Aviation: Issues, Challenges and Solutions* (London: Earthscan, 2009) 347.

Dannenmaier, Eric. "The Role of Non-State Actors in Climate Compliance" in Jutta Brunnée, Meinhard Doelle & Lavanya Rajamani, eds, *Promoting Compliance in an Evolving Climate Regime* (Cambridge: Cambridge University Press, 2012) 149.

Davies, Peter & Jeffrey Goh. "Air Transport and the Environment" (1993) 18 Air & Space L 123.

Dejong, Steven M. "Hot Air and Hot Heads: An Examination of the Legal Arguments Surrounding the Extension of the European Union's Emissions Trading Scheme to Aviation" (2013) 3:1 Asian Journal of International Law 163.

Demailly, Damien & Philippe Quirion. "European Emission Trading Scheme and Competitiveness: A Case Study on the Iron and Steel Industry" (2009) 30 Energy Economics 2027.

Dempsey, Paul. "Compliance & Enforcement in International Law: Achieving Global Uniformity in Aviation Safety" (2004) 30 NC J Int'l L & Com Reg 1.

Deplano, Rossana. "The Fragmentation and Constitutionalisation of International Law: A Theoretical Inquiry" (2013) 6:1 European J Leg Stud 67.

DiMento, Joseph F C. "Process, Norms, Compliance, and International Environmental Law (2003) 19 J Envtl L & Litig 251.

Dinwoodie, Graeme B. "A New Copyright Order: Why National Courts Should Create Global Norms" (2000) 49 U Pa L Rev 469.

Doh, Jonathan P & Hildy Teegen. "Nongovernmental Organizations as Institutional Actors in International Business: Theory and Implications" (2002) 11 International Business Review 665.

Doremus, Holly & W Michael Hanemann. "Of Babies and Bathwater: Why the Clean Air Act's Cooperative Federalism Framework Is Useful for Addressing Global Warming" (2008) 50 Ariz L Rev 799.

Downs, George W, Kyle W Danish & Peter N Barsoom. "The Transformational Model of International Regime Design: Triumph of Hope or Experience?" (2000) 38 Colum J Transnat'l L 465.

Downs, George W & Michael A Jones. "Reputation, Compliance, and International Law" (2002) 31 Journal of Legal Studies 95.

Driesen, David M. "Free Lunch of Cheap Fix? The Emissions Trading Idea and the Climate Change Convention" (1998) 26 BC Envtl Aff L Rev 1.

Driesen, David M. "Is Emissions Trading an Economic Incentive Program? Replacing the Command and Control/Economic Dichotomy (1998) 55 Wash & Lee L Rev 289.

Druzin, Bryan. "Law, Selfishness, and Signals: An Expansion of Posner's Signaling Theory of Social Norms" (2011) 24 Can JL & Juris 5.

Dryzek, John S. "Paradigms and Discourses" in Daniel Bodansky, Jutta Brunnée & Ellen Hey, eds, *The Oxford Handbook of International Environmental Law* (Oxford: Oxford University Press, 2006) 44.

Dunoff, Jeffrey L. "Levels of Environmental Governance" in Daniel Bodansky, Jutta Brunnée & Ellen Hey, eds, *The Oxford Handbook of International Environmental Law* (Oxford: Oxford University Press, 2006) 85.

Dupuy, Pierre-Marie. "The Danger of Fragmentation or Unification of the International Legal System and the International Court of Justice" (1998) 31 NYU J Int'l L & Pol 791.

Dupuy, Pierre-Marie. "The Deficiencies of the Law of State Responsibility Relating to Breaches of 'Obligations Owed to the International Community as a Whole': Suggestions for Avoiding the Obsolescence of Aggravated Responsibility" in Antonio Cassese, ed. *Realizing Utopia: The Future of International Law* (Oxford: Oxford University Press, 2012) 210.

Duthie, Elizabeth. "ICAO Regulation: Meeting Environmental Need?" (2001) 3:3/4 Air & Space Europe 27.

Dworkin, Ronald M. "The Model of Rules" (1967) 35 U Chi L Rev 14.

Economy, Elizabeth. "Chinese Policy-Making and Global Climate Change: Two-Front Diplomacy and the International Community" in Miranda A Schreurs & Elizabeth C Economy, eds, *The Internationalization of Environmental Protection* (Cambridge: Cambridge University Press, 1997) 19.

Günther, Edeltraud. "Accounting for Emission Rights" in Ralf Antes Bernd Hansjürgens & Peter Letmathe, eds, *Emissions Trading and Business* (Heidelberg: Physica-Verlag, 2006) 219.

Eichenberg, Benjamin. "Greenhouse Gas Regulation and Border Tax Adjustments: The Carrot and Stick" (2010) 3 Golden Gate U Envtl LJ 283.

Ekholm, Tommi, Sampo Soimakallio, Sara Moltmann, Niklas Höhne, Sanna Syri & Ilkka Savolainen. "Effort Sharing in Ambitious, Global Climate Change Mitigation Scenarios" (2010) 38 Energy Policy 1797.

Ekins, Paul et al. "Increasing Carbon and Material Productivity through Environmental Tax Reform" (2012) 42 Energy Policy 365.

Ellickson, Robert C. "The Market for Social Norms" (2001) 3 Am L Econ Rev 1.

Ellis, Jaye. "Fisheries Conservation in an Anarchical System: A Comparison of Rational Choice and Constructivist Perspectives" (2007) 3 J Int'l L & Int'l Rel 1.

Ellis, Jaye. "Sustainable Development and Fragmentation in International Society" in Duncan French, ed, *Global Justice and Sustainable Development* (Leiden: Martinus Nijhoff Publishers, 2010) 57.

Ellis, Jaye. "Extraterritorial Exercise of Jurisdiction for Environmental Protection: Addressing Fairness Concerns" (2012) 25:2 Leiden J Int'l L 397.

Engau, Christian & Vokker H Hoffmann."Effects of Regulatory Uncertainty on Corporate Strategy – An Analysis of Firms' Responses to Uncertainty about Post-Kyoto Policy" (2009) 12 Environmental Science & Policy 766.

Eritja, Mar Campins. "Reviewing the Challenging Task Faced by Member States in Implementing the Emissions Trading Directive: Issues of Member State Liability" in

Marjan Peeter & Kurt Deketelaere, eds, *EU Climate Change Policy: The Challenge of New Regulatory Initiatives* (Cheltenham: Edward Elgar Publishing Limited, 2006) 69.

Escorihuela, Alejandro Lorite. "Humanitarian Law and Human Rights Law: The Politics of Distinction" (2011) 19 Mich S J Int'l L 299.

Etchart, Alejo, Begum Sertyesilisik & Greig Mill. "Environmental Effects of Shipping Imports from China and Their Economic Valuation: The Case of Metallic Valve Components" (2012) 21 J Cleaner Production 51.

Evans, Brian. "Principles of Kyoto and Emissions Trading Systems: A Primer for Energy Lawyers" (2004) 42 Alta L Rev 167.

Fahey, Elaine. "The EU Emissions Trading Scheme and the Court of Justice: The High Politics of Indirectly Promoting Global Standards" (2012) 13:11 German L J 1247.

Falkner, Robert. "Private Environmental Governance and International Relations: Exploring the Links" in Peter M Haas, ed, *International Environmental Governance* (Burlington: Ashgate, 2008) 281.

Falkner, Robert, Hannes Stephan & John Vogler. "International Climate Policy after Copenhagen: Toward a Building Blocks Approach" in David Held, Angus Hervey & Marika Theros, eds, *The Governance of Climate Change: Science, Economics, Politics & Ethics* (Cambridge: Polity Press, 2011) 202.

Farnsworth, Nick. "The EU Emissions Trading Directive: Time for Revision" in W Th Douma, L Massai & M Montini, eds, *The Kyoto Protocol and Beyond: Legal and Policy Challenges of Climate Change* (The Hague: TMC Asser Press, 2007) 29.

Finnemore, Martha & Kathryn Sikkink. "International Norm Dynamics and Political Change" (1998) 52:4 International Organization 887.

Fitzgerald, Paul P. "Europe's Emissions Trading System: Questioning Its *Raison d'Être*" (2010) 10:2 Issues Aviation L & Pol'y 189.

Fitzgerald, Warren B, Oliver J A Howitt & Inga J Smith. "Greenhouse Gas Emissions from the International Maritime Transport of New Zealand's Import and Exports" (2011) 39 Energy Policy 1521.

Florini, Ann. "The Evolution of International Norms" (1996) 40 International Studies Quarterly 363.

Fölster, Stefan & Johan Nyström. "Climate Policy to Defeat the Green Paradox" (2010) 39 AMBIO 223.

Forster, Piers M de F, Keith P Shine & Nicola Stuber. "It Is Premature to Include Non-CO_2 Effects of Aviation in Emission Trading Schemes" (2006) 40 Atmospheric Environment 1117.

Forsyth, Peter. "Environmental and Financial Sustainability of Air Transport: Are They Incompatible?" (2011) 17 Journal of Air Transport Management 27.

Fort, Jeffrey C & Cynthia A Faur. "Can Emissions Trading Work Beyond a National Program? Some Practical Observations on the Available Tools" (1997) 18 U Pa J Int'l Econ L 463.

Fossungu, Peter Ateh-Afac. "999 University, Please Help the Third World (Africa) Help Itself: A Critique of Council Elections" (1999) 64:2 J Air L & Com 339.

Foster, Mark Edward. "Making Room for Environmental Trade Measures within the GATT" (1998) 71 S Cal L Rev 393.

Fray, James D. "Remaining Valid: Security Council Resolutions, Textualism, and the Invasion of Iraq" (2007) 15 Tul J Int'l & Comp L 609.

Freestone, David. "The UN Framework Convention on Climate Change, the Kyoto Protocol, and the Kyoto Mechanism" in David Freestone & Charlotte Streck, eds, *Legal Aspects of Implementing the Kyoto Protocol Mechanisms: Making Kyoto Work* (Oxford: Oxford University Press, 2005) 3.

Freyer, Tony A. "Restrictive Trade Practices and Extraterritorial Application of Antitrust Legislation in Japanese-American Trade" (1999) 16 Ariz J Int'l & Comp L 159.

Fry, James D. "International Human Rights Law in Investment Arbitration: Evidence of International Law's Unity" (2007) 18 Duke J Comp & Int'l L 77.

Fukunaga, Yuka. "Civil Society and the Legitimacy of the WTO Dispute Settlement System" (2008) 34 Brooklyn J Int'l L 85.

Gable, Kelly A. "Cyber-Apocalypse Now: Securing the Internet against Cyberterrorism and Using Universal Jurisdiction as a Deterrent" (2010) 43 Vand J Transnat'l L 57.

Gaines, Sanford E. "Rethinking Environmental Protection, Competiveness, and International Trade" (1997) U Chi Legal F 231.

Gaines, Sanford E. "The Problem of Enforcing Environmental Norms in the WTO and What to Do About It" (2003) 26 Hastings Intl & Comp L Rev 321.

Gardner, Allison F. "Environmental Monitoring Undiscovered Country: Developing a Satellite Remote Monitoring System to Implement the Kyoto Protocol's Global Emissions-Trading Program" (2000) 9 NYU Envtl LJ 152.

Garnaut, Ross et al. "Emissions in the Platinum Age: The Implications of Rapid Development for Climate-Change Mitigation" (2008) 24:2 Oxford Rev of Econ Policy 377.

Gattini, Andrea. "Between Splendid Isolation and Tentative Imperialism: The EU's Extension of Its Emission Trading Scheme to International Aviation and the ECJ's Judgment in the ATA Case" (2012) 61 ICLQ 977.

Gehring, Markus W. "Case Note Air Transport Association of America v. Energy Secretary: Clarifying Direct Effect and Providing Guidance for Future Instrument Design for a Green Economy in the European Union" (2012) 2 RECIEL 149.

Gehring, Thomas & Sebastian Oberthür. "Introduction" in Sebastian Oberthür & Thomas Gehring, eds, *Institutional Interaction in Global Environmental Governance: Synergy and Conflict among International and EU Policies* (Cambridge: MIT Press, 2006) 1.

Gehring, Thomas, Sebastian Oberthür & Marc Mühleck. "European Union Actorness in International Institutions: Why the EU Is Recognized as an Actor in Some International Institutions, but Not in Others" (2013) 51:5 Journal of Common Market Studies 849.

Gendzier, Irene. "Just a War: Reflections on the U.S. Invasion of Iraq: Evidence, International Law, and Past Policy" (2004) 9 NEXUS: J Opinion 101.

George, Erika R. "The Place of the Private Transnational Actor in International Law: Human Rights Norms, Development Aims, and Understanding Corporate Self-Regulation as Soft Law" in Roundtable – A Multiplicity of Actors and Transnational Governance (2007) 101 Am Soc'y Int'l L Proc 469.

Geraghty, Anne H. "Universal Jurisdiction and Drug Trafficking: A Tool for Fighting One of the World's Most Pervasive Problems" (2004) 16 Fla J Int'l L 371.

Gerber, David J. "The Extraterritorial Application of the German Antitrust Laws" (1983) 77 Am J Int'l L 756.

Gerlach, Anne. "Sustainability Entrepreneurship in the Context of Emissions Trading" in Ralf Antes Bernd Hansjürgens & Peter Letmathe, eds, *Emissions Trading and Business* (Heidelberg: Physica-Verlag, 2006) 73.

Ghosh, Koushik & Peter Gray. "Rushing to Copenhagen? Is Cap-and-Trade the Answer?" (2010) 53:1 Challenge 5.

Gibney, Mark P. "The Extraterritorial Application of U.S. Law: The Perversion of Democratic Governance, the Reversal of Institutional Roles, and the Imperative of Establishing Normative Principles" (1996) 19 BC Int'l & Comp L Rev 297.

Gibney, Mark & R David Emerick. "The Extraterritorial Application of United States Law and the Protection of Human Rights: Holding Multinational Corporations to Domestic and International Standards" (1996) 10 Temp Int'l & Comp L J 123.

Gillette, Clayton P. "Lock-in Effects in Law and Norms" (1998) 78 BUL Rev 813.

Gilligan, Michael J & Nathaniel H Nesbitt. "Do Norms Reduce Torture" (2009) 38 J Legal Stud 445.

Gilman, Daniel. "Of Fruitcakes and Patriot Games" (2002) 90 Geo LJ 2387.

Gilman, Ryan. "Expanding Environmental Justice after War: The Need for Universal Jurisdiction over Environmental War Crimes" (2011) 22 Colo J Int'l Envtl L & Pol'y 447.

Givoni, Moshe & Piet Rietveld."The Environmental Implications of Airlines' Choice of Aircraft Size" (2010) 16 J Air Transport Management 159.

Glennon, Jason N. "Directive 2008/101 and Air Transport – A Regulatory Scheme Beyond the Limits of the Effects Doctrine" (2013) 78:3 J Air L & Com 479.

Glensy, Rex D. "Quasi-Global Social Norms" (2005) 38 Conn L Rev 79.

Goffman, Joseph. "Title IV of the Clean Air Act: Lessons for Success of the Acid Rain Emissions Trading Program" (2006) 14 Penn St Envtl L Rev 177.

Goldenberg, Evan. "The Design of an Emissions Permit Market for RECLAIM: A Holistic Approach" (1993) 11 UCLA J Envtl L & Pol'y 297.

Goldschein, Perry S. "Going Mobile: Emissions Trading Gets a Boost from Mobile Source Emission Reduction Credits" (1994) 13 UCL J Envtl L & Pol'y 225.

Gopalan, Sandep. "Changing Social Norms and CEO Pay: The Role of Norms Entrepreneurs" (2007) 39 Rut LJ 1.

Gopalan, Sandeep. "Alternative Sanctions and Social Norms in International Law: The Case of Abu Ghraib" (2007) Mich St L Rev 785.

Goodwin, Joshua Michael. "Universal Jurisdiction and the Pirate: Time for an Old Couple to Part" (2006) 39 Vand J Transnat'l L 973.

Gossling, Stefan & Paul Upham. "Introduction: Aviation and Climate Change in Context" in Stefan Gossling and Paul Upham, eds, *Climate Change and Aviation: Issues, Challenges and Solutions* (London: Earthscan, 2009) 1.

Gossling, Stefan, Jean-Paul Ceron, Chislain Dubois & Michael C. Hall. "Hypermobile Travellers" in Stefan Gossling & Paul Upham, eds, *Climate Change and Aviation: Issues, Challenges and Solutions* (London: Earthscan, 2009) 131.

Graham, Brian & Jon Shaw. "Low-Cost Airlines in Europe: Reconciling Liberalization and Sustainability" (2008) 39 Geoforum 1439.

Green, Andrew. "You Can't Pay Them Enough: Subsidies, Environmental Law, and Social Norms" (2006) 30 Harv Envtl L Rev 407.

Griffin, Josepth P. "US Supreme Court Encourages Extraterritorial Application of US Antitrust Laws" (1993) 21 Int'l Bus Law 389.

Grover, Steven. "Blackjack at Thirty Thousand Feet: American's Attempt to Enforce Its Ban on In-flight Gambling Extraterritorially" (1999) 4 Gaming Research & Rev J 2.

Gupta, Aarti. "Global Biosafety Governance: Emergence and Evolution" in Oran R Young et al, eds, *Institutional Interplay* (New York: United Nations University Press, 2008) 19.

Gupta, Joyeeta. "Developing Countries and the Post-Kyoto Regime: Breaking the Tragic Lock-in of Waiting for Each Other's Strategy" in W Th Douma, L Massai, M Montini, eds, *The Kyoto Protocol and Beyond: Legal and Policy Challenges of Climate Change* (The Hague: TMC Asser Press, 2007) 161.

Gupta, Joyeeta. "Climate Change and Shifting Paradigms" in Duncan French, ed, *Global Justice and Sustainable Development* (Leiden: Martinus Nijhoff Publishers, 2010) 167.

Guzman, Andrew T. "A Compliance-Based Theory of International Law" (2002) 90:6 CLR 1823.

Haanappel, PPC. "The Impact of Changing Air Transport Economics on Air Law and Policy: A Short Commentary" (2009) 34 Ann Air & Sp L 519.

Hacking, Lord David. "Appendix No. 1 The Increasing Extraterritorial Impact of U.S. Laws: A Cause for Concern Amongst Friends of America" in Joseph P Griffin, ed, *Perspectives on the Extraterritorial Application of U.S. Antitrust and Other Laws* (Chicago: American Bar Association, 1979) 155.

Hafner, Gerhard. "Pros and Cons Ensuing From Fragmentation of International Law" (2004) 25 Mich J Int'l L 849.

Hahn, Robert W. "Marketable Permits: What's All the Fuss About?" (1982) 2:4 J Pub Pol'y 395.

Hahn, Robert W. "Book Review: Regulatory Reform at EPA: Separating Fact from Illusion Reforming Air Pollution Regulation: The Toil and Trouble of EPA's Bubble by Richard A. Liroff" (1986) 4 Yale J on Reg 173.

Hahn, Robert W. "Climate Policy: Separating Fact from Fantasy" (2009) 33 Harv Envtl L Rev 557.

Hahn, Robert W. "Greenhouse Gas Auctions and Taxes: Some Political Economy Considerations" (2009) 3:2 Rev Env Economics & Pol'y 167.

Hahn, Robert W & Gordon L Hester. "Marketable Permits: Lessons for Theory and Practice" (1989) 16 Ecology LQ 361.

Hahn, Robert W & Robert N Stavins. "Incentive-Based Environmental Regulation: A New Era from an Old Idea" (1991) 18 Ecology LQ 1.

Hammarskjöld, Knut. "About the Need to Bridge a Jurisdictional Chasm" (1983) 8 Ann Air & Sp L 97.

Hans, Monica. "Providing for Uniformity in the Exercise of Universal Jurisdiction: Can Either the Princeton Principles on Universal Jurisdiction or an International Criminal Court Accomplish this Goal" (2002) 15 Transnat'l Law 357.

Hardeman, Andreas. "A Common Approach to Aviation Emissions Trading" (2007) 32:1 Air & Space L 1.

Hardeman, Andreas. "Reframing Aviation Climate Politics and Policies" (2011) 36 Ann Air & Sp L 1.

Hardy, Brettny. "How Positive Environmental Politics Affected Europe's Decision to Oppose and then Adopt Emissions Trading" (2007) 17 Duke Envtl L & Pol'y F 297.

Hares, Andrew, Janet Dickinson & Keith Wilkes. "Climate Change and the Air Travel Decisions of UK Tourists" (2010) 18 Journal of Transport Geography 466.

Harrison, Kathryn. "Talking with the Donkey: Cooperative Approaches to Environmental Protection" in Peter M Haas, ed, *International Environmental Governance* (Burlington: Ashgate, 2008) 339.

Harrison, Kathryn & Lisa McIntosh Sundstrom. "Introduction: Global Commons, Domestic Decisions" in Kathryn Harrison & Lisa Mintosh Sundstrom, eds, *Global Commons, Domestic Decisions: The Comparative Politics of Climate Change* (Cambridge, MIT Press, 2010) 1.

Harrison, Kathryn. "The United States as Outlier: Economic and Institutional Challenges to US Climate Policy" in Kathryn Harrison & Lisa McIntosh Sundstrom, eds, *Global Commons, Domestic Decisions: The Comparative Politics of Climate Change* (Cambridge: MIT Press, 2010) 67.

Hart, Lee. "International Emissions Trading Between Developing Countries: The Solution to the Other Half of the Climate Change Problem" (2008) 20 Fla J Int'l L 79.

Harvard Research. "Part II – Jurisdiction with Respect to Crime" (1935) 29 Am J Int'l L Supp 435.

Hatch, Michael T. "Assessing Environmental Policy Instruments: An Introduction" in Michael T Hatch, ed, *Environmental Policy Making: Assessing the Use of Alternative Policy Instruments* (Albany: State University of New York Press, 2005) 1.

Havel, Brian F & Niels van Antwerpen. "Dutch Ticket Tax and Article 15 of the Chicago Convention" (2009) 34:2 Air & Space L 141.

Havel, Brian F & Niels van Antwerpen. "Dutch Ticket Tax and Article 15 of the *Chicago Convention* (continued)" (2009) 34:6 Air and Space L 447.

Havel, Brian F & Gabriel S Sanchez. "Restoring Global Aviation's Cosmopolitan Mentalité" (2010) 29 BU Int'l LJ 1.

Havel Brian F & John Q Mulligan. "The Triumph of Politics: Reflections on the Judgment of the Court of Justice of the European Union Validating the Inclusion of Non-EU Airlines in the Emissions Trading Scheme" (2012) 37:1 Air & Space L 3.

Havel, Brian F & Gabriel S Sanchez. "Toward a Global Aviation Emissions Agreement" (2012) 36 Harv Envtl L Rev 352.

Hawkings, Slayde. "Skirting Protectionism: A GHG-Based Trade Restriction under the WTO" (2008) 20 Geo Int'l Envtl L Rev 427.

Heitmann, Nadine & Setareh Khalilian. "Accounting for Carbon Dioxide Emissions from International Shipping: Burden Sharing under Different UNFCCC Allocation Options and Regime Scenarios" (2011) 35 Marine Policy 682.

Heller, Thomas. "Climate Change: Designing an Effective Response" in Ernesto Zedillo, ed, *Global Warming: Looking Beyond Kyoto* (Washington: Brookings Institution Press, 2008) 115.

Helm, Dieter. "Climate Change Policy: Why Has So Little Been Achieved?" (2008) 24:2 Oxford Rev Econ Pol'y 211.

Hemingson, Tate L. "Why Airlines Should Be Afraid: The Potential Impact of Cap and Trade and Other Carbon Emissions Reduction Proposals on the Airline Industry" (2010) 75 J Air L & Com 741.

Hepburn, Cameron & Nicolas Stern. "A New Global Deal on Climate Change" (2008) 34:2 Oxford Rev Econ Pol'y 259.

Hermida, Julian. "Crimes in Space: A Legal and Criminological Approach to Criminal Acts in Outer Space" (2006) 31 Ann Air & Sp L 405.

Herold, Anke. "Experiences with 5, 7, and 8 Defining the Monitoring, Reporting, and Verification System under the Kyoto Protocol in an Evolving Climate Regime" in Jutta Brunnée, Meinhard Doelle & Lavanya Rajamani, eds, *Promoting Compliance in an Evolving Climate Regime* (Cambridge: Cambridge University Press, 2012) 122.

Hess, Gerald F. "The Trail Smelter, the Columbia River, and the Extraterritorial Application of CERCLA" (2005) 18 Geo Int'l Envtl L Rev 2.

Hetcher, Steven. "The FTC as Internet Privacy Norm Entrepreneur" (2000) 53 Vand L Rev 204.

Higgins, Rosalyn. "General Course on Public International Law" (1991) Rec des Cours 10.

Hilaire, Max. "International Law and the United States Invasion of Iraq" (2005) 44 Mil L & L War Rev 125.

Hirsch, Robert L, Roger Bezdek & Robert Wendling. "Peaking of World Oil Production and Its Mitigation" in Daniel Sperling & James S Cannon, eds, *Driving Climate Change: Cutting Carbon from Transportation* (Amsterdam: Elsevier, 2007) 9.

Hobley, Anthony. "The UK Emissions Trading System: Some Legal Issues Explored" in Julian Boswall & Robert Lee, eds, *Economics, Ethics and the Environment* (London, Sydney: Cavendish Publishing Limited, 2002) 61.

Hodgkinson, David. "IOSA: The Revolution in Airline Safety Audits" (2005) 30:4/5 Air & Space L 302.

Hof, Andries, Michel den Elzen & Detlef van Vuuren. "Environmental Effectiveness and Economic Consequences of Fragmented versus Universal Regimes: What Can We Learn from Model Studies" in Frank Biermann, Philipp Pattberg & Fariborz Zelli, eds, *Global Climate Governance Beyond 2012: Architecture, Agency and Adaptation* (Cambridge: Cambridge University Press, 2010) 35.

Hofer, Christina, Martin E Dresner & Robert J Windle. "The Environmental Effects of Airline Carbon Emissions Taxation in the US" (2010) 15 Transportation Research Part D 37.

Horwitz, Paul S. "Harnessing the Power of the Montréal Protocol to Deliver Even More Climate Change Benefits" in Donald Kaniaru, ed, *The Montréal Protocol: Celebrating 20 Years of Environmental Progress* (London: Cameron May Ltd, 2007) 187.

Howse, Robert. "The Appellate Body Rulings in the Shrimp/Turtle Case: A New Legal Baseline for the Trade and Environment Debate" (2002) 27:2 Colum J Envtl L 489.

Howse, Robert. "The Political and Legal Underpinnings of Including Aviation in the EU ETS" in Lorand Bartels, *The Inclusion of Aviation in the EU ETS: WTO Law Considerations*, Trade and Sustainable Energy Series, Issue Paper No 6 (Geneva: International Centre for Trade and Sustainable Development Switzerland, 2012) 28.

Howse, Robert. "Fragmentation and Utopia: Towards an Equitable Integration of Finance, Trade, and Sustainable Development" in Antonio Cassese, ed, *Realizing Utopia: The Future of International Law* (Oxford: Oxford University Press, 2013) 427.

Hudson, Blake. "Climate Change, Forests, and Federalism: Seeing the Treaty for the Trees" (2011) 82 U Colo L Rev 363.

Hunter, David, James E Salzman & Durwood Zaelke. "International Trade and Investment Law" in David Hunter, James E Salzman & Durwood Zaelke, eds, *International Environmental Law and Policy*, 3rd edn (New York: Foundation Press, 2007) 1233.

Hurrel, Andrew & Terry MacDonald. "Ethics and Norms in International Relations" in Walter Emmanuel Carlsnaes & Beth A Simmons, eds, *Handbook of International Relations* (London: Sage Publications, 2013) 57.

Janda, Richard. "Passing the Torch: Why ICAO Should Leave Economic Regulation of International Air Transport to the WTO" (1995) 21:1 Ann Air & Sp L 409.

Janzen, Bernd G. "International Trade Law and the 'Carbon Leakage' Problem: Are Unilateral U.S. Import Restrictions the Solution?" (2008) 8 Sustainable Dev L & Pol'y 22

Jenks, C. Wilfred. "The Conflict of Law-Making Treaties" (1953) 30 Brit YB Int'l 401.

Jennings, RY. "Extraterritorial Jurisdiction and the United States Antitrust Laws" (1957) 33 Brit YB Int'l L 146.

Jennison, Michael B. "The Chicago Convention and Safety after 50 Years" (1995) 20:1 Ann Air & Sp L 283.

Jordan, Andrew, Dave Huitema & Harro van Asselt. "Climate Change Policy in the European Union: An Introduction" in Andrew Jordan et al, eds, *Climate Change Policy in the European Union: Confronting the Dilemmas of Mitigation and Adaptation* (Cambridge: Cambridge University Press, 2010) 3.

Jordan, Jon B. "Universal Jurisdiction in a Dangerous World: A Weapon for all Nations against International Crime" (2009) 9 MSU-DCL J Int'l L 1.

Junker, Kirk W. "Ethical Emissions Trading and the Law" (2005) 13 U Balt J Envtl L 149.

Kaleck, Wolfgang. "From Pinochet to Rumsfeld: Universal Jurisdiction in Europe 1998-2008" (2009) 30 Mich J Int'l L 927.

Kaplan, Margo. "Using Collective Interests to Ensure Human Rights: An Analysis of the Articles on State Responsibility" (2004) 79 NYUL Rev 1902.

Karim, Md Saiful. "IMO Mandatory Energy Efficiency Measures for International Shipping: The First mandatory Global Greenhouse Gas Reduction Instrument for an International Industry" (2011) 7 Macquarie J Int'l & Comp Envtl L 111.

Karim, Md Saiful & Shawkat Alam. "Climate Change and Reduction of Emissions of Greenhouse Gases from Ships: An Appraisal" (2011) 1:1 Asian J Int'l L 131.

Karpel, Amy Ann. "The European Commission's Decision on the Boeing-McDonnel Douglas Merger and the Need for Greater U.S.-EU Cooperation in the Merger Field" (1998) 47 Am U L Rev 1029.

Kastenberg, Joshua E. "Universal Jurisdiction and the Concept of a Fair Trial: *Prosecutor v. Fulgence Niyonteze*: A Swiss Military Tribunal Case Study" (2004) 12 U Miami Int'l & Comp L Rev 1.

Kennedy, Kevin C. "The Illegality of Unilateral Trade Measures to Resolve Trade-Environment Disputes" (1998) 22 Wm & Mary Envtl L & Pol'y Rev 375.

Keohane, Robert O. "When Does International Law Come Home" (1998) 35 Hous L Rev 699.

Kete, Nancy. "Air Pollution Control in the United States: A Mixed Portfolio Approach" in Ger Klaassen & Finn R Førsund, eds, *Economic Instruments for Air Pollution Control* (Dordrecht: Kluwer Academic Publishers, 1994) 122.

Kim, Won-Ki. "The Extraterritorial Application of U.S. Antitrust Law and Its Adoption in Korea" (2003) 7 Singapore J Int'l & Comp L 386.

Kirgis, Frederic L Jr. "Standing to Challenge Human Endeavors that Could Change the Climate" (1990) 84 Am J Int'l L 525.

Kirk, J. "Creating an Emissions Trading System for Greenhouse Gases: Recommendations to the California Air Resources Board" (2008) 26 Va Envtl LJ 547.

Kisska-Schulze, Kathryn & Gregory P. Tapis. "Projections for Reducing Aircraft Emissions" (2012) 77 J Air L & Com 701.

Kivits, Robbert, Michael B Charles & Neal Ryan. "A Post-Carbon Aviation Future: Airports and the Transition to a Cleaner Aviation Sector" (2010) 42 Futures 199.

Klepper, Gernot & Sonja Peterson. "The European Emissions Trading Regime and the Future of Kyoto" in Ernesto Zedillo, ed, *Global Warming: Looking Beyond Kyoto* (Washington: Brookings Institution Press, 2008) 101.

Knighton, William. "Britain: Blocking and Claw-Back" in John R Lacey, ed, *Act of State and Extraterritorial Reach: Problems of Law and Policy* (Chicago: American Bar Association, 1983) 52.

Kobrick, Eric S. "The Ex Post Facto Prohibition and Exercise of Universal Jurisdiction Over International Crimes" (1987) 87 Colum L Rev 1515.

Koh, Harold Hongju. "Bringing International Law Home" (1998) 35 Hous L Rev 623.

Koh, Stephanie. "The Case Against Extending the EU Emissions Trading Scheme to International Aviation" (2012) 30 Sing L Rev 125.

Kontorovich, Eugene. "The Piracy Analogy: Modern Universal Jurisdiction's Hollow Foundation" (2004) 45 Harv Int'l LJ 183.

Kontorovich, Eugene. "The Inefficiency of Universal Jurisdiction" (2008) U Ill L Rev 389.

Kontorovich, Eugene. "Beyond the Article I Horizon: Congress Enumerated Powers and Universal Jurisdiction Over Drug Crimes" (2009) 93 Min L Rev 1191.

Kontorovich, Eugene & Steven Art. "Agora: Piracy Prosecution: An Empirical Examination of Universal Jurisdiction for Piracy" (2010) 104 Am J Int'l L 436.

Koremenos, Barbara, Charles Lipson & Duncan Snidal. "The Rational Design of International Institutions" (2001) 53:4 International Organization 761.

Korkeakivi, Antti. "Consequences of 'Higher' International Law: Evaluating Crimes of State and *Erga Omnes*" (1996) 2 J Int'l Legal Stud 81.

Koskenniemi, Matti & Päivi Leino. "Fragmentation of International Law? Postmodern Anxieties" (2002) 15 Leiden J Int'l L 553.

Kruger, Joe & Christian Egenhofer. "Confidence through Compliance in Emission Trading Markets" (2006) 6 Sustainable Dev L & Pol'y 2.

Kübler, Dorothea. "On the Regulation of Social Norms" (2001) 17 J L Econ & Org 449.

Kulovesi, Kati. "Make Your Own Special Song, Even if Nobody Else Sings Along": International Aviation Emissions and the EU Emissions Trading Scheme (2011) 2:4 Climate Law 535.

Kulovesi, Kati. "Addressing Sectoral Emissions Outside the United Nations Framework Convention on Climate Change: What Roles for Multilateralism, Minilateralism and Unilateralism" (2012) 21 RECIEL 193.

Kurkowski, Susan J. "Distributing the Right to Pollute in the European Union Efficiency, Equity, and the Environment" (2006) 14 NYU Envtl LJ 698.

Kysar, Douglas A & James Salzman. "Environmental Tribalism" 87 (2003) Minn L Rev 1099.

Laborde, Isabelle. "EU Regulation of Aviation CO_2 Emissions" (2010) 24 WTR Nat Resources & Env't 54.

Lamendola, Leigh Robin. "The Continuing Transformation of International Antitrust Law and Policy: Criminal Extraterritorial Application of the Sherman Act in *United States* v. *Nippon Paper Industries*" (1998) 22 Suffolk Transnat'l L Rev 663.

Lan, Hua. "Comments on EU Aviation ETS Directive and EU – China Aviation Emission Dispute" (2011) 45 RJT 589.

Lawson, Robert. "UK Air Passenger Duty Held to Be Consistent with the *Chicago Convention*" (2008) 33:1 Air & Space L 1.

Lazar, Fred. "Multilateral Trade Agreement for Civil Aviation" (2011) 36:6 Air & Space L 379.

Leathley, Christian. "An Institutional Hierarchy to Combat the Fragmentation of International Law: Has the ILC Missed an Opportunity" (2007) 40 NYU J Intl'l L & Pol 259.

Lee, DS & R Sausen. "New Directions: Assessing the Real Impact of CO_2 Emissions Trading by the Aviation Industry" (2000) 34 Atmospheric Environment 5337.

Lee, DS et al. "Transport Impacts on Atmosphere and Climate: Aviation" (2010) 44 Atmospheric Environment 4678.

Lee, Jae-Seung. "Coping with Climate Change: A Korean Perspective" in Antonio Marquina, ed, *Global Warming and Climate Change: Prospects and Policies in Asia and Europe* (London: Palgrave Macmillan, 2010) 357.

Legault, Leonard H. "U.S. Assertions of Extraterritorial Jurisdiction: A Canadian Perspective" in John R Lacey, ed, *Act of State and Extraterritorial Reach: Problems of Law and Policy* (Chicago: American Bar Association, 1983) 129.

Liberatore, Angela. "The European Union: Bridging Domestic and International Environmental Policy-Making" in Miranda A Schreurs & Elizabeth C Economy, eds, *The Internationalization of Environmental Protection* (Cambridge: Cambridge University Press, 1997) 188.

Lin, Boqiang & Xuehui Li. "The Effect of Carbon Tax on Per Capita CO_2 Emissions" (2011) 39 Energy Policy 5137.

Lindroos, Anja. "Addressing Norm Conflicts in a Fragmented Legal System: The Doctrine of Lex Specialis" (2005) 74 Nordic J Int'l L 27.

Lindroos, Anja & Michael Mehling. "Dispelling the Chimera of Self-Contained Regime' International Law and the WTO" (2005) 16 Eur J Int'l L 857.

Linton, Suzannah & Firew Kebede Tiba. "The International Judge in an Age of Multiple International Courts and Tribunal" (2009) 9 Chi J Int'l L 407.

Liu, Jin. "The Role of ICAO in Regulating the Greenhouse Gas Emissions of Aircraft" (2011) 4 Carbon & Climate L Rev 417.

Lohmann, Larry. "Toward a Different Debate in Environmental Accounting: The Cases of Carbon and Cost-Benefit" (2009) 34 Accounting, Organizations and Society 499.

Lövbrand, Eva, Teresia Rindefjäll & Joakim Nordqvist. "Closing the Legitimacy Gap in Global Environmental Governance? Lessons from the Emerging CDM Market" (2009) 9:2 Global Environmental Politics 74.

Lowe, Vaughan A. "Blocking Extraterritorial Jurisdiction: The British Protection of Trading Interest Act, 1980" (1981) 75:2 American J Int'l L 257.

Lowe, Vaughan A. "The Problems of Extraterritorial Jurisdiction: Economic Sovereignty and the Search for a Solution" (1985) 34:4 ICLQ 724.

Lowe, Vaughan A. "Extraterritorial Jurisdiction: The British Practice" (1988) 52:1/2 Rabels Zeitschrift für Ausländisches und Internationales Privatrecht 157.

Lowe, Vaughan A. "US Extraterritorial Jurisdiction: The Helms-Burton and D'Amato Acts" (1997) 46:2 Int'l & Comp L Q 378.

Lowe, Vaughan A. "Jurisdiction" in Malcolm D Evans, ed, *International Law*, 1st edn (Oxford: Oxford University Press, 2003) 329.

Lu, Cherie. "The Implications of Environmental Costs on Air Passenger Demand for Different Airline Business Models" (2009) 15 J Air Transport Management 158.

Luna, Erik & Paul G Cassell. "Mandatory Minimalism" (2010) 32 Cardozo L Rev 1.

Mace, MJ. "Comparability of Efforts" in Jutta Brunnée, Meinhard Doelle & Lavanya Rajamani, eds, *Promoting Compliance in an Evolving Climate Regime* (Cambridge: Cambridge University Press, 2012) 286.

Machado-Filho, Haroldo. "Financial Mechanisms under the Climate Regime" in Jutta Brunnée, Meinhard Doelle & Lavanya Rajamani, eds, *Promoting Compliance in an Evolving Climate Regime* (Cambridge: Cambridge University Press, 2012) 216.

Macintosh, Andrew. "Overcoming the Barriers to International Aviation Greenhouse Gases Emissions Abatement" (2008) 33:6 Air & Space L 405.

Macintosh, Andrew & Lailey Wallace. "International Aviation Emissions to 2025: Can Emissions Be Stabilized Without Restricting Demand?" (2009) 37 Energy Policy 264.

Madden, Mary McKleen. "Strengthening Protection of Employees at Home and Abroad: The Extraterritorial Application of Title VII of The Civil Rights Act of 1964 and the Age Discrimination in Employment Act" (1996) 20 Hamline L Rev 739.

Madison, Kristin. "Government, Signaling, and Social Norms" (2001) U Ill L Rev 867.

Magraw, Kendra. "Universally Liable? Corporate-Complicity Liability under the Principle of Universal Jurisdiction" (2009) 18 Minn J Int'l L 458.

Malina, Robert et al. "The Impact of the European Union Emissions Trading Scheme on US Aviation" (2012) 19 J Air Transport Management 36.

Maloy, Bruce. "Extraterritorial Application of United States Criminal Statutes to the Gaming Industry" (1997) 1 Gaming Law Review 491.

Mann, FA. "The Doctrine of Jurisdiction in International Law" (1964)1 11 RCADI 1.

Mann, FA. "Germany's Present Legal Status Revisited" (1967) 16 Intl'l & Comp LQ 760.

Marceau, Gabrielle. "A Call for Coherence in International Law: Praises for the Prohibition against Clinical Isolation in WTO Dispute Settlement" (1999) 33:5 J World Trade 87.

Marceau, Gabrielle. "Conflicts of Norms and Conflicts of Jurisdiction: The Relationship between the WTO Agreement and MEAs and Other Treaties" (2001) 35:6 J World Trade 1081.

Marks, Jonathan H. "Mending the Web: Universal Jurisdiction, Humanitarian Intervention and the Abrogation of Immunity by the Security Council" (2004) 42 Colum J Transnat'l L 445.

Marong, Alhaji BM. "From Rio to Johannesburg: Reflections on the Role of International Legal Norms in Sustainable Development" 16 Geo Int'l Envtl L Rev 21.

Martin-Cejas, Roberto Rendeiro. "Ramsey Pricing Including CO_2 Emission Cost: An Application to Spanish Airports" (2010) 16 Journal of Air Transport Management 45.

Massai, Leonardo. "Legal Challenges in European Climate Policy" in W Th Douma, L Massai, M Montini, eds, *The Kyoto Protocol and Beyond: Legal and Policy Challenges of Climate Change* (The Hague: TMC Asser Press, 2007) 13.

Master, Julie B. "International Trade Trumps Domestic Environmental Protection: Dolphins and Sea Turtles are Sacrificed on the Altar of Free Trade" (1998) 12 Temp Int'l & Comp LJ 423.

Matheson, Michael J. "The Fifty-Sixth Session of the International Law Commission" (2005) 99 Am J Int'l L 211.

Matheson, Michael J. "The Fifty-Seventh Session of the International Law Commission" (2006) 100 Am J Int'l L 416.

Matz-Lück, Nele. "Structural Questions of Fragmentation" (2011) 105 Am Soc'y Int'l L Proc 123.

Mayer, Benoît. "The International Legal Challenges of Climate-Induced Migration: Proposal for an International Legal Framework" (2011) 22 Colo J Int'L Envtl L & Pol'y 357.

Mayer, Benoît. "Case -366/10, Air Transport Association of America and Others v. Secretary of State for Energy and Climate Change" (2012) 49:3 CML Rev 1113.

Mazzochi, Sarah. "The Age of Impunity Using the Duty to Extradite or Prosecute and Universal Jurisdiction to End Impunity for Acts of Terrorism Once and for All" (2011) 32 N Ill UL Rev 75.

McAdams, Richard H. "The Origin, Development, and Regulation of Norms" (1997) 96 Mich L Rev 338.

McGonigle, Sean. "Past Its Use-by Date: Regulation 868 Concerning Subsidy and State Air in International Air Services" (2013) 38:1 Air & Space L 1.

McLure, Charles E Jr. "The GATT-Legality of Border Adjustments for Carbon Taxes and the Cost of Emissions Permits: A Tiddle, Wrapped in a Mystery, Inside an Enigma" (2011) 11 Fla Tax Rev 221.

Meessen, Karl M. "Antitrust Jurisdiction under Customary International Law" (1984) 78 Am J Int'l L 783.

Mehling, Michael. "Enforcing Compliance in an Evolving Climate Regime" in Jutta Brunnée, Meinhard Doelle & Lavanya Rajamani, *Promoting Compliance in an Evolving Climate Regime* (Cambridge: Cambridge University Press, 2012) 194.

Meidan, Michal. "China's Emissions Reduction Policy: Problems and Prospects" in Antonio Marquina, ed, *Global Warming and Climate Change: Prospects and Policies in Asia and Europe* (London: Palgrave Macmillan, 2010) 307.

Mellin, Anna & Hanna Rydhed. "Swedish Port's Attitudes Towards Regulations of the Shipping Sector's Emissions of CO_2" (2011) 38:4 Marti Pol Mgmt 437.

Meltzer, Joshua. "Climate Change and Trade: The EU Aviation Directive and the WTO" (2012) 15 J Intl'l Econ L 111.

Mendelsohn, Robert. "The Policy Implications of Climate Change Impacts" in Ernesto Zedillo, ed, *Global Warming: Looking Beyond Kyoto* (Washington: Brookings Institution Press, 2008) 82.

Mendes de Leon, Pablo. "The Fight Against Terrorism Through Aviation: Data Protection Versus Data Production" (2006) 31:4/5 Air & Space L 320.

Mendes de Leon, Pablo. "ATA and Others v. the UK Secretary of State for Energy and Climate Change" (2009) (2010) 35:2 Air and Space L 199.

Mendes de Leon, Pablo. "Enforcement of the EU ETS: The EU's Convulsive Efforts to Export Its Environmental Values" (2012) 37:4 Air & Space L 287.

Meyer, Jeffrey A. "Dual Illegality and Geoambiguous Law: A New Rule for Extraterritorial Application of U.S. Law" (2010) 95 Minn L Rev 110.

Michaelowa, Axel et al. "The Market Potential of Large-Scale Non-CO_2 CDM Projects" in W Th Douma, L Massai & M Montini, eds, *The Kyoto Protocol and Beyond: Legal and Policy Challenges of Climate Change* (The Hague: TMC Asser Press, 2007) 59.

Miko, Samantha A. "Norm Conflict, Fragmentation and the European Court of Human Rights" (2013) 36 BC Int'l & Comp L Rev 1351.

Milde, Michael. "*Chicago Convention* – 50 Years Later: Are Major Amendments Necessary or 'Desirable'?" (1994) 19:1 Ann Air & Sp L 401.

Milde, Michael. "Aviation Safety Oversight: Audits and the Law" (2001) 26 Ann Air & Sp L 165.

Milde, Michael. "Aviation Safety & Security: Legal Management" (2004) 29 Ann Air & Sp L 1.

Milde, Michael. "The EU Emissions Trading Scheme – Confrontation or Compromise? – A Unilateral Action Outside the Framework of ICAO" (2012) 62:2 ZLW 173.

Miller, Geoffrey P. "Norms and Interests" (2003) 32 Hofstra L Rev 637.

Miller, Heather L. "Civil Aircraft Emissions and International Treaty Law" (1997) 63 J Air L & Com 697.

Mills, Zachary. "Does the World Need Knights Errant to Combat Enemies of All Mankind? Universal Jurisdiction, Connecting Links, and Civil Liability" (2009) 66 Wash & Lee L Rev 1315.

Miola, A, M Marra & B Ciuffo. "Designing a Climate Change Policy for the International Maritime Transport Sector: Market-Based Measures and Technological Options for Global and Regional Policy Actions" (2011) 39 Energy Policy 5490.

Mitchell, Ronald, Moira L McConnell, Alexei Roginko & Ann Barrett. "International Vessel-Source Oil Pollution" in Oran R Young, ed, *The Effectiveness of International Environmental Regimes: Causal Connections and Behavioral Mechanism* (Cambridge: MIT Press, 1999) 33.

Miyoshi, C & K J Mason. "The Carbon Emissions of Selected Airlines and Aircraft Types in Three Geographic Markets" (2009) 15 J Air Transport Management 138.

Moeckli, Daniel. "The Emergence of Terrorism as a Distinct Category of International Law" (2008) 44 Tex Int'l L J 157.

Moghadam, Tanaz. "Revitalizing Universal Jurisdiction: Lessons from Hybrid Tribunals Applied to the Case of Hissene Habre" (2008) 39 Colum Human Rights L Rev 471.

Monjon, Stephanie & Philippe Quirion. "Addressing Leakage in the EU ETS: Border Adjustment or Output-Based Allocation?" (2011) 70 Ecological Economics 1957.

Montgomery, W David. "Markets in Licenses and Efficient Pollution Control Programs" (1972) 5 J Econ Theory 395.

Morgan, Amanda L. "U.S. Officials' Vulnerability to 'Global Justice': Will Universal Jurisdiction over War Crimes Make Traveling for Pleasure Less Pleasurable" (2005) 57 Hasting LJ 423.

Morgan, Jennifer. "The Emerging Post-Cancun Climate Regime" in Jutta Brunnée, Meinhard Doelle & Lavanya Rajamani, eds, *Promoting Compliance in an Evolving Climate Regime* (Cambridge: Cambridge University Press, 2012) 17.

Morrell, Peter. "The Potential for European Aviation CO2 Emissions Reduction Through the Use of Larger Jet Aircraft" (2009) 15 Journal of Air Transport Management 151.

Morrell, Peter. "An Evaluation of Possible EU Air Transport Emissions Trading Scheme Allocation Methods" (2007) 35 Energy Policy 5562.

Morris, Madeline H. “Universal Jurisdiction in a Divided World” (2001) 35 New Eng L Rev 337.

Motaal, Doaa Abdel. “Curbing CO_2 Emissions from Aviation: Is the Airline Industry Headed for Defeat?” (2012) 3 Climate Change 1.

Muchmore, Adam I. “Jurisdictional Standards (and Rules)” (2013) 46 Vand J Transnat’l L 171.

Murphy, Sean D. “Deconstructing Fragmentation: Koskenniemi’s 2006 ILC Project” (2013) 27:2 Temp Int’l & Comp LJ 293.

Muskin, Jerold B. “An Effluent Charge Approach to Aircraft Noise Abatement” (1978) 5 J Env Eco & Manag 333.

Nagle, Luz E. “Terrorism and Universal Jurisdiction: Opening a Pandora’s Box” (2011) 27 Ga St UL Rev 339.

Nazifi, Fatemeth. “The Price Impact of Linking the European Union Emissions Trading Scheme to the Clean Development Mechanism” (2010) 12 Enviro Econo & Pol’y Stud 164.

Neuhold, Hanspeter. “Variations on the Theme of Soft International Law” in Isabelle Buffard et al, eds, *International Law between Universalism and Fragmentation* (Leiden: Martinus Nijhoff Publishers, 2008).

Neuling, Bruce. “The Shrimp-Turtle Case: Implications for Article XX of GATT and the Trade and Environment Debate” (1999) 22 Loy LA Int’l & Comp L Rev 1.

Nilsson, Jan Henrik. “Low-Cost Aviation” in Stefan Gossling & Paul Upham, eds, *Climate Change and Aviation: Issues, Challenges and Solutions* (London: Earthscan, 2009) 113.

Nissel, Alan. “The ILC Articles on State Responsibility: Between Self-Help and Solidarity” (2005) 38 NYU J Int’l L & Pol 355.

Nkuepo, Henri J. “EU ETS Aviation Discriminates Against Developing Countries” (2012) 7 African Trade Law 1.

Nygyren, Emma, Kjell Aleklett & Mikael Hook. “Aviation Fuel and Future Oil Production Scenarios” (2009) 37 Energy Policy 4003.

Oberthür, Sebastian. "The Climate Change Regime: Interactions with ICAO, IMO, and the EU Burden-Sharing Agreement" in Sebastian Oberthür & Thomas Gehring, eds, *Institutional Interaction in Global Environmental Governance: Synergy and Conflict Among International and EU Policies* (Cambridge: MIT Press, 2006) 53.

Oberthür, Sebastian & Thomas Gehring. "Conceptual Foundations of Institutional Interaction" in Sebastian Oberthür & Thomas Gehring, eds, *Institutional Interaction in Global Environmental Governance: Synergy and Conflict among International and EU Policies* (Cambridge: MIT Press, 2006) 19.

Oberthür, Sebastian & Thomas Gehring. "Disentangling the Interaction between the Cartagena Protocol and the World Trade Organization" in Oran R Young, et al, eds, *Institutional Interplay* (New York: United Nations University Press, 2008) 94.

Oberthür, Sebastian & Thomas Gehring. "Institutional Interaction: Ten Years of Scholarly Development" in Sebastian Oberthür & Olav Schram Stokke, eds, *Managing Institutional Complexity: Regime Interplay and Global Environmental Change* (Cambridge: Massachusetts Institute of Technology, 2011) 25.

Oberthür, Sebastian, Claire Dupont & Yasuko Matsumoto. "Managing Policy Contradictions between the Montréal and Kyoto Protocols: The Case of Fluorinated Greenhouse Gases" in Sebastian Oberthür & Olav Schram Stokke, eds, *Managing Institutional Complexity: Regime Interplay and Global Environmental Change* (Cambridge: Massachusetts Institute of Technology, 2011) 115.

Oberthür, Sebastian & Claire Dupont. "The Council, the European Council and International Climate Policy: From Symbolic Leadership to Leadership by Example" in Rüdiger K.W. Wurzel & James Connelly, eds, *The European Union as a Leader in International Climate Change Politics* (London: Routledge, 2011) 74.

Oellers-Frahm, Karin. "Multiplication of International Courts and Tribunals and Conflicting Jurisdiction – Problems and Possible Solutions" (2001) 5 Max Planck UNYB 67.

Opschoor, Hans. "Developments in the Use of Economic Instruments in OECD Countries" in Ger Klaassen & Finn R Førsund, eds, *Economic Instruments for Air Pollution Control* (Dordrecht: Kluwer Academic Publishers, 1994) 75.

Orentlicher, Diane F. "Whose Justice? Reconciling Universal Jurisdiction with Democratic Principles" (2004) 92 Geo LJ 1057.

Paines, A J C. "Extraterritorial Aspects of Mergers and Joint Ventures" (1985) 13 Int'l Bus Law 344.

Palmer, Alice, Beatrice Chaytor & Jacob Werksman. "Interactions between the World Trade Organization and International Environmental Regimes" in Sebastian Oberthür & Thomas Gehring, eds, *Institutional Interaction in Global Environmental Governance: Synergy and Conflict among International and EU Policies* (Cambridge: MIT Press, 2006) 181.

Pan, Jiahua, Jonathan Phillips & Ying Chen. "China's Balance of Emissions Embodied in Trade: Approaches to Measurement and Allocating International Responsibility" (2008) 24:2 Oxford Rev Econ Pol'y 354.

Parchomovsky, Gideon & Philip J Weiser. "Beyond Fair Use" (2010) 96:1 Cornell L Rev 91.

Parker, Andrew. "Emirates: A Perspective on Issues in Canadian Aviation" (2012) 37:6 Air & Space L 419.

Parrish, Austen. "The Effects Test: Extraterritoriality's Fifth Business" (2008) 61 Vand L Rev 1455.

Partsch, Karl Josef. "Discrimination against Individuals and Groups" in R Bernhardt & Max Planck, eds, *Encyclopedia of Public International Law*, vol 1 (Amsterdam: North Holland, 1992) 1079.

Pavel, Carmen. "Normative Conflict in International Law" (2009) 46 San Diego L Rev 883.

Pearse, Peter H. "Developing Property Rights as Instruments of Natural Resources Policy: The Case of the Fisheries" in OECD, *Climate Change: Designing a Tradable Permit System* (Paris: OECD, 1992) 110.

Peck, Brian. "Extraterritorial Application of Antitrust Laws and the U.S.-EU Dispute Over the Boeing and McDonnell Douglas Merger: From Comity to Conflict? An Argument for a Binding International Agreement on Antitrust Enforcement and Dispute Resolution" (1998) 35 San Diego L Rev 1163.

Peeters, Paul & Victoria Williams. "Calculating Emissions and Radiative Forcing" in Stefan Gossling & Paul Upham, eds, *Climate Change and Aviation: Issues, Challenges and Solutions* (London: Earthscan, 2009) 69.

Pendleton, Gregory D. "State Responsibility and the High Seas Marine Environment: A Legal Theory for the Protection of Seamounts in the Global Commons" (2005) 14 Pac Rim L & Pol'y J 485.

Pentelow, Laurel & Daniel J Scott. "Aviation's Inclusion in International Climate Policy Regimes: Implications for the Caribbean Tourism Industry" (2011) 17 J Air Transport Management 199.

Perdan, Slobodan & Adisa Azapagic. "Carbon Trading: Current Schemes and Future Developments" (2011) 39 Energy Policy 6040.

Perez, Oren. "Private Environmental Governance as Ensemble Regulation: A Critical Exploration of Sustainability Indexes and the New Ensemble Politics" (2011) 12 Theoretical Inq L 543.

Peterson, Thomas D & Adam Z Rose. "Reducing Conflicts between Climate Policy and Energy Policy in the US: The Important Role of the States" (2006) 34 Energy Policy 619.

Piera, Alejandro. "Report of the Rapporteur of the Special Sub-Committee on the Preparation of an Instrument to Modernize the Convention on Offences and Certain Other Acts Committed on Board Aircraft of 1963" (2012) 37 Ann Air & Sp L 485.

Piera, Alejandro & Michael Gill. "Unruly & Disruptive Passengers: Do We Need to Revisit the International Legal Regime?" (2010) XXXV Ann Air & Sp L 355.

Piera, Alejandro & Michael Gill. "Will the New ICAO-Beijing Instruments Build a Chinese Wall for International Aviation Security?" (2014) 47:1 Vand J Transnat'l L 145.

Pope, Jeff & Anthony D Owen. "Emission Trading Schemes: Potential Revenue Effects, Compliance Costs and Overall Tax Policy Issues" (2009) 37 Energy Policy 4595.

Posner, Richard A. "Some Economics of International Law: Comment on Conference Papers" (2002) J Legal Stud 321.

Posner, Eric A. "International Law and the Disaggregated State" (2005) 32 Fla St UL Rev 797.

Powell, Catherine. "The Role of Transnational Norm Entrepreneurs in the U.S. 'War on Terrorism'" (2004) 5 Theoretical Inquiries in Law 47.

Pozen, David. "We Are All Entrepreneurs Now" (2008) 43 Wake Forest L Rev 283.

Pronto, Arnold N. "Human-Rightism and the Development of General International Law" (2007) 20 Leiden J Int'l L 753.

Quénivet, Noëlle. "Binding the United Nations to Human Rights Norms by Way of the Laws of Treaties" (2010) 42 Geo Wash Int'l L Rev 587.

Quereshi, Asif H. "Extraterritorial Shrimps, NGOs and the WTO Appellate Body" (1999) 48 ICLQ 199.

Quinn, Elias Leake. "The Solitary Attempt: International Trade Law and the Insulation of Domestic Greenhouse Gas Trading Schemes from Foreign Emissions Credit Markets" (2009) 80 U Colo L Rev 201.

Radetzki, Mariam. "The Fallacies of Concurrent Climate Policy Efforts" (2010) 39 AMBIO 211.

Radgen, Peter, Jane Butterfield & Jürgen Rosenow. "EPS, ETS, Renewable Obligations and Feed in Tariffs: Critical Reflections on the Compatibility of Different Instruments to Combat Climate Change" (2011) 4 Energy Procedia 5814.

Rajamani, Lavanya, Jutta Brunnée & Meinhard Doelle. "Introduction: The Role of Compliance in an Evolving Climate Regime" in Jutta Brunnée, Meinhard Doelle & Lavanya Rajamani, eds, *Promoting Compliance in an Evolving Climate Regime* (Cambridge: Cambridge University Press, 2012) 1.

Randall, Kenneth C. "Universal Jurisdiction under International Law" (1998) 66 Tex L Rev 785.

Ranking, Jennifer K. "U.S. Laws in the Rainforest: Can a U.S. Court Find Liability for Extraterritorial Pollution Caused by a U.S. Corporation? An Analysis of *Aguinada* v. *Texaco, Inc.*"(1995) 18 BC Int'l & Comp L Rev 221.

Rauch, Isabel. “Developing a German and an International Emissions Trading System: Lessons from U.S. Experiences with the Acid Rain Program” (2000) 11 Fordham Envtl LJ 307.

Raustiala, Kal. “The Architecture of International Cooperation: Transgovernmental Networks and the Future of International Law” (2002) 43 Va J Int’l L 1.

Reagan, Daniel B. “Putting International Aviation into the European Union Emissions Trading Scheme: Can Europe Do It Flying Solo?” (2008) 35 BC Envtl Aff L Rev 349.

Redgwell, Catherine. “Facilitation of Compliance” in Jutta Brunnée, Meinhard Doelle & Lavanya Rajamani, ed, *Promoting Compliance in an Evolving Climate Regime* (Cambridge: Cambridge University Press, 2012) 177.

Reuven, S Avi-Yonah & David M Uhlmann. “Combating Global Climate Change: Why a Carbon Tax Is a Better Response to Global Warming than Cap and Trade” (2009) 28 Stan Envtl LJ 3.

Rietvelt, Marc. “Multilateral Failure: A Comprehensive Analysis of the Shrimp/Turtle Decision” (2005) 15 Ind Int’l & Comp L Rev 473.

Roht-Arriaza, Naomi. “The Pinochet Precedent and Universal Jurisdiction” (2001) 35 New Eng L Rev 311.

Rose, Carol M. “Hot Spots in the Legislative Climate Change Proposals” (2008) 102 Nw UL Rev Colloquy 189.

Rosendal, G Kristin. “Impacts of Overlapping International Regimes: The Case of Biodiversity” (2001) 7 Global Governance 95.

Rosendal, G Kristin. “The Convention on Biological Diversity: Tensions with the WTO TRIPS Agreement over Access to Genetic Resources and the Sharing of Benefits” in Sebastian Oberthür & Thomas Gehring, eds, *Institutional Interaction in Global Environmental Governance: Synergy and Conflict among International and EU Policies* (Cambridge: MIT Press, 2006) 79.

Rothengatter, Werner. “Climate Change and the Contribution of Transport: Basic Facts and the Role of Aviation” (2010) 15 Transportation Research Part D5.

Ruckelshaus, William D. "Toward a Global Environmental Policy" in Ruth S DeFries & Thomas F Malone, eds, *Global Change and our Common Future* (Washington: National Academy Press, 1989) 3.

Ryan, Bernard. "Extraterritorial Immigration Control: What Role for Legal Guarantees?" in Bernard Ryan and Valsamis Mitsilegas, eds, *Extraterritorial Immigration Control: Legal Challenges* (The Hague: Martinus Nijhoff Publishers, 2010) 3.

Ryangaert, Cedric. "The International Criminal Court and Universal Jurisdiction: A Fraught Relationship?" (2009) 12 New Crim LR 498.

Sachs, Noah. "Beyond the Liability Wall: Strengthening Tort Remedies in International Environmental Law" (2008) 55 UCLA L Rev 837.

Sadat, Leila Nadya. "Crimes against Humanity in the Modern Age" (2013) 107 Am J Int'l L 334.

Salama, Randa. "Fragmentation of International Law: Procedural Issues Arising in Law of the Sea Disputes" (2005) 19 MLAANZ J 24.

Samaniego, José Luis & Christiana Figueres. "Evolving to a Sector-Based Clean Development Mechanism" in Kevin A Baumert, ed, *Building on the Kyoto Protocol: Options for Protecting the Climate* (United States: World Resources Institute, 2002) 89.

Sampson, Gary P. "WTO Rules and Climate Change: The Need for Policy Coherence" in W Bradnee Chambers, ed, *Inter-Linkages: The Kyoto Protocol & the International Trade & Investment Regimes* (New York: United Nations University Press, 2001) 69.

Sanchez, Gabriel S. "In Defense of Incrementalism for International Aviation Emissions Regulation" (2012) 53:1 Va J Int'l L 1.

Sands, Philippe. "The Greening of International Law: Emerging Principles and Rules" (1994) 1:2 Ind J Global Legal Stud 293.

Sands, Philippe. "Turtles and Torturers: The Transformation of International Law" (2001) 33 NYU J Int'l L & Pol 527.

Sato, Chie. "Extraterritorial Application of EU Competition Law: Is it Possible for Japanese Companies to Steer Clear of EU Competition Law?" 11 J Pol Sci & Soc 23.

Schaffer, Ian L. "An International Train Wreck Caused in Part by a Defective Whistle: When the Extraterritorial Application of Sox Conflicts with Foreign Laws" (2006) 75:3 Fordham L Rev 1829.

Scharf, Michael P. "Application of Treaty-Based Universal Jurisdiction to Nationals of Non-Party States" (2001) 35 New Eng L Rev 363.

Scharf, Michael P. "Universal Jurisdiction and the Crime of Aggression" (2012) 53 Harv Int'l LJ 357.

Scheelhaase, Janina D. "Local Emission Charges – A New Economic Instrument at German Airports" (2010) 16 J Air Transport Management 94.

Scheelhaase, Janina D & Wolfgang G Grimme. "Emissions Trading for International Aviation – An Estimation of the Economic Impact on Selected European Airlines" (2007) 13 J Air Transport Management 253.

Schleich, Joachim, Karl-Martin Ehrhard, Christian Hoppe & Stefan Seifert. "Generous Allocation and a Ban on Banking – Implications of a Simulation Game for EU Emissions Trading" in Ralf Antes Bernd Hansjürgens & Peter Letmathe, eds, *Emissions Trading and Business* (Heidelberg: Physica-Verlag, 2006) 27.

Schlumberger, Charles E. "The Oil Price Spike of 2008" (2009) 34 Ann Air & Sp L 114.

Schlumberger, Charles E. "Are Alternative Fuels an Alternative? A Review of the Opportunities and Challenges of Alternative Fuels for Aviation" (2010) 35-1 Ann Air & Sp L 119.

Schneider, Stepehn H. "Dangerous Climate Change: Key Vulnerabilities" in Ernesto Zedillo, ed, *Global Warming: Looking Beyond Kyoto* (Washington: Brookings Institution Press, 2008) 57.

Schoenbaum, Thomas J. "International Trade and Protection of the Environment: The Continuing Search for Reconciliation" (1997) 91 Am J Int'l L 268.

Schreurs, Miranda A & Yves Tiberghien. "European Union Leadership in Climate Change: Mitigation through Multilevel Reinforcement" in Kathryn Harrison & Lisa McIntosh Sundstrom, *Global Commons, Domestic Decisions: The Comparative Politics of Climate Change* (Cambridge: MIT Press, 2010) 23.

Schroeder, Heike. "Analyzing Biosafety and Trade through the Lens of Institutional Interplay" in Oran R Young et al, eds, *Institutional Interplay* (New York: United Nations University Press, 2008) 49.

Schütze, Robert. "EC Law and International Agreements of the Member States – An Ambivalent Relationship?"(2006) 9 Cambridge YB Eur Legal Stud 387.

Scott, Karen N. "International Environmental Governance: Managing Fragmentation through Institutional Connection" (2011) 12 Melbourne J of Int'l Law 177.

Scott, Robert E. "The Limits of Behavioral Theories of Law and Social Norms" (2000) 86:8 Va L Rev 1603.

Sgouridis, Sgouris, Philippe A Bonnefoy & R John Hansman. "Air Transportation in a Carbon Constrained World: Long-Term Dynamics of Policies and Strategies for Mitigating the Carbon Footprint of Commercial Aviation" (2011) 1 Transportation Research Part A 1077.

Sheeran, Kristen. "Environmental Consequences of Free Trade: Beyond Kyoto: North-South Implications of Emissions Trading and Taxes" (2007) 5 Seattle J Soc Just 697.

Shenefield, John H. "Thoughts on Extraterritorial Application of United States Antitrust Laws" (1983) 52:3 Fordham L Rev 350.

Silversmith, Jol A. "The Long Arm of the DOT: The Regulation of Foreign Air Carriers Beyond US Borders" (2013) 38:3 Air & Space L 3 173.

Simma, Bruno. "Universality of International Law from the Perspective of a Practitioner" (2009) 20:2 EJIL 265 at 270

Singham, Shanker A. "Shaping Competition Policy in the American: Scope for Transatlantic Cooperation" (1998) 24 Brooklyn J Int'l L 363.

Skeel, David A Jr. "Shaming in Corporate Law" (2001) 149 U Pa L Rev 1811.

Skjærseth, Jon Birger & Jørgen Wettestad. "Implementing EU Emissions Trading: Success or Failure?" (2008) 8 Int Environ Agreements 275.

Skjærseth, Jon Birger & Jørgen Wettestad. "Fixing the EU Emissions Trading System? Understanding the Post-2012 Changes" (2010) 10:4 Global Environmental Politics 10.

Slaughter, Anne-Marie, Andrew S Tulumello & Stepan Wood. "International Law and International Relations Theory: A New Generation of Interdisciplinary Scholarship" (1998) 92 Am J Int'l L 367.

Sloane, Robert D. "On the Use and Abuse of Necessity in the Law of State Responsibility" (2012) 106 Am J Int'l L 447.

Smith, Charles E. "Air Transportation Taxation: The Case for Reform" (2010) 75 J Air L & Com 915.

Smith, Inga J & Craig J Rodger. "Carbon Emission Offsets for Aviation-Generated Emissions Due to International Travel To and From New Zealand" (2009) 37 Energy Policy 3438.

Smitherman III, Charles W. "The Future of Global Competition Governance: Lessons from the Transatlantic" (2004) 19 Am U Int'l L Rev 769.

Smithies, Richard. "Regulatory Convergence: Extending the Reach of EU Aviation Law" (2007) 72 J Air L & Com 3.

Snidal, Duncan. "Rational Choice and International Relations" in Walter Emmanuel Carlsnaes & Beth A Simmons, eds, *Handbook of International Relations* (London: Sage Publications Inc, 2013) 85.

Solomon, Diana S & Kenneth F D Hughey. "A Proposed Multi Criteria Analysis Decision Support Tool for International Environmental Policy Issues: A Pilot Application to Emissions Control in the International Aviation Sector" (2007) Environmental Science and Policy 645.

Sprinz, Detlef & Tapani Vaahtoranta. "The Interest-Based Explanation of International Environmental Policy" in Peter M Haas, ed, *International Environmental Governance* (Burlington: Ashgate, 2008) 131.

Staniland, Martin. "Air Transport and the EU's Emissions Trading Scheme: Issues and Arguments" (2008) 8 Issues Aviation L & Pol'y 153.

Stavins, Robert N. "Addressing Climate Change with a Comprehensive US Cap-And-Trade System" (2008) 34:2 Oxford Rev of Econ Pol'y 298.

Stavins, Robert N. "An International Policy Architecture for the Post-Kyoto Era" in Ernesto Zedillo, ed, *Global Warming: Looking Beyond Kyoto* (Washington: Brookings Institution Press, 2008) 145.

Steinberg, Richard H. "Power and Cooperation in International Environmental Law" in Andrew T Guzman & Alan O Sykes, eds, *Research Handbook in International Economic Law* (Northampton: Edward Elgar, 2007) 485.

Steiner, Henry J. "Three Cheers for Universal Jurisdiction – or Is It Only Two?" (2004) 5 Theoretical Inq L 199.

Steppler, Ulrich & Angela Klingmüller. "EU Emissions Trading Scheme and Aviation: Quo Vadis?" (2009) 34:4/5 Air & Space L 253.

Sterk, Wolfgang & Joseph Kruger. "Establishing a Transatlantic Carbon Market" (2009) 9 Climate Policy 389.

Stern, Nicholas. "The Economics of Climate Change" (2008) 98:2 American Economic Review: Papers & Proceedings 1.

Stevenson, Peter. "The World Trade Organization Rules: A Legal Analysis of Their Adverse Impact on Animal Welfare" (2002) 8 Animal L 107.

Stewart, Richard B. "Instrument Choice" in Daniel Bodansky, Jutta Brunnée & Ellen Hey, eds, *The Oxford Handbook of International Environmental Law* (Oxford: Oxford University Press, 2007) 147.

Stokke, Olav Schram & Sebastian Oberthür. "Institutional Interaction in Global Environmental Change" in Sebastian Oberthür & Olav Schram Stokke, eds, *Managing Institutional Complexity: Regime Interplay and Global Environmental Change* (Cambridge: Massachusetts Institute of Technology, 2011) 1.

Stokke, Olav Schram, Lee G. Anderson & Natalia Mirovitskaya. "The Barents Sea Fisheries" in Oran R Young, ed, *The Effectiveness of International Environmental Regimes: Causal Connections and Behavioral Mechanism* (Cambridge: MIT Press, 1999) 91.

Stoltenberg, Clyde et al. "A Comparative Analysis of Post-Sarbanes-Oxley Corporate Governance Developments in the US and European Union: The Impact of Tensions Created by Extraterritorial Application of Section 404" (2005) 53 Am J Comp L 457.

Stone, Christopher D. "Common but Differentiated Responsibilities in International Law" (2004) 98 Am J Int'l L 276.

Strahilevitz, Lior Jacob. "Charismatic Code, Social Norms, and the Emergence of Cooperation on the File-Swapping Networks" (2003) 89 Va L Rev 505.

Streck, Charlotte & David Freestone. "The EU and Climate Change" in Richard Macrory, ed, *Reflections on 30 Years of EU Environmental Law: A High Level of Protection* (Groningen: Europa Law Publishing, 2006) 87.

Stripple, Johannes & Eva Lövbrand. "Carbon Market Governance Beyond the Public-Private Divide" in Frank Biermann, Philipp Pattberg & Fariborz Zelli, eds, *Global Climate Governance Beyond 2012: Architecture, Agency and Adaptation* (Cambridge: Cambridge University Press, 2010) 165.

Summers, Lawrence. "Foreword" in Joseph E Aldy & Robert N Stavins, eds, *Architectures for Agreement: Addressing Global Climate Change in the Post-Kyoto World* (Cambridge: Cambridge University Press, 2007) xviii.

Summers, Mark A. "The International Court of Justice's Decision in *Congo* v. *Belgium*: How Has It Affected the Development of a Principle of Universal Jurisdiction that Would Obligate All States to Prosecute War Criminals?" (2003) 21 BU Int'l LJ 63.

Sunstein, Cass R. "Social Norms and Social Roles" (1996) 96:4 Colum L Rev 903.

Suzannah Linton & Firew Kebede Tiba."The International Judge in an Age of Multiple International Courts and Tribunal" (2009) 9 Chi J Int'l L 407.

Sweeney, Brendan. "Reflections on a Decade of International Law: International Competition Law and Policy: A Work in Progress" (2009) 10 Melbourne J of Int'l Law 58.

Switzer, Stephanie. "Aviation and Emissions Trading in the European Union: Pie in the Sky or Compatible with International Law" (2012) 39:1 Ecology L Currents 1.

Sydnes, Are K. "Overlapping Regimes: The SPS Agreement and the Cartagena Biosafety Protocol" in Oran R Young et al, eds, *Institutional Interplay* (New York: United Nations University Press, 2008) 71.

Teegen, Hildy. "International NGOs as Global Institutions: Using Social Capital to Impact Multinational Enterprises and Governments" (2003) 9 Journal of International Management 271.

Teitel, Rutti & Robert Howse. "Patterns, Possibilities and Problems: Cross Judging: Tribunalization in a Fragmented but Interconnected Global Order" (2009) 41 NYU J Int'l L & Pol'y 959.

Thirlway, Hugh. "Injured and Non-Injured States before the International Court of Justice" in Maurizio Ragazzi, ed, *International Responsibility Today: Essays in Memory of Oscar Schachter* (Leiden: Martinus Nijhoff Publishers, 2005) 311.

Thomas, Daniel C. "Boomerangs and Superpowers: International Norms, Transnational Networks and US Foreign Policy" (2002) 15:1 Cambridge Review of International Affairs 25.

Thompson, Barton H Jr. "Tragically Difficult: The Obstacles to Governing the Commons" (2000) 30 Entl L 241.

Thornton, Robert L. "Governments an Airlines" in Robert O Keohane & Joseph S Nye Jr, eds, *Transnational Relations and World Politics* (Cambridge: Harvard University Press, 1970) 202.

Tien, Lee. "Architectural Regulation and the Evolution of Social Norms" (2004) 7 Yale J L & Tech 1.

Tietenberg, Tom. "The Tradable-Permits Approach to Protecting The Commons: Lessons for Climate" (2003) 19:3 Oxford Rev Econ Pol'y 400.

Tol, Richard S K. "The Impact of a Carbon Tax on International Tourism" (2007) 12 Transportation Research Part D 129.

Tolba, Mostafa K & Iwona Rummel-Bulska. "The Story of the Ozone Layer" in Donald Kaniaru, ed, *The Montréal Protocol: Celebrating 20 Years of Environmental Progress* (London: Cameron May Ltd, 2007) 27.

Torres, Miguel & Paulo Pinho. "Encouraging Low Carbon Policies through a Local Emissions Trading Scheme (LETS)" (2011) 28 Cites 576.

Toy, Alan. "Cross-Border and Extraterritorial Application of New Zealand Data Protection Laws to Online Activity" (2010) 24 New Zealand Universities L Rev 238.

Tsai, Allen Pei-Jan & Annie Petsonk. "Tracking the Skies: An Airline-Based System for Limiting Greenhouse Gas Emissions from International Civil Aviation" (1999) 6 Envtl Law 763.

Turley, Jonathan. "When in Rome: Multinational Misconduct and the Presumption Against Extraterritoriality" (1990) 84 NW UL Rev 598.

Ulen, Thomas S. "Rational Choice and the Economic Analysis of Law" (1994) 49:2 Law & Social Inquiry 487.

Unruh, Gregory C. "Escaping Carbon" (2002) Energy Policy 30 317.

Van Asselt, Harro et al. "Global Climate Change and the Fragmentation of International Law" (2008) 20:4 Law & Pol'y 429.

Van Asselt, Harro. "Emissions Trading: The Enthusiastic Adoption of an 'Alien' Instrument?" in Andrew Jordan et al, eds, *Climate Change Policy in the European Union: Confronting the Dilemmas of Mitigation and Adaptation?*(Cambridge: Cambridge University Press, 2010) 125.

Van Asselt, Harro. "Legal and Political Approaches in Interplay Management: Dealing with the Fragmentation of Global Climate Governance" in Sebastian Oberthür & Olav Schram Stokke, eds, *Managing Institutional Complexity: Regime Interplay and Global Environmental Change* (Cambridge: MIT, 2011) 59.

Van Asselt, Harro. "Managing the Fragmentation of International Environmental Law: Forests at the Intersection of the Climate and Biodiversity Regimes" (2012) 44 NYU J Int'l L & Pol 1205.

Van den Brink, Ryan. "Competitiveness Boarder Adjustments in U.S. Climate Change Proposals Violate GATT: Suggestions to Utilize GATT's Environmental Exceptions" (2010) 21 Colo J Int'l Envtl L & Pol'y 85.

Vandenbergh, Michael P. "Order Without Social Norms: How Personal Norm Activation Can Protect the Environment" (2005) 99 Nw UL Rev 1101.

Varghese, Tracey P. "The WTO's Shrimp-Turtle Decisions: The Extraterritorial Enforcement of U.S. Environmental Policy via Unilateral Trade Embargoes" (2001) 8 Envtl Law 421.

Vesparmann, Jan & Andreas Wald. "Much Ado About Nothing? – An Analysis of Economic Impacts and Ecologic Effects of the EU-Emission Trading Scheme in the Aviation Industry"

Victor, David G. "On the Regulation of Geoengineering" (2008) 24:2 Oxford Rev Econ Pol'y 322.

Vierdag, E W. "The Time of the 'Conclusion' of a Multilateral Treaty: Article 30 of the Vienna Convention on the Law of Treaties and Related Provisions" (1988) 59 Brit YB Int'l L 75.

Viñuales, Jorge E. "Managing Abidance by Standards for the Protection of the Environment" in Antonio Cassese, ed, *Realizing Utopia: The Future of International Law* (Oxford: Oxford University Press, 2012) 326.

Viñuales, Jorge E. "The Contribution of the International Court of Justice to the Development of International Environmental Law: A Contemporary Assessment" (2008) 32 Fordham Int'l LJ 232.

Vogler, John. "The European Contribution to Global Environmental Governance" (2005) 81:4 Int'l Affairs 835.

Vogler, John & Charlotte Bretherton. "The European Union as a Protagonist to the United States on Climate Change" (2006) 7 International Studies Perspectives 1.

Volkman, John M. "Making Change in a New Currency: Incentives and the Carbon Economy" (2009) 29 Pub Land & Resources L Rev 1.

Volkovitsch, Michael John. "Righting Wrongs: Towards a New Theory of State Succession to Responsibility for International Delicts" (1992) 92 Colum L Rev 2162.

Wang, Haifeng. "Economic Cost of CO_2 Emissions Reduction for Non-Annex I Countries in International Shipping" (2010) 14 Energy for Sustainable Development 280.

Warner, Daniel. "Commercial Aviation: An Unsustainable Technology" (2009) 74 J Air L & Com 553.

Weiner, Michael L. "Conflict and Cooperation: Meeting the Challenge of Increasing Globalization" (1997) 12 Antitrust ABA 4.

Weiss, Peter. "The Future of Universal Jurisdiction" in Wolfgang Kaleck et al, eds, *International Prosecution of Human Rights Crimes* (Berlin: Springer, 2007) 29.

Wemaëre, Matthieu. "Legal Nature of Kyoto Units" in W Th Douma, L Massai, M Montini, eds, *The Kyoto Protocol and Beyond: Legal and Policy Challenges of Climate Change* (The Hague: TMC Asser Press, 2007) 71.

Werksman, Jacob. "Compliance and the Use of Trade Measures" in Jutta Brunnée, Meinhard Doelle & Lavanya Rajamani, eds, *Promoting Compliance in an Evolving Climate Regime* (Cambridge: Cambridge University Press, 2012) 262.

Whatstein, Liad. "Extraterritorial Application of EC Competition Law: Comments and Reflections" (1992) 26 Isr L Rev 195.

Whitehead, Charles K. "What's Your Sign? – International Norms, Signals, and Compliance" (2006) 27 Mich J Int'l L 695.

Whytock, Christopher A. "A Rational Design Theory of Transgovernmentalism: The Case of E.U.-U.S. Merger Review Cooperation" (2005) 23 BU Int'l LJ 1.

Wiener, Jonathan B. "Precaution" in Daniel Bodansky, Jutta Brunnée & Ellen Hey, eds, *The Oxford Handbook of International Environmental Law* (Oxford: Oxford University Press, 2006) 597.

Wiener, Jonathan B. "Responding to the Global Warming Problem: Something Borrowed for Something Blue: Legal Transplants and the Evolution of Global Environmental Law" (2001) 21 Ecology LQ 1295.

Wilson, Kathyrn C. "The International Air Quality Management District: Is Emissions Trading the Innovative Solution to the Transboundary Pollution Problem" (1995) 30 Tex Int'l LJ 369.

Winkler, Harald. "An Architecture for Long-term Climate Change: North-South Cooperation Based on Equity and Common but Differentiated Responsibilities" in Frank Bierman, Phillip Pattberg & Fariborz Zelli, eds, *Global Climate Governance Beyond 2012: Architecture, Agency and Adaptation* (Cambridge: Cambridge University Press, 2010) 97.

Winter, Gerd. "On Integration of Environmental Protection into Air Transport Law: A German and EC Perspective" (1996) 21:3 Air & Space L 132.

Witherell, Brendan J. "Trademark Law: The Extraterritorial Application of the Lanham Act: The First Circuit Cuts the Fat from the Vanity Fair Test" (2006) 29 W New Eng L Rev 193.

Wolfrum, Rüdiger. "International Environmental Law: Purpose, Principles and Means of Ensuring Compliance" in Fred. L. Morrison & Rüdiger Wolfrum, eds, *International, Regional and National Environmental Law* (The Hague: Kluwer Law International, 2000) 3.

Worster, William Thomas. "Competition and Comity in the Fragmentation of International Law" (2008) 34 Brooklyn J Int'l L 119.

Yan, Xu. "Green Taxation in China: A Possible Consolidated Transport Fuel Tax to Promote Clean Air?" (2010) 21 Fordham Envtl L Rev 295.

Yifeng, Chen. "Structural Limitations and Possible Future Work of the International Law Commission" (2010) 9 Chinese J Int'l L 473.

Young, Hoong N. "An Analysis of a Global CO_2 Emissions Trading Program" (1998) 14 J Land Use & Envt L 125.

Young, Oran R. "The Rise and Fall of International Regimes" (1982) 36:2 International Organization 277.

Young, Oran R. "Regime Dynamics: The Rise and Fall of International Regimes" (1982) 36:2 International Organization 277.

Young, Oran R. "Regime Effectiveness: Taking Stock" in Oran R Young, ed, *The Effectiveness of International Environmental Regimes: Causal Connections and Behavioral Mechanism* (Cambridge: MIT Press, 1999) 249.

Young, Oran R. "Deriving Insights from the Case of the WTO and the Cartagena Protocol" in Oran R Young et al, eds, *Institutional Interplay* (New York: United Nations University Press, 2008) 131.

Young, Oran R & Marc A Levy. "The Effectiveness of International Environmental Regimes" in Oran R Young, ed, *The Effectiveness of International Environmental Regimes: Causal Connections and Behavioral Mechanisms* (Cambridge: MIT Press, 1999) 1.

Zapatero, Pablo. "Modern International Law and the Advent of Special Legal Systems" (2005) 23 Ariz J Int'l & Comp L 55.

Zelli, Fariborz et al. "The Consequences of a Fragmented Climate Governance Architecture: A Policy Appraisal" in Frank Biermann, Philipp Pattberg & Fariborz Zelli, eds, *Global Climate Governance Beyond 2012: Architecture, Agency and Adaptation* (Cambridge: Cambridge University Press, 2010) 25.

Zelli, Fariborz. "The Fragmentation of the Global Climate Governance Architecture" (2011) 2 Wires Clim Change 255.

Zemach, Ariel. "Reconciling Universal Jurisdiction with Equality before the Law" (2011) 47 Tex Int'l LJ 143.

Zhang, Yue-Jun & Yi-Ming Wei. "An Overview of Current Research on EU ETS: Evidence from Its Operating Mechanism and Economic Effect" (2010) 87 Applied Energy 1804.

Zhang, Anming, Sveinn Vidar Gudmundsson & Tae H Oum. "Air Transport, Global Warming and the Environment" (2010) 15:1 Transportation Research Part D 1.

Zuppi, Alberto Luis. "Immunity v. Universal Jurisdiction: The Yerodia Ndombasi Decision of the International Court of Justice" (2003) 63 La L Rev 309.

Letters

Letter from Sim Kallas (Vice-President of the European Commission) and Connie Hedegaard (Climate Change Commissioner) to Ms. Hillary Rodham Clinton (US Secretary of State) and Mr. Raymond H. LaHood (US Secretary of Transportation) (16 January 2012).

Letter from numerous Nobel Laureates and renowned economists addressed to President Barack Obama (14 March 2012) online: WWF <http://assets.wwf.org.uk/downloads/eu_ets_letter_from_economists_to_obama.pdf>

Letter from Mr. Todd Stern, Special Envoy for Climate Change, inviting some States to discuss these issues in Washington DC (2 July 2012) [unpublished, archived on file with the author].

Letter from 19 NGOs Addressed to Hilary Rodham Clinton, US Secretary of State & Raymond H La Hood, Secretary of Transportation (30 July 2012) online: NBAA <www.nbaa.org/ops/environment/eu-ets/20120731-coalition-letter-us-hosted-aviation-climate-meeting.pdf>.

Letter from 350.org, Center for Biological Diversity, Climate Protection Campaign, Climate Solutions, Earthjustice, Environmental Defense Fund, Environment America, Environment Northeast, Greenpeace USA, Interfaith Power & Light, League of Conservation Voters, Natural Resources Defense Council, Oxfam America, Sierra Club, US Climate Action Network, and World Wildlife Fund to US President, Mr. Barack Obama (3 August 2012) online: EDF <www.edf.org/sites/default/files/NGOs_POTUS_EUETS_Article84_080312.pdf>.

Letter from 25 NGOs Addressed to ICAO (7 November 2012) [unpublished, archived on file with the author].

Letter from Prashant Sukul, Representative of India on the Council of ICAO, to Mr. Roberto Kobeh Gonzalez, President of the ICAO Council, untitled (14 March 2013) [unpublished, archived on file with the author].

Letter from Nicolas E Calio, President and CEO A4A, to the Honorable Duane Woerth, Ambassador to the US Mission to ICAO, Ms. Julie Oettinger, Assistant Administrator, Policy, International Affairs and Environment, FAA, and Mr. Todd Stern, Special Envoy for Climate Change, US Department of State (29 August 2013).

Letter from MA Tao to Raymond Benjamin, "Statement of Reservation of China regarding Resolution 17/2 of the 38th Session of the Assembly Consolidated Statement of Continuing ICAO Policies and Practices related to Environmental Protection: Climate Change" (8 October 2013) online: ICAO <www.icao.int/Meetings/a38/Documents/Resolutions/China_en.pdf>.

Letter from Joseph L Novak to Raymond Benjamin, "Statement of Reservation of China regarding Resolution 17/2 of the 38th ICAO Assembly Resolution: Consolidated Statement of Continuing ICAO Policies and Practices related to Environmental Protection – Climate Change" online: ICAO <www.icao.int/Meetings/a38/Documents/Resolutions/United_States_en.pdf>.

Presentations, Reports and Studies

"The European Union Emission Trading Scheme (EU-ETS) Insights and Opportunities" *Pew Center on Global Climate Change* online: C2ES <www.c2es.org/docUploads/EU-ETS per cent20White per cent20Paper.pdf>.

Annela Anger et al. "Air Transport in the European Union Emissions Trading Scheme, Final Report" *Aviation in a Sustainable World Omega* (December 2008) online: <www.verifavia.com/bases/ressource_pdf/109/AU-OmegaStudy-17-finalreport-AAPMA-2-1-240209.pdf>.

Anthony Aust, "Vienna Convention on the Succession of States in Respect of State Property, Archives and Debts" *United Nations Audiovisual Library of International Law* online: United Nations <http://legal.un.org/avl/pdf/ha/vcssrspad/vcssrspad_e.pdf>.

Mark Bisset & Georgina Crowhurst, "Is the EU's Application of Its Emissions Trading Scheme to Aviation Illegal" *Clyde & Co* (31 March 2011) online: <www.clydeco.com/news/articles/is-the-eus-application-of-its-emissions-trading-scheme-to-aviation-illegal>.

Tomer Broude, "Keep Calm and Carry On: Martti Koskenniemi and the Fragmentation of International Law", *SSRN* online: <http://papers.ssrn.com/sol3/papers.cfm?abstract_id=2297626##>.

Kate Cook, "The Extension of the EU Emissions Trading Scheme to the Aviation Sector Does Not Contravene International Law", *eutopia law* (1 November 2011) online: <http://eutopialaw.com/2011/11/01/the-extension-of-the-eu-emissions-trading-scheme-to-the-aviation-sector-does-not-contravene-international-law>.

Coraline Goron, "The EU Aviation ETS Caught between Kyoto and Chicago: Unilateral Legal Entrepreneurship in the Multilateral Governance System", *GR:EEN-GEM Doctoral Working Papers Series* (2012) online: University of Warwick <www2.warwick.ac.uk/fac/soc/csgr/green/papers/workingpapers/gem/no._2_c._goron.pdf>.

Robert Howse, "EU Aviation Emissions Scheme Opinion", International Economic Law and Policy Blog *World Trade Law* online: <worldtradelaw.typepad.com/ielpblog/2011/10/eu-aviation-emissions-scheme-opinion.html>.

IPCC, *Climate Change 2007: Synthesis Report, IPCC* (2007) online: <www.ipcc.ch/pdf/assessment-report/ar4/syr/ar4_syr.pdf>.

Baris Karapinar & Kateryna Holzer, *Legal Implications of the Use of Export Taxes in Addressing Carbon Leakage: Competing Border Adjustment Measures*, NCCR Trade Regulation Working Paper No 2012/15, Swiss National Centre of Competence in Research (April 2012) online: World Trade Institute <http://goo.gl/ny1jgR>.

Chris Lyle, "Mitigation of Air Transport Emissions which Contribute to Climate Change: A Tourism Perspective" (Presentation delivered at the International Symposium on Sustainable Tourism Development, Quebec City, 18 March 2009) online: Tourisme Québec <www.tourisme.gouv.qc.ca/activites/symposium-developpement-durable/presentations/18-C-14h00-1-Chris-Lyle.pdf>.

Michigan Technological University, "Life Cycle Assessment of Green Jet from Oils and Tallow: Comparison to Petroleum Jet Fuel". February 2009.

Jiahuan Pan, "Common but Differentiated Commitments: A Practical Approach to Engaging Large Developing Emitters under L20" (Presentation delivered at the Commissioned Briefing Notes for the CIGI/CFGS L20 Project, 20-21 September 2004) online: <http://goo.gl/H8TTFz>.

Dirk Pulkowski, "Narratives of Fragmentation: International Law between Unity and Multiplicity" *European Society of International Law* online: ESIL <www.esil-sedi.eu/sites/default/files/Pulkowski_0.PDF>.

Joanne Scott & Lavanya Rajamani, "EU Climate Change Unilateralism: International Aviation in the European Emissions Trading Scheme" *India Environmental Portal* online: <www.indiaenvironmentportal.org.in/files/file/EU per cent20Climate per cent20Change per cent20Unilateralism.pdf>.

Faiz Shakir, "Bush Ignores Science, Claims 'There Is A Debate' Over The Cause of Global Warming", *Think Progress* (26 June 2006) online: <http://thinkprogress.org/politics/2006/06/26/5997/bush-debate-climate/#>.

David Southgate, "Aviation Carbon Footprint: Global Scheduled International Passenger Flights 2012", *Scribd* online: <www.scribd.com/doc/137044034/Aviation-Carbon-Footprint-Global-Scheduled-International-Passenger-Flights-2012>.

W Tuinstra et al., *Aviation in the EU Emissions Trading Scheme: A First Step Towards Reducing the Impact of Aviation on Climate Change, Netherlands Environmental Assessment Agency* (2005) online: RIVM <www.rivm.nl/bibliotheek/rapporten/500043001.pdf>.

Andreas Tuerk, Michael Mehling, Sonja Klinsky & Xin Wang, "Emerging Carbon Markets: Experiences, Trends, and Challenges" *Climate Strategies* (2013) online: <www.climatestrategies.org/research/our-reports/category/63/370.html>.

Nancy Young, "A Look Back & Forward from Rio – Ever – Sustainable Aviation" (Presentation delivered at ATW's 5th Annual Eco-Aviation Conference, 21 June 2012) online: A4A <www.airlines.org/Pages/A-Look-Back-and-Forward-from-Rio-Ever-Sustainable-Aviation.aspx>.

Jennifer Zerk, *Extraterritorial Jurisdiction: Lessons for the Business and Human Rights Spheres from Six Regulatory Areas, Corporate Social Responsibility Initiative Working Paper No 59* (June 2010) (Cambridge; John F Kennedy School of Government, Harvard University) online: <www.hks.harvard.edu/m- rcbg/CSRI/publications/workingpaper_59_zerk.pdf>.

Nongovernmental Organizations Materials

a) IATA

i) Documents

Articles of Association, online: IATA <www.iata.org/about/Documents/articles-of-association.pdf>.

Bisignani, Giovanni, *Words of Change* (Geneva: IATA Corporate Communications, 2011) 53.

Bisignani, Giovanni, "State of the Air Transport Industry" (Presentation delivered at the IATA 66th IATA Annual General Meeting, World Air Transport Summit, Berlin 6-8 June 2010) online: IATA <www.iata.org/pressroom/speeches/pages/2010-06-07-01.aspx>.

IATA, *Resolution on the Implementation of the Aviation CNG2020 Strategy*, *IATA* online: IATA <www.iata.org/pressroom/pr/Documents/agm69-resolution-cng2020.pdf>.

IATA, *World Air Transport Statistics* (Montréal: IATA, 2000).

Jennison, Michael B., "Regulating Greenhouse Gas Emissions: Legal Aspects of Levies and Trading Systems" (Presentation delivered at the IATA Legal Symposium, February 2007) [unpublished, archived on file with the author].

Milde, Michael, "Can Airlines Get Any Relief" (Presentation delivered at the IATA Legal Symposium, Bangkok, 10 February 2009) [unpublished, archived on file with the author].

ii) Websites

IATA, "About US", online: IATA <www.iata.org/about/Pages/index.aspx>.

IATA, "IATA Environmental Assessment (IEnvA)", *IATA* online: <www.iata.org/what wedo/environment/Pages/environmental-assessment.aspx>.

IATA, "IATA Environmental Assessment (IEnvA) Registry", *IATA* online: <www.iata.org/whatwedo/environment/Pages/ienva.aspx?Query=all>.

IATA, "Jet Fuel Price Monitor", *IATA* online: <www.iata.org/publications/economics/fuel-monitor/Pages/index.aspx>.

IATA, "Price Fuel Analysis", *IATA* online: <www.iata.org/publications/economics/fuel-monitor/Pages/price-analysis.aspx>.

IATA, "Strategic Partnerships: Building Relationships", *IATA* online: <www.iata.org/about/sp/Pages/index.aspx>.

IATA, "Striking Oil: Understanding Oil Price Volatility", *IATA* (October 2009) online: <www.iata.org/publications/airlines-international/october-2009/Pages/2009-10-07.aspx>.

IATA, "IATA Economic Briefing Impact of Ash Plume", *IATA* (May 2010) online: <www.iata.org/whatwedo/Documents/economics/Volcanic-Ash-Plume-May2010.pdf>.

IATA, "Remarks of Tony Tyler at the Greener Skies Conference in Hong Kong", *IATA* (26 February 2013) online: <www.iata.org/pressroom/speeches/Pages/2013-02-26-01.aspx>.

IATA, "Remarks of Tony Tyler at the Aviation Fuel Forum in Berlin", *IATA* (8 May 2013) online: <www.iata.org/pressroom/speeches/Pages/2013-05-08-01.aspx

IATA, "Responsibly Addressing Climate Change", *IATA* (June 2013) online: <www.iata.org/policy/environment/Documents/policy-climate-change.pdf>

IATA, "Remarks of Tony Tyler at the Media Round Table, Singapore", *IATA* (16 October 2013) online: <www.iata.org/pressroom/speeches/Pages/2013-10-16-01.aspx>.

IATA, "Remarks of Tony Tyler at the AACO Annual General Meeting in Doha", *IATA* (6 November 2013) online: <www.iata.org/pressroom/speeches/Pages/2013-11-06-01.aspx>.

Point Carbon, "EAU Last 30 Days", online: <www.pointcarbon.com>.

iii) Press Releases

IATA, News Release, PS/12/00, "2000 W.A.T.S. – More Passengers, Less Profits" (19 June 2000) online: IATA <www.asiatraveltips.com/travelnews2000/20June2000IATA.htm>

IATA, "IATA Applauds ICAO Agreement on Aviation and Climate Change" (8 October 2010) online: IATA <www.iata.org/pressroom/pr/pages/2010-10-08-01.aspx>.

IATA, Press Release, "Statement on US Bill on the EU Emissions Trading" (20 July 2011) online: <wwww.iata.org/pressroom/pr/Pages/2011-07-20-01.aspx>.

IATA, Press Release, no 60, "Partnerships, Innovation, Key to Maintaining Rapid Growth in Middle East Aviation" (29 November 2011) online: IATA <wwww.iata.org/pressroom/pr/Pages/2011-11-29-01.aspx>.

IATA, Press Release, no 63, "IATA Disappointed with EU Court Decision on ETS" (21 December 2011) online: IATA <www.iata.org/pressroom/pr/pages/2011-12-21-01.aspx>.

IATA, Press Release, no 25, "IATA Urges ICAO Solution on Economic Measures for Climate Change" (11 June 2012) online: IATA <www.iata.org/pressroom/pr/Pages/2012-06-11-04.aspx>.

IATA, Press Release, no 8, "2012 Best in History of Continuous Safety Improvements" (28 February 2013) online: IATA <www.iata.org/pressroom/pr/Pages/2013-02-28-01.aspx>.

IATA, Press Release, no 34, "Historic Agreement on Carbon-Neutral Growth" (3 June 2013) online: IATA <www.iata.org/pressroom/pr/Pages/2013-06-03-05.aspx>.

IATA, Press Release, no 55, "Landmark Agreement on Climate Change at ICAO: Major Progresson Environment, Safety, Security, Operations and Regulation" (4 October 2013 online: IATA <www.iata.org/pressroom/pr/Pages/2013-10-04-01.aspx>.

b) Airlines for America (A4A)

"Stop the European Union from Unilaterally Applying a Cap-and-Trade Tax to U.S. Airlines and Passengers", *A4A* online: <www.airlines.org/Pages/EU-ETS-Fact-Sheet.aspx>.

"ATA Calls EU ETS Application to U.S. Airlines Illegal", *A4A* (5 July 2011) online: <www.airlines.org/Pages/ATA-Calls-EU-ETS-Application-to-U.S.-Airlines-Illegal.aspx>.

"Airlines for America (A4A) Commends Senate Opposition of EU ETS", *A4A* (7 December 2011) online: A4A <www.airlines.org/Pages/news_12-07-2011_2.aspx>.

"A4A Comment on European Court of Justice Decision", *A4A* (21 December 2011) online: A4A <www.airlines.org/Pages/news_12-21-2011.aspx>.

"A4A EU ETS Lawsuit Defines Clear Path for Government Action", *A4A* (27 March 2012) online: A4A <www.airlines.org/Pages/news_3-27-2012.aspx>.

"A4A Commends U.S. House for Taking Further Action to Protect U.S. Airlines and Their Passengers from Unlawful EU Emissions Trading Scheme", *A4A* (28 June 2012) online: A4A <www.airlines.org/Pages/news_6-28-2012.aspx>.

"A4A Commends Bipartisan Support of Senate EU ETS Bill", *A4A* (7 March 2012) online: A4A <www.airlines.org/Pages/news_3-7-2012.aspx>.

"EU ETS Remains Bad News for Airlines", *A4A* (11 February 2013) online: A4A <www.airlines.org/Pages/EU-ETS-Remains-Bad-News-For-U.S.-Airlines.aspx>.

"A4A Applauds ICAO Climate Change Resolution", *A4A* (4 October 2013) online: www.airlines.org/Pages/A4A-Applauds-ICAO-Climate-Change-Resolution.aspx>.

c) Air Transport Action Group (ATAG)

"Our Members", *ATAG* online: <www.atag.org/membership/our-members.html>.

"The Economic and Social Benefits of Air Transport 2008", *ATAG* online: <www.atag.org/component/downloads/downloads/61.html>.

"Beginners Guide to Bio Fuels", *ATAG* (2 September 2011) online: <www.atag.org/component/downloads/downloads/97.html>.

"The Right Flight Path to Reduce Aviation Emissions", *ATAG* (November 2011) online: <www.atag.org/component/downloads/downloads/121.html>.

d) Association of European Airlines (AEA)

AEA, Position Paper, "AEA Contribution to a Global Approach for International Aviation Emissions", *AEA* (15 December 2008) online: <http://files.aea.be/Downloads/gap_paper_final.pdf>.

AEA, Position Paper, "Striving for an ETS that Supports a Sustainable Aviation Sector, Position", *AEA* (28 April 2011) online: <www.aea.be/assets/documents/positions/ETS%20Paper_April%2028.pdf>.

AEA, Press Release, "AEA Challenges Five Emissions Trading Myths" (16 June 2011) online: AEA <http://files.aea.be/News/PR/Pr11-014.pdf>.

AEA, Information, "AEA Calls on the Commission to Act in Face of Mounting International Pressure on ETS", *AEA* (26 July 2011) online: <http://files.aea.be/News/PR/Pr11-015.pdf>.

AEA, Press Release, "ETS Should Be About Sustainability, Not Politics" (3 November 2011) online: AEA <http://files.aea.be/News/PR/Pr11-020.pdf>.

AEA, Press Release, "Emissions Trading Concerns Will Be Resolved through ICAO" (20 December 2011) online: AEA <http://files.aea.be/News/PR/Pr11-025.pdf>.

AEA, Press Release, "U.S. Customs Facility in Abu Dhabi Harms the Competiveness of the Transatlantic Aviation Market" (17 April 2013) online: AEA <http://files.aea.be/News/PR/Pr13-011.pdf>.

e) *Environmental Defense Fund (EDF)*

Annie Petsonk & Adam Peltz, "Why Europe's Climate Program for Airlines Is Not a Tax", *EDF* (13 February 2012) online: <http://blogs.edf.org/climatetalks/2012/02/13/why-europe per centE2 per cent80 per cent99s-climate-program-for-airlines-is-not-a-tax>.

"Scenario A: Airspace MBM and 1% De Minimis Exemption" (27 September 2013) [unpublished, archived on file with the author].

"ICAO Committee Agrees on Path Forward to Limit Aviation Emissions, but U.S. Leadership Will Be Critical to Delivering Results", *EDF* (3 October 2013) online: <www.edf.org/media/icao-committee-agrees-path-forward-limit-aviation-emissions-us-leadership-will-be-critical>.

"Aviation Emissions Deal: ICAO Takes One Step Forward, Half Step Back", *EDF* (4 October 2013) online: <http://blogs.edf.org/climatetalks/2013/10/04/aviation-emissions-deal-icao-takes-one-step-forward-half-step-back>.

f) *Transport & Environment*

"Briefing: The Billion Euro Aviation Bonanza", *Transport & Environment* (January 2013) online: <www.transportenvironment.org/sites/te/files/publications/Briefing_The_billion_Euro_Aviation_Bonanza.pdf>.

"The Clock Has Stopped: Where Is ICAO Now", *Transport & Environment* (2 May 2013) online: <www.transportenvironment.org/news/clock-has-stopped-where-icao-now>.

"Europe Caves in to Foreign Pressure, Guts Aviation Emission-Reduction Law", *Transport & Environment* (3 April 2014) online: <www.transportenvironment.org/press/europe-caves-foreign-pressure-guts-aviation-emissions-reduction-law>.

"Allocating Aviation CO_2 Emissions: The Airspace-Based Approach and Its Alternatives", *Transport & Environment* online: <www.transportenvironment.org/sites/te/files/publications/201301%20Airspace%20Version%209%20Final%20%281%29_0.pdf>.

"Global Deal or No Deal: Your Free Guide to ICAO's 38th Triennial Assembly?", *Transport & Environment* online: <www.transportenvironment.org/sites/te/files/publications/2013%2009%20 Your%20Guide%20to%20ICAO_final.pdf>.

g) Other Materials

ACI, Media Release, "ACI Welcomes Historic ICAO Assembly Resolution", (9 September 2010) online: ACI <www.aci.aero/aci/aci/file/Press%20Releases/2010/PR_091010%20_ICAO_Assembly_Resolution.pdf>.

Airport Watch, "EU Emissions Trading System", *Airport Watch* online: <www.airport-watch.org.uk/?page_id=8234>.

AWG, "AWG Purpose", *AWG* online: <www.awg.aero/inside/purpose>.

European Parliament, P7_TA-PROV (2014) 0278, Greenhouse Gas Emission Trading (International Aviation Emissions) online: <www.europarl.europa.eu/sides/getDoc.do?type=TA&reference=20140403&secondRef=TOC&language=en>.

European Parliament, MEPs back CO2 Permit Exemption for Long-Haul Flights, <www.europarl.europa.eu/pdfs/news/expert/infopress/20140331IPR41187/20140331IPR41187_en.pdf>.

International Maritime Emissions Reduction Scheme, "Rebate Mechanism for Fair and Global Carbon Pricing of International Transport", *IMERS* online: <http://imers.org/docs/RM_Aviation_Fact_Sheet.pdf>.

The Greens: European Free Alliance in the European Parliament, "MEPs Vote to Let Aviation Off the Hook for Vague Hope of Future Global Action" (3 April 2014) online <www.greens-efa.eu/airline-emissions-12161.html>.

National Snow and Ice Data Center, "Climate Change", *NSIDC* online: <http://nsidc.org/cryosphere/climate-change.html>.

NRDC, "Aviation Biofuel Sustainability Survey", *NRDC* online: <www.nrdc.org/energy/aviation-biofuel-sustainability-survey>.

Jake Schmidt, "US Signs Into Law Bill Calling for Global Solution to Aviation's Carbon Pollution", *Switchboard NRDC* (27 November 2012) online: <http://switchboard.nrdc.org/blogs/jschmidt/ball_in_the_hands_of_proponent.html>.

WCO, "SAFE Framework of Standards to Secure and Facilitate Global Trade", *World Custom Organization* (2012) WCO online: <http://goo.gl/ocZoL6>.

Robert Ireland, "WCO Policy Research Brief – The EU Aviation Emissions Policy and Border Tax Adjustments", WCO Research Paper No 26 (July 2012), *World Custom Organization* online: <http://goo.gl/HvItca>.

WWF, "Aviation Report: Market Based Mechanisms to Curb Greenhouse Gas Emissions from International Aviation", *WWF* (2012) online: <http://awsassets.panda.org/downloads/aviation_main_report_web_simple.pdf>.

Newspapers, Newswires and Other News Sources

a) *Bloomberg*

Andrea Rothman, "EADS Says A330 Boost is Hostage to China Views on Carbon Tax", *Bloomberg* (8 March 2012) online: <www.bloomberg.com/news/2012-03-08/eads-says-a330-boost-is-hostage-to-china-views-on-carbon-tax-1-.html>.

Ewa Krukowska, "Airlines Face Carbon Reduction Verdict on 708 Million Industry", *Bloomberg Businessweek* (24 September 2013) online: <www.businessweek.com/news/2013-09-23/airlines-face-carbon-reduction-verdict-on-708-billion-industry>.

Frederico Tomesco & Ewa Krukowska, "EU to Defend Limited Carbon Market for Airlines' Emissions", *Bloomberg Businessweek* (25 September 2013) online: <www.businessweek.com/news/2013-09-24/eu-to-defend-limited-carbon-market-for-emissions-from-airlines>.

Ewa Krukowska, "First Global Emissions Market for Airlines Wins Support", *Bloomberg* (3 October 2013) online: <www.bloomberg.com/news/2013-10-04/first-global-emissions-market-for-airlines-wins-support.html>.

Ewa Krukowska & Alesandro Vitelli, "EU Lawmaker Liese to Seek Changes to Draft Aviation Carbon Law", *Bloomberg* (20 November 2013) online: <www.bloomberg.com/news/2013-11-20/eu-lawmaker-liese-to-seek-changes-to-draft-aviation-carbon-law.html>.

Alessandro Vitelli, "U.K. Opposes EU Plan to Renew Carbon Curbs on Foreign Flights", *Bloomberg* (29 November 2013) online: <www.bloomberg.com/news/2013-11-29/u-k-opposes-eu-plan-to-renew-carbon-curbs-on-foreign-flights.html>.

Mathew Carr, "EU May Extend Aviation Carbon Freeze to 2020, Adviser Says", *Bloomberg* (5 December 2013) online: <www.bloomberg.com/news/2013-12-05/europe-may-extend-aviation-carbon-freeze-to-20-eu-adviser-says.html>.

b) Green Air Online

"Stiff Challenge Facing ICAO after Unprecedented Number of Reservations on Assembly Climate Change Resolution", *Green Air Online* (29 January 2011) online: <www.greenaironline.com/news.php?viewStory=1043>.

"Sovereignty the Key Issue as Europe, US Airlines and Environmental Groups Argue Their EU ETS Cases Before the CJEU", *Green Air Online* (6 July 2011) online: <www.greenaironline.com/news.php?viewStory=1284>.

"European Airlines Nervous as International Demands for Europe to Exclude Foreign Airlines from EU ETS Continue to Grow", *Green Air Online* (19 August 2011) online: <www.greenaironline.com/news.php?viewStory=1314>.

"China and Russia Join Forces to Oppose EU ETS, Threatening Taxes or Charges on EU Airlines in Retaliation", *Green Air Online* (29 September 2011) online: <www.greenaironline.com/news.php?viewStory=1342>.

"US DOT Official Warns of Damaging Trade War between US and EU over EU ETS as ECJ Announces Date of Ruling on US airline case", *Green Air Online* (6 December 2011) online: <www.greenaironline.com/news.php?viewStory=1381>.

"Following House Passage, Mirror Bill Introduced into US Senate to Block US Airlines from Participating in EU ETS", *Green Air Online* (8 December 2011) online: <www.greenaironline.com/news.php?viewStory=1382>.

"Aviation Must Tackle Its Environmental Impact Before It Can Be Allowed to Grow, Says UK's Aviation Regulator", *Green Air Online* (27 January 2012) online: <www.greenaironline.com/news.php?viewStory=1421>.

"US Legislation to Prohibit Airlines from Joining EU ETS Moves a Step Closer as Senate Bill Receives Bipartisan Support", *Green Air Online* (8 March 2012) online: <www.greenaironline.com/news.php?viewStory=1436>.

"US Airlines Give Up on Legal Case Against Inclusion into the EU ETS But Call on Their Government to Step Up Retaliatory Action", *Green Air Online* (29 March 2012) online: <www.greenaironline.com/news.php?viewStory=1444>.

"The Three Main Obstacles Facing the Introduction of Sustainable Aviation Biofuels: Price, Price and Price", *Green Air Online* (2 April 2012) online: <www.greenaironline.com/news.php?viewStory=1446>.

"UK Climate Advisers Recommend International Aviation Emissions Be Included in National Carbon Budgets", *Green Air Online* (5 April 2012) online: <www.greenaironline.com/news.php?viewStory=1447>.

"Maldives President Calls for a Stronger Will and Leadership in Bringing about the Sustainable Growth of Aviation", *Green Air Online* (3 May 2012) online: <www.greenaironline.com/news.php?viewStory=1458>.

"European Commission Backs Down on EU ETS and Agrees to 'Stop the Clock' on International Aviation Emissions", *Green Air Online* (12 November 2012) online: <www.greenaironline.com/news.php?viewStory=1620>.

"European Parliament Rapporteur Backs 'Stop the Clock' EU ETS Proposal but Calls for Clarity on EU Stance", *Green Air Online* (14 January 2013) online: <www.greenaironline.com/news.php?viewStory=1640>.

"Hedegaard Sets Out Conditions on ICAO Agreement as EU Legislators Approve EU ETS 'Stop the Clock' Measure", *Green Air Online* (17 April 2013) online: <www.greenaironline.com/news.php?viewStory=1681>.

"As 'Stop the Clock' Passes into Law, EU Piles on the Pressure for a Meaningful ICAO Agreement on MBMs", *Green Air Online* (30 April 2013) online: <www.greenaironline.com/news.php?viewStory=1683>.

"Aviation Industry Calls for Global Agreement and Climate Change Leadership by Governments Ahead of ICAO Assembly", *Green Air Online* (20 September 2013) online: <www.greenaironline.com/news.php?viewStory=1745>.

"ICAO Moves Closer to Agreement on Limiting Growth of Aviation Emissions as EU Officials Justify Climb-Down", *Green Air Online* (9 September 2013) online: <www.greenaironline.com/news.php?viewStory=1753>.

"An NGO Message for the ICAO Assembly: Introduce a Global Market-Based Measure Now", *Green Air Online* (17 September 2013) online: <www.greenaironline.com/news.php?viewStory=1754>.

"Climate Researchers Find Even Carbon-Neutral Growth from 2020 Will Not Be Enough to Stave off Climate Impacts", *Green Air Online* (3 October 2013) online: <www.greenaironline.com/news.php?viewStory=1761>.

"ICAO States Reach Agreement on Roadmap Towards a Global MBM but Europe Suffers Defeat over EU ETS", *Green Air Online* (4 October 2013) online: <www.greenaironline.com/news.php?viewStory=1762>.

"ICAO Assembly Climate Change Outcome Hailed by Industry but Seen as a Missed Opportunity by Environmental NGOs", *Green Air Online* (6 October 2013) online: <www.greenaironline.com/news.php?viewStory=1763>.

"Count Us Out of Carbon-Neutral Growth Measures, China and Other Major Emerging Countries Tell ICAO", *Green Air Online* (4 November 2013) online: <www.greenaironline.com/news.php?viewStory=1777>.

"Russia Urges ICAO to Reconsider Unrealistic Carbon Neutrality Goal and Market Measures for International Aviation", *Green Air Online* (18 November 2013) online: <www.greenaironline.com/news.php?viewStory=1786>.

"Proposal by BRIC States to Set Up a New Group on a Global MBM for Aviation Approved by ICAO Council", *Green Air Online* (20 December 2013) online: <www.greenaironline.com/news.php?viewStory=1804>.

Chris Lyle, "Aviation and Climate Change: What Now for a Global Approach", *Green Air Online* (24 January 2011) online: <www.greenaironline.com/news.php?viewStory=1039>.

Chris Lyle, "Rio, Kyoto, Brussels and Chicago: Reconciling Principles Related to International Air Transport Emissions", *Green Air Online* (27 July 2012) online: <www.greenaironline.com/news.php?viewStory=1573>.

Chris Lyle, "Will ICAO States at Last Deliver a Meaningful Global Agreement on Mitigating International Aviation Emissions", *Green Air Online* (2 September 2013) online: <www.greenaironline.com/news.php?viewStory=1732>.

c) The Hill

Keith Laing, "Sen. Isakson: Obama Administration Should Challenge EU Airline Emissions Rules", *The Hill* (8 January 2012): online <http://thehill.com/blogs/transportation-report/aviation/231787-sen-isakson-obama-administration-should-challenge-eu-airline-emission-rules>.

Keith Laing, "LaHood, Senate Committee Hammers 'Lousy' EU airline Emission Trading Rules", *The Hill* (6 June 2012): online <http://thehill.com/blogs/transportation-report/aviation/231295-lahood-senate-committee-hammer-eu-airline-emission-trading-rules>.

Keith Laing, "Sen Thune Pleased with Airline Emission Fees Halt but Not Satisfied", *The Hill* (12 November 2012): online <http://thehill.com/blogs/transportation-report/aviation/267345-sen-thune-pleased-with-airline-emission-fees-halt-but-not-satisfied->.

Keith Laing, "Greens Pressure Obama to Veto Airline Emissions Bill", *The Hill* (14 November 2012): online <http://thehill.com/blogs/transportation-report/aviation/267841-enviros-cast-airline-emissions-bill-as-obamas-first-post-sandy-climate-test>.

The Hill Staff, "The Hill's 2012 Top Lobbyists", *The Hill* (31 November 2012): online <http://thehill.com/business-a-lobbying/264987-2012-top-lobbyists>.

Pete Kasperowicz, "House Members Warn Carbon Tax Would Increase Unemployment Rate", *The Hill* (3 December 2012): online <http://thehill.com/blogs/floor-action/house/270579-house-members-warn-carbon-tax-would-increase-unemployment-rate>.

Zack Colman, "White House Extends Aviation Biofuel Program", *The Hill* (15 April 2013): online <http://thehill.com/blogs/e2-wire/e2-wire/293951-white-house-extends-aviation-biofuel-program#ixzz2SQL8wP3e>.

d) Reuters

"Airlines face CO_2 bill of 300 Million Euros in 2012", *Reuters* (16 February 2012) online: <www.reuters.com/article/2012/02/16/carbon-aviation-barcap-idUSL5E8DG4O520120216>.

"Update 2 – UN Aviation Body Says Emissions Proposal by Year-End", *Reuters* (2 March 2012) online: <www.reuters.com/article/2012/03/02/airlines-emissions-idUSL2E8E2B97

20120302>.

"China Says EU Carbon Rule to Cost $2.8 Billion by 2030", *Reuters* (5 March 2012) online: <www.reuters.com/article/2012/03/05/china-eu-emissions-idUSL4E8E51ME20120305>.

"Update 1 – Commission seeks to speed up EU carbon market remedy", *Reuters* (5 October 2012) online: <www.reuters.com/article/2012/10/05/eu-ets-idUSL6E8L5HJE20121005>.

"Opponents of EU Airline CO_2 Scheme to Meet in Moscow", *Reuters* (6 February 2012) online: <http://uk.reuters.com/article/2012/02/06/uk-russia-aviation-idUKTRE8151QW 20120206>.

"UN Pins Hopes on Govts, Airlines to Bail out CDM", *Reuters* (4 December 2012) online: <www.pointcarbon.com/nes/1.2086199>.

"Trade War 'Unavoidable' If EU Airline Emissions Plan Block – Lawmaker", *Reuters* (20 November 2013) online: <www.reuters.com/article/2013/11/20/eu-airlines-emissions-idUSL5N0J533O20131120>.

"RPT-UK, France, Germany Attack EU Aircraft Carbon Plan", *Reuters* (6 December 2013) online: <www.reuters.com/article/2013/12/06/eu-aviation-idUSL5N0JL1MN20131206>.

Nigam Prusty, "EU Airline Charge Hurts Climate Fight-China, India", *Reuters Africa* (14 February 2012) online: <http://af.reuters.com/article/energyOilNews/idAFL5E8DE8 AS20120214>.

Barbara Lewis & Nina Chestney, "EU Airline Carbon Cash Should Help Fill Climate Fund", *Reuters* (16 May 2012) online: <www.reuters.com/article/2012/05/16/uk-energy-summit-hedegaard-idUSLNE84F01220120516>.

Valerie Volcovici, "Airlines Urge US to Take Stronger Approach Versus EU Carbon Law", *Reuters Africa* (31 July 2012) online: <http://af.reuters.com/article/energyOilNews/idAFL 2E8IUG6C20120731>.

Valerie Volcovici & Barbara Lewis, "EU Sees Progress on UN airline Emissions Deal", *Reuters* (11 November 2012) online: <www.reuters.com/article/2012/11/11/uk-airlines-eu-us-co-idUSLNE8AA00I20121111>.

Valerie Volcovici, "Senate Votes to Shield US Airlines from EU's Carbon Scheme", *Reuters* (24 September 2012) online: <www.reuters.com/article/2012/09/24/uk-usa-carbon-airlines-idUSLNE88N00K20120924>.

Valerie Volcovici & Barbara Lewis, Reuters, "US Offers Airspace-Based Emissions Regime" *Reuters* (22 February 2013) online: <http://uk.reuters.com/article/2013/02/25/uk-eu-icao-climate-idUKLNE91O00K20130225>.

Valerie Volcovici, "Shuttle Diplomacy Under Way on Global Aviation Emissions Deal", *Reuters* (22 July 2013) online: <http://uk.reuters.com/article/2013/07/22/us-eu-aviation-emissions-idUKBRE96L0ZJ2013072>.

Valerie Volcovici & Joshua Schneyer, "EU Stands Firm on Aviation Emissions Position", *Reuters* (27 September 2013) online: <www.reuters.com/article/2013/09/27/us-aviation-climate-eu-idUSBRE98Q17K20130927>.

Stian Reklev, "UN Agrees Multi-Billion Dollar Framework to Tackle Deforestation", *Reuters* (22 November 2013) online: <http://uk.reuters.com/article/2013/11/22/forest-climate-talks-idUKL4N0J72XJ20131122>.

e) The Wall Street Journal

Doug Cameron, "U.S. Airlines Seek Action on EU Carbon Tax", *The Wall Street Journal* (27 March 2012) online: <http://online.wsj.com/article/SB10001424052702304177104 577307980838597906.html?mod=dist_smartbrief>.

Tenille Tracy "Airlines Press White House on EU Emissions Curbs" *The Wall Street Journal* (2 August 2012) online: <http://online.wsj.com/article/SB10000872396390443545 504577565451198051384.html?mod=dist_smartbrief>.

Alessandro Torello, "Airlines Profit from CO_2 Plan?", *The Wall Street Journal* (22 January 2013) online: <http://blogs.wsj.com/brussels/2013/01/22/airlines-profit-from-emissions-plan>.

David Pearson, "Airbus Reaches Deal to Sell Jets to China", *The Wall Street Journal* (25 April 2013) online: <http://online.wsj.com/article/SB10001424127887324474004578444562817850562.html?mod=dist_smartbrief>.

Johanne Chiu, "Air China's Board Approves Plan to Buy 100 Aircraft from Airbus", *The Wall Street Journal* (7 May 2013) online: <http://online.wsj.com/article/BT-CO-20130507-707347.html?mod=dist_smartbrief>.

Daniel Michaels, "Airlines: EU Plans Put Global Emissions Deal at Risk", *The Wall Street Journal* (17 October 2013) online: <http://blogs.wsj.com/brussels/2013/10/17/airlines-eu-plans-put-global-emissions-deal-at-risk>.

Susan Carey & Daniel Michaels, "Rise of Middle East Airlines Does not Fly with U.S. Rivals", *The Wall Street Journal* (29 October 2013) online: <http://online.wsj.com/news/articles/SB10001424052702304384104579141732219208604>.

f) Other Newspapers, Newswires and News Sources

Michael Murray, "Airlines Made 5.7 Billion in Revenues in 2010 from Fees" *ABC News* (13 June 2011) online: <http://abcnews.go.com/Travel/airlines-made-57-billion-profits-2010-fees/story?id=13832011>.

"Air Passenger Duty", *ABTA* (24 March 2014) online: <http://abta.com/news-and-views/policy-zone/more/air-passenger-duty>.

"US Airlines Report USD 2.9 Billion Profit for 2010 – But There Is More to the Story", *Airline Leader* online: <www.airlineleader.com/this-months-highlights/us-airlines-report-a-usd2-9-billion-profit-for-2010-but-there-is-more-to-the-story>.

Curt Epstein, "Alternative Fuels Still Face Hurdles in Environment", *Aviation International News* (27 August 2010) online: <www.ainonline.com/sites/ainonline.com/files/filead-min/template/main/pdfs/Bivav_Environment.pdf> 30.

AFP, "EU 'Open' to Talks on Airlines Tax, but Won't Change Law", *Business Recorder* (4 March 2012) online: <www.brecorder.com/top-news/1-front-top-news/48229-eu-open-to-talks-on-airlines-tax-but-wont-change-law-.html>.

Ian Goold, "EU's Emissions Trading Scheme Could Cancel A330 Production Hike", *Aviation International News* (2 July 2012) online: <www.ainonline.com/aviation-news/2012-07-02/eus-emissions-trading-scheme-could-cancel-a330-production-hike>.

Gregory Polek, "Airbus Eyes Year-End Horizon To Solve ETS Row with China", *Aviation International News* (16 July 2012) online: <www.ainonline.com/aviation-news/ain-air-transport-perspective/2012-07-16/airbus-eyes-year-end-horizon-solve-ets-row-china>.

Gregory Polek, "Climate Experts Challenge Anti-ETS 'Myths'", *Aviation International News* (15 October 2012) online: <www.ainonline.com/aviation-news/ain-air-transport-perspective/2012-10-15/climate-experts-challenge-anti-ets-myths>.

Chad Trauvetter, "Aviation Groups Back ICAO Aircraft Emissions Framework", *Aviation International News* (8 October 2013) online: <www.ainonline.com/aviation-news/ainalerts/2013-10-08/aviation-groups-back-icao-aircraft-emissions-framework>.

Karen Walker, "US State Views EU ETS as a Challenge to ICAO Process", *Air Transport World* (23 March 2012) online: <http://atwonline.com/airline-finance-data/news/us-state-views-ets-challenge-icao-process-0323>.

Aaron Karp, "IATA: 2012 Western-Built Jet Accident Rate Lowest in History", *Air Transport World* (28 February 2012) online: <http://atwonline.com/operations-maintenance/news/iata-2012-western-built-jet-accident-rate-lowest-history-0228>.

"Obama Administration Pressed On EU ETS", *Airwise* (28 March 2012) online: <http://news.airwise.com/story/view/1332974498.html>.

"US House OKs Bill to Shield Airlines from ETS", *Airwise* (14 November 2012) online: <http://news.airwise.com/story/view/1352872192.html>.

"Economics Turned EU Powers against ETS", *Airwise* (9 December 2012) online: <http://news.airwise.com/story/view/1355097725.html>.

"Biofuel Flights Signal Demand from U.S. Airlines", *Aviation Week & Space Technology* (14 November 2011) online: <http://aviationweek.com/awin/biofuel-flights-signal-demand-us-airlines>.

Jen DiMascio, "LaHood Won't Endorse Senate Ban On EU ETS Compliance", *Aviation Week & Space Technology* (6 June 2012) online: <http://aviationweek.com/business-aviation/lahood-wont-endorse-senate-ban-eu-ets-compliance>.

Rachel Brewster, "US-Europe Fight over Airline Emissions Could Help Talks on Climate Change", *The Christian Science Monitor* (15 January 2013) online: <www.csmonitor.com/lay-

out/set/print/Commentary/Opinion/2013/0115/US-Europe-fight-over-airline-emissions-could-help-talks-on-climate-change>.

"Nobel Economists Urge Obama to Support EU Aviation Carbon Scheme", *Clean Technica* (15 March 2012) online: <http://cleantechnica.com/2012/03/15/nobel-economists-urge-obama-to-support-eu-aviation-carbon-scheme>.

"Will India Follow the EU in Capping Aviation Emissions?", *Clean Technica* online: <http://cleantechnica.com/2012/02/01/will-india-follow-the-eu-in-capping-aviation-emissions>.

Patrick Goodenough, "Sen. Kerry Blames U.S. for Dispute Over E.U. Airline Carbon Tax: 'We Dragged Our Feet'", *CNS News* (7 June 2012) online: <http://cnsnews.com/news/article/sen-kerry-blames-us-dispute-over-eu-airline-carbon-tax-we-dragged-our-feet>.

Charlie Leocha, "Airlines Embrace the Environment: It Is Good for Their Bottom Line", *Consumer Traveler* (13 September 2013) online: <www.consumertraveler.com/today/airline-embrace-the-environment-its-good-for-their-bottom-line>.

"EU Carbon Tax Will Have Significant Impact on Air-Fares: Sources", *The Economic Times* (2 September 2012) online: <http://articles.economictimes.indiatimes.com/2012-09-02/news/33548735_1_eu-ets-carbon-tax-emission-data>.

"Green Taxes – A Nice Little Earner for Some", *The Economist* (6 February 2012) online: <www.economist.com/blogs/gulliver/2012/02/airlines-and-emissions-permits>.

"IATA: Europe Should Abandon 'Misguided' ETS Plans", *Global Travel Industry News* (28 September 2011) online: <www.eturbonews.com/25447/iata-europe-should-abandon-misguided-ets-plans>.

Mavis Toh, "Airbus and Partners Urge EU Leaders to Stop ETS Trade Conflict", *Flight Global* (12 March 2012) online: <www.flightglobal.com/news/articles/airbus-and-partners-urge-eu-leaders-to-stop-ets-trade-conflict-369357/>.

Margaret Wente, "Whatever Happened to Global Warming", *The Globe and Mail* (6 January 2014) online: <www.theglobeandmail.com/commentary/whatever-happened-to-global-warming/article7725145/>.

Amy Westervelt, "How Airlines Are Fighting Carbon Trading", *Greenbiz.com* (8 August 2012) online: <www.greenbiz.com/news/2012/08/08/how-airlines-fighting-carbon-trading>.

"Iceland Volcano", *The Guardian* online: <www.guardian.co.uk/world/iceland-volcano>.

Vikram Dodd, "Cargo Plane Bomb 'Timed to Detonate over US'", *The Guardian* (10 November 2010) online: <www.theguardian.com/world/2010/nov/10/cargo-plane-bomb-us-alqaida>.

Robert "Bo" Van Valkenburg, "Directive 2008/101: The EU Emissions Trading System", *International Relations* (13 January 2013) online: <http://internationalreports.wordpress.com/2013/01/13/directive-2008101>.

"Cargo Plane Bombs Were Wired to Explode, Officials Say", *Los Angeles Times* (31 October 2010) online: <http://articles.latimes.com/2010/oct/31/nation/la-na-cargo-planes-20101031>.

"Airbus Worried About Fallout of E.U. Plan to Charge for Airline Emissions", *The New York Times* (13 February 2012) online: <www.nytimes.com/2012/02/14/business/global/airbus-worried-about-fallout-of-eu-plan-to-charge-for-airline-emissions.html?_r=1>.

Nancy Young, Letter to the Editor, "Airlines, Getting Greener", *The New York Times* (3 February 2013) online: <www.nytimes.com/2013/02/04/opinion/airlines-getting-greener.html?_r=1&>.

Elizabeth Rosenthal, "Your Biggest Carbon Sin May Be Air Travel", *The New York Times* (26 January 2013) online: <www.nytimes.com/2013/01/27/sunday-review/the-biggest-carbon-sin-air-travel.html>.

Kathryn A Wolfe, "Europe's New Airline Rules Anger Congress", *Politico* (3 April 2012) online: <www.politico.com/news/stories/0412/74804_Page2.html>.

Sbishophall, "EU Flies Solo to Reduce Aviation Emissions", *Reporting the EU* (16 November 2011) online: <http://reportingeu.mediajungle.dk/2011/11/16/eu-flies-solo-to-reduce-aviation-emissions>.

Nitin Sethi, "Trade War: Cabinet to Consider Steps against EU Carbon Tax", *The Times of India* (9 July 2012) online: <http://timesofindia.indiatimes.com/home/environment/devel-

opmental-issues/Trade-war-Cabinet-to-consider-steps-against-EU-carbon-tax/articleshow/14753183.cms>.

Cenap Çakmak, "Assessing Turkey's Performance as Norm Entrepreneur in Syrian Crisis", *Todays Zaman* (8 August 2013) online: <www.todayszaman.com/newsDetail_openPrintPage.action?newsId=324772>.

"The Carbon Caper", *Travel Weekly* (9 January 2012) online: <www.travelweekly.com/Editorials/The-carbon-caper/>.

Bart Jansen, "FAA Official Raps EU over Airline Carbon Tax", *USA Today* (13 April 2012) online: <http://travel.usatoday.com/flights/post/2012/04/faa-official-wraps-eu-over-airline-carbon-tax/670800/1>.

Nicholas E Calio & Lee Moak, "Why U.S. Airlines Oppose EU's Emissions Tax Scheme", *USA Today* (27 March 2012) online: <www.usatoday.com/news/opinion/forum/story/2012-03-27/airlines-eu-planes-emissions-tax/53805612/1>.

Nicholas E Calio, "EU's Emissions Tax Scheme Kills American Jobs", *Politico* (4 November 2012) online: <www.politico.com/news/stories/1112/83284.html>.

Eckhard Pache, "On the Compatibility with International Legal Provisions of Including Greenhouse Gas Emissions from International Aviation in the EU Emission Allowance Trading Scheme as a Result of the Proposed Changes to the EU Emission Allowance Trading Directive Legal Opinion Commissioned by the Federal Ministry for the Environment, Nature Conservation and Nuclear", online: <http://grist.files.wordpress.com/2011/06/aviation_emission_trading.pdf>

Pete Sepp, "EU Energy Tax Would Hike U.S. Airfares Even Higher", *US News* (6 June 2012) online: <www.usnews.com/opinion/blogs/on-energy/2012/06/06/national-energy-tax-would-hike-airfares-even-higher>.

Peter Finn & Greg Miller, "Cargo Plane Bombs More Lethal than Christmas Day Attempt; Yemen Charges Aulaqi in Absentia", *The Washington Post* (2 November 2010) online: <http://goo.gl/n4Q1MG>.

Tim Devaney, "Leader of U.S. Aviation Group Slams 'Flawed' EU Emissions Scheme", *The Washington Times* (17 May 2012) online: <www.washingtontimes.com/news/2012/may/17/leader-of-us-aviation-group-slams-flawed-eu-emissi/>.

Magazines and Blogs

Jacques Hartmant, "The European Emissions Trading System and Extraterritorial Jurisdiction", *EJIL: Talk* (23 April 2012) online: <www.ejiltalk.org/the-european-emissions-trading-system-and-extraterritorial-jurisdiction>.

Annie Petsonk, "Aviation on the Flight Path to Success", *Carbon Finance*, (Spring 2013), online: EDF <www.edf.org/sites/default/files/CFSpring2013_On_the_flightpath_to_success_Petsonk.pdf> 22.

Joseph E Stiglitz, "A New Agenda for Global Warming", *The Economists' Voice* (July 2006) online: De Gruyter <www.degruyter.com/view/j/ev.2006.3.7/ev.2006.3.7.1210/ev.2006.3.7.1210.xml?format=INT>.

Michel Wachenheim. "Interview with Michel Wachenheim – President of the 38th Session of the ICAO Assembly" in *The European Civil aviation Conference Magazine, ECAC News* 50 (Winter 2013) 10.

Corporate Websites

Airbus, "Global Market Forecast: Future Journey 2013-2032", *Airbus* (2013) online: <www.airbus.com/company/market/forecast/?eID=dam_frontend_push&docID=3375>.

Boeing, "Current Market Outlook 2009-2028", *Boeing* online: <www.boeing.com/commercial/cmo/pdf/Boeing_Current_Market_Outlook_2009_to_2028.pdf at 8>.

Boeing, "Current Market Outlook 2013-2032", *Boeing* online: <www.boeing.com/boeing/commercial/cmo>.

Emirates Group, *Annual Report 2012-2013, Emirates Group* online: <http://content.emirates.com/downloads/ek/pdfs/report/annual_report_2013.pdf>.

LATAM, "LATAM Airlines Group", *LATAM* online: <www.latamairlinesgroup.net/phoenix.zhtml?c=81136&p=irol-home>.

Other Materials

C-SPAN, "Airlines Fees", video interview with John Heimlich, vice-president & Chief Economist of the Air Transport Association, (19 June 2011) online: <www.c-span-video.org/program/AirlineFees>.

Open Secrets.org, "John Micca", online: <www.opensecrets.org/politicians/summary.php?cid=N00002793&cycle=2012>.

Open Secrets.Org, "Sen. John Thune 2007 /2012", online: <www.opensecrets.org/politicians/industries.php?cycle=2012&cid=N00004572&type=I&newmem=N>.

Essential Air and Space Law (Series Editor: Marietta Benkö)

Volume 1: Natalino Ronzitti & Gabriella Venturini (eds.), The Law of Air Warfare – Contemporary Issues, ISBN 978-90-77596-14-2
Volume 2: Marietta Benkö & Kai-Uwe Schrogl (eds.), Space Law: Current Problems and Perspectives for Future Regulations, ISBN 978-90-77596-11-1
Volume 3: Tare Brisibe, Aeronautical Public Correspondence by Satellite, ISBN 978-90-77596-10-4
Volume 4: Michael Milde, International Air Law and ICAO, ISBN 978-90-77596-54-8
Volume 5: Markus Geisler & Marius Boewe, The German Civil Aviation Act, ISBN 978-90-77596-72-2
Volume 6: Ulrich Steppler & Angela Klingmüller, EU Emissions Trading Scheme and Aviation, ISBN 978-90-77596-79-1
Volume 7: Heiko van Schyndel (ed.), Aviation Code of the Russian Federation, ISBN 978-90-77596-80-7
Volume 8: Zang Hongliang & Meng Qingfen, Civil Aviation Law in the People's Republic of China, ISBN 978-90-77596-91-3
Volume 9: Ronald M. Schnitker & Dick van het Kaar, Aviation Accident and Incident Investigation. Concurrence of Technical, ISBN 978-94-90947-01-9
Volume 10: Michael Milde, International Air Law and ICAO, second edition, ISBN 978-90-90947-35-4
Volume 11: Ronald Schnitker & Dick van het Kaar, Safety Assessment of Foreign Aircraft Programme. A European Approach to Enhance Global Aviation Safety, ISBN 978-94-9094-793-4
Volume 12: Marietta Benkö & Engelbert Plescher, Space Law: Reconsidering the Definition/Delimitation Question and the Passage of Spacecraft through Foreign Airspace, ISBN 978-94-6236-076-1
Volume 13: Heiko van Schyndel (ed.), Aviation Code of the Russian Federation, second edition, ISBN 978-94-6236-433-2
Volume 14: Alejandro Piera Valdés, Greenhouse Gas Emissions from International Aviation: Legal and Policy Challenges, ISBN 978-94-6236-467-7